Basic

algebra

for College Students

Basic algebra for College Students

Second edition

Lawrence G. Gilligan
University of Cincinnati

Robert B. Nenno
Monroe Community College

Anthony S. Pruchnicki, Jr.
Mattatuck Community College

D. C. Heath and Company
Lexington, Massachusetts Toronto

Cover photograph: ''Helioptix,'' designed by Henry Ries

International Standard Book Number: 0-669-05693-6

Library of Congress Catalog Card Number: 84-80451

This text is dedicated

with love

to our parents

and to the loving memory of

Muriel M. Gilligan (1920–1983).

Preface

Like the first edition, *Basic Algebra for College Students,* Second Edition, is a textbook designed to be used in several different classroom settings by a variety of college students. The text may be used in a traditional lecture situation; there are over 440 worked-out examples to help illustrate the topics being covered and over 3750 carefully graded exercises for homework assignments. In a self-paced laboratory, students will find the step-by-step approach to solving examples helpful. The first ten odd-numbered exercises in each section and the first ten odd-numbered supplementary exercises have solutions at the back of the text to aid students in solving problems on their own.

Prerequisites

This text was written for students in any of the following three categories:

1. those who have never had a course (in high school or elsewhere) in elementary algebra,
2. those who have had an unsuccessful experience with algebra,
3. those who had algebra several years ago, but need to "brush up" when they return to school.

Although a working knowledge of arithmetic is helpful, the text provides a review of two of the most troublesome areas: fractions (the review is in Chapter 7) and decimals (the review is in Appendix A).

Course Coverage

The amount of the text covered depends on many factors, of course. Not the least of those factors is the length of time available. In a sixteen-week semester, three-credit-hour course, all eleven chapters *could* be covered. We feel that Chapters 1 through 8 offer the core of material the student will need to go on to courses in intermediate algebra, business math, or technical math. After those eight chapters are finished, several alternatives may be possible. We offer some below:

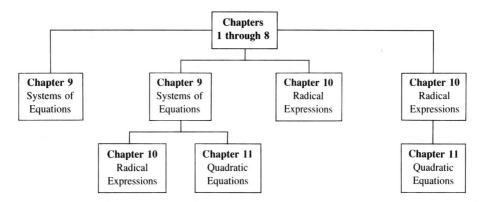

Features of the Second Edition

- The number of exercises has drastically increased, including the addition of Supplementary Exercises for each section.
- Chapter 4 is entirely new and deals with solving applied ("word") problems.

- Factoring is discussed in its own chapter, Chapter 6.
- The material on graphing has been split into two chapters: Chapter 8, which introduces graphing, and Chapter 9, which deals with systems of equations.
- The quadratic formula chapter (Chapter 11) has been expanded to include more work with the method of completing the square.
- Extensive review material is found at the end of each chapter: a summary; a vocabulary matching quiz; review exercises; a practice test; and now cumulative review exercises, which reinforce concepts learned in earlier chapters.
- At the end of the text there are answers to odd-numbered section exercises and supplementary exercises, all vocabulary quiz questions, all chapter review exercises, all cumulative review exercises, and all practice test questions. The *complete solutions* for the first ten odd-numbered exercises and supplementary exercises of each section are included.
- As in the first edition, the very successful approach to problem solving—"problem translation . . . equation solution . . . check"—is stressed throughout the text.
- Calculator exercises have been added throughout the text.
- Functional use of a second color is made throughout the text.
- An abundance of applications has been included to help show the *reason* algebra is used, not merely its abstract nature. An index listing these applications appears on the inside front cover of the text.

Supplement

An Instructor's Guide includes a diagnostic placement test, five alternative forms of each chapter test with answer keys, and answers to even-numbered section exercises.

Lawrence G. Gilligan
Robert B. Nenno
Anthony S. Pruchnicki, Jr.

Acknowledgments

This book has benefited from the valuable contributions of many First Edition users. A talented committee of reviewers offered many helpful suggestions for the text and its companion, *Intermediate Algebra;* we appreciate their input. They are: Ann S. Bretscher, University of Georgia; Ted Clinkenbeard, Des Moines Area Community College; Michael Detlefsen, Slippery Rock State College; Susan Forman, Bronx Community College; Pamela E. Matthews, Chemeketa Community College; and Betty S. Weissbecker, J. Sargeant Reynolds Community College. Special recognition is in order for Ann Smallen, Mohawk Valley Community College, for her valuable contribution to the exercises. Thanks, Ann.

In preparing the text manuscript and the Instructor's Guide, the expert typing skills of Susan Gilligan and Christine Suleski were invaluable. Mary Keefe did a great job verifying each of the answers. Thank you all for such fine work.

Our editor, Mary Lu Walsh, has been an inspiration on this project and has worked closely with us on it for three years. Her talents, too numerous to list here, are greatly appreciated and we are convinced that without her patience and guidance, the project would not have been. Thanks, Mary Lu.

To the staff who worked on the production of the text, Nancy Blodget, Peggy J. Flanagan, Mike O'Dea, and Libby Van de Kerkhove, we would like you to know that we are extremely pleased with your work and appreciate your talents.

Finally but foremost, Nenno and Gilligan would like to say "welcome aboard" to Tony Pruchnicki, the latest addition to this writing team. His contributions and enthusiasm for this project have exceeded all our expectations.

Contents

Chapter 5 Exponents and polynomials   *159*

Chapter 6 Factoring   *195*

Chapter 7 Fractions   *227*

Chapter 8 Linear equations in two variables 271

Chapter 9 Systems of equations in two variables 307

Chapter 10 Radical expressions 331

Chapter 11 Quadratic equations 363

Basic

for College Students

chapter
one

The real
numbers

1.1 Introduction to the real numbers

Arithmetic involves the study of numbers using the operations of addition, subtraction, multiplication, and division. The ability to deal with the arithmetic of whole numbers, fractions, and decimals provides the foundation for the study of algebra. Algebra extends our ability to use numbers by introducing symbols to represent some of the numbers.

One way we use numbers in arithmetic is for counting. The **natural numbers** (or **counting numbers**) can be written as follows:

$$\{1, 2, 3, \ldots\}$$

Because there is no last or largest number, we say that there are **infinitely many** natural numbers and use three dots to indicate this.

Example 1

Write the natural numbers between 3 and 8.

Solution

The word "between" does not allow us to include 3 and 8. Thus we have $\{4, 5, 6, 7\}$.

Example 2

Write the natural numbers larger than 100.

Solution

This collection never ends; so we write $\{101, 102, 103, \ldots\}$.

Example 3

Write the natural numbers used to count the atoms in the universe.

Solution

Assuming the number of atoms in the universe is infinitely large, we have $\{1, 2, 3, \ldots\}$.

The whole numbers consist of the natural numbers and the number 0; they can be listed as

$$\{0, 1, 2, 3, \ldots\}$$

The whole numbers can be illustrated graphically by the use of a **number line:**

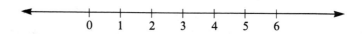

The point represented by 0 is chosen arbitrarily while the equally spaced markings to the right of 0 are labeled by the natural numbers 1, 2, 3, ... The arrow on the right end of the number line emphasizes the fact that the natural numbers are infinite and have no largest number.

If equally spaced markings are placed to the left of 0, we can label these successive numbers by -1, -2, -3, and so on, where the minus sign indicates that these values are located to the left of 0. The number line can now be pictured as:

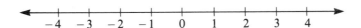

This group of numbers comprises the **integers** $\{\ldots -3, -2, -1, 0, 1, 2, 3, \ldots\}$. The numbers to the right of 0, the natural numbers, will be called the **positive integers,** while the numbers to the left of 0 will be called the **negative integers.** Zero itself is an integer, but is neither positive nor negative. Since negative integers are preceded by a minus sign, positive integers should be preceded by a plus sign. However, we will assume that any nonzero number without a sign is understood to be positive. For example, 6 means $+6$; 20 means $+20$.

Example 4

Picture the numbers $\{-4, -1, 0, 3, 7\}$ on a number line.

Solution

Draw a number line and locate each number with a solid dot.

There are other numbers besides integers. In arithmetic the fraction $\frac{1}{2}$ represents the number halfway between 0 and 1. Also, the numbers $2\frac{2}{3}$ and $-\frac{3}{4}$ are not integers. These and several others are pictured below.

Each of the numbers in the figure above can be obtained by dividing one integer by another integer. Each quotient formed in this manner is called a **rational number.** Rational numbers can be described as quotients of two integers, with the divisor integer not equal to 0. Note that every integer is a rational number since an integer can be written as the quotient of itself and 1. That is, $6 = \frac{6}{1}$, $-5 = \frac{-5}{1}$, and so on.

There are still other numbers that do not fall into any of the previous categories. These numbers are called **irrational** numbers and are defined as numbers on the number line that are not rational. One example of an irrational number is the number π whose value can be approximated by the fraction $\frac{22}{7}$ or the decimal 3.141592654. The symbol "$\approx$" is read "is approximately equal to," so we can write $\pi \approx \frac{22}{7}$ or $\pi \approx 3.14$. More examples of irrational numbers will be given when square roots are discussed in Section 1.7.

The **real numbers** consist of both rational and irrational numbers. Every point on the number line corresponds to a real number, and every real number can be located as a specific point on the number line.

The following diagram summarizes the discussion of real numbers.

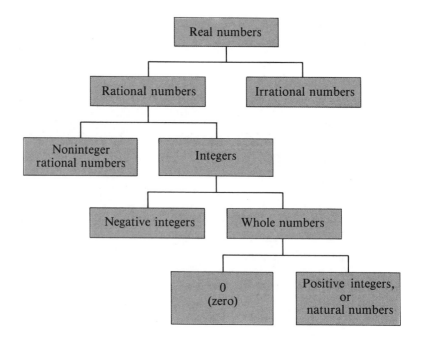

In algebra we will be working with real numbers. In doing so we will need to establish **order relations** between the numbers.

One such relationship is indicated by the equality symbol, $=$, which states that two numbers are equal. Obvious examples are given by the mathematical statements: $6 = 6$ or $\frac{8}{4} = 2$. Any symbol with a diagonal through it negates the relation involved. Thus the symbol $\neq$ means "is not equal to." For example, $4 \neq 7$ states that 4 is not equal to 7.

When two numbers are not equal, one of them must be larger and the other smaller. The symbol $<$ means "is less than." To say 4 is less than 7, we write $4 < 7$. The symbol $>$ means "is greater than." Since 6 is larger than 3, we write $6 > 3$. Each of the symbols, $<$ and $>$, are referred to as **inequality** symbols.

To keep the symbols $>$ and $<$ straight, notice that either symbol will always point to the smaller number. Also note that the order relation between two numbers can be expressed using either of the two symbols. For example, $5 < 8$ and $8 > 5$ both give the true relation between the two numbers. In both cases above the arrow points to the smaller number, namely, toward the 5. As another example, notice that $7 < 9$ can also be written with the symbol $>$ as $9 > 7$.

Example 5

Insert $<$ or $>$ to make the following statements true.

(a) 6 _____ 11   (b) 8 _____ 6   (c) 11 _____ 13

Solution

(a) $6 < 11$ Since 6 is smaller than 11

(b) $8 > 6$ Since 8 is larger than 6

(c) $11 < 13$ Since 11 is smaller than 13

Example 6

Write an equivalent statement for each of the following by reversing the inequality symbol.

(a) $7 > 5$ (b) $3 < 8$ (c) $0 < 4$

Solution

(a) $5 < 7$ (b) $8 > 3$ (c) $4 > 0$

Two other symbols, $\geq$ and $\leq$, also represent the idea of inequality. The symbol $\geq$ means "is greater than or equal to." The $\geq$ statement is true if either the $>$ part is true or the $=$ part is true.

For example, $7 \geq 6$ is true since $7 > 6$ is true. Also, $8 \geq 8$ is true since $8 = 8$ is true, but $9 \geq 10$ is not true since $9 \neq 10$ and $9 \not> 10$. (*Note*: $\not>$ is read "is not greater than.")

Similarly, the symbol $\leq$ means "is less than or equal to." The $\leq$ statement is true if either the $<$ part is true or the $=$ part is true. Thus $3 \leq 5$ is true since $3 < 5$ is true. $9 \leq 9$ is true since $9 = 9$ is true. But $5 \leq 4$ is not true since $5 \neq 4$ and $5 \not< 4$.

Example 7

Determine whether each statement is true or false.

(a) $8 \geq 6$ (b) $4 \leq 2$ (c) $0 \leq 0$

Solution

(a) True since $8 > 6$ is true.

(b) False since $4 \neq 2$ and $4 \not< 2$.

(c) True since $0 = 0$.

Note that as we move to the right on the number line, the numbers get bigger. Thus, the number farther to the right on the number line must be the larger number.

Each of the examples involving inequality symbols used only positive integers to illustrate them. Now let's look at some negative integers. Remember that the number to the right is still larger than the one on the left.

Example 8

Show that $-2 > -3$ is true.

Solution

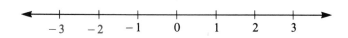

Since -2 lies to the right of -3 on the number line, -2 is larger than -3.

Example 9

Is $-4 < -1$?

Solution

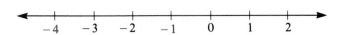

Yes, -1 lies to the right of -4 on the number line. Thus -1 is greater than -4, which is the same as saying -4 is less than -1.

Notice that the $<$ symbol can be interpreted as "to the left of" on the number line. That is, the statement $-4 < -1$ is true because -4 is to the left of -1 on the number line. Similarly, $-3 < -2$ is true because -3 is to the left of -2.

Also note that the $>$ symbol can be interpreted as "to the right of" on the number line. Thus, $2 > -3$ is true because 2 is to the right of -3 on the number line. Similarly, $-1 > -4$ is true because -1 is to the right of -4.

Example 10

Insert $\geq$ or $\leq$ to make the following statements true.

(a) 0 _____ -3 (b) $-1\frac{1}{4}$ _____ -1 (c) π _____ 3

Solution

(a)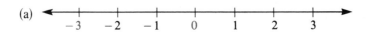

Since 0 is to the right of -3,

$$0 \geq -3$$

(b)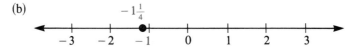

Since -1 is to the right of $-1\frac{1}{4}$,

$$-1 \geq -1\frac{1}{4} \quad \text{or equivalently} \quad -1\frac{1}{4} \leq -1$$

(c)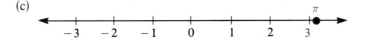

Since π is to the right of 3,

$$\pi \geq 3$$

The following diagram summarizes the inequality symbols introduced.

Symbol	Meaning
$<$	is less than
$>$	is greater than
$\leq$	is less than or equal to
$\geq$	is greater than or equal to

Section 1.1 Exercises

1. Write the natural numbers between 7 and 12 inclusive.
2. Write the natural numbers between 5 and 9.
3. Write the integers evenly divisible by 2.
4. Write the integers evenly divisible by 5.

In Exercises 5 through 10, graph the numbers on the number line.

5. $\{0, 3, -7, -2\}$

6. $\{-4, 5, 0, -3\}$

7. $\left\{-\dfrac{1}{2}, 4, \dfrac{5}{9}, -3\dfrac{1}{2}\right\}$

8. $\left\{-2\dfrac{1}{4}, 1, \dfrac{3}{5}, -\dfrac{7}{8}\right\}$

9. $\left\{-\dfrac{3}{4}, \pi, \dfrac{5}{2}, -1\dfrac{1}{4}\right\}$

10. $\left\{\dfrac{2}{3}, -\pi, \dfrac{6}{7}, 3\dfrac{1}{6}\right\}$

Given the numbers $\left\{0, -3, 2, \dfrac{3}{4}, -5\dfrac{1}{8}, \pi\right\}$, list

11. The natural numbers

12. The integers

13. The rational numbers

14. The irrational numbers

15. The real numbers

Given the numbers $\left\{-\dfrac{3}{4}, -5, 0, 3, 4\dfrac{1}{4}, -\pi\right\}$, list

16. The natural numbers

17. The integers

18. The rational numbers

19. The irrational numbers

20. The real numbers

21. Complete the following chart by checking off which sets of numbers each of the following belong to.

	Natural numbers	Integers	Rational numbers	Irrational numbers	Real numbers
-4					
0					
$\dfrac{2}{3}$					
600					

In Exercises 22 through 35, determine whether each of the statements is true or false.

22. $7 < -2$

23. $-3 < -4$

24. $-7 < -5$

25. $3 \geq 7$

26. $-4 \leq -4$

27. $-8 \leq -6$

28. $-7 > 3$

29. $-2 < -1\dfrac{1}{2}$

30. $-\pi \geq -3$

31. $\pi \geq 3$

32. $7 \neq 6\dfrac{1}{2}$

33. $4 \neq \dfrac{8}{2}$

34. $-3 \nless -2$

35. $4 \ngtr 4$

In Exercises 36 through 41, rewrite each of the following so that the inequality symbol is reversed.

36. $-4 < -2$

37. $-6 \geq -7$

38. $0 > -3$

39. $0 \leq 4$

40. $-\dfrac{5}{2} \leq -2$

41. $\dfrac{3}{4} > \dfrac{1}{2}$

In Exercises 42 through 45, place the symbol $<$ or the symbol $\geq$ in the blank to make a true statement.

42. -8.0375 _____ -8.0127

43. $(1.076 + 3.502 + 8.1271)$ _____ $(6.003 + 6.7115 - 0.0997)$

44. $(1.014 + 2.020 + 3.123)$ _____ $(9.8715 - 3.7124)$

45. $(25.01) \times (24.99)$ _____ $625.000 - 0.0001$

Supplementary Exercises

1. Complete the following chart by checking off which sets of numbers each of the following belong to.

	Natural numbers	Integers	Rational numbers	Irrational numbers	Real numbers
-10					
$\dfrac{4}{9}$					
$-\pi$					
12					
$6+9$					

2. Determine if each of the following is true or false.
   (a) Every whole number is an integer.
   (b) No integer is a real number.
   (c) 0 is the only whole number which is not a natural number.
   (d) Some fractions are integers.
   (e) The real numbers are made up of the rational numbers and the irrational numbers.

In Exercises 3 through 10, insert $<$, $>$, or $=$ to make the following statements true.

3. 129 ____ 149

4. 15 ____ 8

5. -3 ____ 9

6. -3 ____ -8

7. 0 ____ -12

8. -2 ____ 0

9. -21 ____ -26

10. $4 - 2$ ____ $0 + 2$

In Exercises 11 through 20 insert $<$ or $\geq$ to make the following statements true.

11. -39 ____ -43

12. -48 ____ -48

13. -2 ____ -1

14. 13 ____ -4

15. 0 ____ $9 - 9$

16. $5 + 6$ ____ $3 + 9$

17. -3 ____ 8

18. 12 ____ -12

19. -23 ____ -24

20. 12198 ____ 12199

1.2   Addition of real numbers

To discuss addition of real numbers, we will first present the concept of the absolute value of a number.

The **absolute value** of a number is defined as the distance between 0 and the number on the number line. Since the absolute value of a number represents distance, the absolute value of a number can never be negative. We symbolize the absolute value of a number by placing vertical bars before and after the number. For example, the absolute value of 3 is written as $|3|$. Since the distance between 0 and 3 on the number line is 3 units, we see that

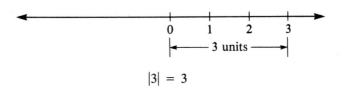

$$|3| = 3$$

Also, since the distance between 0 and -3 is 3 units we see that

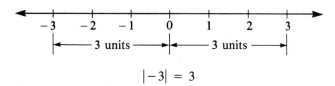

$$|-3| = 3$$

In a similar fashion we know that $|10| = 10$ and $|-10| = 10$, since both 10 and -10 are a distance of 10 units from 0 on the number line. Since 0 is a distance of 0 units from 0, we know $|0| = 0$.

The absolute value of a number is always positive or zero.

Example 1

Find the value of each of the following.

(a) $|-13|$ (b) $|4|$ (c) $\left|-\dfrac{1}{2}\right|$ (d) $|\pi|$

Solution

(a) $|-13| = 13$ Since -13 is *13* units from 0

(b) $\quad|4| = 4$ Since 4 is *4* units from 0

(c) $\left|-\dfrac{1}{2}\right| = \dfrac{1}{2}$ Since $-\frac{1}{2}$ is $\frac{1}{2}$ units from 0

(d) $\quad|\pi| = \pi$ Since π is π units from 0

Example 2

In (a) through (d) below, determine which of the two numbers has the larger absolute value.

(a) 5 or 3   (b) -5 or 3   (c) 5 or -3 (d) -5 or -3

Solution

(a) $|5| = 5$ and $|3| = 3$. Since $5 > 3$, we know that $|5| > |3|$. That is, of the two numbers, 5 and 3, the one with the larger absolute value is 5.

(b) $|-5| = 5$ and $|3| = 3$. Thus $|-5| > |3|$ and for the two numbers, -5 and 3, the one with the larger absolute value is -5.

(c) $|5| = 5$ and $|-3| = 3$. So $|5| > |-3|$ and it follows that of the two numbers, 5 and -3, 5 has the larger absolute value.

(d) $|-5| = 5$ and $|-3| = 3$. Thus $|-5| > |-3|$. That is, of the two numbers, -5 and -3, the one with the larger absolute value is -5.

Example 2 illustrates that when comparing two numbers to see which has the larger absolute value, the one that is further from the origin is the one with the larger absolute value.

With an understanding of absolute value, let us proceed to the addition of real numbers. We will again use the number line as an aid.

Example 3

Add 3 and 2.

Solution

We start at 0 on the number line. Since the first number in the addition is 3, we count 3 units to the right of 0 and from there we count 2 additional units to the right since the second number is 2.

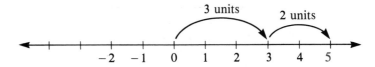

We end up at 5. Thus $3 + 2 = 5$.

Example 4

Add 4 and -7.

Solution

Start at 0 and count 4 units to the right. From there we count 7 units *to the left* (since the second number is negative).

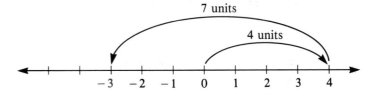

We end up at -3; so $4 + (-7) = -3$.

Note the use of the parentheses around -7; they clearly identify the second number as negative.

Example 5

Add -3 and -6.

Solution

We start at 0 and count 3 units to the left. From that point we count another 6 units to the left.

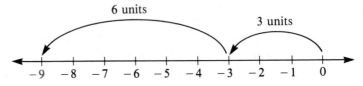

We finish at -9. Thus $(-3) + (-6) = -9$.

Example 6

Add -8 and 6.

Solution

Begin at 0 and count 8 units to the left. From that point we count 6 units to the right.

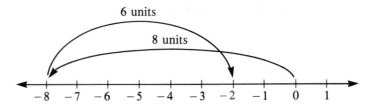

We finish at -2; so $(-8) + 6 = -2$.

Example 7

Add 6 and -8.

Solution

The number line shows the addition.

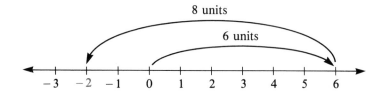

Thus, $6 + (-8) = -2$.

The last two examples involved the addition of the same two integers, 6 and -8, in two different orders. Note that both resulted in the *same sum*.

The use of a number line to add numbers is a time-consuming task. The results in the five previous examples can be obtained without the number line by applying the following rules.

Rules for adding two real numbers

1. If both numbers have the *same sign*, add their absolute values. The sign of this result is the same as the sign of the original numbers. (Adding two positive numbers yields a positive number; adding two negative numbers yields a negative number.)
2. If the numbers have *different signs*, subtract *the smaller absolute value from the larger absolute value*. The sign of this result is the same as the sign of the original number whose absolute value is larger.

The following examples illustrate the use of these rules in adding real numbers.

Example 8

Find $4 + 7$.

Solution

Since 4 and 7 have the same sign, we add their absolute values: $4 + 7 = 11$. Their sum will have the same sign, which is $+$. Therefore, the sum is $+11$, which we write as 11. Thus $4 + 7 = 11$.

Example 9

Find $(-5) + (-4)$.

Solution

-5 and -4 have the same sign (both negative), so we add their absolute values. $|-5| = 5$, $|-4| = 4$, and $5 + 4 = 9$. The sum will have the same sign, which is $-$. Thus $(-5) + (-4) = -9$.

Example 10

Find $(-6) + 4$.

Solution

-6 and 4 have different signs. $|-6| = 6$. $|4| = 4$. The difference in absolute values is $6 - 4 = 2$ and, since -6 has the larger absolute value, the sign of the sum is $-$. Thus $(-6) + 4 = -2$.

Example 11

Find $(-7) + 12$.

Solution

-7 and 12 have different signs. $|-7| = 7$ and $|12| = 12$. The difference in absolute values is $12 - 7 = 5$ and, since 12 has the larger absolute value, the sign of the sum is $+$. Thus $(-7) + 12 = 5$.

Example 12

Find the following sums.

   (a) $(-7) + 4$ (b) $(-9) + (-6)$ (c) $(-6) + 10$

Solution

   (a) $(-7) + 4 = -3$
   (b) $(-9) + (-6) = -15$
   (c) $(-6) + 10 = 4$

When more than two numbers are to be added, we add the numbers in pairs from left to right. We add the first two numbers, then add that result to the third number, and so on.

Example 13

Add $(-3) + 7 + (-8)$.

Solution

$$(-3) + 7 + (-8) = 4 + (-8) \quad \text{Since } (-3) + 7 = 4$$
$$= -4$$

Example 14

Add $4 + (-6) + (-8) + 3$.

Solution

$$4 + (-6) + (-8) + 3 = (-2) + (-8) + 3 \quad \text{Since } 4 + (-6) = -2$$
$$= (-10) + 3 \quad \text{Since } (-2) + (-8) = -10$$
$$= -7$$

Section 1.2 Exercises

In Exercises 1 through 40, find each indicated sum.

1. $(-3) + 4$
2. $(-5) + 7$
3. $7 + 3$
4. $5 + 8$
5. $9 + (-5)$
6. $8 + (-3)$
7. $(-11) + (-7)$
8. $(-8) + (-6)$
9. $(-7) + 3$
10. $(-9) + 2$
11. $2 + (-9)$
12. $(-2) + 9$
13. $(-9) + (-2)$
14. $(-2) + (-9)$
15. $9 + (-2)$
16. $(-11) + (-20)$
17. $(-15) + (-12)$
18. $(-23) + 28$
19. $23 + (-28)$
20. $(-23) + (-28)$
21. $22 + (-14)$
22. $(-18) + 25$
23. $36 + (-43)$
24. $48 + (-39)$
25. $(-205) + 185$
26. $(-78) + 149$
27. $(-5) + 9 + 10$
28. $5 + (-9) + (-10)$
29. $(-6) + (-5) + 7$
30. $(-11) + 8 + (-5)$
31. $12 + (-7) + (-8)$
32. $(-7) + (-8) + (-12)$
33. $(-3) + (-19) + (-1)$
34. $(-18) + (-14) + 19$
35. $16 + (-14) + (-20) + (-15) + 12$
36. $(-6.315) + (-8.937)$
37. $14.316 + (-18.053)$
38. $(-4638) + 7427 + (-8154) + 2147$
39. $(-5.0012) + 8.1476 + (-3.00147)$
40. $(\$1576.14) + (\$876.15) + (-\$3135.95)$

Supplementary Exercises

In Exercises 1 through 18, perform the indicated operations.

1. $24 + 46$
2. $52 + 15$
3. $(-23) + (-42)$
4. $(-139) + (-21)$
5. $(-3) + 28$
6. $(-81) + 25$
7. $37 + (-32)$
8. $29 + (-28)$
9. $(-32) + (-45) + (-21)$
10. $24 + 31 + 36$
11. $(-32) + (-24) + (-4)$
12. $(-21) + (32) + (-12)$
13. $(-7) + (-3) + 12$
14. $92 + (-23) + 21$
15. $94 + (-94) + 12$

16.
$$\begin{array}{r} 23 \\ -12 \\ +\ \ \ 9 \\ \hline \end{array}$$

17.
$$\begin{array}{r} -32 \\ -22 \\ +\ -73 \\ \hline \end{array}$$

18.
$$\begin{array}{r} -32 \\ 26 \\ +\ -54 \\ \hline \end{array}$$

19. Find the sum of -21, 43, and -32.
20. Find the sum of -32, -53, and -67.

1.3   Subtraction of real numbers

Before we get into the subtraction of real numbers, we will first define the **opposite** of a number. Two numbers with different signs are opposites of one another if each number is the same distance from 0 on the number line. For example, 4 is the opposite of -4 since both numbers are 4 units from 0 on the number line. Similarly, -4 is the opposite of 4.

Example 1

State the opposite of each of the following positive numbers.

(a) 7   (b) 12   (c) $\frac{3}{4}$ (d) π

Solution

(a) -7 (b) -12 (c) $-\frac{3}{4}$ (d) $-\pi$

Example 2

State the opposite of each of the following negative numbers.

(a) -6 (b) -9 (c) $-\frac{1}{2}$ (d) $-\pi$

Solution

(a) 6   (b) 9   (c) $\frac{1}{2}$ (d) π

Example 3

How can we represent the opposite of b?

Solution

If b represents a number, then $-b$ represents its opposite. Similarly, we say b is the opposite of $-b$.

Note that the opposite of a number changes the sign of the number but not its absolute value. Also, the opposite of 0 is 0.

Another idea we need to look at in the subtraction process is the idea of order. If we wish to subtract 7 *from* 10, 7 is the number we are subtracting and 10 is the number from which we are subtracting. We write this as $10 - 7$, where the minus sign is used to indicate the operation of subtraction. If we wish to subtract 10 *from* 7, 10 is the number we are subtracting and 7 is the number from which we are subtracting. In this case we write $7 - 10$. How these two examples are ordered is most important to the answer obtained.

Example 4

Write the following subtraction of numbers in symbols.

(a) Subtract -6 from 3.

(b) Subtract -5 from -7.

(c) Subtract the number b from the number a.

Solution

(a) $3 - (-6)$ Note the use of parentheses to indicate that -6 is the number being subtracted.

(b) $(-7) - (-5)$

(c) $a - b$

We actually *define* the operation of subtraction *in terms of addition*. In other words, to subtract the number b from the number a (written $a - b$), we add the opposite of b to a. Symbolically,

$$a - b = a + (-b)$$

The following procedure will be used:

Rules for subtracting two real numbers

1. Change the subtraction sign to an addition sign and change the number being subtracted to its opposite.
2. Complete the addition by using the rules for adding two real numbers in Section 1.2.

Example 5

Subtract 7 from 10.

Solution

Subtract 7 from 10 means $10 - 7$.

opposites

$$10 - 7 = 10 + (-7)$$

Subtraction changes
to addition

$$10 + (-7) = 3$$

Example 6

Subtract 10 from 7.

Solution

Subtract 10 from 7 means $7 - 10$.

opposites

$$7 - 10 = 7 + (-10)$$

Subtraction changes
to addition

$$7 + (-10) = -3$$

Example 7

Find $3 - (-6)$.

Solution

$$3 - (-6) = 3 + (+6)$$ Subtracting -6 is the same
as adding $+6$.

$$= 3 + 6 = 9$$

Example 8

Find $-7 - (-5)$.

Solution

$$-7 - (-5) = -7 + (+5)$$ Subtracting -5 is the same
as adding $+5$.

$$= -7 + 5 = -2$$

Example 9

Find $-16 - 14$.

Solution

$$-16 - 14 = -16 + (-14) = -30 \qquad \text{Subtracting 14 is the same as adding } -14.$$

Example 10

Subtract 16 from 0.

Solution

Subtract 16 from 0 means

$$0 - 16 = 0 + (-16) = -16 \qquad \text{Subtracting 16 is the same as adding } -16.$$

Example 11

Find $-7 - 0$.

Solution

$$-7 - 0 = -7 + 0 = -7 \qquad \text{0 is its own opposite.}$$

Section 1.3 Exercises

In Exercises 1 through 8, state the opposite of the given number.

1. 8
2. 17
3. -6
4. -7
5. $\dfrac{3}{8}$
6. $-\dfrac{4}{5}$
7. 0
8. $-\pi$

In Exercises 9 through 32, find the differences.

9. $7 - 12$
10. $6 - 14$
11. $(-7) - 6$
12. $(-8) - 3$
13. $(-4) - (-5)$
14. $(-8) - (-9)$
15. $16 - 7$
16. $21 - 15$
17. $(-20) - 6$
18. $(-15) - 9$
19. $(-14) - (-10)$
20. $(-12) - (-15)$
21. $6 - 20$
22. $(-16) - 24$
23. $8 - (-9)$
24. $(-9) - (-16)$
25. $17 - 0$
26. $0 - (-3)$
27. $(-13) - 8$
28. $14 - 18$
29. $(-19) - (-12)$
30. $(-25) - 45$
31. $112 - (-73)$
32. $98 - (-35)$
33. Subtract -10 from 32.
34. Subtract 16 from -24.
35. Subtract -73 from 110.
36. Subtract 83 from 65.
37. A share of GNP Associates, Inc., common stock was worth $83 a share when the stock market closed on Tuesday. At the closing bell on Wednesday, a share of that stock was worth $79. What number represents the change in value of a share of GNP stock at the close on Wednesday?
38. Subtract -573.014 from 257.150.
39. Find $87.9304 - (-678.4102)$.
40. Find $(-38.4076) - 3.9041$.

Supplementary Exercises

In Exercises 1 through 20 find the indicated differences.

1. $15 - 8$
2. $54 - 21$
3. $86 - 94$
4. $35 - 72$
5. $(-25) - 21$
6. $(-89) - 54$
7. $(-23) - (-24)$
8. $(-35) - (-21)$
9. $(-45) - (-24)$
10. $0 - (-3)$
11. $(-9) - 0$
12. $(-19) - (-19)$
13. $(-8) - 2 - 15$
14. $(-21) - (-21) - 32$
15. $12 - 23 - (-2)$
16. $(-27) - 13 - (-12)$

17. $\begin{array}{r} -245 \\ - 134 \\ \hline \end{array}$

18. $\begin{array}{r} 231 \\ - -84 \\ \hline \end{array}$

19. $\begin{array}{r} 24 \\ - 9 \\ \hline \end{array}$

20. $\begin{array}{r} -34 \\ - -25 \\ \hline \end{array}$

1.4 Multiplication and division of real numbers

From previous experience in arithmetic, we know that multiplication is simply a process of *repeated addition*. For example, to multiply 4 by 3, we write $3 \times 4 = 4 + 4 + 4 = 12$ where the "$\times$" indicates the operation of multiplication. The numbers 4 and 3 used in the multiplication are called **factors,** and the result of that multiplication, 12, is called the **product.** Note that if we wish to multiply 3 by 4 we would write

$$4 \times 3 = 3 + 3 + 3 + 3 = 12$$

The same product, 12, is obtained as in the first example. Changing the order of the factors will not change the resulting product.

Multiplication may be indicated in a variety of ways. Here is a list of possibilities for indicating multiplication.

$$3 \times 4 = 12$$
$$3 \cdot 4 = 12$$
$$3(4) = 12$$
$$(3)4 = 12$$
$$(3)(4) = 12$$

Note the following about writing a product:

1. We rarely use the symbol "$\times$" for multiplication since it can easily be confused with the letter x, a letter frequently used in algebra.
2. Be careful to distinguish the multiplication dot from a decimal point. $3 \cdot 4$ represents the product, whereas 3.4 is the decimal.
3. The use of parentheses around either or both of the numbers is understood to signify multiplication as long as no other operation is written between the numbers.

Now let's look at products where at least one of the numbers involves a negative number. The following examples illustrate the idea.

Example 1

Find the product of 3 and -4.

Solution

We use parentheses around -4 to identify that number as negative. Then we use the repeated addition principle.

$$3(-4) = (-4) + (-4) + (-4) = -12$$

The usual value of 12 is obtained in multiplying the values involved, but note that the product is *negative*. If we assume the order of the factors will not change the resulting product, we have

$$-4(3) = -12$$

From this example we see that when a positive number is multiplied by a negative number, the resulting product is negative.

Example 2

Find the product of -3 and -4.

Solution

To determine this product, we will examine the following pattern:

$$\text{Decrease by 1} \begin{cases} 3(-4) = -12 \\ 2(-4) = -8 \\ 1(-4) = -4 \\ 0(-4) = 0 \end{cases} \text{Increase by 4}$$

Note that if the first factor decreases by 1, the resulting product increases by 4. Continuing the pattern, we must have

$$-1(-4) = 4$$
$$-2(-4) = 8$$
$$-3(-4) = 12$$

We conclude that $-3(-4) = 12$. Again, by changing the order of the factors we would have $-4(-3) = 12$.

From this example we see that when two negative numbers are multiplied together, the resulting product is positive. We summarize these ideas as follows:

Rules for multiplying two real numbers

1. Multiply the absolute values of the numbers.
2. The product is positive if both numbers are positive or both negative. (Both numbers have the same sign.) The product is negative if one number is positive and the other one is negative. (The numbers have different signs.)

Example 3

Multiply $6(-3)$.

Solution

$$6(-3) = -18$$

$|6| = 6$ and $|-3| = 3$ so $6 \cdot 3 = 18$; the product is negative since one number is positive and the other negative.

Example 4

Multiply $-5(7)$.

Solution

$-5(7) = -35$ $\qquad$ $|-5| = 5$ and $|7| = 7$ so we have $5 \cdot 7 = 35$; the product is negative since one number is positive and the other negative.

Example 5

Multiply $(-6)(-5)$.

Solution

$(-6)(-5) = 30$ $\qquad$ $|-6| = 6$ and $|-5| = 5$ and thus $6 \cdot 5 = 30$; the product is positive since both numbers are negative.

Example 6

Multiply $(-2.5)(6.4)$.

Solution

$(-2.5)(6.4) = -16$ $\qquad$ $|-2.5| = 2.5$ and $|6.4| = 6.4$ so $(2.5)(6.4) = 16$; the product is negative since one number is negative and the other positive.

We will define division of real numbers in terms of multiplication. We first present a variety of ways in which division can be indicated. The division of 18 by 3 can be written as

$$\frac{18}{3}, \qquad 18 \div 3, \qquad 18/3, \qquad \text{or} \qquad 3\overline{)18}$$

In each of these forms, the number 18 is called the **dividend,** the number 3 is called the **divisor,** and the result of the division is called the **quotient.**

$$\frac{\text{Dividend}}{\text{Divisor}} = \text{Quotient}$$

We know that $\frac{18}{3} = 6$ since the product of the divisor 3 and the quotient 6 equals the dividend 18. This is the principle that relates division to multiplication.

$$(\text{Divisor})(\text{Quotient}) = \text{Dividend}$$

The following examples use this relationship to investigate division that involves combinations of positive and negative real numbers.

Example 7

$$\frac{18}{3} = 6 \qquad \text{Since } 3 \cdot 6 = 18$$

$$\frac{-18}{3} = -6 \qquad \text{Since } 3 \cdot (-6) = -18$$

$$\frac{18}{-3} = -6 \qquad \text{Since } (-3)(-6) = 18$$

$$\frac{-18}{-3} = 6 \qquad \text{Since } (-3) \cdot 6 = -18$$

Note that the rules governing the sign of a quotient are the same as those for the sign of a product.

Also note that when division is indicated by a fraction form, such as $\frac{18}{3}$, the dividend, 18, is called the **numerator,** and the divisor, 3, is called the **denominator.**

Rules for dividing two real numbers

1. Divide the absolute values of the numbers.
2. The quotient is positive if both numbers are positive or both numbers are negative. (Both numbers have the same sign.) The quotient is negative if one number is positive and the other one is negative. (The numbers have different signs.)

Example 8

Divide $\frac{-27}{9}$.

Solution

$$\frac{-27}{9} = -3$$ $|-27| = 27$ and $|9| = 9$, so $\frac{27}{9} = 3$; since one of the numbers is negative and the other positive, the quotient must be negative.

Example 9

Divide $\frac{24}{-6}$.

Solution

$$\frac{24}{-6} = -4$$ $|24| = 24$ and $|-6| = 6$ so $\frac{24}{6} = 4$; since one number is positive and the other negative, the quotient is negative.

Example 10

Find $(-48) \div (-4)$.

Solution

$$(-48) \div (-4) = 12$$ $|-48| = 48$ and $|-4| = 4$ so $\frac{48}{4} = 12$; since both numbers are negative, the quotient is positive.

Example 11

Find $(-52) \div 8$.

Solution

$$(-52) \div 8 = -6.5$$ $|-52| = 52$ and $|8| = 8$ so $\frac{52}{8} = 6.5$; since one number is positive and the other negative, the quotient is negative.

Section 1.4 Exercises

In Exercises 1 through 28, find each indicated product.

1. $5 \cdot 7$
2. $8 \cdot 4$
3. $(-6) \cdot 5$
4. $(-7) \cdot 4$
5. $(-8)(-6)$
6. $(-6)(-9)$
7. $5(-9)$
8. $8(-5)$
9. $(-3)(-7)$
10. $(-4)(-9)$
11. $-9(9)$
12. $-8(7)$
13. $(-12) \cdot 10$
14. $(-10) \cdot 9$
15. $15(-25)$
16. $35(-12)$
17. $(-8)(-40)$
18. $(-9)(-30)$
19. $(-14)(-18)$
20. $(-21)(-16)$
21. $(-14) \cdot 12$
22. $(-24) \cdot 8$
23. $(-4.63)(10)$
24. $(-5.76)(-10)$
25. $(-4.7)(2.8)$
26. $(-5.3)(2.4)$
27. $(3.1416)(-6.8)$
28. $(-12.0478)(-193.02)$

In Exercises 29 through 52, find the indicated quotient.

29. $\dfrac{28}{2}$
30. $\dfrac{36}{3}$
31. $(-15) \div 3$

32. $(-24) \div 8$
33. $\dfrac{-42}{-6}$
34. $\dfrac{-32}{-16}$

35. $21 \div (-7)$
36. $18 \div (-2)$
37. $\dfrac{-56}{8}$

38. $\dfrac{-63}{7}$
39. $(-54) \div (-6)$
40. $(-72) \div (-8)$

41. $\dfrac{25}{-5}$
42. $\dfrac{28}{-4}$
43. $30 \div (-3)$

44. $(-45) \div 5$
45. $\dfrac{180}{-10}$
46. $\dfrac{-500}{-50}$

47. $(-200) \div 25$
48. $360 \div (-15)$
49. $\dfrac{-576}{10}$

50. $\dfrac{-873.47}{1006.5}$
51. $\dfrac{67.438}{-5.24}$
52. $\dfrac{-86.746}{3.1416}$

Supplementary Exercises

In Exercises 1 through 20, perform the indicated operations.

1. $12(-2)$
2. $-15(-3)$
3. $-3(5)$
4. $-9(8)(2)$
5. $-2(-3)(3)$
6. $12(-2)(-3)$
7. $2(3)(5)(0)$
8. $235765(1)$
9. $-3(-2)(-4)$
10. $-12(21)(-1)$
11. $(-45) \div (-9)$
12. $35 \div (-7)$
13. $-121 \div (-11)$
14. $330 \div (-10)$
15. $-85 \div 5$
16. $\dfrac{-135}{-5}$
17. $\dfrac{200}{10}$
18. $\dfrac{-144}{12}$
19. $7\overline{)-56}$
20. $-12\overline{)96}$
21. Find the product of -12, -5, and 2.

22. Find the product of -5, 12, and 3.
23. Find the product of -2, -5, and -7.
24. Find the quotient when you divide -9 by -3.
25. Find the quotient when you divide 990 by -9.
26. Find the quotient when you divide -56 by -8.

1.5   Grouping symbols and properties of the real numbers

Grouping symbols are frequently used when more than two numbers are involved in operations on real numbers. The most often used grouping symbols are:

(	)	parentheses
[	]	brackets
{	}	braces
_____		bar (as used in division)

Certain numerical expressions require more than one operation to be completed for a given result. Grouping symbols indicate that the operation inside of that symbol is to be completed first. For those cases in which more than one set of grouping symbols appears within an expression, the operation within the *innermost* set of grouping symbols should be completed first.

Example 1

Find a value for the expression $8 + (-5) + 4$.

Solution

We can find the value in two different ways depending on how we group pairs of numbers to complete the addition.

(a) $[8 + (-5)] + 4 = 3 + 4$
$= 7$

Note that when we complete the addition of 8 and (-5), we drop the brackets since we know that $[3] + 4 = 3 + 4$.

(b) $8 + [(-5) + 4] = 8 + [-1]$
$= 7$

Note that in this case we keep the brackets so we can clearly identify the negative number to be added to 8.

In both solutions to Example 1, the same result, 7, was obtained. How the three numbers were grouped did not affect their sum. Using the letters a, b, and c to represent any real numbers, we state the general property exemplified here as follows.

Associative Property for Addition

If a, b, and c represent any real numbers, then

$$a + (b + c) = (a + b) + c$$

We now examine the other operations to see if they satisfy the associative property.

Example 2

Evaluate (a) $[7 - (-3)] - (-5)$ and (b) $7 - [(-3) - (-5)]$ and (c) determine whether the two expressions are equal.

Solution

(a) $[7 - (-3)] - (-5) = [7 + 3] - (-5)$ Change subtraction to addition.

$\qquad\qquad\qquad\quad\; = 10 - (-5)$ Add in brackets first and then drop brackets.

$\qquad\qquad\qquad\quad\; = 10 + 5 = 15$ Change subtraction to addition.

(b) $7 - [(-3) - (-5)] = 7 - [(-3) + 5]$ Change subtraction to addition.

$\qquad\qquad\qquad\quad\; = 7 - 2$

$\qquad\qquad\qquad\quad\; = 7 + (-2) = 5$ Change subtraction to addition.

(c) Thus $[7 - (-3)] - (-5) \neq 7 - [(-3) - (-5)]$.

From this example we see that *subtraction is not associative*. What about multiplication?

Example 3

Evaluate (a) $[(-3) \cdot 4](-5)$ and (b) $(-3)[4 \cdot (-5)]$ and (c) determine whether the two expressions are equal.

Solution

(a) $[(-3) \cdot 4] \cdot (-5) = [-12] \cdot (-5)$ Multiply in brackets first; we keep the brackets since the result is negative.

$\qquad\qquad\qquad\quad\; = 60$

(b) $(-3) \cdot [4 \cdot (-5)] = (-3)[-20]$

$\qquad\qquad\qquad\quad\; = 60$

(c) Thus $[(-3) \cdot 4](-5) = (-3)[4 \cdot (-5)]$.

As was the case for addition, the method of grouping had no affect on the product. This means that *multiplication is associative*. In general terms:

Associative Property for Multiplication

If a, b, and c represent any real numbers, then

$$a(bc) = (ab)c$$

Example 4

Evaluate (a) $[36 \div (-12)] \div 3$ and (b) $36 \div [(-12) \div 3]$ and (c) determine whether the two expressions are equal.

Solution

(a) $[36 \div (-12)] \div 3 = [-3] \div 3$ Divide in brackets first; we keep the brackets since the result is negative.

$\qquad\qquad\qquad\quad\; = -1$

(b) $36 \div [(-12) \div 3] = 36 \div [-4]$
$$= -9$$

(c) Thus $[36 \div (-12)] \div 3 \neq 36 \div [(-12) \div 3]$.

As we saw for subtraction, *division is not associative*. The associative properties for addition and multiplication allow us to group numbers in any manner to complete those operations. The key word here is "grouping."

Another idea we investigate is that of *order*. Does it make a difference in the end result if the order of the numbers is changed? The following examples show that order does not matter when adding or multiplying but does make a difference when subtracting or dividing.

Example 5

(a) $7 + (-5) = 2$
$(-5) + 7 = 2$
Thus $7 + (-5) = (-5) + 7$.

(b) $(-8) + (-6) = -14$
$(-6) + (-8) = -14$
Thus $(-8) + (-6) = (-6) + (-8)$.

This example shows that real numbers can be added in any order. This property is known as the **commutative property.**

Commutative Property for Addition

If a and b represent any real numbers, then

$$a + b = b + a$$

Example 6

(a) $7 - (-4) = 7 + 4$
$$= 11$$

(b) $(-4) - 7 = (-4) + (-7)$
$$= -11$$

(c) Thus $7 - (-4) \neq (-4) - 7$.

As we see, *subtraction is not commutative.*

Example 7

(a) $(-7) \cdot 4 = -28$
$4(-7) = -28$

(b) $(-5)(-8) = 40$
$(-8)(-5) = 40$

Numbers may be multiplied in any order. In other words, *multiplication is commutative.* In general terms:

Commutative Property for Multiplication

If a and b represent any real numbers, then

$$a \cdot b = b \cdot a$$

Example 8

(a) $20 \div 4 = 5$

(b) $4 \div 20 = \dfrac{1}{5}$

(c) Thus $\dfrac{20}{4} \neq \dfrac{4}{20}$.

Division is not commutative.

Example 9

State whether each of the following statements are true or false and indicate the reason.

(a) $[(-7) \cdot 5] \cdot 12 = (-7)[5 \cdot 12]$
(b) $[(-7) + 5] + 10 = (-7) + [5 + 10]$
(c) $6 \cdot (-7) = (-7) \cdot 6$
(d) $6 - 7 = 7 - 6$
(e) $\dfrac{3}{4} + \pi = \pi + \dfrac{3}{4}$

Solution

(a) True by the Associative Property of Multiplication.
(b) True by the Associative Property of Addition.
(c) True by the Commutative Property of Multiplication.
(d) False because $-1 \neq 1$. Remember, subtraction is *not* commutative.
(e) True by the Commutative Property of Addition.

Another property of real numbers involves both addition and multiplication. Known as the **distributive property,** it is stated as follows:

Distributive Property for Multiplication over Addition

If a, b, and c represent any real numbers, then

$$a(b + c) = ab + ac$$

Note: The use of the parentheses on the lefthand side of the equals sign indicates that the addition within the parentheses is to be completed first, and the result is to be multiplied by the number a. On the righthand side of the expression, the multiplications are to be done first and then those results are to be added. To make the order of these operations clearer, we could have written the multiplications in parentheses, but the meaning is the same either way.

In general, we will assume that multiplication is to be done before addition in any expression in which parentheses are not used. The order of operations for numbers will be discussed in more detail in Section 1.8. Right now let's look at some examples of the distributive property.

Example 10

Show that $7(6 + 5) = 7 \cdot 6 + 7 \cdot 5$.

Solution

$$7(6 + 5) = 7(11)$$ Do the addition first.
$$= 77$$

$$7 \cdot 6 + 7 \cdot 5 = 42 + 35$$ Do the multiplication first.
$$= 77$$

Therefore $7(6 + 5) = 7 \cdot 6 + 7 \cdot 5$.

Example 11

Does $(-3)[(-4) + 7] = (-3)(-4) + (-3)(7)$?

Solution

It does since the expression illustrates the distributive property. To verify this, we evaluate each expression.

$$(-3)[(-4) + 7] = (-3)3$$ Do the addition first.
$$= -9$$

$$(-3)(-4) + (-3)(7) = 12 + (-21)$$ Do the multiplication first.

$$= -9$$

Thus $(-3)[(-4) + 7] = (-3)(-4) + (-3)(7)$.

The distributive property is very useful in rewriting algebraic expressions. Certain situations require eliminating the parentheses in a given expression, whereas others call for the introduction of parentheses. In either case the expression should be examined closely to see whether the distributive property can be applied.

Example 12

Assuming x and y represent real numbers, use the distributive property to write $3 \cdot (x + y)$ without the parentheses.

Solution

$$3 \cdot (x + y) = 3 \cdot x + 3 \cdot y$$

Note: The distributive property applies since the number 3 outside the parentheses is a factor (multiplier) and the operation inside the parentheses is addition.

Example 13

Can we apply the distributive property to the expression $5(10 - 3)$ to eliminate the parentheses?

Solution

Yes. Since subtraction has been defined as a type of addition, the original expression could be rewritten as follows:

$$5(10 - 3) = 5[10 + (-3)]$$

Note the use of brackets here in place of the original parentheses so that the expression is easier to read.

$$= 5 \cdot 10 + 5(-3)$$

Note that parentheses are used only to make it easier to identify -3 as the second factor in this product.

Example 14

Can the distributive property be used to rewrite the expression $5 \cdot (7 \cdot 10)$?

Solution

No! The distributive property requires addition inside the parentheses. This example involves multiplication within the parentheses. We can, however, *evaluate* the expression:

$$5 \cdot (7 \cdot 10) = 5 \cdot 70$$
$$= 350$$

In addition to the properties presented so far, there are two additional properties unique to the numbers 0 and 1. Since each requires some discussion, we will cover them in the next section.

Section 1.5 Exercises

In Exercises 1 through 22, state whether each statement is true or false and give the reason for your choice.

1. $(7 + 6) + 5 = 7 + (6 + 5)$

2. $3 \cdot (4 \cdot 6) = (3 \cdot 4) \cdot 6$

3. $(-3)(4 + 9) = (-3) \cdot 4 + (-3) \cdot 9$

4. $5[(-3) + 5] = 5(-3) + 5 \cdot 5$

5. $18 + (-5) = (-5) + 18$

6. $(-24) \div 3 = 3 \div (-24)$

7. $16 - (-3) = (-3) - 16$

8. $18 \div 6 = 18 - 6$

9. $5 \cdot [24 \cdot (-8)] = [5 \cdot 24] \cdot (-8)$

10. $[(-7) + (-8)] + 6 = (-7) + [(-8) + 6]$

11. $(-6) \cdot [8 + (-10)] = (-6) \cdot 8 + (-6)(-10)$

12. $25 \div [6 + (-1)] = (25 \div 6) + [25 \div (-1)]$

13. $(5 \cdot x) \cdot y = 5 \cdot (x \cdot y)$

14. $(-6) + (x + y) = [(-6) + x] + y$

15. $x + 3 = 3 + x$

16. $a + (-4) = (-4) + a$

17. $(-6)(3) + (-6) \cdot y = (-6)[3 + y]$

18. $x \cdot (-4) + x \cdot y = x[(-4) + y]$

19. $5(a \cdot 3) = (5 \cdot a) \cdot 3$

20. $8 \cdot (4 \cdot x) = (8 \cdot 4) \cdot x$

21. $6(x - 4) = 6 \cdot x + 6 \cdot (-4)$

22. $(-9)(y - 3) = (-9)y + (-9)(-3)$

In Exercises 23 through 34, use the indicated property to write a new expression that is equal to the given expression.

23. $(-5) + 8$ Commutative Property of Addition

24. $(-6) \cdot (-9)$ Commutative Property of Multiplication

25. $(6 \cdot x) \cdot (-3)$ Associative Property of Multiplication

26. $(17 + y) + (-8)$ Associative Property of Addition
27. $(-6)(4 + 12)$ Distributive Property
28. $x[y + (-5)]$ Distributive Property
29. $6 + (17 + y)$ Associative Property of Addition
30. $[(-3) \cdot (-6)] \cdot x$ Associative Property of Multiplication
31. $(-3) + x$ Commutative Property of Addition
32. $y \cdot z$ Commutative Property of Multiplication
33. $(-4) \cdot x + (-4)(6)$ Distributive Property
34. $5(-3) + 5(-7)$ Distributive Property

Supplementary Exercises

In Exercises 1 through 17, simplify each side of the expression to see whether the statement is true or false. Tell what property you are using if it is true.

1. $[-3 + 5] = [5 + (-3)]$
2. $[-4 + (-8)] = [-8 + (-4)]$
3. $8 - 9 = 9 - 8$
4. $-10 \div 21 = 21 - 10$
5. $4 + (-5 + 23) = [4 + (-5)] + 23$
6. $6 - (8 - 9) = (6 - 8) - 9$
7. $[3(-2)](-5) = 3[(-2)(-5)]$
8. $(12 \div 2) \div 2 = 12 \div (2 \div 2)$
9. $0 - 9 = 9 - 0$
10. $4(6 - 9) = 4(6) - 4(9)$
11. $-6[5 + (-2)] = -6(5) + (-6)(-2)$
12. $13(-2) = -2(13)$
13. $2(8 \div 4) = 2(8) \div [2(4)]$
14. $2(-5 + 6) = 2(-5) + 2(6)$
15. $-7(5 - 9) = -7(5) - (-7)(9)$
16. $8 \div (-2) = -2 \div 8$
17. $0 + 21 = 21 + 0$

1.6   Properties of zero and one

As we said at the close of the last section, there are two properties unique to the numbers 0 and 1. We will discuss these properties in this section.

If we add 0 to any real number, the number remains unchanged. As a result, we say that 0 *preserves the identity* of other numbers under addition. Thus, 0 is called the **identity element for addition** or, simply, the **additive identity.**

Identity property for addition

If a represents any real number, the number 0 is such that

$$a + 0 = 0 + a = a$$

Since subtraction is not commutative, we must consider two cases when dealing with subtractions involving zero.

Example 1

Find the following differences.

(a) $8 - 0$ (b) $(-17) - 0$

Solution

(a) $8 - 0 = 8 + 0$ Recall that 0 is its own opposite.

          $= 8$

(b) $(-17) - 0 = -17 + 0$

              $= -17$

In this example subtracting 0 does not change either number it is subtracted from.

Example 2

Find the following differences.

   (a) $0 - 8$ (b) $0 - (-17)$

Solution

(a) $0 - 8 = 0 + (-8)$ Definition of subtraction

          $= -8$

(b) $0 - (-17) = 0 + 17$ Definition of subtraction

              $= 17$

In Example 2 notice that the subtraction of a number *from* zero yields the opposite of the number. We summarize our results here.

> If a represents any real number, then
>
> 1. $a - 0 = a$
> 2. $0 - a = 0 + (-a) = -a$

Since multiplication *is* repeated addition, we know that multiplying a number by 0 will always result in 0.

Example 3

(a) $2 \cdot 0 = (0 + 0) = 0$

(b) $6 \cdot 0 = (0 + 0 + 0 + 0 + 0 + 0) = 0$

Example 4

(a) $(-7) \cdot 0 = 0$

(b) $0 \cdot \left(-\dfrac{3}{4}\right) = 0$

(c) $(-\pi) \cdot 0 = 0$

The commutative property of multiplication would also give us $0 \cdot 2 = 0$ and $0 \cdot (-\pi) = 0$. So, in general:

> If a represents any real number, then
>
> $$a \cdot 0 = 0 \cdot a = 0$$

Division is not a commutative operation, so we must be careful in considering 0 under this operation. We first consider an example where 0 is the dividend (or numerator) and a nonzero number appears in the divisor (or denominator).

Example 5

(a) $\dfrac{0}{7} = 0$ since $7 \cdot 0 = 0$ Definition of division

(b) $\dfrac{0}{-12} = 0$ since $(-12) \cdot 0 = 0$ Remember that 0 multiplied by any number is 0.

(c) $\dfrac{0}{\pi} = 0$ since $\pi \cdot 0 = 0$.

This example shows that dividing 0 by a nonzero number always results in 0.

Example 6

What is the quotient of $0 \div 0$?

Solution

Suppose we use a ''?'' to represent the quotient of $0 \div 0$. In other words, if we write

$$\frac{0}{0} = ?$$

by our understanding of division, we can write

$$0 = (?) \cdot 0$$

What number should replace the question mark to make the last statement true? Since the product of *any number* and 0 is 0, we see that the ''?'' can be *any* real number. Since no *one* real number represents the quotient of 0 by 0, we conclude that $\dfrac{0}{0}$ **cannot be determined.**

Now let us examine a third possibility that exists for quotients involving 0, namely, those cases in which the dividend is a nonzero number and 0 is the divisor.

Example 7

What is the quotient of $\dfrac{8}{0}$?

Solution

Again we use a question mark to represent the result of this quotient.

$$\frac{8}{0} = ?$$

By the definition of division, this means $0 \cdot (?) = 8$, but 0 times any number will result in 0, not 8. There is no real number that will make the statement true. So we say that any nonzero number divided by 0 **cannot be defined.**

We summarize these results as follows:

If a represents any nonzero real number, then

1. $\dfrac{0}{a} = 0$.

2. $\dfrac{a}{0}$ is undefined.

3. $\dfrac{0}{0}$ cannot be determined.

In the beginning of this section, we stated that 0 is the identity element under addition. There is also an **identity element under multiplication,** namely 1. The number 1 *preserves the identity* of other numbers under multiplication. For this reason, the number 1 is called the **multiplicative identity.**

Identity property for multiplication

If a represents any real number, the number 1 is such that

$$a \cdot 1 = 1 \cdot a = a$$

Section 1.6 Exercises

In Exercises 1 through 27, determine a value for each of the following, if possible. Identify those that involve identity properties.

1. $(-13) + 0$ 2. $0 + (-6)$ 3. $\dfrac{0}{7}$

4. $\dfrac{0}{-12}$ 5. $8 \cdot 1$ 6. $1 \cdot (-14)$

7. $16 - 0$ 8. $0 - 7$ 9. $0 + (-10)$

10. $(-6) + 0$ 11. $\dfrac{-16}{0}$ 12. $\dfrac{18}{0}$

13. $\dfrac{0}{0}$ 14. $\dfrac{0}{6}$ 15. $1 \cdot (-15)$

16. $7 \cdot 1$ 17. $\pi + 0$ 18. $0 + (-\pi)$

19. $\left(\dfrac{3}{4}\right) \cdot 1$ 20. $1 \cdot \left(-\dfrac{5}{8}\right)$ 21. $\dfrac{5}{8} \div 0$

22. $0 \div \dfrac{5}{8}$ 23. $0 \div \left(-\dfrac{1}{2}\right)$ 24. $\left(-\dfrac{1}{2}\right) \div 0$

25. $(-126) + 0$ 26. $0 + (1971.3)$ 27. $\dfrac{-212.6}{0}$

 28. On your calculator, try to "find" $15 \div 0$. What is the result?

29. On your calculator, try to "find" $0 \div 0$. What is the result?

30. (a) Using your calculator, complete the following chart.

Expression	Value of Expression
15 ÷ 0.5	30
15 ÷ 0.02	
15 ÷ 0.0001	
15 ÷ 0.0000001	

(b) Although $15 \div a$ is undefined if $a = 0$, what happens to the value of $15 \div a$ if a is a positive number that gets closer and closer to 0?

Supplementary Exercises

In Exercises 1 through 15, determine the value of each of the following, if possible.

1. $0 + 181$
2. $-29 + 0$
3. $0 - 9$
4. $(-12) + 0$
5. $0(5)$
6. $2(7)(-3)(0)$
7. $\dfrac{5}{0}$
8. $\dfrac{0}{9}$
9. $\dfrac{0}{0}$
10. $5(1)$
11. $1(-9)$
12. $18 \div 1$
13. $-9 \div 1$
14. $\dfrac{0}{-12}$
15. $1573(-4)(23)(7)0$

1.7 Powers and roots

We have seen that multiplication is a shorthand notation for repeated addition. We will now examine a shorthand notation for repeated multiplication. For example, consider

$$2 \cdot 2 \cdot 2 \cdot 2 \cdot 2 = 32$$

The factor (multiplier) 2 appears 5 times. We introduce the symbol 2^5 to represent the above product, that is

$$2^5 = 2 \cdot 2 \cdot 2 \cdot 2 \cdot 2 = 32$$

The expression 2^5 is read "2 to the fifth power" or more simply "2 to the fifth." The 2 is called the **base** and the 5 is called the **exponent.**

$$\overset{\text{exponent}}{2^5} = 32$$

base ⟶   ⟵ fifth power of 2

Example 1

Rewrite each of the following using exponent notation.

(a) $3 \cdot 3 \cdot 3 \ 3$ (b) $5 \cdot 5 \cdot 5 \cdot 5 \cdot 5 \cdot 5$ (c) $(-4) \cdot (-4) \cdot (-4)$

Solution

(a) $3 \cdot 3 \cdot 3 \cdot 3 = 3^4$ 3 is used as a factor 4 times.

(b) $5 \cdot 5 \cdot 5 \cdot 5 \cdot 5 \cdot 5 = 5^6$ 5 is used as a factor 6 times.

(c) $(-4) \cdot (-4) \cdot (-4) = (-4)^3$ -4 is used as a factor 3 times. Note that the parentheses are a necessary part of the notation since the base is a negative number.

Example 2

Find the values of each of the following.

(a) 7^3 (b) 8^2 (c) 5^4 (d) $(-3)^4$ (e) -3^4 (f) 0^3

Solution

(a) $7^3 = 7 \cdot 7 \cdot 7$ *Note:* 7^3 can be read as "7 to the third power" or "7 *cubed.*"

$ = 343$

(b) $8^2 = 8 \cdot 8$ *Note:* 8^2 is read "8 to the second power" or "8 *squared.*"

$ = 64$

(c) $5^4 = 5 \cdot 5 \cdot 5 \cdot 5$ 5 is used as a factor 4 times.

$ = 625$

(d) $(-3)^4 = \underbrace{(-3)(-3)}(-3)(-3)$ *Note:* The parentheses around the number -3 clearly identify -3 as the base.

$ = \underbrace{9(-3)}(-3)$

$ = -27(-3)$

$ = 81$

(e) $-3^4 = -(3 \cdot 3 \cdot 3 \cdot 3)$ *Note:* The exponent 4 applies only to the quantity immediately preceding it. Thus the base is 3.

$ = -81$

(f) $0^3 = 0 \cdot 0 \cdot 0$

$ = 0$

If a represents any **positive** real number, then

$$0^a = 0$$

Parts (d) and (e) of Example 2 point out that the parentheses are crucial in distinguishing $(-3)^4$ from -3^4. The next example further illustrates this distinction.

Example 3

Find the value of each expression:

(a) $(-5)^4$ (b) -5^4 (c) $(-2)^3$ (d) -2^3 (e) $-(-2)^3$

Solution

(a) $(-5)^4 = (-5)(-5)(-5)(-5)$ -5 is the base; it must be raised to the fourth power.

$ = 625$

(b) $-5^4 = -(5 \cdot 5 \cdot 5 \cdot 5)$ Think of -5^4 as "the opposite of 5^4."

$ = -625$

(c) $(-2)^3 = (-2)(-2)(-2)$
$= -8$

(d) $-2^3 = -(2 \cdot 2 \cdot 2)$
$= -8$

(e) $-(-2)^3 = -[(-2)(-2)(-2)]$ Here, we find the opposite of $(-2)^3$.
$= -(-8)$
$= 8$

The operations of addition and subtraction are called **inverse operations** since, if we start with a number, then add and subtract the same number, the result is the original number. For example,

$$(7 + 3) - 3 = 7$$

Adding and subtracting 3 leaves us with the original number, 7.

The operations of multiplication and division are also called inverse operations. For example,

$$(12 \div 2) \cdot 2 = 12$$

Dividing 12 by 2 and then multiplying by 2 results in 12, the original number.

There is also an inverse operation for "raising to a power." This operation is called "taking a root." The inverse of squaring a number is called *taking a square root*. The positive square root of a number is indicated by using the symbol $\sqrt{}$.

Suppose we wish to determine the square root of 16, written $\sqrt{16}$. We want to find the positive number that, when squared, is equal to 16. That number is 4 since $4^2 = 16$.

Actually, there are two square roots for every positive number. The number 25 has two square roots, 5 and -5, since $(5)^2 = 25$ and $(-5)^2 = 25$. In this book, however, we will only be concerned with the *positive* square root of 25, which is also called the **principal square root.** Thus we write $\sqrt{25} = 5$.

Example 4

Find the value of each of the following square roots.

(a) $\sqrt{9}$ (b) $\sqrt{49}$ (c) $\sqrt{100}$ (d) $\sqrt{1}$ (e) $\sqrt{0}$

Solution

(a) $\sqrt{9} = 3$ Since $3^2 = 9$

(b) $\sqrt{49} = 7$ Since $7^2 = 49$

(c) $\sqrt{100} = 10$ Since $10^2 = 100$

(d) $\sqrt{1} = 1$ Since $1^2 = 1$

(e) $\sqrt{0} = 0$ Since $0^2 = 0$

Every positive number has a principal (positive) square root and a negative square root. The symbol $\sqrt{}$ always indicates the principal (positive) square root of a positive number. We will use the symbol $-\sqrt{}$ to designate the negative square root of a number. For example, $\sqrt{16} = 4$ since $4^2 = 16$. But since $(-4)^2$ is also 16, we can write $-\sqrt{16} = -4$.

Example 5

Find (a) $-\sqrt{25}$ and (b) $-\sqrt{81}$.

Solution

(a) $-\sqrt{25} = -5$ Since $5^2 = 25$ and the opposite of 5 is -5

(b) $-\sqrt{81} = -9$ Since $9^2 = 81$ and the opposite of 9 is -9

Example 6

Express $\sqrt{3}$ as a rational number, if possible.

Solution

$\sqrt{3}$ is the principal square root of 3. When squared, it gives us 3, that is, $(\sqrt{3})^2 = 3$. But 3 itself is not a *perfect* square. (By *perfect* square, we mean the square of an integer.) This means that $\sqrt{3}$ is not an integer nor can it be written as a rational number, which is the quotient of two integers. $\sqrt{3}$ is an *irrational* number; $-\sqrt{3}$ is also an irrational number.

Thus we have seen that the square root of a positive integer that is not a perfect square represents an irrational number. Such irrational numbers do correspond to a particular point on the real number line and their location can be approximated by using a table or a hand-held calculator. Most calculators express irrational numbers to six or seven decimal places. The next example shows some approximations for irrational numbers; they have been rounded off to three decimal places.

Example 7

Use a calculator or Table I on page 401 to approximate the following.

(a) $\sqrt{3}$ (b) $\sqrt{40}$ (c) $\sqrt{1024}$ (d) $-\sqrt{57}$ (e) $-\sqrt{378}$

Solution

(a) $\sqrt{3} \approx 1.732$ Recall that the symbol $\approx$ is read "approximately equal to."

(b) $\sqrt{40} \approx 6.325$

(c) $\sqrt{1024} = 32.000$ Note that $\sqrt{1024}$ is not an irrational number; it represents another way of writing the integer 32.

(d) $-\sqrt{57} \approx -7.545$

(e) $-\sqrt{378} \approx -19.442$

Example 8

Can $\sqrt{-4}$ be written as a real number?

Solution

The only possibilities would seem to be 2 and -2. But $(2)^2 = 4$ and $(-2)^2 = 4$. We see that there is no real number that is the principal square root of -4. In fact, since the squaring of a number always results in a nonnegative number, we can make the following general statement:

The square root of a negative number is not real.

Example 9

(a) $\sqrt{-16}$ is not real.

(b) $-\sqrt{16} = -4$ *Note:* Since the negative symbol is not *inside* the square root symbol, we take the square root first and then find the opposite of the value obtained.

(c) $-\sqrt{-16}$ is not real.

(d) $-\sqrt{-247}$ is not real.

We end this introduction to roots by discussing the "inverse operations" concept for higher powers. The inverse of raising to the third power (cubing) is called the *cube root* and is symbolized by $\sqrt[3]{}$. The inverse of raising to the fourth power is called the *fourth root* and is symbolized by $\sqrt[4]{}$. The inverse of raising to the fifth power is called the *fifth root* and is symbolized by $\sqrt[5]{}$. We could continue this process and define $\sqrt[6]{}$, $\sqrt[7]{}$, and so on.

Example 10

(a) $4^3 = 64$, so we can write $\sqrt[3]{64} = 4$.

(b) $2^4 = 16$, so we can write $\sqrt[4]{16} = 2$.

(c) $3^5 = 243$, so we can write $\sqrt[5]{243} = 3$.

(d) $(-4)^3 = -64$, so we can write $\sqrt[3]{-64} = -4$.

(e) $(-2)^5 = -32$, so we can write $\sqrt[5]{-32} = -2$.

Note that certain higher roots for negative numbers *do* exist. In particular, if the root corresponds to an odd-numbered power ($\sqrt[3]{}$, $\sqrt[5]{}$, and so on), those roots are real numbers. However, roots corresponding to even powers ($\sqrt{}$, $\sqrt[4]{}$, $\sqrt[6]{}$, and so on) of negative numbers are *not* real numbers.

Example 11

Find each of the indicated roots.

(a) $\sqrt[3]{27}$ (b) $\sqrt[3]{-27}$ (c) $\sqrt[4]{16}$ (d) $\sqrt[4]{-16}$ (e) $-\sqrt[3]{-8}$ (f) $\sqrt[7]{-1}$

Solution

(a) $\sqrt[3]{27} = 3$ Since $3^3 = 27$

(b) $\sqrt[3]{-27} = -3$ Since $(-3)^3 = -27$

(c) $\sqrt[4]{16} = 2$ Since $2^4 = 16$

(d) $\sqrt[4]{-16}$ is not real.   There is no number that can be raised to the fourth power (an even power) to yield a negative number.

(e) $-\sqrt[3]{-8} = -(-2) = 2$

(f) $\sqrt[7]{-1} = -1$ Since $(-1)^7 = -1$

The following is a list of the most commonly used principal roots.

Square roots	Cube roots	Fourth roots	Fifth roots
$\sqrt{1} = 1$	$\sqrt[3]{1} = 1$	$\sqrt[4]{1} = 1$	$\sqrt[5]{1} = 1$
$\sqrt{4} = 2$	$\sqrt[3]{8} = 2$	$\sqrt[4]{16} = 2$	$\sqrt[5]{32} = 2$
$\sqrt{9} = 3$	$\sqrt[3]{27} = 3$	$\sqrt[4]{81} = 3$	$\sqrt[5]{243} = 3$
$\sqrt{16} = 4$	$\sqrt[3]{64} = 4$	$\sqrt[4]{256} = 4$	$\sqrt[5]{1024} = 4$
$\sqrt{25} = 5$	$\sqrt[3]{125} = 5$	$\sqrt[4]{625} = 5$	
$\sqrt{36} = 6$	$\sqrt[3]{216} = 6$		
$\sqrt{49} = 7$			
$\sqrt{64} = 8$			
$\sqrt{81} = 9$			
$\sqrt{100} = 10$			
$\sqrt{121} = 11$			
$\sqrt{144} = 12$			

Appendix B can also be used to find approximate values (to three decimal places) of square roots between 1 and 100.

Section 1.7 Exercises

In Exercises 1 through 28, find the value of each expression, if possible.

1. 3^3

2. 4^3

3. 9^2

4. 7^2

5. $(-6)^3$

6. $(-5)^3$

7. $(-1)^5$

8. $(-1)^4$

9. 0^5

10. 0^3

11. 3^5

12. 4^4

13. $(-3)^4$

14. $(-2)^5$

15. $(-10)^3$

16. $(-10)^4$

17. -6^3

18. -5^3

19. -2^4

20. -4^3

21. $(-2)^4$

22. $(-4)^3$

23. -9^3

24. -8^3

25. $(46.3)^2$

26. $(83.7)^2$

27. $(17)^3$

28. $(21)^3$

In Exercises 29 through 64, find the indicated roots, if possible.

29. $\sqrt{64}$

30. $\sqrt{100}$

31. $\sqrt[3]{64}$

32. $\sqrt[3]{125}$

33. $\sqrt[3]{-8}$

34. $\sqrt[3]{-27}$

35. $\sqrt{81}$

36. $\sqrt{36}$

37. $\sqrt[4]{81}$

38. $\sqrt[4]{16}$

39. $\sqrt[3]{-64}$

40. $\sqrt[3]{-125}$

41. $\sqrt[5]{-1}$

42. $\sqrt[5]{-32}$

43. $\sqrt[4]{-16}$

44. $\sqrt[4]{-81}$

45. $-\sqrt{81}$

46. $-\sqrt{25}$

47. $\sqrt[3]{-216}$

48. $-\sqrt[3]{-125}$

49. $-\sqrt[4]{256}$

50. $-\sqrt[4]{625}$

51. $\sqrt[4]{-256}$

52. $\sqrt[4]{-625}$

53. $\sqrt[3]{1000}$

54. $\sqrt{10000}$

55. $-\sqrt{144}$

56. $-\sqrt{49}$

57. $\sqrt{676}$

58. $\sqrt{729}$

59. $\sqrt{396}$

60. $\sqrt{513}$

61. $\sqrt{1327}$

62. $\sqrt{941}$

63. $\sqrt{86.73}$

64. $\sqrt{112.6}$

Supplementary Exercises

In Exercises 1 through 32, evaluate each of the following, if possible.

1. 5^3

2. $(-5)^3$

3. -5^3

4. 2^4

5. $(-2)^4$

6. -2^4

7. 0^2

8. 0^4

9. 25^1

10. $(-25)^2$

11. $(-7)^3$

12. 6^3

13. $(-6)^2$

14. -6^2

15. 1^5

16. -1^2

17. $\sqrt{121}$

18. $\sqrt{49}$

19. $\sqrt{100}$

20. $\sqrt{1}$

21. $\sqrt{-9}$

22. $\sqrt[3]{216}$

23. $\sqrt[3]{-8}$

24. $\sqrt[3]{-1}$

25. $\sqrt[3]{1}$

26. $-\sqrt{36}$

27. $-\sqrt{25}$

28. $-\sqrt[3]{8}$

29. $-\sqrt[3]{-8}$

30. $-\sqrt[4]{16}$

31. $-\sqrt[4]{1}$

32. $-\sqrt[5]{32}$

1.8   The priority of operations

When an expression contains more than one operation, there may be confusion as to which operation is to be completed first. For example, to determine $5 + 3 \cdot 4$, we might calculate either

$$5 + 3 \cdot 4 = 5 + 12 = 17$$

where the multiplication is done first (followed by the addition), or

$$5 + 3 \cdot 4 = 8 \cdot 4 = 32$$

where the addition is completed first (followed by the multiplication). To avoid the possibility of different answers, there is a standard order in which to operate. This priority of operations will be followed according to these rules:

Order of operations

1. If grouping symbols appear, work within each pair of grouping symbols *from the innermost outward*.

2. If no grouping symbols are present, or if the evaluation is to begin within the innermost grouping symbols:

    (a) Powers and roots are done first.

    (b) Multiplication and division are done in order from left to right.

    (c) Addition and subtraction are done in order from left to right.

According to the rules, $5 + 3 \cdot 4$ must have the value 17. (Since there are no grouping symbols, the multiplication must precede the addition.)

Example 1

Evaluate $5^2 - 6 + 3$.

Solution

$$\begin{aligned} 5^2 - 6 + 3 &= 25 - 6 + 3 && \text{Evaluate the power first.} \\ &= 19 + 3 && \text{Subtract from left to right.} \\ &= 22 \end{aligned}$$

Example 2

Evaluate $12 \cdot 3 + 6 - 10$.

Solution

$$\begin{aligned} 12 \cdot 3 + 6 - 10 &= 36 + 6 - 10 && \text{Multiply first.} \\ &= 42 - 10 && \text{Add next from left to right.} \\ &= 32 \end{aligned}$$

Example 3

Evaluate $24 \div (-3) \cdot 2 + 12$.

Solution

$$24 \div (-3) \cdot 2 + 12 = (-8) \cdot 2 + 12$$ Parentheses are used to denote the divisor -3. Divide first in the left-to-right order.

$$= (-16) + 12$$ Multiply before adding.

$$= -4$$

Note: It is best to keep parentheses around any negative number obtained in the evaluation process.

Example 4

Evaluate $14 - 6(3 + 4)$.

Solution

$$14 - 6(3 + 4) = 14 - 6(7)$$ Add in the parentheses first.
$$= 14 - 42$$ Multiply next.
$$= 14 + (-42)$$ Change subtraction to addition and add.
$$= -28$$

Example 5

Evaluate $2 \cdot 3^2 - 4\sqrt{25} + (-4)(-3)$.

Solution

$$2 \cdot 3^2 - 4\sqrt{25} + (-4)(-3) = 2 \cdot 9 - 4 \cdot 5 + (-4)(-3)$$ Evaluate powers and roots first.

$$= 18 - 20 + 12$$ Multiply from left to right.

$$= 18 + (-20) + 12$$ Change subtraction to addition.

$$= (-2) + 12$$ Add from left to right.

$$= 10$$

Example 6

Evaluate $2[4^2 + 3(5 - 12)]$.

Solution

$$2[4^2 + 3(5 - 12)] = 2[4^2 + 3(-7)]$$ Subtract in the innermost grouping symbol (parentheses inside brackets) first.

$$= 2[16 + 3(-7)]$$ Evaluate powers in brackets next.
$$= 2[16 + (-21)]$$ Multiply in the brackets next.
$$= 2[-5]$$ Add in the brackets.
$$= -10$$ Multiply.

Example 7

Evaluate $4^3 - 3[2\sqrt{36} - (6 \cdot 8 \div 12)]$.

Solution

$$4^3 - 3[2\sqrt{36} - (6 \cdot 8 \div 12)] = 4^3 - 3[2\sqrt{36} - (48 \div 12)]$$ Multiply left to right inside inner-most parentheses.

$$= 4^3 - 3[2\sqrt{36} - 4]$$ Divide next in innermost parentheses.

$$= 4^3 - 3[2 \cdot 6 - 4]$$ Evaluate root in brackets next.

$$= 4^3 - 3[12 - 4]$$ Multiply in the brackets.

$$= 4^3 - 3[8]$$ Subtract in the brackets.

$$= 64 - 3[8]$$ Evaluate power next.

$$= 64 - 24$$ Multiply.

$$= 40$$ Subtract.

The fraction bar (which indicates division) acts as a grouping symbol and requires that we perform any operations in the numerator and denominator before the division.

Example 8

Evaluate

$$\frac{7(-8) + 12 \cdot 3}{3 - [4(-3) + 10]}$$

Solution

$$\frac{7(-8) + 12 \cdot 3}{3 - [4(-3) + 10]} = \frac{-56 + 36}{3 - [(-12) + 10]}$$ Multiply in the numerator; multiply within the brackets in the denominator.

$$= \frac{-20}{3 - [-2]}$$ Add in numerator; add in brackets in the denominator.

$$= \frac{-20}{3 + 2}$$ Change subtraction to addition.

$$= \frac{-20}{5}$$ Add.

$$= -4$$ Divide.

Example 9

Evaluate

$$\frac{(-3)^2\sqrt{16} + 4(-7)}{2(3 \cdot 8 - 5^2)}$$

Solution

$$\frac{(-3)^2\sqrt{16} + 4(-7)}{2(3 \cdot 8 - 5^2)} = \frac{9 \cdot 4 + 4(-7)}{2(3 \cdot 8 - 25)}$$ Evaluate powers and roots first in both numerator and denominator.

$$= \frac{36 + (-28)}{2(24 - 25)}$$ Multiply in numerator; multiply within parentheses in the denominator.

$$= \frac{8}{2(-1)}$$ Add in numerator; subtract inside parentheses in the denominator.

$$= \frac{8}{-2}$$ Multiply.

$$= -4$$ Divide.

The square root symbol is also treated as a grouping symbol and requires that we evaluate the expression under it before we take the square root.

Example 10

Evaluate $\sqrt{10^2 - 8^2}$.

Solution

$$\sqrt{10^2 - 8^2} = \sqrt{100 - 64} \qquad \text{Evaluate powers first under the square root.}$$
$$= \sqrt{36} \qquad \text{Subtract.}$$
$$= 6$$

Section 1.8 Exercises

In Exercises 1 through 40, evaluate each expression.

1. $36 \div 4 \cdot 3$

2. $16 \cdot 2 \div 8$

3. $2 \cdot \sqrt{49} - 6$

4. $3\sqrt{16} + 5$

5. $12 - 6 + 3 - 8$

6. $(-14) + 4 + 7 - 9$

7. $8 - 3 \cdot 6$

8. $10 - 6 \cdot 4$

9. $5 \cdot 4 + 21 \div 3$

10. $6 \div 3 + 16 \div 4$

11. $-16 + 4(-2) \div 2^3$

12. $-18 + 6(-4) \div (-2)^2$

13. $(-9 + 6)^2 \div 3 + 5\sqrt{4}$

14. $(-8 + 3)^2 \div 5 - 4\sqrt{9}$

15. $(-12) \cdot 4 \div (-4)^2 - 5 \cdot 3^2$

16. $(-16) \cdot 3 \div (-2)^2 - 4 \cdot 2^3$

17. $16 - [8 - (5 - 7)]$

18. $24 - [12 - (6 - 9)]$

19. $10 - 3[(18 \div 3) - 12]$

20. $15 - 4[(24 \div 8) - 6]$

21. $(-3)(-6) - [(8 - 5)^3 - \sqrt{49}]$

22. $4(-5) - [\sqrt{64} - (6 - 10)^2]$

23. $(-4)(-3) - [7 - 3(6 - 8)^2]$

24. $6(-5) - [8 - 4(7 - 10)^2]$

25. $\dfrac{16 - 2(-4)}{9 - 3(2)}$

26. $\dfrac{-24 - 3(-6)}{-21 + (-4)(-5)}$

27. $\dfrac{3 \cdot 2^3 - [4(-6) + 8]}{2 \cdot 3^2 - [3(-7) + 31]}$

28. $\dfrac{2(-4)^2 - [12 - 4(-7)]}{3 \cdot (-2)^2 - [16 - 3(4)]}$

29. $\sqrt{4^2 + 3^2}$

30. $\sqrt{5^2 + 12^2}$

31. $\sqrt{13^2 - 12^2}$

32. $\sqrt{5^2 - 3^2}$

33. $\sqrt{(16.3)^2 + (20.4)^2}$

34. $\sqrt{(9.72)^2 + (7.04)^2}$

35. $17.35 + (4.26)(8.73)$

36. $21.63 + (5.06)(7.83)$

37. $(\sqrt{1.69} - 1.8)^2 \div (.05)$

38. $(\sqrt{5.76} - 3.2)^2 \div (-.04)$

39. $(7.29 - 5.68)^2 \div \sqrt{40.96}$

40. $(9.36 - 6.89)^2 \div \sqrt{10.24}$

41. Calculators use, for the most part, either of two different types of logic: "algebraic" logic or "left-to-right" logic. Algebraic logic calculators (including most scientific calculators) obey the same order of operations as we do here. So, in algebraic logic calculators, $5 + 3 \cdot 4$ has the value of 17. In the left-to-right types, all operations are performed from left to right, with no priority. In left-to-right calculators, $5 + 3 \cdot 4$ would produce 32:

$$5 + 3 \cdot 4 = 8 \cdot 4$$
$$= 32$$

Which kind of calculator do you have?

Supplementary Exercises

In Exercises 1 through 20, evaluate each of the following.

1. $16 \div 2(8)$
2. $-8 + 3(4)$
3. $5 - (2 + 3^2)$
4. $-4 + 5^2$
5. $-(5 - 9) - 7 \div (-1)$
6. $5 - 9 + 6 \div (-3)$
7. $8 + 5(3 - 7)$
8. $2^2 \div (-1) + 4$
9. $-3^2 + 7$
10. $-15 - 12 \div 4$
11. $1 - 2^2 + 5$
12. $3 - [2 - 4(7 - 9)]$
13. $6 + 5(9 - 10 \div 5)$
14. $12 \div (-2) - 10$
15. $2^2 - 3^2$
16. $(2 - 3)^2$
17. $4\sqrt{9} - 12$
18. $-7 + 12 \div (-4) - 2(-1)$
19. $15 - 3^2 \div (-1) + 4$
20. $-[6 - 2(3 - 1)^2]$

1.9   Chapter review

Summary

We began this chapter with a study of numbers. Natural numbers, whole numbers, integers, rational numbers, irrational numbers, and real numbers were defined. Then we discussed the order relations between numbers, establishing the meaning of the following symbols: $=$, $<$, $>$, $\leq$, and $\geq$.

In Sections 1.2, 1.3, and 1.4, the rules for addition, subtraction, multiplication, and division of real numbers were presented. Through the use of many examples involving these operations on real numbers, we formalized seven basic properties:

Associative Property for Addition

Commutative Property for Addition

Associative Property for Multiplication

Commutative Property for Multiplication

Distributive Property for Multiplication over Addition

Identity Property for Addition

Identity Property for Multiplication

These properties are useful when evaluating numerical expressions, which we did throughout the chapter. In Section 1.7 we defined powers and roots of numbers and established a priority of operations. By the rules of the priority of operations, we learned first to complete the operations within the innermost grouping symbols, working outward. The order to follow within the grouping symbols or when no grouping symbols were present is: (1) evaluate powers and roots in any order, (2) do multiplication and division from left to right, and, (3) do addition and subtraction from left to right.

Vocabulary Quiz

Match the expression in Column I with the phrase in Column II that best describes it.

Column I

1. The Distributive Property for Multiplication over Addition says that
2. The Commutative Property for Addition says that
3. The Commutative Property for Multiplication says that
4. The Associative Property for Addition says that
5. The Associative Property for Multiplication says that
6. The Identity Property for Addition says that
7. The Identity Property for Multiplication says that
8. The natural numbers include
9. The integers include
10. The rational numbers include
11. The irrational numbers include
12. The absolute value of a number is

Column II

a. when three numbers are multiplied, the first two numbers can be grouped together or the last two numbers can be grouped together—the product will be the same.

b. 1, 2, 3, . . .

c. when a number is multiplied by 1, the result is the same as the original number.

d. we can rewrite $3(4 + 2)$ as $3 \cdot 4 + 3 \cdot 2$.

e. two numbers can be multiplied in either order—the product will be the same.

f. . . . $-3, -2, -1, 0, 1, 2, 3, \ldots$

g. the quotients of two integers so long as the divisor (denominator) integer is not 0.

h. the positive distance between 0 and that number on the number line.

i. when three numbers are added, the first two numbers can be grouped together or the last two numbers can be grouped together—the sum will be the same.

j. two numbers can be added in either order—the sum will be the same.

k. when 0 is added to the number, the sum is the same as the original number.

l. numbers such as π and $\sqrt{3}$, which are not rational numbers.

Chapter 1 Review Exercises

1. Write the natural numbers that are less than 7.
2. Write the integers evenly divisible by 3.
3. Given the numbers $\left\{ -\dfrac{3}{4}, 0, -5, 7, \dfrac{1}{2}, \sqrt{3}, -\sqrt{2} \right\}$, list
   (a) the integers
   (b) rational numbers
   (c) irrational numbers
   (d) real numbers

Determine whether each of the statements is true or false.

4. $-5 \geq -4$
5. $-\sqrt{2} \leq -\sqrt{2}$
6. $3 + (-4) \neq 1$
7. $2(-6) > -10$
8. $2\pi > 6$
9. $-5 \not> 0$

In Exercises 10 through 58, evaluate the expressions, if possible.

10. $6 - 15$
11. $6 - (-15)$
12. $(-12) + (-8)$
13. $(-3) - 7$
14. $(-3) - (-7)$
15. $(-8) + 3 + (-7)$
16. $6 - 9 - 4$
17. $5 \cdot 2 - 21$
18. $(-4)(-7) + 5$
19. $(-7)(-1)$
20. $6(-12)$
21. $(-24) \div (-12)$
22. $36 \div (-4)$
23. $[8 \cdot (-6)] \cdot (-1)$
24. $[7 \cdot (-5)] \cdot (-2)$

25. $(-48) \div (-8)$

26. $(-60) \div (-5)$

27. $5 \cdot (-3) + 2(-4)$

28. $(-4)(-6) - 3(-9)$

29. $(-5)^2$

30. $(-7)^2$

31. $4^2 - 3(-5)$

32. $8^2 + 4(-3)$

33. $[4^2 - 3](-5)$

34. $[8^2 + 4](-3)$

35. $-5 + 4(-3)$

36. $-7 - 4(6)$

37. $24 - [3^3 - 4(5)]$

38. $36 - [(-5)^2 + 3(-6)]$

39. $\sqrt[3]{-125}$

40. $\sqrt[3]{-216}$

41. $\sqrt{-25}$

42. $\sqrt{-36}$

43. $\sqrt{36} - [(5)(4) \div (-10)]$

44. $\sqrt{81} - [(-6)(8) \div 3 \cdot 2]$

45. $5(-6) - [18 - (6 - 9)^3]$

46. $8(-4) - [15 - (3 - 7)^2]$

47. $\dfrac{(-3)8 + 6(-4)}{3(-1) + 5(3)}$

48. $\dfrac{(-36) \div 4 \cdot (-5)}{56 \div 7 - 5}$

49. $\dfrac{3^3 - [4(-8) + 3(5)]}{5^2 - [(-3)(-9) + 2]}$

50. $\dfrac{(-2)^4 - [6(-5) + (-4)]}{2^3 - [4(6) - 11]}$

51. $36 \div \sqrt[3]{27} \cdot (-2)^3$

52. $(-24) \div \sqrt{36} \cdot (-3)^2$

53. $\sqrt{8^2 + 6^2}$

54. $\sqrt{10^2 - 6^2}$

📱 55. $\dfrac{-16.45 + 8.37}{9.45 - 6.32}$

📱 56. $\dfrac{17.43 - (-5.21)}{4.73 + 6.49}$

📱 57. $(17.39 \div 4.27)^2 - (-6.34)$

📱 58. $(6.48 \div 2.37)^2 - (-10.46)$

In Exercises 59 through 68, state whether each statement is true or false and give the reason for your choice.

59. $7[8 + (-10)] = 7 \cdot 8 + 7(-10)$

60. $(-14) + 9 = 9 + (-14)$

61. $(-17)[4(5)] = [(-17) \cdot 4] \cdot 5$

62. $a \cdot b - c = a - b \cdot c$

63. $(-5) \cdot x = x \cdot (-5)$

64. $x + 0 = x$

65. $1 \cdot (-16) = -16$

66. $6(x - y) = 6x - 6y$

67. $48 \div [6 + x] = (48 \div 6) + (48 \div x)$

68. $(x + 5) + y = x + (5 + y)$

Chapter 1 test

Take this test to determine how well you have mastered the concept of real numbers and their properties; check your answers with those found at the end of the book.

Given the numbers $\left\{ -7, \frac{4}{5}, -\pi, 0, 5, -\frac{2}{3} \right\}$, list the following.   *(Section 1.1)*

1. The natural numbers

2. The integers

1. _____

2. _____

3. The rational numbers

4. The real numbers

3. _____

4. _____

In each of the following, perform the indicated operation. If the operation cannot be done, give a reason.   *(Sections 1.2, 1.3, 1.4, and 1.6)*

5. $5 - 12$

6. $(-6) + (-7)$

5. _____

6. _____

7. $14 \div (-7)$

8. $(-3)(-2)$

7. _____

8. _____

9. $8 - 3$

10. $(-15) + 7$

9. _____

10. _____

11. $-3(4)$

12. $18 \div (-6)$

11. _____

12. _____

13. $0 - (-11)$

14. $(-4) + 0$

13. _____

14. _____

15. $\dfrac{-16}{0}$

16. $\dfrac{0}{-16}$

15. _____

16. _____

17. $(-24) \div (-2)$

18. $(-5)8$

17. _____

18. _____

19. $\dfrac{0}{0}$

20. Subtract -5 from -3.

19. _____

20. _____

In Problems 21–24, state whether each statement is true or false and give the reason for your choice. *(Section 1.5)*

21. $3 + [(-4) + 5] = [3 + (-4)] + 5$

22. $1 \cdot (-6) = -6$

21. _____

22. _____

23. $-15 \div 5 = -15 + 5$

24. $6[(-3) + 7] = 6(-3) + 6(7)$

23. _____

24. _____

In Problems 25 through 40, evaluate each expression.   *(Section 1.7, 25–32; Section 1.8, 33–40)*

25. 3^3

26. -2^2

25. _____

26. _____

27. $(-5)^2$

28. $(-1)^5$

27. _____

28. _____

29. $\sqrt{49}$

30. $-\sqrt{36}$

29. _____

30. _____

31. $\sqrt{-8}$

32. $\sqrt[3]{-8}$

33. $16 - 6 \cdot 3$

34. $24 \div 4 \cdot 3$

35. $[(-3) - (-7)] \cdot 6$

36. $3^2 \cdot (-4) + 5\sqrt{16}$

37. $\dfrac{-18 + 6(-3)}{4^2 - 7}$

38. $(-8 + 2)^2 \div (-4) - 3(-5)$

39. $\sqrt{6^2 + 8^2}$

40. $3(-5) - [\sqrt{81} - (7 - 10)^3]$

31. _____

32. _____

33. _____

34. _____

35. _____

36. _____

37. _____

38. _____

39. _____

40. _____

chapter two

The language of algebra

2.1 Representing numbers by the alphabet

Algebra is simply a generalization of arithmetic. In this chapter we begin a study of that generalization. For instance, consider the expression $2 + 5$. Its value is definite because there is only one number that is the sum of 2 and 5. On the other hand, consider the expression $? + 5$. Think of it as "to some *unknown* number (?), we add 5." When $? = 2$, the value of $? + 5$ is 7. When $? = 100$, $? + 5 = 105$. In this chapter we will examine expressions where one or more of the quantities on which we operate is not a specific number; its value is **unknown.** What we will do is to allow a quantity to have the privilege to *vary.*

Example 1

(a) Find the value of $? + 3$ if $? = 8$.

(b) If $? = 10$, find the value of $6 \cdot ?$.

(c) Evaluate $(? \cdot 4) - 5$ when $? = -3$.

Solution

(a) When $? = 8$, $? + 3$ becomes $8 + 3$. The value is 11.

(b) When $? = 10$, the value of $6 \cdot ?$ is $6 \cdot 10 = 60$.

(c) We work within parentheses first:

$$
\begin{aligned}
(? \cdot 4) &- 5 \\
= (-3 \cdot 4) &- 5 \\
= (-12) &- 5 \\
= &-17
\end{aligned}
$$

The varying quantity, called the **variable,** can take the written form of a question mark (as above) or a blank space. But in algebra it has become the convention to use a letter of the alphabet for the variable. Thus, we will write $a + 6$, $3 \cdot x - 17$, or $5 \cdot (R - 7)$ to represent varying quantities.

Another algebraic convention is to drop the multiplication "dot." Thus $25 \cdot x$ will be written as $25x$, and when no operation symbol appears, the operation will be understood to mean multiplication. We also agree to write $(-5) \cdot x$ as $-5x$ with the same understanding.

Example 2

Rewrite $(-3) \cdot x \cdot (2 \cdot x - 6)$ without multiplication dots.

Solution

We write

$$-3x(2x - 6)$$

and respect the fact that three multiplication symbols have been dropped, but multiplication is what we mean.

We *evaluate* expressions (determine their value) in the next three examples. Notice that we obey the **priority of operations** discussed in Section 1.8.

Example 3

(a) Evaluate $a(b + 7)$ when $a = 4$ and $b = -5$.

(b) If $p = 6$ and $q = -3$, find the value of $pq(p + q)$.

Solution

(a) Recall $a(b + 7)$ means multiply a by the value of $b + 7$:

$$a(b + 7) = 4[(-5) + 7]$$
$$= 4(2)$$
$$= 8$$

(b) $pq(p + q)$ means $p \cdot q \cdot (p + q)$.

$$pq(p + q) = (6)(-3)[6 + (-3)]$$
$$= (6)(-3)[3]$$
$$= -54$$

Example 4

Assume $x = 2$, $y = -3$, and $z = 5$. Evaluate each of the following:

(a) $x[3(y - 1) + z]$ (b) $(x - y)(x + y)$ (c) $x - 3(z - 1)$

Solution

(a) Work within innermost grouping symbols first:

$$x[3(y - 1) + z] = 2[3(-3 - 1) + 5] \qquad \text{Substitute values for } x, y, z.$$
$$= 2[3(-4) + 5] \qquad \text{Work within parentheses first.}$$
$$= 2[-12 + 5] \qquad \text{Multiply before adding within brackets.}$$
$$= 2[-7]$$
$$= -14$$

(b) $(x - y)(x + y) = [2 - (-3)][2 + (-3)]$ Substitute 2 for x and -3 for y and work within brackets first.

$$= 5 \cdot [-1]$$
$$= -5$$

(c) $x - 3(z - 1) = 2 - 3(5 - 1)$

$$= 2 - 3(4) \qquad \text{Work within parentheses first.}$$
$$= 2 - 12 \qquad \text{Multiply before subtracting.}$$
$$= -10$$

In the next example, we evaluate an algebraic expression that involves absolute value notation.

Example 5

Evaluate the expression $|2x - 1| + |5 - y|$ assuming $x = 4$ and $y = 10$.

Solution

Each pair of absolute value symbols acts as a grouping symbol; we work within each separately.

$$|2x - 1| + |5 - y| = |2 \cdot 4 - 1| + |5 - 10| \quad \text{Substitute 4 for } x \text{ and 10 for } y.$$
$$= |7| + |-5| \quad \text{Work within each pair of symbols.}$$
$$= 7 + 5 \quad \text{Evaluate each absolute value before adding.}$$
$$= 12$$

We conclude this section by showing three examples in which we translate sentences into **algebraic expressions.** Algebraic expressions are expressions involving at least one variable.

Example 6

Walter earns $5 per hour. What algebraic expression represents his gross earnings if h represents the number of hours he worked?

Solution

When Walter works 6 hours, he earns $5 \cdot 6 = \$30$; when he works 40 hours, he earns $5 \cdot 40 = \$200$. In other words, he earns $5 *times* the number of hours worked. So, *in general*, he can expect $5h$ as his gross earnings.

$$\underset{\substack{\downarrow \\ \$5 \text{ per hour}}}{5} \quad \underset{\substack{\downarrow \\ \text{times}}}{\cdot} \quad \underset{\substack{\downarrow \\ \text{number of hours worked}}}{h} = 5h$$

Example 7

The gram is a unit of mass in the metric system. There are 454 grams in one pound.

  (a) What *numerical* expression represents the number of grams in 10 pounds?
  (b) What *algebraic* expression represents the number of grams in x pounds?

Solution

  (a) There are 454 grams in one pound, so in 10 pounds there would be $454 \cdot 10$ grams.
  (b) To express the number of grams in x pounds, we multiply:

$$454 \cdot x = 454x \text{ grams}$$

Example 8

Andrea is 28 years less than twice Gloria's age. If g represents Gloria's age, what algebraic expression represents Andrea's age?

Solution

"28 less than twice Gloria's age" is a complex statement. First, let's examine how to represent "twice Gloria's age." "Twice" means double, so we multiply Gloria's age, g, by 2. "Twice Gloria's age," then, is $2g$. Now, "28 less than $2g$" means to subtract 28 from $2g$. So

$$2g - 28$$

represents Andrea's age.

Section 2.1 Exercises

1. Rewrite $3 \cdot j - 7$ using algebraic conventions.

2. Rewrite $(3 \cdot x - y) \cdot (2 \cdot x + y)$ using algebraic conventions.

In Exercises 3 through 18, assume $x = 3$, $y = 0$, $d = -3$, and $n = 5$. Evaluate each expression.

3. $x + n$

4. $n - x$

5. $x - n$

6. $dx - 5$

7. $-3(x + 2)$

8. $-3(x + 2) - y(d + n)$

9. $x[d - 5(n + d)]$

10. $x(d + 2)$

11. $2x + dx$

12. $y - x(3 \div d)$

13. $nx(x - d)$

14. $3[x - y(d + 2)] - 10[x - d(n + 3)]$

15. $x \cdot x \cdot x$

16. $(nx + dn)[5 - x(d + 2)]$

17. $y \div n$

18. $n \div y$

In Exercises 19 through 33, assume $p = 2$, $q = -5$, and $r = 6$. Evaluate each of the algebraic expressions.

19. $p + 8$

20. $p - 8$

21. $8 - p$

22. $p + q$

23. $(p + q) + r$

24. $p + (q + r)$

25. $8(p + q)$

26. $8p + 8q$

27. $(p \cdot r) + 7$

28. $p \cdot p \cdot p$

29. $3p - 1$

30. $p[6(5 - q) + r]$

31. $(p - q) - 2(3 - r)$

32. $p - 2(r - q)$

33. $(p + q)(r + 10)$

34. $|2p - 3q|$

35. $|2p| - |3q|$

36. $|p + q| - 5r$

37. $|p + q - 5r|$

38. A certain shirt pattern calls for 2 meters (a metric unit equal to about 39 inches) of fabric. If s represents the number of shirts to be made, what algebraic expression represents the number of meters of fabric needed?

39. In figuring his income tax, John has determined his adjusted gross income to be $25,000 plus $2000 for each bank account he has. If n represents the number of John's bank accounts, what algebraic expression represents John's adjusted gross income?

40. If, in addition to the condition of Exercise 39, John may subtract $1000 for each child to determine his taxable income, what algebraic expression represents John's taxable income if c represents the number of children he has?

41. Mary is 5 years older than twice Jill's age. If j represents Jill's age, what algebraic expression represents Mary's age?

42. Use your answers to Exercises 39 and 40 to determine John's taxable income when $n = 1$ and $c = 8$.

43. Use your answer to Exercise 41 to determine Mary's age when $j = 21$.

44. George makes $16 per hour. If g represents the number of hours he works, represent algebraically the amount George earns.

45. Samantha is going to buy 300 grams of potassium nitrate for her chemistry class. If c represents the cost per gram, write an algebraic expression for the total cost.

46. Dom sells used cars. His annual salary is $5000. In addition he makes $25 for each car he sells. If x represents the number of cars Dom sells in a year, express algebraically his annual income.

47. Ken's car gets 11 miles per gallon. If k represents the number of gallons Ken has, write an algebraic expression for how far he can travel.

48. A man willed all of his estate as follows: $3595 to his wife, $2630 of his estate to each of his daughters, and $4647 to each of his sons. Let d represent the number of daughters and s represent the number of sons.

    (a) Represent the total amount of his estate algebraically.

    (b) If he had two sons and four daughters, what was the value of his estate?

In Exercises 49 through 60 write each of the following algebraic expressions in words.

Example: $5 - 2x$ means subtract two times x from 5

49. $x + 2$ 50. $r + t$ 51. $3x - y$

52. $4xy$ 53. $x \div y$ 54. $3(x + y)$

55. $6 - (2x + 4)$ 56. $\dfrac{3xy}{4}$ 57. $|4 + 2s|$

58. $(3 + t)(6 + t)$ 59. $100x - 3y$ 60. $6 - 2(4 - x)$

Supplementary Exercises

In Exercises 1 through 20 evaluate each of the algebraic expressions assuming that $x = -6$, $t = -8$, $q = 0$, $r = 2$, and $s = 3$.

1. $-x$ 2. t 3. $5x - 2t + r$

4. $\dfrac{5x}{2}$ 5. $r(x + t)$ 6. $5q - r$

7. $x \div q$ 8. $4(2r - t)$ 9. $-2rst$

10. $6[x - (r + t)]$ 11. $4(r + s)$ 12. $2s + x$

13. sss 14. $s - 3t$ 15. $(t + s) - (t + x)$

16. $5 - 6(2r + s)$ 17. $|x - s| + 3$ 18. $-|2r + t|$

19. $s[2r - (t + 4s)]$ 20. $x \div rs$

21. Carol earns $1000 more than three times Joan's salary. If s represents Joan's salary, what algebraic expression represents Carol's salary?

22. Chris purchased four less pounds of apples as bananas and one-half as many pounds of peaches as bananas. If x represents the number of pounds of bananas Chris purchased, what algebraic expressions represent the number of pounds of apples and peaches he bought?

23. Use your answer to Exercise 21 to determine Carol's salary if Joan's salary was $4000.

24. Use your answer to Exercise 22 to determine the number of pounds of apples and peaches Chris purchased if the number of pounds of bananas purchased was 12.

25. Suzanne sews dresses in a factory that pays a weekly salary of $200. In addition, she is paid $5 for each dress she completes. If d represents the number of dresses she completes, write an algebraic expression that represents her weekly salary.

26. Use your answer to Exercise 25 to determine Suzanne's salary for the week if she completes 8 dresses in the week.

In Exercises 27 through 30, write the algebraic expressions in words.

27. $5t - q$ 28. $3t \div s$ 29. $q - (2r + 3)$

30. $4rs + t$

2.2   The arithmetic of letters: simplification

As we learned in Section 2.1, algebraic expressions are just arithmetic expressions with letters substituted for numbers. Therefore, the seven properties of Chapter 1 apply. We repeat them here.

The commutative property for addition

$$a + b = b + a$$

The associative property for addition

$$a + (b + c) = (a + b) + c$$

The commutative property for multiplication

$$ab = ba$$

The associative property for multiplication

$$a(bc) = (ab)c$$

The distributive property for multiplication over addition

$$a(b + c) = ab + ac$$

The identity property for addition

$$a + 0 = 0 + a = a$$

The identity property for multiplication

$$a \cdot 1 = 1 \cdot a = a$$

In this section we will use the seven properties to simplify algebraic expressions. First, however, we need a few definitions.

In this chapter we will use the word "term" to indicate a part of an algebraic expression separated from other parts by " + " or " − " symbols. Each + and − sign is part of the term that follows it. Thus, the expression $4y - 3x + 17$ consists of three terms:

$$\boxed{4y} \; + \; \boxed{-3x} \; + \; \boxed{17}$$

$$\downarrow \qquad\quad \downarrow \qquad\quad \downarrow$$

term 1   term 2   term 3

Notice that even though they are not marked with + signs, the first and third terms are understood as $+4y$ and $+17$, respectively. In general, we understand expressions to be positive unless they are clearly identified with a minus sign as negative.

Like terms are terms that have identical literal (letter) components. The terms $6x$, $-10x$, and x are like terms since each term has x as the only letter. The terms $8x$ and $-3y$ contain different letters; we refer to these terms as **unlike terms.**

One *simplification process* is called **collecting like terms** and essentially involves writing an expression using fewer symbols. For example, the expression

$$(3a + 5) + (4a + 7)$$

can be rewritten by using the following steps:

$(3a + 5) + (4a + 7)$	$= 3a + (5 + 4a) + 7$	Associative Property for Addition ("Regrouping Property")
	$= 3a + (4a + 5) + 7$	Commutative Property for Addition
	$= (3a + 4a) + (5 + 7)$	Regroup again. (Associative Property)
	$= a(3 + 4) + 12$	Distributive Property; also add $5 + 7$.
	$= a(7) + 12$	Add $3 + 4$.
	$= 7a + 12$	Commutative Property for Multiplication (It is customary to write numbers to the left of letters.)

Thus, $(3a + 5) + (4a + 7)$

is simplified to

$$7a + 12$$

We call 7 the **coefficient** (or multiplier) of a; 12 is called the **constant term.**

Example 1

Simplify $6x - 2y + 5x + 4y$.

Solution

We regroup these four terms in the steps below with the strategy of placing like terms together. We place the reason for each step to the right for thoroughness.

$6x - 2y + 5x + 4y$	$= 6x + 5x - 2y + 4y$	Commutative Property for Addition (Think of $-2y$ as adding a negative $2y$.)
	$= x(6 + 5) + y(-2 + 4)$	Distributive Property
	$= x \cdot 11 + y \cdot 2$	Add $6 + 5$ and $-2 + 4$.
	$= 11x + 2y$	Commutative Property for Multiplication

Before proceeding to additional examples, we emphasize three things:

Guidelines for simplification

1. Although the properties of arithmetic were used to *justify* our steps in the simplification above, it is bothersome to write them all down. When $(3a + 5) + (4a + 7)$ was simplified, we **added** the **like terms** of $3a$ and $4a$ and got $7a$. (Don't forget, "a" stands for some quantity, and 3 "quantities" plus 4 "quantities" are 7 "quantities."
2. Remember that the sign immediately preceding a term is part of it. Thus, in $3t - 4(t + 2)$ the -4 will multiply both the t and 2. (See Example 3.)
3. Simplified results will be written with terms in alphabetical order and with the constant terms (numbers) last. This is another algebraic convention.

Example 2

Simplify $(6r - 4s) + (3r + 6s)$.

Solution

First, notice that there is really no need for parentheses since there is no quantity preceding each set of parentheses that will change the value of the expression inside. We could write

$$6r - 4s + 3r + 6s$$

Now, the "like" terms of $6r$ and $3r$ can be "combined" (added) and the like terms of $-4s$ and $6s$ can be added as follows:

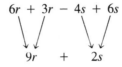

$$6r + 3r - 4s + 6s$$
$$9r + 2s$$

Example 3

Simplify $3t - 4(t + 2)$.

Solution

The parentheses *cannot* be dropped here because each term in parentheses (the "t" and "2") must be multiplied by -4:

$$3t - 4(t + 2) = 3t - 4t - 8 \qquad \text{The Distributive Property}$$
$$= -1t - 8 \qquad \text{Add } 3t \text{ and } -4t.$$

One further convention: we will not write coefficients of 1 or -1. Since 1 is the multiplicative identity and, thus, $1 \cdot x = x$, the coefficient is said to be understood. Thus

$$-1t - 8$$

becomes

$$-t - 8$$

Example 4

Simplify $6x - 2(x - 5) + 25$.

Solution

First we eliminate parentheses by applying the Distributive Property.

$$6x - 2(x - 5) + 25 = 6x - 2x + 10 + 25 \qquad \textit{Note: } (-2) \cdot (-5) \text{ is } +10.$$
$$= 4x + 35 \qquad \text{Collect like terms.}$$

Example 5

Simplify $[3(x + 7) - 2x] + 3[x + 2(x - 1)]$.

Solution

Before we collect like terms, we must eliminate parentheses and brackets.

$$[3(x + 7) - 2x] + 3[x + 2(x - 1)]$$
$$= [3x + 21 - 2x] + 3[x + 2x - 2] \qquad \text{Eliminate parentheses by using the distributive property.}$$
$$= [x + 21] + 3[3x - 2] \qquad \text{Collect terms within brackets.}$$
$$= x + 21 + 9x - 6 \qquad \text{Eliminate brackets.}$$
$$= 10x + 15$$

The final two examples of this section show how to eliminate parentheses when they are preceded by a minus sign.

Example 6

Simplify $2x - (3x - 4)$.

Solution

Notice that $(3x - 4)$ is identical to $1(3x - 4)$. So, we can write

$$2x - (3x - 4)$$

as

$$2x - 1(3x - 4)$$

Now, multiply by -1:

$$2x - 1(3x - 4) = 2x - 3x + 4$$
$$= -x + 4$$

Example 7

Simplify $13(x + 3y) - 5(2x - 4y) - (x + 2y) + 4$.

Solution

$$13(x + 3y) - 5(2x - 4y) - (x + 2y) + 4$$
$$= 13x + 39y - 10x + 20y - x - 2y + 4$$
$$= 2x + 57y + 4$$

Finally, a term can be eliminated in a simplification process when two terms add up to 0.

$$15(x - 2) - 3(5x + 7) = 15x - 30 - 15x - 21$$
$$= 15x - 15x - 51 \qquad \text{\textit{Note:} } 15x - 15x \text{ is } 0 \cdot x \text{ or } 0.$$
$$= -51$$

Section 2.2 Exercises

In Exercises 1 through 28, simplify each algebraic expression, if possible.

1. $(t + 4) + 9$
2. $8 + (3x - 47)$
3. $6 + (2x + 5)$
4. $2 + (y - 10)$
5. $10 + (3 - 6x) + 5$
6. $5t - 2t$
7. $7 + (4p - 2)$
8. $43 + (7 - 2x) + 8x$
9. $(v - 9) + 27$
10. $13 - 7(5u - 11) + 55u$
11. $-2y - 4(y + 3)$
12. $20p + 10q - 21r$
13. $3(a + 20) - 2(a + 6)$
14. $3(a - 20) - 2(a - 6)$
15. $5a + 4j - 2a - j$
16. $8k - 20k + 1$
17. $6(x - 4) + 24$
18. $4(2x - 3) - 8x + 12$
19. $8(12x + 1) + 2(x + 5)$
20. $8(12x - 1) - 2(x - 5)$
21. $3(2x + y) + 3[x - (y + 2z)]$
22. $-7(2x - 9) + 8(x - 1)$
23. $6(x - 4) - 3(2x - 18)$
24. $7(2x - 2) - 2(7x + 5)$
25. $-2[x - 2(3x - 4)] - 5(x + 6)$
26. $3[x + 2(x - 3)] - 2[3 - 2(x - 1)]$

27. $3(2x + y) - 13(x - y) + 2(x + z)$

28. $10[a - 2(b + 3c)] - 4[a + 7(b - c)] + 8[a + 6(b + 1)]$

In Exercises 29 and 30, keep in mind that although we are rewriting algebraic expressions, the *value* of the expressions will remain unchanged.

29. $2(3x - 4) + 7$ is simplified as $6x - 1$.

   (a) If $x = 5$, what is the value of $2(3x - 4) + 7$? What is the value of $6x - 1$?

   (b) If $x = -1$, what is the value of $2(3x - 4) + 7$? What is the value of $6x - 1$?

30. In Example 7, $13(x + 3y) - 5(2x - 4y) - (x + 2y) + 4$ was simplified to $2x + 57y + 4$.

   (a) If $x = 2$ and $y = 0$, find the value of $13(x + 3y) - 5(2x - 4y) - (x + 2y) + 4$. Find the value of $2x + 57y + 4$.

   (b) Repeat part (a) if $x = -2$ and $y = -5$.

31. The Distributive Property for Multiplication over Addition as written for algebraic expressions,

$$a(b + c) = ab + ac$$

   can be interpreted geometrically.

   (a) First, find the area of each rectangle below

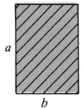 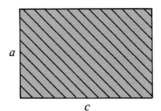

   (b) Now find the area of the new rectangle that was made by adjoining the two rectangles above:

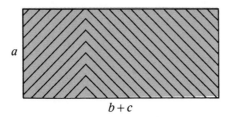

   (c) Compare your results in parts (a) and (b).

32. The Distributive Property can be extended to three terms in parentheses:

$$a(b + c + d) = ab + ac + ad$$

   Simplify each expression below.

   (a) $3(x - y + 6) - 2(x + y - 7)$

   (b) $14(r - 2s + 10) - 7(2r - 4s + 21)$

   (c) $14(r - 2s + 10) - 7(2r - 4s + 20)$

33. Charlene purchased x 13-cent stamps, $(2x - 1)$ 15-cent stamps and $(15 - x)$ 20-cent stamps. Write a simplified algebraic expression to represent the total number of stamps Charlene purchased in terms of x.

34. Pierre worked $x + 5$ hours on Monday, x hours on Tuesday, $x - 1$ hours on Wednesday, $2x$ hours on Thursday, and $x + 3$ hours on Friday. Write a simplified expression to represent the total number of hours Pierre worked during the week.

35. Refer back to Exercise 33. If Charlene purchased seven 13-cent stamps, how many 15-cent stamps and 20-cent stamps did she purchase? (*Hint*: Using your expression for the total number of stamps Charlene purchased, evaluate this expression for $x = 7$.)

36. Referring to Exercise 34, find the total number of hours Pierre worked if $x = 6$. (*Hint*: Use your algebraic expression for the total number of hours Pierre worked. Then check your answer by evaluating for each day and then adding.)

37. Mary sells her own coffee blend using Brazilian beans and Colombian beans by mixing large bags of these beans together and then repackaging her mix in small quantities. The Brazilian beans she buys are packed in bags that contain x pounds each and the Colombian beans are packed in bags that contain y pounds each. If her blend contains 5 bags of Brazilian beans and 4 bags of Colombian beans, write a simplified expression for the number of pounds of beans she uses in terms of x and y.

38. If each bag of Brazilian beans in Exercise 37 weighs 3 pounds and each bag of Colombian beans weighs 4 pounds, how many pounds of beans does Mary use in her blend?

39. Refer again to Exercise 37. Suppose Mary decides to triple her blend. Write an algebraic expression to represent the total number of pounds of beans used in terms of x and y.

40. Ann purchases $2x + 3$ pounds of apples, $3x + y$ pounds of bananas, and $3y + 4$ pounds of grapes. What simplified algebraic expression, in terms of x and y, represents the total number of pounds of fruit Ann purchased?

Supplementary Exercises

In Exercises 1 through 16, simplify each of the following, if possible.

1. $(3t - 2) + (4 - 5t)$
2. $4(x - 1) - 5(3 - x)$
3. $6 - 2(3 + a)$
4. $5a - (3 + 5a)$
5. $(2x + y - 5) - (x - y - 3) + (3y - 2x + 1)$
6. $4(x + 2y) - 3(x + y)$
7. $3[x - 2(x + 1)]$
8. $6 - [x - (2 - x)]$
9. $2(3x + y - 1) - (x - 2y - 1)$
10. $(a + b + c) - (a - b - c)$
11. $3x - [2(x - 1) - 4x]$
12. $5y + 3(2y + 4) - 1$
13. $3 - [2 - (3x - 1)]$
14. $(a + 2b) - (a + 2b)$
15. $-2 - 5(a + 1) + 4a$
16. $5(x + y - 1) - 2(y + 2x + 1) + 3(2y + x)$

17. Karl purchased $3x + y$ pounds of nuts and $x - 3y$ pounds of dried fruit. Write a simplified expression for the total number of pounds of nuts and dried fruit purchased by Karl.

18. If $x = 10$ and $y = 1$ in Exercise 17, what was the total number of pounds of nuts and dried fruit purchased by Karl?

19. Kit bought $3x + 1$ pounds of fruit, which cost 39 cents per pound. She also bought $x - 1$ pounds of nuts, which cost 75 cents per pound. Write a simplified expression for the total cost of the fruit and nuts.

20. If $x = 5$ in Exercise 19, what was Kit's total cost for the fruit and nuts?

2.3   Exponent Rules 1 and 2

In Chapter 1 we saw how exponents were used in the evaluation of arithmetic expressions. Here, we are concerned with extending the use of exponents to *algebraic* expressions.

$$b^5$$

means

$$b \cdot b \cdot b \cdot b \cdot b$$

In b^5, b is called the **base** and 5 is the **exponent.** We say b is used as a **factor** five times. A factor is an expression that is part of a product.

In the next two examples, we evaluate algebraic expressions involving exponents.

Example 1

Evaluate each expression. Assume $x = 4$ and $y = 2$.

(a) x^3 (b) $(2y)^2$ (c) $(-3y)^3$ (d) x^y

Solution

(a) $x^3 = 4^3$ Substitute 4 for x.

$\quad = 4 \cdot 4 \cdot 4$

$\quad = 64$

(b) $(2y)^2 = (2 \cdot 2)^2$ Substitute 2 for y.

$\quad = 4^2$

$\quad = 16$

(c) $(-3y)^3 = (-3 \cdot 2)^3$ Substitute 2 for y.

$\quad = (-6)^3$ Multiply within parentheses first.

$\quad = (-6)(-6)(-6)$

$\quad = -216$

(d) $x^y = 4^2$ Substitute 2 for y and 4 for x.

$\quad = 16$ 4 is **squared** (that is, raised to the second power.)

Before proceeding to the next example, we point out again the difference between the symbols $(-2)^4$ and -2^4. In the expression $(-2)^4$, the -2 is in parentheses and thus -2 must be used as a factor four times:

$$(-2)^4 = (-2)(-2)(-2)(-2) = 16$$

On the other hand, in -2^4, the "$-$" sign is used to denote "the opposite sign of" 2^4. So,

$$-2^4 = -(2 \cdot 2 \cdot 2 \cdot 2) = -16$$

Example 2

Evaluate the following expressions.

(a) $(-2)^3$ (b) -3^2 (c) $(-3)^2$

Solution

(a) $(-2)^3 = (-2)(-2)(-2) = -8$

(b) $-3^2 = -(3 \cdot 3) = -9$

(c) $(-3)^2 = (-3)(-3) = 9$

Example 3

Assume $p = -4$ and $q = 3$. Find the value of each expression.

(a) p^3 (b) $(p + q)^2$ (c) $p^2 + q^2$ (d) $-q^2$

Solution

(a) When we substitute -4 for p, we use parentheses to indicate the *entire quantity*, -4, is to be **cubed** (raised to the third power):

$$p^3 = (-4)^3 = (-4)(-4)(-4) = -64$$

(b) $(p + q)^2 = (-4 + 3)^2$ Substitute -4 for p and 3 for q.

$= (-1)^2$ $(-1)^2 = (-1)(-1)$

$= 1$

(c) $p^2 + q^2 = (-4)^2 + 3^2$ Substitute values for p and q. Notice, there is no need to put parentheses around ''3'' but we *must* put them around -4.

$= 16 + 9$

$= 25$

(d) $-q^2 = -(3^2)$

$= -9$

Example 4

Assume $x = -2$ and $y = 3$. Evaluate each expression.

(a) $x^3 - y^3$ (b) $|x^3 - y^3|$ (c) $|x^3| - |y^3|$

Solution

(a) After substituting in -2 for x and 3 for y, we have

$$x^3 = (-2)^3 = -8 \quad \text{and} \quad y^3 = 3^3 = 27$$

So,

$$x^3 - y^3 = -8 - 27 = -35$$

(b) In $|x^3 - y^3|$, the absolute value symbols act as parentheses to group $x^3 - y^3$. So,

$$|x^3 - y^3| = |-8 - 27|$$
$$= |-35| = 35$$

(c) $|x^3| = |(-2)^3| = |-8| = 8$

$|y^3| = |3^3| = |27| = 27$

Hence, $|x^3| - |y^3| = 8 - 27 = -19$.

The examples above involved *evaluating* expressions with exponents. The *simplification* of exponential expressions will be explored in the remainder of this section and in the next section.

Consider what happens when we multiply two one-term expressions that have the same base, such as

$$x^4 \cdot x^5$$

Using the definition of exponent, we can write x^4 as $x \cdot x \cdot x \cdot x$ and x^5 as $x \cdot x \cdot x \cdot x \cdot x$:

$$\overbrace{x \cdot x \cdot x \cdot x}^{x^4} \cdot \overbrace{x \cdot x \cdot x \cdot x \cdot x}^{x^5}$$

Since we have a string of nine factors of x, we see that

$$x^4 \cdot x^5 = x^9$$

This leads us to the following rule:

Exponent Rule 1

$$x^a \cdot x^b = x^{a+b}$$

That is, when multiplying two expressions with the *same base*, add the exponents.

Example 5

Use Exponent Rule 1 to rewrite each of the following, if possible.

(a) $x^5 \cdot x^6$ (b) $y \cdot y^4$ (c) $z \cdot z^2 \cdot z^4$

Solution

(a) $x^5 \cdot x^6 = x^{5+6} = x^{11}$

(b) Notice that "y" could also be written as "y^1," even though the exponent 1 is rarely written. However, we will write it below to show our computation:

$$y \cdot y^4 = y^1 \cdot y^4 = y^{1+4} = y^5$$

(c) Exponent Rule 1 can be extended to involve any number of factors. Thus,

$$z \cdot z^2 \cdot z^4 = z^{1+2+4} = z^7$$

To simplify a term raised to a power, such as $(x^2)^3$, observe that the "3" means use x^2 as a factor three times.

$$(x^2)^3 = x^2 \cdot x^2 \cdot x^2 \qquad \text{Use } x^2 \text{ as a factor 3 times.}$$
$$= x^{2+2+2} = x^{2 \cdot 3}$$
$$= x^6$$

Notice that $(x^2)^3 = x^{2 \cdot 3} = x^6$. This leads to Exponent Rule 2.

Exponent Rule 2

$$(x^a)^b = x^{ab}$$

That is, when raising an expression involving a base with an exponent to another power, multiply the exponents.

Example 6

Simplify each expression.

(a) $(y^4)^2$ (b) $(a^2)^{100}$ (c) $[(2x)^3]^4$

Solution

(a) $(y^4)^2 = y^{4 \cdot 2} = y^8$

(b) $(a^2)^{100} = a^{2 \cdot 100} = a^{200}$

(c) $[(2x)^3]^4 = (2x)^{3 \cdot 4} = (2x)^{12}$ *Note:* The parentheses cannot be dropped!

Example 7

Simplify each expression, if possible.

(a) $(-4x^2)(-3x^4)$ (b) $(-5y^2)(y^2)^3$ (c) $(8xy^2)(-3xy^3)$

Solution

(a) $(-4x^2)(-3x^4) = (-4)(-3)x^2x^4$
$$= 12x^6 \qquad \text{Note that } x^2 \cdot x^4 = x^{2+4} = x^6.$$

(b) $(-5y^2)(y^2)^3 = -5y^2y^6 \qquad \text{First rewrite } (y^2)^3 \text{ as } y^{2 \cdot 3} = y^6.$
$$= -5y^8$$

(c) $(8xy^2)(-3xy^3) = 8(-3) \cdot x \cdot x \cdot y^2 \cdot y^3$
$$= -24x^2y^5$$

Example 8

Simplify each expression.

(a) $(-5y^2)^3$ (b) $(6ab)^2(2a)^2$

Solution

(a) Note that neither of the two rules given in this section can be applied directly to the expression $(-5y^2)^3$. We can use the definition of exponent, however, to simplify this expression as follows:

$$(-5y^2)^3 = (-5y^2)(-5y^2)(-5y^2)$$ $-5y^2$ is used as a factor three times.

$$= (-5)(-5)(-5) \cdot y^2 \cdot y^2 \cdot y^2$$
$$= -125y^6$$

(b)

$$(6ab)^2(2a)^2 = (6ab)(6ab)(2a)(2a)$$
$$= (6)(6)(2)(2) \cdot ab \cdot ab \cdot a \cdot a$$
$$= 144a \cdot a \cdot a \cdot a \cdot b \cdot b$$
$$= 144a^4b^2$$

Section 2.3 Exercises

In Exercises 1 through 10, evaluate each expression. Assume $x = 1$, $y = 2$, and $z = 3$.

1. y^4
2. x^4
3. $x^4 + y^4$
4. $(x + y)^4$
5. $(-3y)^2$
6. y^z
7. $(xyz)^2$
8. $(3y - z)^2$
9. $2(3y - 4z)^3$
10. $2(3y - 4z)^3 + x$

In Exercises 11 through 20, evaluate each expression. Assume $a = -2$, $b = -4$, and $c = 5$.

11. $a^2 + b^2$
12. $a^2 - b^2$
13. $(a + b)^2$
14. $(a - b)^2$
15. $5a^2 - 3b^2$
16. $b^2 - a^3$
17. $|a^3 + c^2|$
18. $|a^3 - c^2|$
19. $|a^3| - c^2$
20. $|a^3| + c^2$

In Exercises 21 through 35, simplify each expression, if possible.

21. $(2x^3)(3x^2)$
22. $(-2y)(7y^2)$
23. $(-10xy^2)(-4xy)$
24. $(7a^2)(8b^2)$
25. $(x^2y^3z^4)(xyz)$
26. $(4x^2y^3z^4)(x^4y^3z)$
27. $-11(a^2b^2)(-abc)$
28. $(10ax)^2(4x)^2$
29. $10(x^4)^4$
30. $(10x^4)^4$
31. $[(3x - 2y)^3]^2$
32. $x^a \cdot x^b$
33. $x^a + x^b$
34. $x^a \cdot y^b$
35. $(3a^2b^3c)(-2ab)(-a^2b^4)$
36. $5^4 5^a$
37. $3^2 3^6 3^4$
38. $(5.8671x^2)(-13.0015x^3)$
39. $(5.8671x^2)^2(-13.0015x^3)$
40. $(5.8671x^2)^2(-13.0015x^3)^2$

Supplementary Exercises

In Exercises 1 through 5, evaluate each expression. Assume that $t = -1$, $s = 2$, and $r = -2$.

1. t^2
2. $-s^2$
3. $-r^3$
4. $(4rs)^3$
5. $s^3 \div r^2$

In Exercises 6 through 10, evaluate each expression for $a = -1$, $b = 2$, and $c = -2$. You may wish to simplify each expression first.

6. $(2a^2b)(-ab^2)$
7. $(2bc)^2$
8. $-[3a^3(b^2)^2]$
9. $a^2 - 2b^2$
10. aa^5a^3

In Exercises 11 through 20, simplify each expression using Exponent Rules 1 and 2, if possible.

11. $(-3a^2)(5a^7)$
12. $(x^3)^4$
13. $(2x^3)(x^5)^2$
14. $(5x^2y^3)(-3xy)(4x^4y^4)$
15. $(-2x^7y)(5x^3y^2)(3y^2x)$
16. $6(x^3)^3$
17. $x^ax^bx^c$
18. $4^54^64^7$
19. $x^5x^3x^y$
20. $x^2 + x^3$

2.4   Simplifying algebraic expressions

Now that we have Exponent Rules 1 and 2, let's look again at the simplification process of *collecting like terms* as discussed in Section 2.2. Equipped with these rules, we can simplify many more algebraic expressions.

Example 1

Using the distributive property, simplify $5x(x^2 - 2)$.

Solution

$$5x(x^2 - 2) = 5x \cdot x^2 - 5x \cdot 2$$
$$= 5x^3 - 10x \qquad \text{Use Exponent Rule 1.}$$

Notice that $5x^3$ and $-10x$ are *unlike* terms; there are three x's in $5x^3$, but only one in $-10x$. No further simplification is possible.

Example 2

Simplify by collecting like terms:

$$3x^2(x - 5) - 2x(x^2 + 7x)$$

Solution

$$3x^2(x - 5) - 2x(x^2 + 7x) = 3x^3 - 15x^2 - 2x^3 - 14x^2 \qquad \text{Eliminate parentheses and use exponent rules.}$$

$$= x^3 - 29x^2 \qquad \text{Collect like terms:}$$

$\boxed{3x^3 - 2x^3}$ and

$\boxed{-15x^2 - 14x^2}$

Example 3

Simplify $y^2(2y - 1) - y^3(y - 1) + 6$.

Solution

$$y^2(2y - 1) - y^3(y - 1) + 6 = 2y^3 - y^2 - y^4 + y^3 + 6$$
$$= -y^4 + 2y^3 + y^3 - y^2 + 6 \qquad \text{The only ''like'' terms are } 2y^3 \text{ and } y^3.$$
$$= -y^4 + 3y^3 - y^2 + 6$$

The example above demonstrates that it is **customary** to write such an algebraic expression in **descending powers of the variable.** That is, the term with the largest exponent is written first, then the term with the next largest exponent, and so on. Thus, $7 + 2x^2 - 20x$ would be written as $2x^2 - 20x + 7$.

Often, especially in applied problems, it is convenient to use the same letter of the alphabet to represent two *different* variables. This is done by distinguishing the letter with a **subscript.** For example, if a physicist was doing an experiment involving velocity (v), she might use v_1 and v_2 to represent two different velocities. The symbol v_1 is read ''v sub 1'' and v_1 and v_2 are treated as totally different quantities (just like ''x'' and ''y''). Some examples follow.

Example 4

Simplify $2x_1(3x_1 - 5x_2) + 5x_1(x_1 - x_2)$.

Solution

$$2x_1(3x_1 - 5x_2) + 5x_1(x_1 - x_2)$$
$$= 6x_1^2 - 10x_1x_2 + 5x_1^2 - 5x_1x_2 \qquad \text{Multiply to eliminate parentheses.}$$
$$= 11x_1^2 - 15x_1x_2 \qquad \text{Collect like terms:}$$

$$\boxed{6x_1^2 + 5x_1^2} \quad \text{and} \quad \boxed{-10x_1x_2 - 5x_1x_2}$$

Example 5

Simplify:

$$x_1(x_1 + x_2 - 5) - 2x_2(2x_1 - 3x_2 - 2) + 5(x_1 + x_2 + 6)$$

Solution

$$x_1(x_1 + x_2 - 5) - 2x_2(2x_1 - 3x_2 - 2) + 5(x_1 + x_2 + 6)$$
$$= x_1^2 + x_1x_2 - 5x_1 - 4x_1x_2 + 6x_2^2 + 4x_2 + 5x_1 + 5x_2 + 30$$
$$= x_1^2 + 6x_2^2 - 3x_1x_2 + 9x_2 + 30$$

Example 6

Simplify $2r^2s(5rs^2 - 4rs + 6) - 5rs(3r^2s^2 - r^2s + 2r)$.

Solution

$$2r^2s(5rs^2 - 4rs + 6) - 5rs(3r^2s^2 - r^2s + 2r)$$
$$= 10r^3s^3 - 8r^3s^2 + 12r^2s - 15r^3s^3 + 5r^3s^2 - 10r^2s$$
$$= -5r^3s^3 - 3r^3s^2 + 2r^2s$$

Example 7

Simplify by collecting like terms:

$$-3[2t(t^2 - 7t + 1) - 5t(t^2 - 7)]$$

Solution

We must work within brackets first.

$$-3[2t(t^2 - 7t + 1) - 5t(t^2 - 7)]$$
$$= -3[2t^3 - 14t^2 + 2t - 5t^3 + 35t] \quad \text{Collect like terms within brackets}$$
before multiplying by -3.
$$= -3[-3t^3 - 14t^2 + 37t]$$
$$= 9t^3 + 42t^2 - 111t$$

Section 2.4 Exercises

In Exercises 1 through 36, simplify each algebraic expression.

1. $3x(x^2 - 2)$
2. $5y(y^2 - 1)$
3. $5z^2(3z - 6)$
4. $-2a^2(a^2 - 5a)$
5. $-3t^2(t^2 - 4t + 1)$
6. $6r(r^2 - 5r + 11)$
7. $30b^3(b^2 - 3b + 1)$
8. $-19c^3(1 - 5c^2)$
9. $6a(a^2 - 2) + 7(a - 2)$
10. $9t(t^2 - 5) + 4(t - 5)$
11. $3x(x^2 - x) + 5(x - 4)$
12. $-5x(x^2 - 7x) + 3x^2(9x - 1)$
13. $7x(x^2 - 5x + 1) + 3x(x^2 - 2)$
14. $-2y(y^2 - 7y) + 2y(2y^2 - y + 1)$
15. $2a(a^2b - b) + 4b(a^3 + 9a)$
16. $3x(y^2 - 5x) - 2x(3y^2 - 7x)$
17. $3c^2(2c - 5) - 4c(c^2 + 7c)$
18. $3z^2(4z - 8) - 4z(3z^2 - 6z - 1)$
19. $4x - x(2x - 1) + 10$
20. $4y - 3y(4y - 7) - (y^2 - 7y + 9)$
21. $3x - x(2x - 5) - 3x(x - 7)$
22. $13a^2b - 5a(ab - b) + 7ab(3a - 4)$
23. $20x^2(3x^2 - 7x - 1) - 4x^3(2x - 3) - 5x^2(-11x^2 - 7x + 2)$
24. $3y^2(4y^2 - 5y + 2) - 12y^3(y - 5) - 6y^2(y^2 + y - 9)$
25. $3x_1(2x_2 - 5) + 4x_2(x_1 - 3)$
26. $-x_1(2x_1 - x_2 + 7) + 3x_2(x_1 - x_2 - 1)$
27. $3T_F(5 - 6T_C) - 9T_C(T_F + 7)$
28. $x_1^2(2x_1 - 3x_2) - 5x_1x_2(x_1 - 7)$
29. $-2[3(x_1^2 - 5x_1) - 2x_1(3x_1 - 7)]$
30. $-5t[3(2t - 7) + 6(t - 9)] + 4t(6t - 7)$
31. $(x + y)^2$
32. $(2x - y)^2$
33. $(a + b)^2 - (a^2 - 3ab - 4b^2)$
34. $3(a^2 - 2ab + b^2) - 4(5b^2 - ab)$
35. $6x(2x + y - 1) + 5y(x - y - 1)$
36. $x^3y^2(4xy - x^2y^2) - 3x^2y^3(x^3y - x^2)$

Supplementary Exercises

In Exercises 1 through 16, simplify by collecting like terms, if possible.

1. $15y(2 - 3y^2)$
2. $4x^2y(5x - 3y)$
3. $-4x^3(5 - x + 2x^2 - x^3)$
4. $6x(2x^2 - x + 1) - x(x^2 + 6x + 6)$
5. $5 - 2x - (3x - 2)$
6. $-(2x + y + 4) + 3(y - 4) - 2(3 - x)$
7. $15a - (b - a) + (4b - 7a)$
8. $3(x^3)^2 - 5x^2(2x^4 + 1)$
9. $-5(x^2)^4 + 2x^4(3 - x^4)$
10. $-2[(a + 3b - 1) - 3(a - b + 2)] + 4a$
11. $2t\{4 - [(4t - 3) - (t - 8)]\} - 6t$
12. $5x - x(2x + 1) - 3(x^2 + 2x - 1)$
13. $6 - 4(2x - 8)$
14. $(2xy + 5t)3xyt$
15. $(9x - 3y + 7)2xy^2$
16. $6x^2y + 4xy^2$

In Exercises 17 through 20, evaluate the given expressions for $a = 2$, $b = -1$, and $c = -2$ by first simplifying the expressions.

17. $5(a - 2b) - 2(a + b)$

18. $5a^2(b + a) - 3ab(a + 1)$

19. $(a^2 - b^2) - 3(a^2 + b^2)$

20. $c(2c + 1) + 4(c^2 + c + 1)$

2.5   Geometric formulas

In this section we will examine some formulas from geometry. Volume, area, and perimeter will be described in algebraic formulas.

Example 1

The formulas for the area (A) and perimeter (P) of a rectangle are:

width
(w)

$A = \ell w$

$P = 2\ell + 2w$

length (ℓ)

Find the area and perimeter of a rectangle whose width is 2 inches and whose length is 5 inches.

Solution

We apply the area formula and keep careful track of units:

$A = \ell w$

$A = 5 \text{ in.} \cdot 2 \text{ in.}$ Substitute 5 in. for ℓ and 2 in. for w.

$A = 10 \text{ in}^2$ $5 \cdot 2$ is 10; in. $\cdot$ in. is in^2.
or 10 square inches   (Think of "in." as a variable.)

The unit for area is *square inches:*

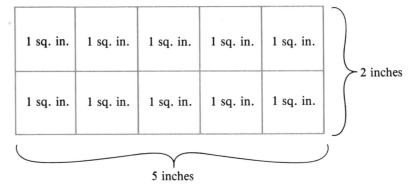

The perimeter, which is the straight-line distance around the rectangle is:

$P = 2\ell + 2w$

$P = 2 \cdot 5 \text{ in.} + 2 \cdot 2 \text{ in.}$ Substitute 5 in. for ℓ and 2 in. for w.

$P = 10 \text{ in.} + 4 \text{ in.}$

$P = 14 \text{ inches}$

Example 2

The formulas for the area (A) and circumference (distance around, C) of a circle with radius (r) are:

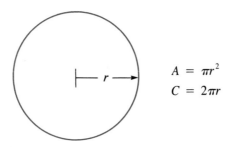

$$A = \pi r^2$$
$$C = 2\pi r$$

where π is an irrational number approximately equal to 3.14159265.

Find the area and circumference of a circle whose radius is 3 meters. (Use 3.14 as an approximation to π.)

Solution

$A = \pi r^2$

$A \approx 3.14(3 \text{ m})^2$ Substitute 3.14 for π and 3 m for r.

$A \approx 3.14 \cdot 9 \text{ m}^2$ $(3 \text{ m})^2$ is $3 \text{ m} \cdot 3 \text{ m}$ or 9 square meters.

$A \approx 28.26 \text{ m}^2$

The circumference is:

$C = 2\pi r$

$C \approx 2 \cdot 3.14 \cdot 3 \text{ m}$

$C \approx 18.84 \text{ m}$

Example 3

The formula for the area (A) of a triangle with base (b) and height (h) is:

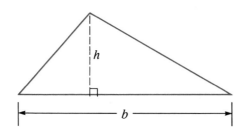

$$A = \frac{1}{2}bh$$

Notice h is perpendicular to b.

Find the area of a triangle whose height is 3 inches and whose base length is 2 feet.

Solution

Be careful! Units should match. We first convert 2 feet to 24 inches.

$$A = \frac{1}{2}bh$$

$$A = \frac{1}{2} \cdot 24 \text{ in.} \cdot 3 \text{ in.}$$

$$A = 36 \text{ in}^2 \quad \text{(or 36 square inches)}$$

Example 4

A rectangular box (sometimes called a **parallelepiped**) has volume (V) given by the formula:

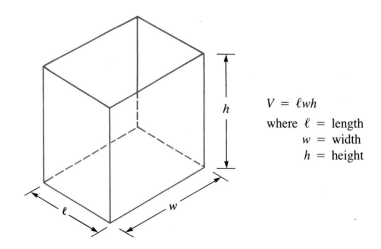

$V = \ell wh$

where ℓ = length
w = width
h = height

Find the volume of a box whose dimensions are 30 cm by 40 cm by 45 cm.

Solution

$A = \ell wh$

$A = 30 \text{ cm} \cdot 40 \text{ cm} \cdot 45 \text{ cm}$

$A = 54{,}000 \text{ cm}^3$ (or 54,000 cubic centimeters)

Example 5

The volume (V) of a right circular cylinder of height (h) and radius (r) is given by

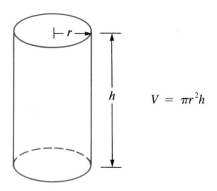

$V = \pi r^2 h$

Find the volume if $r = 30$ cm and $h = 1$ m. (*Note:* 1 m = 100 cm.)

Solution

$V = \pi r^2 h$

$V \approx 3.14 \cdot (30 \text{ cm})^2 \cdot 100 \text{ cm}$

$V \approx 282{,}600 \text{ cm}^3$ (or 282,600 cubic centimeters)

Section 2.5 Exercises

In Exercises 1 through 6, find (a) the perimeter and (b) the area of a rectangle with dimensions listed.

1. length = 2 inches
   width = 4 inches

2. length = 2 meters
   width = 3 meters

3. length = 5 feet
   width = 4 feet

4. length = 12 centimeters
   width = 9 centimeters

5. length = 120 centimeters
   width = 2 meters

6. length = 14 inches
   width = 2 feet

In Exercises 7 through 12, find (a) the circumference and (b) the area of the circle with radius listed. Use 3.14 as an approximation to π.

7. radius = 4 inches

8. radius = 20 centimeters

9. radius = 7 feet

10. radius = 5 meters

11. radius = 10 kilometers

12. radius = 3 miles

In Exercises 13 through 16, find the area of the triangle with the dimensions listed.

13. base = 4 inches
    height = 5 inches

14. base = 10 centimeters
    height = 4 centimeters

15. base = 3 meters
    height = 5 meters

16. base = 12 feet
    height = 30 inches

17. A triangle has base 18 inches and height 1 foot.
    (a) Find the area expressed in square inches.
    (b) Find the area expressed in square feet.

In Exercises 18 through 21, find the volume of the rectangular box with dimensions listed.

18. length = 5 inches
    width = 4 inches
    height = 2 inches

19. length = 10 centimeters
    width = 8 centimeters
    height = 4 centimeters

20. length = 1 meter
    width = 2 meters
    height = 50 centimeters

21. length = 3 feet
    width = 10 feet
    height = 6 inches

In Exercises 22 through 26, find the volume of the right circular cylinder with radius and height given below. Use 3.14 as an approximation to π.

22. radius = 3 feet
    height = 4 feet

23. radius = 4 feet
    height = 3 feet

24. radius = 5 centimeters
    height = 1 centimeter

25. radius = 1 centimeter
    height = 5 centimeters

26. radius = 18 inches
    height = 2 feet

In Exercises 27 through 30 use the fact that the volume (V), of the right circular cone pictured below with radius r and height h, is given by

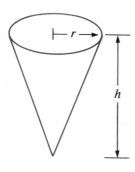

$$V = \frac{1}{3}\pi r^2 h$$

Use 3.14 as an approximate value of π.

27. radius = 5 inches
    height = 4 inches

28. radius = 4 inches
    height = 5 inches

29. radius = 2 feet
    height = 2 inches

30. radius = 2 centimeters
    height = 2 meters

The parallelogram (a four-sided figure with both pairs of opposite sides parallel) pictured below with sides a and b and height h has formulas for area (A) and perimeter (P) given below:

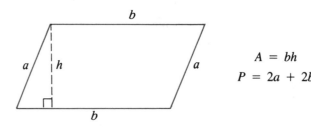

$$A = bh$$
$$P = 2a + 2b$$

In Exercises 31 through 34, find the area and perimeter for each set of dimensions.

31. $a = 2$ inches
    $b = 4$ inches
    $h = 5$ inches

32. $a = 7$ centimeters
    $b = 8$ centimeters
    $h = 9$ centimeters

33. $a = 4$ inches
    $b = 1$ foot
    $h = 6$ inches

34. $a = 4$ centimeters
    $b = 1$ meter
    $h = 6$ centimeters

35. Find the area of a circle of radius 5.25 inches. Use $\pi \approx 3.141592654$.

36. The volume V of a sphere of radius r is given by the formula

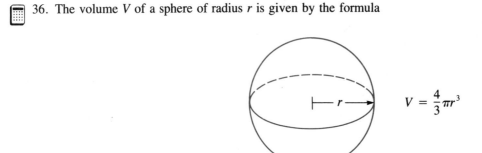

$$V = \frac{4}{3}\pi r^3$$

Using $\pi \approx 3.141592654$, find the volume of a sphere of radius 5.7532 centimeters.

37. Find the volume of a right circular cylinder of radius 10.508 inches and height = 30.508 inches. Use $\pi \approx 3.141592654$.

38. Find the area of a square whose perimeter is 12.8374 meters.

The number π is an irrational number. That is, its decimal expansion is nonterminating and non-repeating. Actually, π is defined to be the ratio of the circumference of a circle to its diameter. In Exercises 39 through 42 use a string to measure the circumference (C) and diameter (d) of four circular objects. Complete the table below.

Object	Circumference, C	Diameter, d	Radius $= \frac{1}{2}d$	$\frac{C}{d}$
39.				
40.				
41.				
42.				

43. Before the new tenants move in, the landlord wishes to carpet the living room floor, which measures 9 feet by 12 feet. How many square feet of carpeting will he need?

44. John wishes to enclose a rectangular cow pasture with fencing. If the field measures 200 feet by 150 feet, find the amount of fencing he will need to enclose the field. (*Hint:* Find the perimeter.)

45. A circular pond with a radius of 7 feet will have a circular fence built around it. What is the amount of fencing needed if the fence's radius is one foot more than the radius of the pond? (*Hint:* Find the circumference and use $\pi \approx 3.14$.)

46. If a cover is to be made to fit the pond in Exercise 45 snugly, what is its area? Use $\pi \approx 3.14$.

47. Jose has a box measuring 4 feet by 3 feet by 5 feet. What is its volume?

48. Carol is filling a can 10 cm tall with a radius of 6 cm with nuts. What is the volume of the can? Use $\pi \approx 3.14$. (*Hint:* A "can" is a right circular cylinder.)

49. Kevin is making a triangular-shaped sail for his boat. If the base of the triangle is 6 feet and the height is 8 feet, what is the area of the sail?

50. Find the volume of a cube (a rectangular box that has all of its sides equal) if each side measures 10 cm.

Supplementary Exercises

In Exercises 1 through 4 find (a) the perimeter and (b) the area of each rectangle with dimensions listed.

1. length = 4 yards
   width = 5 yards

2. length = 30 inches
   width = 12 inches

3. length = 4 feet
   width = 18 inches

4. length = 9 meters
   width = 3 meters

In Exercises 5 through 8 find (a) the circumference and (b) the area of the circle with radius listed. Use 3.14 as an approximation to π.

5. radius = 15 yards

6. radius = 6 feet

7. radius = 9 kilometers

8. radius = 18 miles

In Exercises 9 through 12 find the area of the triangle with the dimensions listed.

9. base = 9 feet
    height = 8 feet

10. base = 12 yards
    height = 8 yards

11. base = 9 feet
    height = 2 yards

12. base = 10 cm.
    height = 8 cm.

In Exercises 13 through 16, find the volume of the rectangular box with dimensions listed.

13. length = 4 feet
    width = 3 feet
    height = 2 feet

14. length = 7 yards
    width = 4 yards
    height = 3 yards

15. length = 2 feet
    width = 3 feet
    height = 1 yard

16. length = 2 meters
    width = 4 meters
    height = 5 meters

In Exercises 17 through 20, find the volume of the right circular cylinder with radius and height given below. Use 3.14 as an approximation for π.

17. radius = 4 feet
    height = 7 feet

18. radius = 2 yards
    height = 50 yards

19. radius = 60 inches
    height = 40 feet

20. radius = 2 meters
    height = 4 meters

21. If the volume V of a right circular cone is $V = \frac{\pi}{3}r^2h$, find the volume when $r = 4$ feet and $h = 9$ feet. Use $\pi \approx 3.14$.

22. Using the formula in Exercise 21, find the volume when $r = 20$ inches and $h = 3$ feet. Use $\pi = 3.14$.

23. The volume of a sphere, V, is given by the formula $V = \frac{4\pi}{3}r^3$. Find the volume if $r = 6$ feet. Use 3.14 for π.

24. Using the formula in Exercise 23, find the volume when $r = 3$ meters. Use 3.14 for π.

25. Find the area of a circle whose diameter is 8 feet. (*Note:* The radius of a circle is one-half the diameter of the circle.)

2.6   Chapter review

Summary

We began this chapter by showing how letters could be used to represent numbers. Evaluating and simplifying these letters, called *variables,* is an integral part of the study of algebra. In Section 2.2 we saw that the seven properties of the real number system can be used to simplify and collect like terms.

Two rules of exponents were studied in Section 2.3:

$$Rule\ 1.\quad x^a \cdot x^b = x^{a+b}$$
$$Rule\ 2.\quad (x^a)^b = x^{ab}$$

Algebraic expressions were simplified in Section 2.4, where we also learned that it is customary to write many terms in descending powers of the variable. Finally, in Section 2.5 we evaluated certain types of algebraic expressions, formulas from geometry.

Vocabulary Quiz

Match the expression in Column I with the phrase in Column II that best describes it.

Column I

1. A variable is
2. $a + b = b + a$ illustrates
3. $a + (b + c) = (a + b) + c$ illustrates
4. According to Exponent Rule 1, $x^2 \cdot x^3 =$
5. According to Exponent Rule 2, $(x^2)^3 =$
6. The formula for the area of a triangle is $A =$
7. The formula for the area of a rectangle is $A =$

Column II

a. the Associative Property for Addition
b. a quantity, usually represented by a letter, that may take on several values
c. The Commutative Property for Addition
d. lw
e. $\frac{1}{2}bh$
f. x^6
g. x^5

Chapter 2 Review Exercises

In Exercises 1 through 11, let $a = 4$, $b = 3$, $x = -5$, $y = 0$, and $z = -11$. Evaluate each expression.

1. $a + 2b$

2. $(a + b)(a - b)$

3. $x^2 - b^2$

4. $2x + 3y - z$

5. $x - y - 7z$

6. $2abx - 3z$

7. $-4(2x - 3z)$

8. $(x + z) \div a$

9. $8[2(a - b) - 4(y - x)]$

10. $|5a| - |4x|$

11. $\dfrac{3a + 2b + 3}{x}$

12. Claudia is 23 years younger than twice Kit's age. If k represents Kit's age, what algebraic expression represents Claudia's age?

In Exercises 13 through 20, simplify each expression.

13. $(2x - 4) + 7$

14. $(2x - 3) + (5x + 12)$

15. $-2y - 3(y + 4)$

16. $-2y - 3(y - 4)$

17. $4(2t - 5) - 3t + 7$

18. $6(x - 2) - 3(2x - 9)$

19. $3(2a + b) + 4[a - 3(a - 2b)]$

20. $3[x + 2(x + 1)] - 4[1 - 2(3x - 4)]$

In Exercises 21 through 28, evaluate each expression if $a = -3$, $b = 2$, and $c = -5$.

21. $a^2 + b^2$

22. $a^2 - b^2$

23. $(a + b)^2$

24. $2a^2 + b^2 + c^2$

25. $(b + c)^2$

26. $b^2 + c^2$

27. a^b

28. $(a + 3b)^3$

In Exercises 29 through 32, simplify each expression.

29. $(-10xy^2)(-2xy)$

30. $(7a^2b)(5ab^2)$

31. $-2(x^2y^3z)(3xy^2)$

32. $(5x^3)^2$

In Exercises 33 through 38, simplify each expression.

33. $3x(x^2 + 4x)$

34. $-2a^2(a^2 - 5a + 7)$

35. $-14c^3(1 - 6c^2)$

36. $-2y(y^2 - 4) + 3y(2y^2 - y + 5)$

37. $2z^2(3z - 1) - 5z(z^2 - 7)$

38. $3[2z - 1(z^2 + 7)] - 4[3z(z^2 - 5) + 1]$

39. Find the perimeter and area of a rectangle whose length is 25 centimeters and whose width is 10 centimeters.

40. Find the circumference and area of a circle whose diameter is 20 inches. Use 3.14 as an approximation to π.

Cumulative review exercises

1. Write the integers.   *(Section 1.1)*

2. Find the sum $(-6) + (-5) + 4$.   *(Section 1.2)*

3. Find the difference $(-19) - (-3)$.   *(Section 1.3)*

4. Find the product $(-6)(5)(-2)$.   *(Section 1.4)*

5. What property is illustrated in the equation $5(10 + 4) = 50 + 20$?   *(Section 1.5)*

6. What is $6 \div 0$?   *(Section 1.6)*

7. Evaluate -9^2.   *(Section 1.7)*

8. Evaluate $\sqrt[3]{-8}$.   *(Section 1.7)*

9. Evaluate $20 - 2[(12 \div 3) + 6]$.   *(Section 1.8)*

10. Evaluate $\sqrt{4^2 + 3^2} - \dfrac{5 + 7}{\sqrt{4}}$.   *(Section 1.8)*

Chapter 2 test

Take this test to determine how well you have mastered the language of algebra; check your answers with those found at the end of the book.

In Problems 1 through 6, evaluate each expression. Assume $a = -2$, $b = 3$, and $c = -4$.

1. $a + 4b$ *(Section 2.1)*

2. $-3(b + 4) - c(b + 1)$ *(Section 2.1)*

1. _____

2. _____

3. $2a - 5b + c[a - b(a + 2)]$ *(Section 2.1)*

4. $(abc)^2$ *(Section 2.3)*

3. _____

4. _____

5. $(-3a)^2$ *(Section 2.3)*

6. $(2b + 3a)^4 - 5(2a + b)^3$ *(Section 2.3)*

5. _____

6. _____

In Problems 7 through 12, simplify each expression.

7. $7x - 2(x + 3)$ *(Section 2.2)*

8. $4(2x + 7) + 3(x - 4)$ *(Section 2.2)*

7. _____

8. _____

9. $-2(3y - 5) - 7(3y - 1)$ *(Section 2.2)*

10. $(3x^2y^2z)(4xy^2)$ *(Section 2.3)*

9. _____

10. _____

11. $-5x(x^2 - 2x) + 3x^2(x - 5)$
    (Section 2.4)

12. $3a^2(4a - 5) - 2a(a^2 - 5a + 1)$
    (Section 2.4)

11. _____

12. _____

13. Michelle makes $18 per hour. If m represents the number of hours she works, represent algebraically the amount Michelle earns. *(Section 2.1)*

14. Find the circumference and area of a circle with radius 3 feet. Use 3.14 as an approximation to π.   *(Section 2.5)*

13. _____

14. _____

15. Find the volume of a parallelepiped with length 8 centimeters, width 10 centimeters, and height 0.5 meters.   *(Section 2.5)*

15. _____

three

Linear equations and inequalities

3.1 The equation—an algebraic balance

We used the symbol of equality, "=", in Chapter 2 when we simplified algebraic expressions and formulas. In this chapter we will be concerned with *solving* **equations,** statements of equality between two quantities or expressions.

When we write
$$7 + 1 = 8$$

we mean that the quantity $7 + 1$ and the quantity 8 "carry the same weight" or are "in balance." We can make this comparison more visual with a *balance scale:*

In the case of the equation $7 + 1 = 8$, the scale is balanced because an equal "amount" is on both sides of it. That is, both sides of the equation are equal.

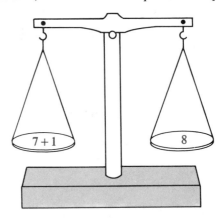

When an equation is false, the scale will not balance. For example, $7 + 1 = 9$ is false. Pictorially, we have:

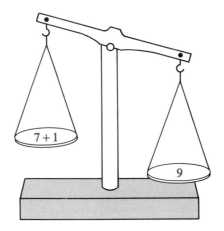

Our object in this chapter will be to solve algebraic equations containing one variable. To do so, we'll employ the balance scale to demonstrate how we "work" on equations.

Example 1

Solve $x - 3 = 2$.

Solution

To solve an equation means to find the replacement(s) for the variable that makes the statement of equality true.

Now, notice that *the scale will still balance if we add the same "weight" to both sides.* In other words, we could add 6 to both sides, or we could add 25 to both sides, or we could add 10,000 to both sides—the scale will still balance. We choose to add the number 3 to both sides.

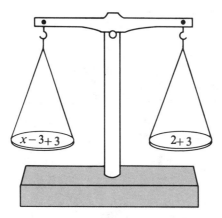

or

We want just "x" here. That's why we chose the number 3 to add to both sides. The expression $x - 3 + 3$ becomes just x when simplified.

So $x = 5$ is our solution.

$$\text{Check:} \quad x - 3 = 2$$
$$5 - 3 = 2$$
$$2 = 2 \ \checkmark$$

Notice in Example 1 that the equation $x - 3 = 2$ became an equivalent equation, $x = 5$. **Equivalent equations** are equations that have the same solution. In this case, the solution is 5.

Our basic goal in solving equations is to try to produce equivalent equations until we can isolate the variable. In this chapter we will learn certain rules for obtaining equivalent equations. By following these rules, it will be possible to isolate the variable. Once that is done, the solution to the equation will be obvious.

The preceding example leads us to the first principle we will use to solve equations.

The Addition Rule for Solving Equations

The same number may be added to both sides of an equation. The result is an equivalent equation. In symbols, if $x - a = b$, then

$$x - a + a = b + a$$

or $$x = b + a$$

In $x - a = b$, the correct number to add is the number opposite in sign to $-a$. Several examples follow.

Example 2

Solve $t - 11 = 17$.

Solution

We want to isolate the t on one side of the equals sign. To accomplish this, we add the number opposite in sign to -11. That is, we add 11 to the lefthand side of the equals sign, remembering to balance it by adding 11 to the righthand side of the equals sign also.

$$t - 11 = 17$$
$$\underline{+ 11 = +11}$$
$$t + 0 = 28$$
$$t = 28$$

$$\text{Check:} \quad t - 11 = 17$$
$$28 - 11 = 17$$
$$17 = 17 \ \checkmark$$

Example 3

Solve $y - 14 = -10$.

Solution

We add 14 to both sides (because that will make the lefthand side of the equation equal to just y).

$$y - 14 = -10$$
$$\underline{+ 14 = +14}$$
$$y + 0 = 4$$
$$y = 4$$

Check: $y - 14 = -10$

$4 - 14 = -10$

$-10 = -10$ $\checkmark$

Example 4

Solve $x - 2 = -5$.

Solution

We add 2 to both sides.

$$x - 2 = -5$$
$$+ 2 = +2$$
$$\overline{x + 0 = -3}$$
$$x = -3$$

Check: $x - 2 = -5$

$-3 - 2 = -5$

$-5 = -5$ $\checkmark$

Since we know from Chapter 1 that subtracting a number is the same as adding its opposite, the Addition Rule really includes the procedure of *subtracting* the same number from both sides. Of course, the balance scale bears this out too:

When we subtract 4 from both sides we get

Now we apply this idea in the next three examples.

Example 5

Solve $x + 4 = 9$.

Solution

Here we'll subtract 4 from both sides because, as before, we are interested in "isolating" x on the lefthand side of the equation.

$$
\begin{array}{rcr}
x + 4 &=& 9 \\
- 4 &=& -4 \\
\hline
x + 0 &=& +5 \\
x &=& 5
\end{array}
$$

$$
\begin{aligned}
\text{Check:} \quad x + 4 &= 9 \\
5 + 4 &= 9 \\
9 &= 9 \; \checkmark
\end{aligned}
$$

Example 6

Solve $y + 13 = 2$.

Solution

We subtract 13 from both sides of the equation.

$$
\begin{array}{rcr}
y + 13 &=& 2 \\
- 13 &=& -13 \\
\hline
y + 0 &=& -11 \\
y &=& -11
\end{array}
$$

$$
\begin{aligned}
\text{Check:} \quad y + 13 &= 2 \\
-11 + 13 &= 2 \\
2 &= 2 \; \checkmark
\end{aligned}
$$

Example 7

Solve $x + 3 = -6$.

Solution

$$
\begin{array}{rcr}
x + 3 &=& -6 \\
- 3 &=& -3 \\
\hline
x + 0 &=& -9 \\
x &=& -9
\end{array}
$$

$$
\begin{aligned}
\text{Check:} \quad x + 3 &= -6 \\
-9 + 3 &= -6 \\
-6 &= -6 \; \checkmark
\end{aligned}
$$

We conclude this section by showing examples in which the same equation can be presented in different forms.

Example 8

Solve $-3 + x = 2$.

Solution

Our goal is to *solve for x* (that is, to isolate x). In order to do so, we must "convert" the lefthand side of the equation to "x." We do this by adding 3 to both sides.

$$
\begin{aligned}
-3 + x &= \ \ 2 \\
\underline{+3 \qquad = +3} \\
0 + x &= \ \ 5 \\
x &= 5
\end{aligned}
$$

$$
\begin{aligned}
\text{Check:} \quad -3 + x &= 2 \\
-3 + 5 &= 2 \\
2 &= 2 \ \checkmark
\end{aligned}
$$

Example 9

Solve $2 = -3 + x$.

Solution

Note that this equation is the same as the one in Example 8 except that the lefthand and righthand sides of the equation have been switched. We now **isolate x on the righthand side of the equation,** which we do again by adding 3 to both sides.

$$
\begin{aligned}
2 &= -3 + x \\
\underline{+3 = +3} \\
5 &= \ \ 0 + x \\
5 &= x
\end{aligned}
$$

Notice, of course, that we get the same answer as in Example 8.

$$
\begin{aligned}
\text{Check:} \quad 2 &= -3 + x \\
2 &= -3 + 5 \\
2 &= 2 \ \checkmark
\end{aligned}
$$

Each example in this section was solved by adding or subtracting the same number from both sides of the equation. Recall that in Chapter 1 addition and subtraction were called inverse operations since, if the same number was added to and subtracted from a given number, the result was the given number. Notice how we used this idea of inverse operations to isolate the variable in the equations in this section. If one side of an equation consists of a number **added** to the variable, we simply **subtract** that number from both sides of the equation to isolate the variable. If one side of an equation consists of a number **subtracted from** the variable, we simply **add** that number to both sides of the equation to isolate the variable.

Section 3.1 Exercises

In Exercises 1 through 45 solve each of the equations by using the Addition Rule developed in this section.

1. $x - 1 = 2$
2. $x + 1 = 2$
3. $x - 7 = 9$
4. $x - 7 = -9$
5. $x - 9 = 7$
6. $x - 9 = -7$
7. $y - 1 = 0$
8. $k - 3 = -11$
9. $m - 9 = 99$
10. $x - 2 = -2$
11. $x - 15 = 10$
12. $x + 15 = 10$

13. $y - 7 = 7$

14. $y + 7 = 7$

15. $m - 4 = -5$

16. $n + 4 = -5$

17. $x + 3 = 8$

18. $x + 1 = 0$

19. $a + 9 = 1$

20. $x + 6 = -6$

21. $q - 10 = -2$

22. $q + 10 = -2$

23. $x - 19 = 31$

24. $x - 19 = -31$

25. $b + 20 = -29$

26. $x + 1 = 1$

27. $x + 1 = -11$

28. $p + 11 = -1$

29. $x + 19 = 31$

30. $x + 19 = -31$

31. $2 = x - 1$

32. $-1 + x = 2$

33. $x - 1 = 9$

34. $y + 7 = 11$

35. $x - 2 = 6$

36. $z - 5 = 1$

37. $x + 12 = 4$

38. $t + 1 = -10$

39. $6 = x - 2$

40. $-1 = z - 5$

41. $x - 102 = 35$

42. $x + 102 = 35$

43. $5 + x = 6$

44. $10 + x = 10$

45. $y + 11 = 7$

In Exercises 46 through 52, use a calculator to solve each equation.

46. $x - 31.587 = 46.719$

47. $y + 3.8704 = 7.8703$

48. $z - 6.7193 = -9.4501$

49. $t + 6.7193 = -9.4501$

50. $a + 0.01138 = 0.001409$

51. $5.6376 = x - 2.7315$

52. $-7.2143 = 3.1147 + y$

Supplementary Exercises

In Exercises 1 through 20, solve each of the equations by using the Addition Rule developed in this section.

1. $m - 9 = 7$

2. $t + 8 = 12$

3. $w - 6 = -9$

4. $x + 15 = 7$

5. $p + 3 = -9$

6. $t + 3 = 3$

7. $m - 15 = 15$

8. $q + 20 = 41$

9. $t - 17 = -63$

10. $x + 25 = -60$

11. $5 + x = 8$

12. $-7 + p = 10$

13. $-8 + y = -8$

14. $15 + w = 15$

15. $12 + x = 49$

16. $9 = x + 36$

17. $10 = t - 14$

18. $-31 = p + 14$

19. $-19 = x - 7$

20. $-15 = m - 21$

3.2   Solving equations using multiplication and division

In this section we will use the operations of multiplication and division to solve equations. In the equation

$$\frac{x}{2} = 10$$

we are saying that some number divided by 2 is 10. To isolate the variable, we need to remove the divisor 2 from the lefthand side of the equation. Since the inverse operation of dividing by 2 is multiplying by 2, we multiply both sides of the equation by 2 to obtain

$$2 \cdot \frac{x}{2} = 2 \cdot 10$$

$$x = 20$$

$$\text{Check:} \quad \frac{x}{2} = 10$$

$$\frac{20}{2} = 10$$

$$10 = 10 \; \checkmark$$

In other words, we can multiply both sides of an equation by the same nonzero quantity. We call this the Multiplication Rule:

The Multiplication Rule for Solving Equations

The same nonzero number may be used to multiply both sides of an equation. The result is an equivalent equation. In symbols, if $\frac{x}{a} = b$, then

$$a \cdot \frac{x}{a} = a \cdot b$$

$$x = a \cdot b$$

provided $a \neq 0$.

Example 1

Solve $\frac{x}{7} = 6$.

Solution

$$\frac{x}{7} = 6$$

$$7 \cdot \frac{x}{7} = 7 \cdot 6 \qquad \text{To remove the divisor 7, use the inverse operation of multiplying both sides by 7.}$$

$$x = 42$$

$$\text{Check:} \quad \frac{x}{7} = 6$$

$$\frac{42}{7} = 6$$

$$6 = 6 \; \checkmark$$

Example 2

Solve $\frac{x}{4} = -5$.

Solution

$$\frac{x}{4} = -5$$

$$4 \cdot \frac{x}{4} = 4 \cdot (-5) \qquad \text{Multiply both sides by 4. (Multiplying by 4 is the inverse of dividing by 4.)}$$

$$x = -20$$

$$\text{Check:} \qquad \frac{x}{4} = -5$$

$$\frac{-20}{4} = -5$$

$$-5 = -5 \; \checkmark$$

Example 3

Solve $\dfrac{y}{-3} = -7$.

Solution

$$\frac{y}{-3} = -7$$

$$(-3) \cdot \frac{y}{-3} = (-3) \cdot (-7) \qquad$$ Multiply both sides by -3 since the variable is divided by -3.

$$y = 21$$

$$\text{Check:} \qquad \frac{y}{-3} = -7$$

$$\frac{21}{-3} = -7$$

$$-7 = -7 \; \checkmark$$

In the equation

$$3x = 30$$

we are saying that 3 times some number is 30. To isolate the variable, we need to remove the multiplier 3 from the lefthand side of the equation. The inverse operation of multiplying by 3 is dividing by 3 and, thus, we choose to divide both sides of the equation by 3 to obtain

$$\frac{3x}{3} = \frac{30}{3}$$

$$x = 10$$

$$\text{Check:} \qquad 3x = 30$$

$$3 \cdot 10 = 30$$

$$30 = 30 \; \checkmark$$

We now state the Division Rule:

The Division Rule for Solving Equations

The same nonzero number may be used to divide both sides of an equation. The result is an equivalent equation. In symbols, if $ax = b$, then

$$\frac{ax}{a} = \frac{b}{a}$$

$$x = \frac{b}{a}$$

provided $a \neq 0$.

Example 4

Solve $8x = -48$.

Solution

$$8x = -48$$

$$\frac{8x}{8} = \frac{-48}{8}$$

To isolate x, we need to remove the multiplier 8; we use the inverse operation of dividing both sides by 8.

$$x = -6$$

Check:
$$8x = -48$$
$$8(-6) = -48$$
$$-48 = -48 \ \checkmark$$

Example 5

Solve $-x = -7$.

Solution

We understand $-x$ to be $(-1) \cdot x$. So to isolate x, we divide both sides by -1.

$$-x = -7$$

$$\frac{-x}{-1} = \frac{-7}{-1}$$

$$x = 7$$

Check:
$$-x = -7$$
$$-(7) = -7$$
$$-7 = -7 \ \checkmark$$

Example 6

Solve $-3x = 15$.

Solution

$$-3x = 15$$

$$\frac{-3x}{-3} = \frac{15}{-3}$$

Divide both sides by -3.

$$x = -5$$

Check:
$$-3x = 15$$
$$-3(-5) = 15$$
$$15 = 15 \ \checkmark$$

Section 3.2 Exercises

In Exercises 1 through 28, solve each equation for the variable. Check your solution.

1. $5x = 30$

2. $4x = 16$

3. $-6x = 48$

4. $-5x = 45$

5. $\dfrac{x}{3} = 7$

6. $\dfrac{x}{5} = 6$

7. $\dfrac{x}{-4} = 7$

8. $\dfrac{x}{-7} = 3$

9. $2t = 28$

10. $2t = -28$

11. $-2t = 28$

12. $-2t = -28$

13. $\dfrac{x}{5} = -4$

14. $\dfrac{x}{6} = -7$

15. $\dfrac{x}{-3} = -8$

16. $\dfrac{x}{-6} = -6$

17. $8y = 0$

18. $-6y = 0$

19. $-5x = -35$

20. $-7x = -35$

21. $\dfrac{x}{-10} = 30$

22. $\dfrac{x}{20} = -30$

23. $\dfrac{x}{-5} = -50$

24. $\dfrac{x}{-7} = -30$

25. $(-3.654)x = 23.273$

26. $(2.407)x = -17.364$

27. $\dfrac{y}{3.536} = 7.913$

28. $\dfrac{y}{-5.637} = -8.471$

Supplementary Exercises

In Exercises 1 through 20, solve each equation for the variable.

1. $4x = 12$

2. $-5x = 25$

3. $-t = 9$

4. $-30 = -6p$

5. $100 = -25t$

6. $-64 = 16x$

7. $-121 = -11t$

8. $6x = 78$

9. $\dfrac{x}{8} = 11$

10. $\dfrac{x}{-4} = 12$

11. $21 = \dfrac{x}{5}$

12. $-4 = \dfrac{y}{-4}$

13. $6 = \dfrac{-t}{5}$

14. $-8 = \dfrac{-x}{5}$

15. $\dfrac{5m}{2} = 10$

16. $\dfrac{5x}{4} = 20$

17. $\dfrac{4x}{3} = -12$

18. $\dfrac{-6x}{5} = 120$

19. $\dfrac{-x}{-7} = 8$

20. $\dfrac{x}{-7} = -8$

3.3   Solving equations using more than one rule

Each example of the preceding two sections involved applying one basic rule to determine the solution to an equation. We now consider examples where the rules must be combined to find the solution. The objective is still the same—to isolate the variable on one side of the equation. In solving equations that require using more than one rule, we begin by *isolating the term containing the variable*. The following examples illustrate this type of solution.

Example 1

Solve $3x + 1 = 19$.

Solution

$$3x + 1 = 19$$
$$\underline{-1 = -1}$$
$$3x + 0 = 18$$
$$3x = 18$$

To isolate the term involving the variable, $3x$, we need to remove $+1$ from the lefthand side of the equation. So we first subtract 1 from both sides.

$$\frac{3x}{3} = \frac{18}{3}$$
$$x = 6$$

To isolate x, divide both sides by 3.

Check:
$$3x + 1 = 19$$
$$3 \cdot 6 + 1 = 19$$
$$18 + 1 = 19$$
$$19 = 19 \ \checkmark$$

Example 2

Solve $\dfrac{2x}{3} = -12$.

Solution

The variable appears in the numerator of the fraction on the lefthand side of the equation. We want to isolate that numerator on the lefthand side of the equation and then proceed as before.

$$\frac{2x}{3} = -12$$

$$3 \cdot \frac{2x}{3} = 3 \cdot (-12)$$

Multiply both sides by 3 to isolate the numerator containing the variable.

$$2x = -36$$

$$\frac{2x}{2} = \frac{-36}{2}$$

Divide both sides by 2 to isolate x.

$$x = -18$$

Check:
$$\frac{2x}{3} = -12$$

$$\frac{2(-18)}{3} = -12$$

$$\frac{-36}{3} = -12$$

$$-12 = -12 \ \checkmark$$

Example 3

Solve $5 - 3x = -16$.

Solution

$$5 - 3x = -16$$

Subtract 5 from each side to isolate the *term* involving x.

$$\frac{-5 \quad\quad = \quad -5}{0 - 3x = -21}$$

$$\frac{-3x}{-3} = \frac{-21}{-3}$$

Divide both sides by -3 to isolate x.

$$x = 7$$

Check:
$$5 - 3x = -16$$
$$5 - 3(7) = -16$$
$$5 - 21 = -16$$
$$5 + (-21) = -16$$
$$-16 = -16 \ \checkmark$$

Example 4

$$-6 = \frac{3x}{4} - 15$$

Solution

$$-6 = \frac{3x}{4} - 15$$

Add 15 to both sides to isolate the term (in this case a fraction) containing the variable.

$$\frac{+15 = \quad\quad + 15}{9 = \frac{3x}{4}}$$

$$4 \cdot 9 = 4 \cdot \frac{3x}{4}$$

Multiply both sides by 4 to isolate the numerator.

$$36 = 3x$$

$$\frac{36}{3} = \frac{3x}{3}$$

Divide both sides by 3 to isolate x.

$$12 = x$$

Check:
$$-6 = \frac{3x}{4} - 15$$

$$-6 = \frac{3(12)}{4} - 15$$

$$-6 = \frac{36}{4} - 15$$

$$-6 = 9 - 15$$

$$-6 = 9 + (-15)$$

$$-6 = -6 \ \checkmark$$

We now consider equations in which the variable appears on *both* sides of the equations. Our first step in solving this type of equation will be to remove the variable from one side of the equation (it doesn't matter which side) so that the variable only appears on the other side. We then proceed with our goal of isolating the variable to determine the solution.

Example 5

Solve $2x - 7 = x + 9$.

Solution

Since x appears on both sides of the equation, we begin by subtracting x from both sides so that the variable only appears on the lefthand side of the equation.

$$
\begin{aligned}
2x - 7 &= x + 9 \\
-x &= -x \\
\hline
x - 7 &= 9 \\
+ 7 &= +7 \\
\hline
x &= 16
\end{aligned}
$$

Add 7 to both sides to isolate x.

Check:
$$
\begin{aligned}
2x - 7 &= x + 9 \\
2(16) - 7 &= 16 + 9 \\
32 - 7 &= 25 \\
25 &= 25 \;\checkmark
\end{aligned}
$$

We could also have gotten the variable on just one side of the equation in the initial step by subtracting $-2x$ from each side. Then we would have continued as follows:

$$
\begin{aligned}
2x - 7 &= x + 9 \\
-2x &= -2x \\
\hline
-7 &= -x + 9 \\
-9 &= -9 \\
\hline
-16 &= -x
\end{aligned}
$$

Subtract 9 from each side to isolate the variable on one side of the equation.

$$
\frac{-16}{-1} = \frac{-x}{-1}
$$

Divide both sides by -1 to isolate x.

$$
16 = x
$$

In either procedure the same solution is obtained.

Example 6

Solve $5 - 2y = 11 + 4y$.

Solution

$$
\begin{aligned}
5 - 2y &= 11 + 4y \\
+ 2y &= + 2y \\
\hline
5 &= 11 + 6y
\end{aligned}
$$

Add $2y$ to both sides to get the variable only on the righthand side.

$$
\begin{aligned}
5 &= 11 + 6y \\
-11 &= -11 \\
\hline
-6 &= 6y
\end{aligned}
$$

Subtract 11 from both sides to isolate y on one side of the equation.

$$
\frac{-6}{6} = \frac{6y}{6}
$$

Divide both sides by 6 to isolate y.

$$
-1 = y
$$

$$\text{Check:}\quad 5 - 2y = 11 + 4y$$
$$5 - 2(-1) = 11 + 4(-1)$$
$$5 + 2 = 11 + (-4)$$
$$7 = 7 \;\checkmark$$

Example 7

Solve $3t + 6 = 4t - 10$.

Solution

$$
\begin{array}{rl}
3t + 6 = 4t - 10 & \\
\underline{-4t \qquad = -4t} & \text{Subtract } 4t \text{ from both sides.} \\
-t + 6 = \qquad -10 & \\
\underline{\quad - 6 = \qquad -6} & \text{Subtract 6 from each side.} \\
-t \quad = \quad -16 & \\
\dfrac{-t}{-1} = \dfrac{-16}{-1} & \text{Divide both sides by } -1. \\
t = 16 &
\end{array}
$$

$$\text{Check:}\quad 3t + 6 = 4t - 10$$
$$3(16) + 6 = 4(16) - 10$$
$$48 + 6 = 64 - 10$$
$$54 = 54 \;\checkmark$$

We conclude this section with an example of an equation whose solution is not an integer.

Example 8

Solve $6x + 10 = 4 - 3x$.

Solution

$$
\begin{array}{rl}
6x + 10 = 4 - 3x & \\
\underline{3x \qquad = \qquad 3x} & \text{Add } 3x \text{ to both sides.} \\
9x + 10 = 4 & \\
\underline{\quad - 10 = -10} & \text{Subtract 10 from both sides.} \\
9x \qquad = \quad -6 & \\
x = -\dfrac{6}{9} = -\dfrac{2}{3} & \text{Divide both sides by 9.}
\end{array}
$$

$$\text{Check:}\quad 6x + 10 = 4 - 3x$$
$$6\left(-\frac{2}{3}\right) + 10 = 4 - 3\left(-\frac{2}{3}\right)$$
$$-4 + 10 = 4 + 2$$
$$6 = 6 \;\checkmark$$

You should keep in mind that there may be more than one strategy for isolating the variable in an equation. Different approaches may lead to the same correct solution. In the previous example, for instance, we could have begun by subtracting 10 from both sides, then we might have added $3x$ to both sides, and so on. In fact, we could have subtracted $6x$ from both sides first!

Section 3.3 Exercises

In Exercises 1 through 40, solve each equation for the variable. Check your solution.

1. $12 - x = 19$

2. $8 - x = 3$

3. $5 + 2x = 11$

4. $5 - 2x = 11$

5. $\dfrac{3z}{8} = -9$

6. $\dfrac{3z}{-8} = -9$

7. $\dfrac{3x}{2} = 15$

8. $\dfrac{3x}{-2} = -15$

9. $3z + 1 = 22$

10. $5 + 2x = 11$

11. $7 - 3y = -11$

12. $8 - 5y = -17$

13. $7 = \dfrac{3x}{4} - 2$

14. $10 = \dfrac{2x}{3} - 4$

15. $\dfrac{5x}{3} + 2 = -8$

16. $\dfrac{4x}{3} + 2 = -6$

17. $17 = 3x - 19$

18. $-12 = 5x + 18$

19. $5t + 18 = -32$

20. $6t + 14 = 68$

21. $\dfrac{4x}{3} + 8 = 0$

22. $\dfrac{3x}{2} + 15 = 0$

23. $6x + 24 = 0$

24. $-3x + 18 = 0$

25. $8z - 1 = 4z + 19$

26. $14 + 3y = 6y - 16$

27. $7x - 9 = 16x$

28. $5x - 10 = 3x$

29. $5 - z = 3z + 81$

30. $5x - 7 = x - 11$

31. $3y - 1 = 4y - 3$

32. $3x + 10 = x - 8$

33. $1 - y = y - 1$

34. $7x - 1 = 4x - 3$

35. $3y - 5 = 8y + 10$

36. $3y - 5 = 8y + 9$

37. $2t - 7 = 5t$

38. $7 - 2t = 5t$

39. $6t + 3 = 7 - 2t$

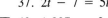 40. $1.897y - 3.147 = 3.916y + 11.001$

Supplementary Exercises

In Exercises 1 through 20, solve each equation for the variable. Check your solution.

1. $2x + 5 = 19$

2. $5y - 1 = 39$

3. $\dfrac{2x}{3} = 12$

4. $\dfrac{2x}{3} = -6$

5. $\dfrac{4x}{5} = 16$

6. $\dfrac{4x}{5} = -16$

7. $2 - 4x = 18$

8. $2 - 4x = -18$

9. $\dfrac{2x}{3} - 5 = 1$

10. $4z = 3z - 11$

11. $2x + 1 = x - 7$

12. $2x + 1 = x + 7$

13. $3x - 9 = x + 7$

14. $3x - 9 = x - 7$

15. $5y - 2 = 4y + 11$

16. $2y - 5 = 11 + 4y$

17. $3x - 10 = x + 7$

18. $3t + 1 = 4t - 15$

19. $6x + 9 = 3 - 3x$

20. $2x + 9 = 5x - 6$

3.4 Simplification and equation-solving

This section combines the rules we have learned so far in this chapter with the Chapter 2 notion of simplifying algebraic expressions.

Example 1

Solve $3(2x + 1) = 7x + 5$.

Solution

First we *simplify* the lefthand side by eliminating the parentheses:

$$3(2x + 1) = 7x + 5$$
$$6x + 3 = 7x + 5$$

The Distributive Property was used to eliminate the parentheses on the lefthand side.

Now we proceed with the methods for *solving* that have been developed in this chapter.

$$
\begin{array}{rcr}
6x + 3 = & & 7x + 5 \\
-7x = & & -7x \\
\hline
-x + 3 = & & 5 \\
-3 = & & -3 \\
\hline
-x = & & 2 \\
x = -2 & &
\end{array}
$$

Check: $3(2x + 1) = 7x + 5$

Note that we check the solution $x = -2$ in the *original* equation.

$$3(2 \cdot -2 + 1) = 7 \cdot -2 + 5$$
$$3(-4 + 1) = -14 + 5$$
$$3 \cdot -3 = -9$$
$$-9 = -9 \ \checkmark$$

Example 2

Solve $4(y - 1) = 5(2y - 8)$.

Solution

We must simplify *both* sides.

$$4(y - 1) = 5(2y - 8)$$
$$4y - 4 = 10y - 40$$

As we said at the end of Section 3.3, there is more than one possible way to find the solution. Of course, it doesn't matter which method we use as long as we are performing the same algebraic operation on both sides of the equation. Here are two possible ways of proceeding.

Method I	Method II

Method I

$$4y - 4 = 10y - 40$$
$$+ 4 = \quad + 4$$
$$4y = 10y - 36$$
$$-10y = -10y$$
$$-6y = -36$$
$$\frac{-6 \cdot y}{-6} = \frac{-36}{-6}$$
$$y = 6$$

Method II

$$4y - 4 = 10y - 40$$
$$-4y = -4y$$
$$-4 = 6y - 40$$
$$+ 40 = \quad + 40$$
$$36 = 6y$$
$$\frac{36}{6} = \frac{6 \cdot y}{6}$$
$$6 = y$$

Check: $4(y - 1) = 5(2y - 8)$
$4(6 - 1) = 5(2 \cdot 6 - 8)$
$4(5) = 5(12 - 8)$
$20 = 5(4)$
$20 = 20 \checkmark$

Check: $4(y - 1) = 5(2y - 8)$
$4(6 - 1) = 5(2 \cdot 6 - 8)$
$4(5) = 5(12 - 8)$
$20 = 5(4)$
$20 = 20 \checkmark$

Example 3

Solve $2[3(x + 1) + 1] + x = 22$.

Solution

First we *simplify* the lefthand side of the equation.

$$2[3(x + 1) + 1] + x = 22$$
$$2[3x + 3 + 1] + x = 22$$
$$2[3x + 4] + x = 22$$
$$6x + 8 + x = 22$$
$$7x + 8 = 22$$

Now we proceed with *solving* the equation.

$$7x + 8 = 22$$
$$- 8 = -8$$
$$7x = 14$$
$$\frac{7x}{7} = \frac{14}{7}$$
$$x = 2$$

Check: $2[3(x + 1) + 1] + x = 22$
$2[3(2 + 1) + 1] + 2 = 22$
$2[3 \cdot 3 + 1] + 2 = 22$
$2[9 + 1] + 2 = 22$
$2 \cdot 10 + 2 = 22$
$20 + 2 = 22$
$22 = 22 \checkmark$

Example 4

Solve $x + 1 = 3[2 + (x + 1)]$.

Solution

First we *simplify* the righthand side of the equation.

$$x + 1 = 3[2 + (x + 1)]$$
$$x + 1 = 3[3 + x]$$
$$x + 1 = 9 + 3x$$

Now we proceed to *solve* the equation.

$$
\begin{array}{rcl}
x + 1 &=& 9 + 3x \\
\underline{x} &=& -x \\
1 &=& 9 + 2x \\
\underline{-9} &=& \underline{-9} \\
-8 &=& 2x
\end{array}
$$

$$\frac{-8}{2} = \frac{2x}{2}$$

$$-4 = x$$

Check: $x + 1 = 3[2 + (x + 1)]$
$$-4 + 1 = 3[2 + (-4 + 1)]$$
$$-3 = 3(2 + (-3))$$
$$-3 = 3[-1]$$
$$-3 = -3 \ \checkmark$$

Example 5

Solve $3(-2y - 5) - 1 = 6 - 8[3 - 2(y - 4)]$.

Solution

$$3(-2y - 5) - 1 = 6 - 8[3 - 2(y - 4)]$$
$$-6y - 15 - 1 = 6 - 8[3 - 2y + 8]$$
$$-6y - 15 - 1 = 6 - 8[-2y + 11]$$
$$-6y - 15 - 1 = 6 + 16y - 88$$

Eliminate grouping symbols.

$$-6y - 16 = 16y - 82$$

Collect like terms.

$$-22y = -66$$
$$y = 3$$

So far the equations studied involve only one variable. They are called **first-degree equations in one variable** or **linear equations in one variable** because no term has an exponent greater than 1. Later in this text we will examine equations in which there are two variables as well as equations involving variables raised to a higher degree.

We summarize the procedure for solving equations as follows:

Procedure for solving linear equations in one variable

1. If appropriate, simplify either side or both sides by eliminating grouping symbols or collecting like terms.
2. Add (or subtract) appropriate quantities to both sides of the equation to isolate the term containing the variable.
3. Multiply or divide both sides of the equation by an appropriate quantity to isolate the variable.
4. Check your solution!

Section 3.4 Exercises

In Exercises 1 through 30, solve each equation and check your solution.

1. $3(4x + 1) = 6x + 9$
2. $5y = 3(y + 2)$
3. $-2(3 - t) = t - 7$
4. $x - 7 = 4 - 3(x - 1)$
5. $3(2x + 8) = 30$
6. $5 = -2(x - 6) + 1$
7. $-(x - 3) = 10$
8. $2(x - 1) = 8$
9. $3(x + 2) - x = 4$
10. $x - (2 + 2x) = 6$
11. $5 - 2(x + 4) = 7$
12. $2 - (x + 1) = 3 - 4x$
13. $3(x + 1) - 4 = 3 - x$
14. $(2 - x) - 3x = 4x - 6$
15. $4 - (x + 2) = 2(x + 1)$
16. $10[x - (x + 1)] = -5x$
17. $a + 2(1 - a) = 3$
18. $8(2t - 5) = t - 40$
19. $9(3y + 8) = 3(y + 8)$
20. $-2(x - 1) = 5 - 3(x + 3)$
21. $4(z - 1) = 5(2z - 8)$
22. $2(z + 1) + 3(z - 1) = 4$
23. $2x + 3[x - 2(x + 2)] = 0$
24. $2x + 3[x - 2(x + 2)] = 4$
25. $a - 1 = 2[2 - (a + 1)]$
26. $x + 3[2x + (1 - x)] = 11$
27. $x + 3[2x - (1 - x)] = 7$
28. $-4(x - 9) + 5 = 21 - 5[16 - 2(x - 4)]$
29. $4(3x + 7) + 1 = 75 - 10[16 - 3(2x + 7)]$
30. $4 - [4 - 4(4 - x)] = 3 + [3 + 3(3 - x)]$

Supplementary Exercises

In Exercises 1 through 15, solve each equation and check your solution.

1. $3 - 2(a + 1) = 5$
2. $5(b + 1) = -10$
3. $4(x - 2) - 7 = 3x + 1$
4. $6 - 3(3 - a) = 2a + 1$
5. $4(a + 5) - (13 + 2a) = a - 9$
6. $16 = 4(3 - x)$
7. $\dfrac{x + 2}{3} = 5$
8. $5 - (x + 2) = 2x + 9$
9. $2x - \{1 - (3 - x)\} = 2 + 2x$
10. $8 - a = 2(5 - 2a) - 1$
11. $\dfrac{6 - x}{2} = 7$
12. $\dfrac{3a + 2}{4} = a - 1$
13. $5 - [6 - (x - 2)] = 4x + 3$
14. $2x - (5 - x) = x - 9$
15. $\dfrac{a + 1}{4} = \dfrac{2a + 3}{4}$

3.5 Using algebraic models to solve problems

Often when an algebraic problem arises, it is in verbal or written form—not in the form of an equation. When we represent the problem in algebraic notation, we are forming an **algebraic model** for the problem. Many algebraic models result in an algebraic equation whose solution can be determined by the methods of this chapter. The following examples illustrate the concept.

Example 1

The sum of Dom's age and Matt's age is 30 years. If Matt is 2 years old, how old is Dom?

Solution

To answer the question, we do three things:

1. Translate the problem into an algebraic equation. (That is, an *algebraic model.*)
2. Solve the equation.
3. Check our answer in the original statement of the problem.

Step 1 Problem translation

We must decide what the *unknown* is. The unknown is what is asked for in the shaded part of the problem—Dom's age. Let d be Dom's age. Next, we know that

$$\underbrace{\text{Dom's age}}_{d} \quad \underbrace{\text{plus}}_{+} \quad \underbrace{\text{Matt's age}}_{2} \quad \underbrace{\text{is}}_{=} \quad \underbrace{30.}_{30} \leftarrow \text{This is our algebraic model.}$$

Step 2 Equation solution

$$\begin{array}{rcr} d + 2 &=& 30 \\ -2 &=& -2 \\ \hline d &=& 28 \end{array}$$

Step 3 Check

The answer of 28 does, in fact, check because "the sum of Dom's age (28) and Matt's age (2) is 30 years."

Example 2

Pamela spent $800 on her new stereo system. This is $500 more than Jeanne spent. How much did Jeanne spend?

Solution

Step 1 Problem translation

We are asked to find the amount Jeanne spent. Let s equal the amount Jeanne spent.

$$\underbrace{800}_{800} \quad \underbrace{\text{is}}_{=} \quad \underbrace{500}_{500} \quad \underbrace{\text{more than}}_{+} \quad \underbrace{s.}_{s} \leftarrow \text{This is our algebraic model.}$$

Step 2 Equation solution

$$\begin{array}{rcr} 800 &=& 500 + s \\ -500 &=& -500 \\ \hline 300 &=& 0 + s \\ 300 &=& s \end{array}$$

So Jeanne spent $300 on her stereo system.

Step 3   Check

Pam spent $800. Jeanne spent $300.

800 *is* 500 more than 300, so our answer is correct.

Example 3

Leland made $83 more than Dwight made last month. If their total earnings were $723, how much did Dwight make?

Solution

Step 1   Problem translation

We are asked to find the amount Dwight made. Let *d* represent the amount Dwight made. Since Leland earned $83 more than Dwight, Leland's earnings can then be represented by $(d + 83)$.

$$
\underbrace{\text{Dwight's earnings}}_{d} \quad \underbrace{\text{and}}_{+} \quad \underbrace{\text{Leland's earnings}}_{(d + 83)} \quad \underbrace{\text{were}}_{=} \quad \underbrace{\$723.}_{\$723}
$$

Step 2   Equation solution

$$
\begin{aligned}
d + d + 83 &= 723 \\
2d + 83 &= 723 \\
-83 &= -83 \\
\hline
2d &= 640 \\
\frac{2d}{2} &= \frac{640}{2} \\
d &= 320
\end{aligned}
$$

So Dwight's earnings were $320.

Step 3   Check

Dwight's earnings were $320.

Leland's earnings were $403 ($83 more than Leland's).

Together they made $320 + $403 = $723. This checks.

In the next four examples, we are asked to find more than one quantity or thing.

Example 4

A nurse is going to administer a dose of phenobarbital to two patients, an adult and a child. The combined amount of the drug needed is 40 mg. The adult is to receive 4 times as much as the child. How much is *each* to receive?

Solution

Step 1   Problem translation

We are asked to find the dosage for each person. The child receives the smaller amount— we let *p* represent the child's dosage. The adult dosage is four times as much—we then have 4*p* representing the adult's dosage.

$$
\underbrace{\text{The child's dosage}}_{p} \quad \underbrace{\text{and}}_{+} \quad \underbrace{\text{the adult's dosage}}_{4p} \quad \underbrace{\text{total}}_{=} \quad \underbrace{40 \text{ mg.}}_{40}
$$

Step 2   Equation solution

$$p + 4p = 40$$
$$5p = 40$$
$$\frac{5p}{5} = \frac{40}{5}$$
$$p = 8$$

So the child receives 8 mg.
The adult receives 32 mg.

Step 3   Check

The combined dosage must total 40 mg.

$$8 + 32 = 40$$
$$40 = 40 \ \checkmark$$

Example 5

Connie, Oksana, and Penny are photographers for a city police department. One day the assignment was to ''shoot'' pictures of a fire of suspicious origin. Penny ''shot'' twice as many pictures as Oksana did. Connie ''shot'' three more pictures than Oksana did. In total they took 103 pictures. How many pictures did *each* person take?

Solution

Step 1   Problem translation

From the information given, Penny and Connie each shot more pictures than Oksana. So let n represent the number of pictures shot by Oksana. Then Penny shot $2n$ pictures and Connie shot $n + 3$ pictures.

Connie's pictures	and	Oksana's pictures	and	Penny's pictures	total	103.
$(n + 3)$	$+$	n	$+$	$2n$	$=$	103

Step 2   Equation solution

$$(n + 3) + n + 2n = 103$$
$$n + 3 + n + 2n = 103$$
$$4n + 3 = 103$$
$$\underline{ -3 = -3}$$
$$4n = 100$$
$$n = 25$$

So Oksana shot 25 pictures.
Connie shot 28 pictures.
Penny shot 50 pictures.

Step 3   Check

The combined number of pictures must be 103.

$$28 + 25 + 50 = 103$$
$$53 + 50 = 103$$
$$103 = 103 \ \checkmark$$

Example 6

Joan and Ron invested money at 9% simple annual interest. If the total amount of the investment is $9265 after one year, how much was invested originally? How much interest did they receive?

Solution

Step 1   Problem translation

Let a represent the amount invested originally. Since the amount a is invested for only one year, the interest earned on that amount is strictly a product of the amount a times the rate of interest (as a decimal). Thus the interest in this case is given by $.09a$ and we have

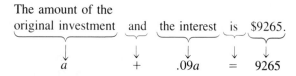

The amount of the original investment   and   the interest   is $9265.

$$a + .09a = 9265$$

Step 2   Equation solution

$$a + .09a = 9265$$
$$1.09a = 9265$$
$$a = 8500$$

Thus the amount invested originally is $8500.
The interest, $.09 \cdot 8500$, is $765.

Step 3   Check

The total, $8500 + $765, *is* $9265. $\checkmark$

Example 7

The sum of two consecutive even integers is 282. Find the integers.

Solution

Step 1   Problem translation

Let x equal the first integer. The next *even* integer is then represented by $x + 2$. (Consecutive even integers must be two units apart.)

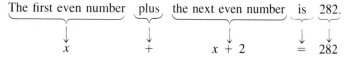

The first even number   plus   the next even number   is   282.

$$x + x + 2 = 282$$

Step 2   Equation solution

$$x + x + 2 = 282$$
$$2x + 2 = 282$$
$$2x = 280$$
$$x = 140$$

The smaller number is 140; the next consecutive even integer ($x + 2$) is 142.

Step 3   Check

140 and 142 are obviously consecutive even integers.
Their sum is $140 + 142 = 282$. $\checkmark$

Section 3.5 Exercises

1. The Ace Trucking Company has six employees. Together with the Carlin Hauling Corporation, they employ a total of fifteen people. How many people are there in the Carlin Hauling Corporation?

2. In March, the House of Guitars sold 600 record albums. This is 72 less than the expected sales for that month. What were the expected sales for March?

3. A number is 23 more than −9. What is the number?

4. A number is 32 more than −9. What is the number?

5. Liz can keypunch (on the average) 10 cards per minute. This is 5 less than the number that José can punch per minute. How many cards can José punch per minute?

6. Larry and Sue are, respectively, 67 in. and 69 in. tall. They each buy a pair of platform shoes and find that they are equally tall with the shoes on. Larry's shoes are twice the height of Sue's. What is the height of Sue's shoes?

7. Linda and Paul are recording artists. Last month their combined income was $13,000. Linda made $1,000 more than five times what Paul made. How much did Paul make?

8. Two identical automobiles, except for engines, each drive the same test course. Auto A, equipped with a standard-type internal combustion engine, uses 3 liters (a unit of volume in the metric system that is slightly larger than a quart) less than twice the amount of gas that auto B uses. (Auto B has a newly designed engine.) Together they consume 9 liters of gasoline.
   (a) How much gas does auto A use?
   (b) How much gas does auto B use?
   (c) Which auto gets better mileage?

9. Erin, Sean, and Philip are each subjects in a psychology experiment in which they are asked to identify colors of blocks while blindfolded. In a certain test run, Erin identified twice the number that Philip did. Sean identified three times the number that Philip did. In total they made 12 identifications. How many did *each* identify?

10. Suppose Erin, Sean, and Philip attempt to make color identifications during another test run. If Erin identifies twice the number that Philip does and Sean identifies one more than twice what Erin does, how many identifications does each make? Assume that, in total, they made 15.

11. A sociologist noted the number of months that four different families were on welfare. Collectively, they were on welfare for 21 months. Individually, Families 1 and 2 spent the same number of months on welfare. Family 3 was on welfare three fewer months than Family 1, and Family 4 was on welfare three times as long as Family 1. How long was each family on welfare?

12. The accompanying pictorial model shows the compositional layering (in miles) of the atmosphere based on information of the *Explorer* satellite program of The National Aeronautics and Space Administration (NASA).

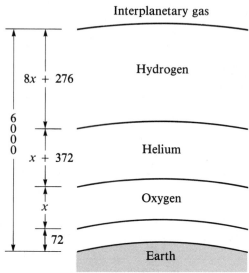

(a) Determine the three unknown distances.

(b) At what height above earth does the layer of hydrogen gas begin?

13. The accompanying pictorial model shows an iceberg. Let h represent the height above water level. The amount below water level is $8h - 3$, and the total height is 888 feet. How much of the iceberg is below water level?

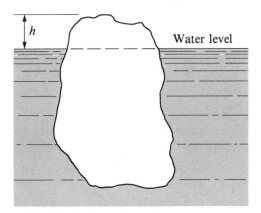

14. Bobby burns up twice as many calories jogging as he does walking the same distance. During an exercise period in which he jogs 1 km (kilometer, a unit of length in the metric system; one km is slightly more than $\frac{1}{2}$ mile) and walks 8 km, he burns up 1200 calories. How many calories does he burn up while jogging 1 km?

15. The sum of two consecutive even integers is 30. What are the two integers?

16. The sum of three consecutive odd integers is 39. What are the three integers?

17. Louise and Ken both recycle all their paper products. Louise recycled 10 more pounds than twice what Ken recycled. If their combined paper for recycling was 100 pounds, how much paper did each recycle?

18. A certain city has three different ambulance dispatch stations, A, B, and C. The A and B stations have an equal average weekly number of calls, but the C station has five less than twice the number of calls of station A. How many calls does each have per week if the total for A, B, and C is 95 calls per week?

19. Paul and Eileen invested money in an account that yielded $12\frac{1}{4}\%$ simple annual interest. How much did they invest originally if their account amounted to $33,675?

20. A coat is on sale with the following ticket on it:

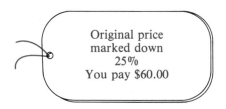

Original price
marked down
25%
You pay $60.00

How much did the coat sell for originally? (*Hint:* A markdown of 25% means 25% of the original price was subtracted from the original price.)

21. A contract between the school board and teachers' union was just negotiated. The new 1986 salary is determined by taking 5% of the 1985 salary, adding it to the 1985 salary, and then adding $800. What was Joe's 1985 salary if his new 1986 salary is $19,175?

22. Lee has 16 coins in dimes and nickels. He has two more nickels than dimes. How many nickels and dimes does he have?

23. Mike sells pens and pencils. He sold twice as many pens as pencils last week, and he sold a total of 27 items. How many pens did he sell?

24. John mixes 20 pounds of fruit that contain peaches, pears, and apples. He mixes 3 more pounds of apples than pears and 2 more pounds of peaches than pears. How many pounds of each did he mix?

25. The sum of two numbers is 56, and one number is 7 times the other. Find the numbers.

26. Five times a number added to 6 is the same as 8 less three times the number. Find the number.

27. Mark works 5 hours more per week than John does, and Joe works 3 hours less per week than John does. If their combined hours per week worked is 38, find how many hours each worked.

28. Kristen and Jack clean chimneys. Kristen cleaned four less than twice the number cleaned by Jack. If 14 chimneys were cleaned by them, how many were cleaned by each?

29. The sum of three consecutive integers is 51. Find the integers.

30. The sum of three consecutive even integers is 66. Find the integers.

31. The sum of three consecutive odd integers is 339. Find the integers.

32. The sum of an integer and four times the opposite of the integer is 12. Find the integer.

33. The sum of two numbers is 35. One number is 5 more than the other. Find the numbers.

34. Sarah has two savings accounts. The combined interest from the two accounts is $400. If one account earned $130 more than the other, find the interest earned by each account.

35. Carla buys a dress on sale for $15. This price had been marked down from the original price by 25%. What was the original price?

36. Judy invests $2000 in an account that yields $8\frac{3}{4}\%$ simple annual interest. How much interest did she earn?

Supplementary Exercises

Solve each of the following word problems.

1. Four times a number less 6 is the same as the number plus 12. Find the number.

2. John has 5 fewer 13-cent stamps than 15-cent stamps. If he has 17 stamps altogether, how many of each does he have?

3. The sum of four consecutive integers is 90. Find the integers.

4. The sum of five times a number and the number less 8 is 64. Find the number.

5. The sum of three consecutive even integers is 114. Find the integers.

6. Dennis is 5 years older than Alice, and Chris is 4 years younger than Alice. Their combined age is 70 years. How old is each?

7. If a number is subtracted from 17, the difference is -10. Find the number.

8. Joe delivered three times as many suits as Frank did. Together they delivered 120 suits. How many did each deliver?

9. Mike invests $2000 in an account that yields $8\frac{1}{2}\%$ simple annual interest. How much interest did he earn?

10. An auditorium holds 500 people on the main floor and balcony. If the main floor holds three times as many people as the balcony, how many can sit in each area?

11. John's parking garage has three floors and can hold 50 more cars on the first floor than on the second floor. It can only hold half as many cars on the top floor as on the second floor. If the garage holds 400 cars, how many will it hold on each floor?

12. One-half a number less 5 is 12. Find the number.

13. Harry sells two kinds of computers. Last month he sold two more than twice the number of Brand A as Brand B. If he sold 44 altogether, how many of each brand did he sell?

14. Natalie needs three times as many units of vitamin C as she does vitamin A. If she needs a combined dosage of 24 units, how much of each type of vitamin does she need?

15. Two-thirds of some number added to 27 is 37. Find the number.

3.6 Formulas and literal equations

So far we have solved equations, each of which was composed of exactly one variable. In this section we will examine **literal equations,** which are equations composed of two or more variables.

We begin by using a familiar example—the rectangle. The area of a rectangle is obtained by multiplying its two dimensions, length and width.

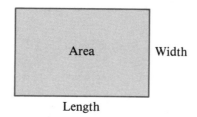

If we call the area "A," the width "w," and the length "ℓ," we can express the area relationship algebraically as

$$A = \ell w$$

This is a literal equation. Furthermore, we call it a **formula,** which means that it is a general fact, rule, or principle expressed in symbols. We can find the area pictorially when we know the length is 4 inches and the width is 3 inches, as follows:

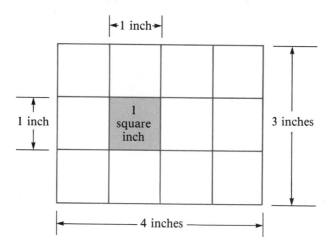

As we see, the area is 12 square inches. Algebraically, we find the area by evaluating the area formula when $\ell = 4$ and $w = 3$, as in Chapter 2:

$$A = \ell w$$
$$A = 4 \text{ inches} \cdot 3 \text{ inches}$$
$$A = 12 \text{ in}^2$$

On the other hand, suppose we know the area of a certain rectangular region (for example, the floor), and we also know its length. Could we find its width? The answer is yes, and one very appropriate means for finding w is to *solve the area formula for w.* That is, take

$$A = \ell w$$

an equation, which right now is *solved for A,* and solve it for w.

To solve for w, we must first isolate it on one side of the equation. Since it is multiplied by ℓ, we divide both sides of the equation by ℓ:

$$\frac{A}{\ell} = \frac{\ell w}{\ell}$$

$$\frac{A}{\ell} = \frac{\cancel{\ell} w}{\cancel{\ell}}$$

$$\frac{A}{\ell} = w \quad \text{or} \quad w = \frac{A}{\ell} \leftarrow \left\{ \begin{array}{l} \text{This is the area formula solved} \\ \text{for } w. \text{ It is now a formula for} \\ \text{the width } w \text{ of a rectangle.} \end{array} \right.$$

So if a rectangle has an area of 48 square feet ($A = 48$) and a length of 12 feet ($\ell = 12$), we can find the width w:

$$w = \frac{A}{\ell}$$

$$w = \frac{48}{12}$$

$$w = 4$$

Thus the width is 4 feet.

If a rectangle has an area of 24 square miles ($A = 24$) and a length of 4 miles ($\ell = 4$), we can find w:

$$w = \frac{A}{\ell}$$

$$w = \frac{24}{4}$$

$$w = 6$$

So the width is 6 miles.

For any values of A (area) and ℓ (length) that a rectangle might have, the formula $w = \frac{A}{\ell}$ can be used to find the width, w. Thus we can see the importance of having a formula. Besides being a symbolic way of expressing a general fact, rule, or principle, a formula can be used to determine the value of one of the variables if the other variables' values are known.

Example 1

The equation $d = rt$ is a formula for the distance (d) traveled by an object in uniform motion. The rate of travel (or speed) is r and the time of travel is t. Assume d is measured in feet, r is measured in feet per second, and t is measured in seconds.

(a) What distance is traveled by an object moving at a speed of 44 ft/sec for 5 seconds?

(b) What distance is traveled by an object moving at a speed of 1100 ft/sec for 5 seconds?

Solution

(a) The question asks: "What distance is traveled . . ." So we are asked to find d. To do so, we *evaluate* the distance formula with $r = 44$ and $t = 5$.

$$d = rt$$
$$d = 44 \cdot 5$$
$$d = 220$$

The distance traveled is 220 feet.

(b) $d = rt$

$$d = 1100 \cdot 5$$
$$d = 5500$$

The distance traveled is 5500 feet.

Example 2

(a) Solve the distance formula for r.

(b) At what uniform speed must an auto move to cover 440 feet in 5 seconds?

Solution

(a) The distance formula $d = rt$ must be solved for r, which means we must isolate r all by itself. Since r is multiplied by t in the formula $d = rt$, we will divide both sides of the distance formula by t.

$$d = rt$$

$$\frac{d}{t} = \frac{r\cancel{t}}{\cancel{t}}$$

$$\frac{d}{t} = r \quad \text{or} \quad r = \frac{d}{t} \leftarrow \left\{ \begin{array}{l} \text{This is the distance formula} \\ \text{solved for } r. \text{ It is now a formula} \\ \text{for the rate of travel (speed).} \end{array} \right.$$

(b) The question asks: "At what uniform speed . . ." So we are asked to find r.

$$r = \frac{d}{t}$$

$$r = \frac{440}{5}$$

$$r = 88$$

The speed must be 88 ft/sec.

Example 3

(a) Solve the distance formula for t.

(b) How long does it take an auto to travel 300 feet if it is traveling with a speed of 30 ft/sec?

Solution

(a) We desire to solve $d = rt$ for t. To isolate t, we will divide both sides of the distance formula by r.

$$d = rt$$

$$\frac{d}{r} = \frac{\cancel{r}t}{\cancel{r}}$$

$$\frac{d}{r} = t \quad \text{or} \quad t = \frac{d}{r} \leftarrow \left\{ \begin{array}{l} \text{This is the distance formula} \\ \text{solved for } t. \text{ It is now a} \\ \text{formula for the time of} \\ \text{travel } t. \end{array} \right.$$

(b) The question asks, "How long does it take . . ." So we are asked to find t.

$$t = \frac{d}{r}$$

$$t = \frac{300}{30}$$

$$t = 10$$

The time would be 10 seconds.

The previous three examples all dealt with the distance formula $d = rt$, which contains three variables. We have seen that the value of any one of these unknowns can be determined if the values of the other two are known. When we want to solve a literal equation for one of the unknowns, we can use the methods developed in this chapter. That is, we can add or subtract the same quantity on *both sides* of the equation. Furthermore, we can multiply or divide both sides of an equation by the same nonzero quantity.

The next two examples illustrate this.

Example 4

(a) Solve the literal equation $2x + 3y = 1$ for y.

(b) Find the value of y when $x = 5$.

Solution

(a) We must isolate y in the equation $2x + 3y = 1$. First we subtract $2x$ from both sides.

$$
\begin{array}{rcl}
2x + 3y &=& 1 \\
-2x &=& -2x \\
\hline
3y &=& 1 - 2x
\end{array}
$$

Next we divide both sides by 3.

$$3y = 1 - 2x$$

$$\frac{3y}{3} = \frac{1 - 2x}{3}$$

$$y = \frac{1 - 2x}{3} \;\;\leftarrow\left\{\begin{array}{l}\text{This is the result of solving} \\ \text{the original equation for } y.\end{array}\right.$$

(b) We want the value of y when $x = 5$.

$$y = \frac{1 - 2x}{3}$$

$$y = \frac{1 - 2 \cdot 5}{3}$$

$$y = \frac{-9}{3}$$

$$y = -3$$

Example 5

(a) Solve the literal equation $3ab + 5ap = 6$ for p.

(b) Determine the value of p, when $a = 1$ and $b = 7$.

Solution

(a) We must isolate p in the equation $3ab + 5ap = 6$. First we subtract $3ab$ from both sides.

$$
\begin{array}{rcl}
3ab + 5ap &=& 6 \\
-3ab &=& -3ab \\
\hline
5ap &=& 6 - 3ab
\end{array}
$$

Next we divide both sides by $5a$; we assume $a \neq 0$.

$$5ap = 6 - 3ab$$

$$\frac{\cancel{5a}p}{\cancel{5a}} = \frac{6 - 3ab}{5a}$$

$$p = \frac{6 - 3ab}{5a} \leftarrow \left\{ \begin{array}{l} \text{This is the result of solving} \\ \text{the original equation for } p. \end{array} \right.$$

(b) We want the value of p when $a = 1$ and $b = 7$.

$$p = \frac{6 - 3ab}{5a}$$

$$p = \frac{6 - 3 \cdot 1 \cdot 7}{5 \cdot 1}$$

$$p = \frac{6 - 21}{5}$$

$$p = \frac{-15}{5}$$

$$p = -3$$

The next four examples contain formulas from a variety of different life situations.

Example 6

A car rental agency determines the rental cost of a car by using the formula

$$c = \frac{500d + 10m}{100}$$

where c represents the cost of rental in dollars,
d represents the number of rental days, and
m represents the number of miles.

(a) If a car is rented for 2 days and 50 miles are traveled, what is the cost?
(b) If a car is rented for 5 days and 300 miles are traveled, what is the cost?

Solution

(a) We are asked to find c when $d = 2$ and $m = 50$.

$$c = \frac{500d + 10m}{100}$$

$$c = \frac{500 \cdot 2 + 10 \cdot 50}{100}$$

$$c = \frac{1000 + 500}{100}$$

$$c = \frac{1500}{100}$$

$$c = 15$$

So the cost of rental is $15.

(b) We are asked to find c when $d = 5$ and $m = 300$.

$$c = \frac{500d + 10m}{100}$$

$$c = \frac{500 \cdot 5 + 10 \cdot 300}{100}$$

$$c = \frac{2500 + 3000}{100}$$

$$c = \frac{5500}{100}$$

$$c = 55$$

The cost of rental is $55.

Example 7

Intelligence quotient, or IQ, is the relationship between mental age and chronological age. For children, it is given by the formula:

$$I = \frac{100M}{C}$$

where I represents IQ,
M represents mental age, and
C represents chronological age.

What is an 8-year-old child's IQ if, as a result of intelligence testing, she has a mental age of 10 years?

Solution

We are asked to find IQ when $M = 10$ and $C = 8$.

$$I = \frac{100M}{C}$$

$$I = \frac{100 \cdot 10}{8}$$

$$I = \frac{1000}{8}$$

$$I = 125$$

The child's IQ is 125.

Example 8

Suppose a 10-year-old child has an IQ of 90. What is the child's mental age?

Solution

We are asked to find the mental age M. So we solve the IQ formula for M.

$$I = \frac{100M}{C}$$

First multiply both sides by C.

$$C \cdot I = C \cdot \frac{100M}{C}$$

$$C \cdot I = 100M$$

Now divide both sides by 100.

$$\frac{C \cdot I}{100} = \frac{\cancel{100}M}{\cancel{100}}$$

$$\frac{C \cdot I}{100} = M \quad \text{or} \quad M = \frac{C \cdot I}{100} \leftarrow \begin{cases} \text{This is the IQ formula} \\ \text{solved for } M. \text{ It is now a} \\ \text{formula for mental age.} \end{cases}$$

Continuing, we now evaluate the mental age formula when $C = 10$ and $I = 90$.

$$M = \frac{C \cdot I}{100}$$

$$M = \frac{10 \cdot 90}{100}$$

$$M = \frac{90}{100}$$

$$M = 9$$

The child's mental age is 9 years.

Example 9

When the air temperature is 0°C, the speed of sound in air is 1087 ft/sec. For each degree rise in temperature (Celsius) the speed of sound increases by 2 ft/sec. The following speed of sound formula results:

$$s = 1087 + 2T$$

where s represents the speed of sound, and
T represents the Celsius temperature.

(a) If the Celsius temperature is 20°C, what is the speed of sound?

(b) If the speed of sound is 1105 ft/sec, what must be the Celsius temperature?

Solution

(a) We must determine s when $T = 20$.

$$s = 1087 + 2T$$
$$s = 1087 + 2 \cdot 20$$
$$s = 1087 + 40$$
$$s = 1127$$

The speed of sound would be 1127 ft/sec.

(b) We are asked to find T when $s = 1105$. First we solve the speed formula for T.

$$\begin{array}{rcl} s & = & 1087 + 2T \\ -1087 & = & -1087 \\ \hline s - 1087 & = & 2T \end{array}$$

$$\frac{s - 1087}{2} = \frac{2T}{2}$$

$$\frac{s - 1087}{2} = T \quad \text{or} \quad T = \frac{s - 1087}{2} \leftarrow \begin{cases} \text{This is the speed of} \\ \text{sound formula} \\ \text{solved for } T. \text{ Now it} \\ \text{is a formula for the} \\ \text{temperature } T. \end{cases}$$

Now we proceed to find T.

$$T = \frac{s - 1087}{2}$$

$$T = \frac{1105 - 1087}{2}$$

$$T = \frac{18}{2}$$

$$T = 9$$

The Celsius temperature is 9°C.

Section 3.6 Exercises

In Exercises 1 through 23, solve for the indicated variable.

1. $x + y = 1$ Solve for y.
2. $t - 2k = 0$ Solve for t.
3. $t - 2k = 0$ Solve for k.
4. $5x - 3y = 1$ Solve for y.
5. $-2x - 3y = 4$ Solve for y.
6. $5x + 3a = 4$ Solve for x.
7. $5x + 3a = 4$ Solve for a.
8. $L = a + (n - 1)d$ Solve for d.
9. $3ps - ax = y$ Solve for p.
10. Solve $V = lwh$ for h.
11. Solve $P = 2l + 2w$ for l.
12. Solve $P = a + b + c$ for a.
13. Solve $Y + 3x = l$ for Y.
14. Solve $A = \frac{1}{2}bh$ for h.
15. Solve $C = 2\pi r$ for r.
16. Solve $s = 2h^2 + 4hb$ for b.
17. Solve $D = rt$ for r.
18. Solve $A = P + PRT$ for T.
19. Solve $Q = \frac{P}{R}$ for R.
20. Solve $V = 4\pi r^2 h$ for h.
21. Solve $Q = \frac{P}{R}$ for P.
22. Solve $3xy - 2t = 5$ for x.
23. Solve $5abc - 7 = 8$ for b.
24. In Exercise 1, find the value of y when $x = 4$.
25. In Exercise 2, find the value of t when $k = -3$.
26. In Exercise 3, find the value of k when $t = 8$.
27. In Exercise 4, find the value of y when $x = 2$.
28. In Exercise 5, find the value of y when $x = 1$.
29. In Exercise 7, find the value of a when $x = 2$.
30. The relationship between Fahrenheit (F) and Celsius (C) temperatures is given by the formula

$$5F - 9C = 160$$

   (a) Solve the formula for F.
   (b) Solve the formula for C.
   (c) Using the appropriate formula [either (a) or (b)], find the Fahrenheit equivalent of 100°C.
   (d) Using the appropriate formula, find the Celsius equivalent of 68°F.

31. Given the equation $7x + 4y = 21$:
   (a) Solve the equation for y.
   (b) Solve the equation for x.
   (c) What is y when x is 7?
   (d) What is x when y is 0?

In this section our use of the distance formula has assumed that d is measured in feet, r is measured in ft/sec, and t is measured in seconds. It is not necessary to maintain that restriction. In the next two exercises, assume that d is measured in kilometers, r is measured in km/hr, and t is measured in hours.

32. A meteor is traveling at the uniform speed of 2000 km/hr. How far does it travel in 5 hours?

33. A car traveled 960 km in 10 hours. Assuming uniform speed, how fast did it go?

34. Hurtz Car Rental Agency charges $8 per day plus 5¢ per mile. A formula that gives the cost C_H of rental (in dollars) from Hurtz is:

$$C_H = \frac{800d + 5m}{100}$$

Smartz Car Rental Agency charges $5 per day plus 20¢ per mile. A formula that gives the cost C_S of rental (in dollars) from Smartz is:

$$C_S = \frac{500d + 20m}{100}$$

In each formula, d = number of days and m = number of miles. Suppose you desire to rent a car for 3 days and you estimate that you'll travel 500 miles.

  (a) Which company offers the better deal?

  (b) How much will it cost you?

  (c) How much will you save by choosing the better deal?

35. Rework Exercise 34 assuming that you want to rent a car for 5 days and estimate that you'll drive 75 miles.

36. Children are not able to tolerate adult doses of drugs. One formula used to calculate the child's dose C is Young's Rule, which states:

$$C = \frac{aD}{a + 12}$$

where D = adult dose and a = age of child.

Find the correct dose of phenobarbital for a 6-year-old child if the adult dose is 30 mg.

37. Sue is training to run her first marathon. Each Saturday she jogs a 16-mile course at the rate of 8 mi/hr, and then she walks a 4-mile course at the rate of 2 mi/hr. How long does she spend exercising each Saturday?

38. What is the IQ of a 10-year-old child with a mental age of 8 years? (Recall $I = \dfrac{100M}{C}$ from Example 7 in this section.)

39. The amount of *simple interest*, I, paid to the lender of money by the receiver, is given by $I = PRT$ where P represents the principal (amount loaned), R represents the rate of interest, and T represents the time period of the loan. Solve the simple interest formula for R.

40. The formula $N = 4T - 160$ is an approximate algebraic model for the number of chirps per minute, N, made by a cricket when the Fahrenheit temperature is T degrees.

  (a) Solve the formula for T.

  (b) If a cricket makes 100 chirps per minute, what is the Fahrenheit temperature?

Supplementary Exercises

In Exercises 1 through 10, solve for the indicated variable.

1. $m - n = 11$ Solve for m.

2. $t + v = 21$ Solve for v.

3. $2t + p = 7$ Solve for t.

4. $k - 3m = y$ Solve for m.

5. $6a + 4b = 12$ Solve for b.

6. $2x + 3y = 5$ Solve for y.

7. $A = \frac{1}{2}bh$ Solve for b.   8. $A = P + PRT$ Solve for R.

9. $P = 2r + b_1 + b_2$ Solve for b_1.   10. $P = 2r + b_1 + b_2$ Solve for r.

11. In Exercise 1, find the value of m when $n = 9$.

12. In Exercise 2, find the value of v when $t = -9$.

13. In Exercise 3, find the value of t when $p = 15$.

14. In Exercise 4, find the value of m when $k = 6$ and $y = 9$.

15. In Exercise 5, find the value of b when $a = -1$.

16. In Exercise 6, find the value of y when $x = 4$.

17. In Exercise 7, find the value of b when $A = 12$ and $h = 3$.

18. In Exercise 8, find the value of R when $A = 1800$, $P = 1000$, and $T = 1$.

19. In Exercise 9, find the value of b_1 when $P = 50$, $r = 5$, and $b_2 = 15$.

20. In Exercise 10, find the value of r when $P = 100$, $b_1 = 40$, and $b_2 = 32$.

21. The perimeter of a rectangle is given by the formula $P = 2b + 2h$ where P represents the perimeter, b represents the length of the base, and h represents the height of the rectangle.

   (a) Solve the formula for b.

   (b) Find b when $P = 30$ feet and $h = 5$ feet.

22. The circumference of a circle is given by the formula $C = 2\pi r$, where C represents the circumference, r represents the radius, and π is a constant approximated by 3.14.

   (a) Solve the formula for r.

   (b) Find r when $C = 37.68$ inches.

23. The volume of a rectangular box is given by the formula $V = lwh$, where V represents the volume, l represents the length, w represents the width, and h represents the height.

   (a) Solve the formula for w.

   (b) Find w when $V = 5400$ cubic feet, $l = 20$ feet, and $h = 9$ feet.

24. The formula for simple interest is given by $I = PRT$, where I represents the interest earned, P represents the principal, R represents the interest rate, and T represents the time period (in the same interval as the interest rate).

   (a) Solve for R.

   (b) Find R if $I = \$24$, $P = \$300$, and $T = 2$ years.

25. Referring to question 24:

   (a) Solve for P.

   (b) Find P if $I = \$84$, $R = 7\%$ per year, and $T = 3$ years.

3.7   Linear inequalities

Recall the four order relations we introduced in Chapter 1:

$<$	"less than"
$>$	"greater than"
$\leq$	"less than or equal to"
$\geq$	"greater than or equal to"

We say that the **sense** or **direction** of $<$ (or $\leq$) is opposite to $>$ (or $\geq$).

When we compare one variable expression with another using one of the order relations, we have an **inequality**. In fact, the inequalities we will solve in this section resemble the linear equations in one variable studied earlier in this chapter except that the "=" symbols will be replaced with one of the four order relations. Such inequalities are called **linear inequalities in one variable.** Some examples are:

$$2x + 1 < 5$$
$$3(2 - 4x) \geq 18$$
$$2(3x - 7) + 1 \leq 3(4x - 9) + 2$$

Linear inequalities in one variable

The procedure we will use is similar to the procedure we used for solving equations: to isolate the variable. Then the solution to the inequality will be obvious. Our first rule is similar to the Addition Rule for Solving Equations.

The Addition Rule for Solving Inequalities

The same number or expression can be added to both sides of an inequality. The inequality obtained will have the same solution as the original inequality.

Example 1

Solve $x - 5 < 6$.

Solution

Using the Addition Rule, we have

$$
\begin{array}{rl}
x - 5 < & 6 \\
+ 5 & + 5 \\
\hline
x < & 11
\end{array}
$$
Add 5 to both sides to isolate x.

Notice that the solution is the collection of all real numbers **less than 11.** Since there are infinitely many such numbers, we represent the solution by *graphing* all points on a number line that correspond to real numbers less than 11.

We circle 11 to indicate it is not part of the solution set.

The Multiplication/Division Rule for Solving Inequalities differs from its equation-solving counterpart when multiplying or dividing by a negative number.

The Multiplication/Division Rule for Solving Linear Inequalities

1. Both sides of an inequality may be multiplied or divided by the same *positive* quantity.
2. If both sides of an inequality are multiplied or divided by the same *negative* quantity, the direction of the inequality must change.

In both of these cases, the inequality obtained will have the same solution as the original inequality.

Some examples follow.

Example 2

Solve $6x \geq 12$ and graph the solution on a real number line.

Solution

$$6x \geq 12$$

$$\frac{6x}{6} \geq \frac{12}{6} \qquad \text{Divide both sides by 6.}$$

$$x \geq 2 \qquad \text{Note that the direction of the inequality symbol is not changed since we are dividing by a positive number, 6.}$$

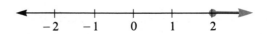

We fill in the circle to indicate that 2 *is* in the solution set.

Example 3

Solve $-3x > 12$ and graph the solution on a real number line.

Solution

$$-3x > 12$$

$$\frac{-3x}{-3} < \frac{12}{-3} \qquad \text{Divide both sides by } -3 \text{ and change the direction of the inequality.}$$

$$x < -4$$

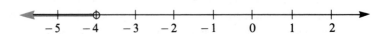

The remaining examples of this section involve solving more complicated linear inequalities.

Example 4

Solve $5 - 2x \leq 7$ and graph the solution on a real number line.

Solution

$$
\begin{array}{rr}
5 - 2x \leq & 7 \\
-5 \qquad & -5 \\
\hline
-2x \leq & 2 \\
x \geq & -1
\end{array}
$$

Subtract 5 from both sides.

Divide by -2; change the direction of the inequality.

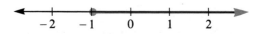

Example 5

Solve $4(3 - x) > 2 - 3x$.

Solution

$$4(3 - x) > 2 - 3x$$

$12 - 4x > 2 - 3x$	Eliminate parentheses.
$\underline{ + 3x \qquad + 3x}$	Add $3x$ to both sides.
$12 - x > 2$	
$\underline{-12 - 12}$	Subtract 12 from both sides.
$-x > -10$	
$x < 10$	Divide by -1 and change direction.

Example 6

Solve $2(3x - 7) + 11 < 3(4x + 1)$ and graph the solution on a real number line.

Solution

$$2(3x - 7) + 11 < 3(4x + 1)$$

$6x - 14 + 11 < 3(4x + 1)$	Eliminate parentheses on the left.
$6x - 14 + 11 < 12x + 3$	Eliminate parentheses on the right.
$6x - 3 < 12x + 3$	Combine like terms.
$-3 < 6x + 3$	Subtract $6x$ from both sides.
$-6 < 6x$	Subtract 3 from both sides.
$-1 < x$	Divide by 6.

Notice $-1 < x$ means "-1 is less than any real number solution, x." That is equivalent to saying "any solution x must be greater than -1", or $x > -1$.

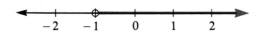

We conclude this section with one example of solving inequalities in an applied setting.

Example 7

Dom, Ken, and John work as solderers in an electronics plant. In one day, Dom solders 10 boards and Ken does 13. How many boards must John solder so that the average for the three of them for that day is at least 14?

Solution

Step 1   Problem translation

Let x equal the number of boards John solders. The threesome's average can be represented algebraically by adding the production of each and dividing by 3:

$$\frac{10 + 13 + x}{3}$$

Since "at least" means equal to or greater than, the following inequality represents the statement of the problem:

$$\frac{10 + 13 + x}{3} \geq 14$$

Step 2   Solving the inequality

$$\frac{10 + 13 + x}{3} \geq 14$$

$$10 + 13 + x \geq 42 \qquad \text{Multiply both sides by 3.}$$

$$23 + x \geq 42$$

$$x \geq 19 \qquad \text{Subtract 23 from both sides.}$$

Hence, John must solder at least 19 boards.

Step 3   Check

If John does 19 boards, the trio's average would be

$$\frac{10 + 13 + 19}{3} = \frac{42}{3} = 14$$

If he does *more than* 19, their average would be *more than* 14. ✓

Section 3.7 Exercises

In Exercises 1 through 22, solve each inequality and graph the solution on a real number line.

1. $x - 2 < 4$
2. $x + 2 > 4$
3. $2 + t \geq 9$
4. $y - 11 \leq -8$
5. $2x + 3 > 7$
6. $2x - 7 \leq 3$
7. $5 - y < 3$
8. $y - 5 < 3$
9. $3a - 4 \geq 8$
10. $4b - 2 \leq 14$
11. $8 - 7y > 29$
12. $2 - 5x \geq 7$
13. $3(2 - x) < 12 - x$
14. $8(1 - 2x) > 5(4 - 3x)$
15. $8 - 7y > y$
16. $3(x + 6) - 10 \geq 2(3x + 4) + 3$
17. $4(z - 1) \leq 5(2z - 8)$
18. $x - 1 < 2[2 - (x + 1)]$
19. $-2(3 - 2y) \geq 2y - 7$
20. $-4(b - 9) + 5 \leq 21 - 5[16 - 2(b - 4)]$
21. $5(2 - 3x) \leq 12(1 - x)$
22. $\dfrac{1 - 3x}{4} > 5$

23. Hank, Carol, and Richie work on an assembly line. In one month, Hank assembles 1400 pieces and Carol assembles 2000 pieces. How many pieces must Richie assemble that month so that the three-some's average is at least 2200 pieces?

24. Ann Marie is on a diet. On Monday she consumed 900 calories and on Tuesday she consumed 1300 calories. How many calories can she have on Wednesday if her three-day average is not to exceed 1200 calories?

25. Elaine received grades of 78, 79, and 91 on three exams in History 101. What grade must she receive on the fourth exam so that she has at least an 80 average?

26. Rework Exercise 25 with the fourth exam counting double.

In Exercises 27 through 30, use a calculator to solve the stated inequality.

27. $2y - 1.57 \leq 3.871$
28. $1.871x + 4.71 \geq 3.2x - 1.5$
29. $5.87 - 3.145x < 5.1(3.23 - 4.78x)$
30. $3.5t - 5(1.5 - 3.2t) > 6.10 - 7.1t$

Supplementary Exercises

In Exercises 1 through 13, solve each inequality and graph the solution on a real number line.

1. $a + 6 \leq 4$

2. $4 > b - 7$

3. $x - 4 \geq -12$

4. $2x < 12$

5. $-3y > 9$

6. $-2y \leq -14$

7. $3a + 1 < 10$

8. $4 - 2y > -10$

9. $3(x - 1) \leq 2x - 4$

10. $5 - 2(1 - x) > 3(x + 2)$

11. $5(x + 2) \leq 2 - (3 - 4x)$

12. $\dfrac{x + 1}{2} \leq 5$

13. $\dfrac{3x - 1}{4} > 2$

14. John needs to have an average of at least 75 in the five courses he takes to be eligible to play basketball. If he has scores of 80, 65, 78, and 85 in four of the five courses, what must he have in the fifth course to be eligible to play?

15. Carol needs to average running at least 10 miles a day for 5 days to prepare for her race. If she has already run 7 miles, 11 miles, 8 miles, and 13 miles for 4 days, how many miles must she run on the fifth day to achieve her goal?

3.8   Chapter review

Summary

In this chapter we examined the techniques for solving linear equations and inequalities in one variable. We usd the addition rule, the multiplication rule, and the division rule, and the concept of simplification to solve equations.

We also looked at some applied problems from various life situations. We used a three-part format to solve them:

problem translation–equation solution–check

In Section 3.6 formulas and literal equations were presented. We saw that it was possible to solve an equation like $P = 2\ell + 2w$ for any of the variables ℓ, w, or P.

The most significant difference between solving linear inequalities (Section 3.7) and solving linear equations (Sections 3.1–3.4) is that when multiplying or dividing an inequality by a negative number, the sense, or direction, changes. That is,

$$-2x \leq 12 \qquad \text{becomes} \qquad x \geq -6$$

Vocabulary Quiz

Match the expression in Column I with the phrase in Column II that best describes it.

Column I

1. The Addition Rule for Solving Equations is illustrated by

2. The Multiplication Rule for Solving Equations is illustrated by

3. A literal equation is

4. A formula is

Column II

a. if $z - 7 = 8$, then $z = 15$.

b. an equation composed of two or more variables.

c. a general fact, rule, or principle expressed in symbols.

d. if $\dfrac{x}{4} = -6$, then $x = -24$.

Chapter 3 Review Exercises

In Exercises 1 through 20, solve each equation.

1. $x - 4 = 7$
2. $x + 4 = 7$
3. $-4 + y = 3$
4. $m - 21 = 21$
5. $2 = 5 - x$
6. $2x + 3 = 3x - 1$
7. $3p + 1 = 10$
8. $4x - 3 = x$
9. $409x = 411x - 4$
10. $2(3x - 2) = 5x + 2$
11. $5 - x = 25$
12. $m + 5 = -17$
13. $4t = -16$
14. $-5p = 30$
15. $7t + 6 = 8t - 1$
16. $4m + 3 = 19$
17. $\frac{m}{9} = 72$
18. $-6 = \frac{t}{9}$
19. $3x - 1 = 7 - x$
20. $2(x - 2) - 3 = 2 - x$

21. Larry mailed two packages to Idaho. One of them weighed 13 ounces less than twice the weight of the other. Together they weighed 38 ounces. How much did each weigh?

22. A rectangular room has a width that is four feet less than its length. A carpenter uses 49 feet of floor molding to go around the entire edge of the floor except for the only doorway in the room, which is 3 feet wide. What are the dimensions of the room?

23. JoAnn received grades of 71, 83, and 52 in Accounting 101. What grade must she get on her fourth exam so that her average will be at least 70?

24. Three times a number less 7 is 11. Find the number.

25. Four times some number added to 1 is at most 17. Find the numbers that make this statement true.

26. A field is 4 feet longer than it is wide. Carl enclosed the field with 48 feet of fencing. What are the dimensions of the field?

27. The sum of two consecutive odd integers is 32. What are the integers?

28. (a) Solve $3x + 2y = 1$ for y.
    (b) Determine the value of y when $x = 3$.

29. Given $2x - 3y = 5$, solve for y. Find y when $x = 10$.

In Exercises 30 through 39, solve each inequality and graph the solution on a real number line.

30. $3x - 7 \leq 17$
31. $26 - 3y > 11$
32. $5(6 - x) < 2(3 - 2x)$
33. $6 - 2x \geq -4(3x + 6)$
34. $3 - 5[2 - (x - 3)] > 7x - 3(x - 2)$
35. $\frac{x}{2} \leq -4$
36. $2x + 3 > 9$
37. $5 - y \leq 12$
38. $4(x - 7) \geq 2 - x$
39. $x - 2 < 9$

40. Carol wishes to average at least 150 in her bowling. If she plays 3 games and scores 128 and 160 in 2 of the three games, what must she score on the third game to achieve her goal?

Cumulative review exercises

1. Find the difference: $-19 - 6$ *(Section 1.3)*

2. Find the quotient: $\frac{-56}{-7}$ *(Section 1.4)*

3. If $x = \dfrac{y}{z}$,

   (a) What value(s) is z prohibited to be?

   (b) For what value of y will x be 0?   *(Section 1.6)*

4. Which of the following does *not* represent a real number?   *(Section 1.7)*

   (a) -5^2 (b) $(-2)^4$ (c) $\sqrt[3]{-8}$ (d) $\sqrt{-2}$

5. Find the value of $\dfrac{2(-3)^2 - [2 - (-3)]}{4^2 - (5 - 2)}$ *(Section 1.8)*

6. Suppose $x = 4$, $y = -5$, and $z = -1$. Evaluate $|x + y| + xz$.   *(Section 2.1)*

7. Simplify $4(3x - 1) - 7x + 5(x - 2)$.   *(Section 2.2)*

8. Simplify $2x(x^2 - 5x + 6) - 3x(x^2 + 7x - 9)$.   *(Section 2.4)*

9. Find the area of a circle of radius 4 in. Use 3.14 as an approximation to π.   *(Section 2.5)*

10. Solve for x: $x + 7 = 19$ *(Section 3.1)*

Chapter 3 test

Take this test to determine how well you have mastered solving linear equations and inequalities; check your answers with those found at the end of the book.

In Problems 1 through 10, solve each equation for x.

1. $x - 7 = 8$ *(Section 3.1)*

2. $4 = x - 9$ *(Section 3.1)*

1. _____

2. _____

3. $5x = -35$ *(Section 3.2)*

4. $\frac{x}{4} = -9$ *(Section 3.2)*

3. _____

4. _____

5. $4x - 1 = 3x - 45$ *(Section 3.3)*

6. $5 - x = -1$ *(Section 3.3)*

5. _____

6. _____

7. $8 - x = 2x - 1$ *(Section 3.3)*

8. $\frac{2x}{3} - 12$ *(Section 3.3)*

7. _____

8. _____

9. $5(2x - 1) = x - 50$ *(Section 3.4)*

10. $x - 5[2x - 3(1 - x)] = 63$ *(Section 3.4)*

9. _____

10. _____

11. The Jones Brothers are professional wrestlers whose combined weight is 800 pounds. If one weighs 50 pounds more than twice the other, how much does each weigh? *(Section 3.5)*

12. Alice is twice Barbara's age, and Kevin is two years younger than Barbara. The total of their ages is 46. How old is each? *(Section 3.5)*

11. _____

12. _____

13. Using the distance formula, $d = rt$, determine the time it takes a runner to travel 300 feet if she is traveling at the uniform speed of 10 ft/sec. *(Section 3.6)*

14. Solve $4x + 7y = 11$ for y. *(Section 3.6)*

13. _____

14. _____

15. The relationship between Fahrenheit (F) and Celsius (C) temperatures is given by the formula

$$5F - 9C = 160$$

Solve the formula for C. *(Section 3.6)*

15. _____

In Problems 16 through 20, solve each inequality and graph the solution on a real number line. *(Section 3.7)*

16. $4 - 2x \le 8$

17. $1 - 3x > -8$

18. $2(x - 7) \ge x - 8$

19. $5(1 - 3x) < -7(x + 2)$

20. $3[1 - 2(1 - x)] \ge 5x + 11$

16. _____

17. _____

18. _____

19. _____

20. _____

chapter four

Word problems

4.1   Number and monetary problems

The three-step procedure of (1) problem translation, (2) equation solution, and (3) check, first introduced in Chapter 3, will be further examined in this chapter.

Number Problems

"Number" problems are so named because we are asked in each problem to find a number with a certain property. Three examples follow.

Example 1

When a number is added to five times itself, the result is 42. What is the number?

Solution

Step 1   Problem translation

We are asked to find a number.

Let n equal the number.

"When
a number   is added to   five times itself   the result is   42"

$$n \qquad + \qquad 5n \qquad = \qquad 42 \leftarrow \text{This is our algebraic model.}$$

Step 2   Equation solution

$$n + 5n = 42$$
$$6n = 42$$
$$n = \frac{42}{6}$$
$$n = 7$$

So the number we set out to find is 7.

Step 3   Check

The number is 7. Five (5) times the number (7) is 35. Their sum is 42, so our answer is correct. $\checkmark$

Example 2

Find two consecutive even integers such that twice their sum is 108.

Solution

Step 1   Problem translation

Let n equal the first even integer. The next consecutive *even* integer is then $n + 2$.

"Twice   their sum   is   108"

$$2 \qquad \cdot \qquad [n + (n + 2)] \qquad = \qquad 108 \leftarrow \text{This is our algebraic model.}$$

Step 2   Equation solution

$$2[n + (n + 2)] = 108$$
$$2(n + n + 2) = 108$$
$$2(2n + 2) = 108$$
$$4n + 4 = 108$$
$$4n = 104 \qquad \text{Subtract 4 from both sides.}$$
$$n = 26 \qquad \text{Divide by 4.}$$

The first even integer, n, is 26, and the next consecutive even integer is 28.

Step 3   Check

The sum of 26 and 28 is 54. Twice that sum is 108 and 26 and 28 *are* consecutive even integers. $\checkmark$

Example 3

Find three consecutive odd integers such that the sum of the first two plus three times the largest is 99.

Solution

Step 1   Problem translation

Let $\qquad x =$ the first integer

$x + 2 =$ the next consecutive odd integer

$x + 4 =$ the largest consecutive odd integer

"Sum of first two   plus   three times the largest   is   99"

$$[x + (x + 2)] \quad + \quad 3(x + 4) \quad\quad = \quad 99 \leftarrow \text{This is our algebraic model.}$$

Step 2   Equation solution

$$[x + (x + 2)] + 3(x + 4) = 99$$
$$+ 2 + \quad + 12 = 99$$
$$+ 14 = 99 \qquad \text{Combine like terms.}$$
$$5x = 85$$
$$x = \frac{85}{5}$$
$$x = 17$$

The numbers are 17, 19, and 21.

Step 3   Check

The sum of 17 and 19 is 36. Three times the largest is 63 added to 36 is 99. $\checkmark$

Monetary Problems

Problems that involve value of some kind, usually money, are explored in the next three examples.

Example 4

Marc has $2.20 in nickels and dimes. If he has twice as many nickels as dimes, how many coins of each type does he have?

Solution

Step 1   Problem translation

Let x equal the number of dimes Marc has. So $2x$ equals the number of nickels. The total *value* of all the dimes is $10x$ cents and the total *value* of all the nickels is $5(2x)$ cents, because the total value of each coin is the product of the number of coins with the value of each coin.

$$
\underbrace{10x}_{\substack{\text{Total value of} \\ \text{dimes in cents}}} + \underbrace{5(2x)}_{\substack{\text{total value of} \\ \text{nickels in cents}}} = \underbrace{220}_{\substack{\text{total value} \\ \text{in cents}}} \leftarrow \text{This is our algebraic model.}
$$

Step 2   Equation solution

$$
\begin{aligned}
10x + 5(2x) &= 220 \\
10x + 10x &= 220 \\
20x &= 220 \\
x &= 11
\end{aligned}
$$

Hence he has 11 dimes and 22 nickels.

Step 3   Check

The value of 11 dimes is $1.10. The value of 22 nickels is $1.10. The total value is $1.10 + $1.10 = $2.20. So our answer checks. √

Example 5

The Gonzales family attended the movies one evening. The cost of an adult ("adult" means "over 12 years of of age") ticket was $3.50, while a child's ticket cost $1.75. If there are 6 people in the Gonzales family, and the total cost for all of them was $19.25, how many adults and children are there in the Gonzales family?

Solution

Step 1   Problem translation

Let x equal the number of adults. Since there are 6 people in all, 6 minus the number of adults, or $6 - x$, represents the number of children. The cost for adults' tickets is $3.50 \cdot x$. The cost for children's tickets is $1.75 \cdot (6 - x)$. So the total cost, $19.25, is equal to the cost of adults' tickets plus the cost of children's tickets.

$$
19.25 = 3.50x + 1.75(6 - x) \leftarrow \text{This is our algebraic model. Note that we}
$$
use dollars as the common unit.

Step 2   Equation solution

$$
\begin{aligned}
19.25 &= 3.50x + 1.75(6 - x) \\
19.25 &= 3.50x + 10.50 - 1.75x \\
8.75 &= 1.75x \\
\frac{8.75}{1.75} &= x \\
5 &= x
\end{aligned}
$$

So there are 5 adults and 1 child in the Gonzales family.

Step 3   Check

$$
\begin{aligned}
\text{5 adults' tickets cost } 5(\$3.50) &= \$17.50 \\
\text{1 child's ticket costs } 1(\$1.75) & \underline{1.75} \\
\text{Total cost} &= \$19.25 \; \checkmark
\end{aligned}
$$

The next example of a monetary problem involves the concept of **simple annual interest.** The simple annual interest is given by the formula

$$I = PR$$

where I is the simple annual interest, P is the amount invested (called the **principal**), and R is the rate of interest (expressed as a percent). For example, if \$400 is invested at an annual rate of interest of 5%, the annual amount of interest is:

$$I = P \cdot R$$
$$I = \$400 \cdot 0.05$$
$$I = \$20$$

Note that I and P are usually expressed as dollars, and R is written as a decimal. The symbol % means "hundredths" so 5% means 5 hundredths, or 0.05. The reader unfamiliar with decimals or percents may refer to Appendix A or Section 4.5.

Example 6

Laverne has \$1200 dollars to invest, part at 12% and part at 10%. If the annual interest from the 12% investment exceeded the interest from the 10% investment by \$56, how much did she invest at the 12% rate?

Solution

Step 1   Problem translation

We are asked to find the principal for the 12% investment. Call it x. Since x is the amount invested at 12%, then $1200 - x$ is the amount invested at 10%. The interest for the 12% investment, which we will call I_{12}, is given by

$$I = P \cdot R$$
$$I_{12} = x \cdot 0.12$$
$$I_{12} = 0.12x$$

The interest for the 10% investment, which we will call I_{10}, is given by

$$I = P \cdot R$$
$$I_{10} = (1200 - x) \cdot 0.10 = 0.10(1200 - x)$$

Finally, $I_{12} - I_{10} = 56$ because "the interest from the 12% investment exceeded the interest from the 10% investment by \$56."

$$I_{12} - I_{10} = 56$$

$$0.12x - 0.10(1200 - x) = 56 \leftarrow \text{This is our algebraic model.}$$

Step 2   Equation solution

$$0.12x - 0.10(1200 - x) = 56$$
$$0.12x - 120 + 0.10x = 56$$
$$0.22x - 120 = 56$$
$$0.22x = 176$$
$$x = \frac{176}{0.22}$$
$$x = 800$$

So the principal for the 12% investment is \$800.

Step 3   Check

$800 at 12% interest yields $96.

$400 at 10% interest yields $40.

Since $96 exceeds $40 by $56, our answer checks. $\checkmark$

Section 4.1 Exercises

1. When a number is added to four times itself, the result is 60. What is the number?

2. When twice a number is subtracted from five times the number, the result is 18. Find the number.

3. Twice a number minus 6 is 19. What is the number?

4. Find two consecutive odd integers whose sum is 56.

5. Find two consecutive even integers whose sum is 302.

6. Find three consecutive even integers such that the sum of the first two added to twice the largest is 234.

7. When 7 is subtracted from three times a number, the result is 11. Find the number.

8. Consider two consecutive integers. Three times the first plus twice the second is 87. Find the integers.

9. Find three consecutive even integers such that one-half their sum is 183.

10. When two consecutive integers are added and their sum is divided by 3, the result is 27. Find the numbers.

11. Twice the sum of a number and 4 is -12. Find the number.

12. Suppose we have two consecutive even integers so that one-half the first added to three times the second is 97. Find the numbers.

13. When three times a number is subtracted from 9 and that result is divided by 2, the result is 18. Find the number.

14. An envelope has twice as many 21-cent stamps as 10-cent stamps on it. If the total postage is $2.60, how many of each type of stamp is there?

15. A group of six friends between the ages of 10 and 14 went to the movies. Those of ages 12 and over each paid $2.25 to get in; those under 12 each paid $1.00. If, in all, they paid $9.75, how many of them were under 12?

16. At a small Minnesota college, out-of-state tuition is $400 per semester, while tuition for residents of the state is $250. One semester the school received $304,000 from 1000 students. How many students did not live in Minnesota?

17. Sam invested $3000 in two bank accounts. One paid interest at a 6% annual rate and the other at a $7\frac{1}{4}\%$ annual rate. If the annual interest from the 6% account exceeded the interest Sam made on the $7\frac{1}{4}\%$ account by $157, how much did he invest in each account?

18. When Juan invested his savings of $10,000 in two businesses (call them A and B) he lost at an annual rate of 4% on A and gained at a 7% annual rate on B.

    (a) If he invested an equal amount in each business, how much income did he have from his investment after one year?

    (b) If he invested $1000 more in Business A than in Business B, how much income did he have from his investment after one year?

19. Bob mailed a package that cost $1.80 for postage. He had only 13¢ and 10¢ stamps and wanted to use twice as many 13¢ stamps as 10¢ stamps. How many of each did he use?

20. At the university basketball game student admission was 50¢ and nonstudent admission was $1.25. There were five times as many students as nonstudents at the game, and the total income at the gate was $1500. How many students attended?

21. Tom's total annual interest income when he invested $900 at 7% and another amount at 6% was $99. How much did he invest at 6%?

22. Nastasia invested $30,000 in two stocks. One yielded annual interest at a rate of 11%, and the other yielded annual interest at a rate of 5.5%. Find the amounts invested at each rate if:

    (a) The amounts of annual income from each stock were equal.

    (b) The amount of annual income from the 11% stock was four times that of the annual income from the 5.5% stock.

    (c) The annual income on the 11% investment exceeded the annual income on the 5.5% investment by $3300.

23. There are two types of tickets sold for the college play: balcony and orchestra. Balcony tickets sell for $3.50 and orchestra tickets sell for $5.00. If there are twice as many orchestra seats as balcony seats and a "sellout" brings in $2025, how many orchestra seats are there?

24. Andrea bought five times as many 20¢ stamps as 15¢ stamps and seven more 35¢ stamps than 15¢ stamps. If she spent $17.45 for the stamps, how many of each did she purchase?

25. Jake invests a total of $25,000 in two accounts. One pays 6% and the other pays $7\frac{1}{2}$% simple annual interest. If Jake earns $1725 in interest for one year, how much does he invest in each account?

Supplementary Exercises

1. Seven times a number added to 9 is 72. Find the number.

2. Three times the sum of a number and -7 is 42. Find the number.

3. One-half a number subtracted from 21 is 17. Find the number.

4. The sum of three consecutive integers divided by 9 is 12. Find the numbers.

5. The sum of a number and 7 is the same as one-half the number less 5. Find the number.

6. Joan purchased x pounds of apples, which cost 39¢ per pound, and 5 more pounds of grapes than apples, which cost 79¢ per pound. If Joan paid $15.75 for the fruit, how many pounds of each did she buy?

7. Kim mailed a package to Alaska that required two more 20¢ stamps than 15¢ stamps. If Kim spent $2.85 to send the package, how many of each stamp did she use?

8. The eight members of the Jenkins family went to a movie that charged $4.75 for each adult ticket and $2.50 for each child's ticket. If they spent $26.75, how many of each type of ticket did they buy?

9. Pam spends twice as much annually on her rent as she does on food. She spends the same amount on her clothing as she spends on food. If she spends $10,000 on these three items, how much does she spend on each?

10. The local baseball stadium holds 5 times as many bleacher seats as box seats. If a box seat costs $4.00 each and a bleacher seat costs $2.25, how many of each seat are in the stadium if, when every seat is filled, the stadium raises $1525?

11. Mary Lu has $15,000 divided between two accounts. One pays 8% annual interest and the other pays 7% annual interest. If the total interest earned for one year is $1140, how much does she have deposited in each account?

12. Carol deposits a total of $9000 in two savings accounts. One pays 7% annual interest and the other pays 10% annual interest. If she receives a check for $780 for the combined annual interest on the two accounts, how much does she have deposited in each?

13. John won $800 in a raffle. He loaned his brother some of the winnings, and his brother promised to pay him back at the end of the year with 5% interest. John deposited the rest of the winnings in an account that paid 9% annual interest. If the total interest from the two investments was $64, how much did he loan his brother and how much did he deposit in the account?

14. Chris invests a total of $1200 in stocks and bonds. The bonds pay 14% annual interest and the stocks pay 7% annual interest. His interest for the year was $140. How much does he have invested in each?

4.2   Motion problems

In Chapter 3 we saw that the literal equation

$$d = rt$$

is a formula for the distance (d) traveled by an object at a uniform rate (r) for a time (t). We saw that this formula can also be written as

$$r = \frac{d}{t} \quad \text{or} \quad t = \frac{d}{r}$$

Now we look at some examples of word problems involving those formulas.

Example 1

Eileen and Denise are runners, and they have decided to enter a race. After giving Denise a 30-minute head start, Eileen runs at a pace of 8 mph. If they finish in a tie after Denise ran for 1 hour and 15 minutes, how fast did Denise run?

Solution

Step 1   Problem translation

Let x = Denise's rate, the quantity asked for.

Facts about Denise:

$$\text{Rate} = x$$

$$\text{Time} = 1 \text{ hour} + 15 \text{ minutes} \left(\text{or } \frac{1}{4} \text{ hour}\right) = \frac{5}{4} \text{ hour}$$

$$\text{Distance} = rt = \frac{5}{4}x = \frac{5x}{4}$$

Facts about Eileen:

$$\text{Rate} = 8$$

$$\text{Time} = \frac{5}{4} \text{ hour} - 30 \text{ minutes} \left(\text{or } \frac{2}{4} \text{ hour}\right) = \frac{3}{4} \text{ hour}$$

$$\text{Distance} = rt = 8 \cdot \frac{3}{4} = 6$$

Since they finished in a tie after having run the same distance, we can find the equation by setting the distances equal:

$$\frac{5x}{4} = 6$$

Step 2   Equation Solution

$$\frac{5x}{4} = 6$$

$$5x = 24$$

$$x = \frac{24}{5} = 4.8 \text{ mph}$$

Step 3   Check

If Denise ran at a rate of 4.8 mph for 1.25 hours, she ran 6 miles. Eileen ran at a rate of 8 mph for .75 hours. She also ran 6 miles. $\checkmark$

Example 2

Two bicyclists, Katie and Andy, leave their house at the same time. Katie travels south at a uniform rate that is 10 km/hr (kilometers per hour) faster than Andy, who is traveling north. After 5 hours they are 160 kilometers apart. Find the rate of travel for each cyclist.

Solution

Step 1   Problem translation

In solving motion problems involving two travelers, it is important to find the one variable (d or r or t) that is the same for both. In this example, it is time; they both travel for 5 hours.

It will be handy to use subscripts so that each cyclist is easily "recognizable":

$$t_k = \text{the time Katie travels (5 hours)}$$
$$r_k = \text{Katie's uniform rate of speed}$$
$$d_k = \text{the distance Katie travels}$$
$$t_a = \text{the time Andy travels (also 5 hours)}$$
$$r_a = \text{Andy's uniform rate of speed}$$
$$d_a = \text{the distance Andy travels}$$

The pictorial model below summarizes some of the information for this example.

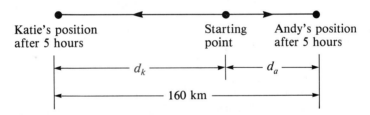

We can set up the general formula for each cyclist as follows:

$$d_k = r_k t_k \qquad \text{and} \qquad d_a = r_a t_a$$

Since we know the amount of time that has passed for both, we know that

$$d_k = 5r_k \qquad \text{and} \qquad d_a = 5r_a$$

Furthermore, since Katie's rate is 10 km/hr faster than Andy's,

$$r_k = r_a + 10$$

The algebraic model is obtained by realizing that the total distance traveled by the two ($d_k + d_a$) is 160 kilometers.

$$d_k + d_a = 160$$
$$5r_k + 5r_a = 160$$

Now, if we substitute $5(r_a + 10)$ for $5r_k$, we can obtain an equation with only one variable.

$$5(r_a + 10) + 5r_a = 160 \leftarrow \text{This is our algebraic model.}$$

Step 2   Equation solution

$$5(r_a + 10) + 5r_a = 160$$
$$5r_a + 50 + 5r_a = 160$$
$$10r_a = 110$$
$$r_a = 11$$

So Andy's rate of travel is 11 km/hr and Katie's rate of travel ($r_k = r_a + 10$) is 21 km/hr.

Step 3   Check

Katie, after traveling at 21 km/hr for 5 hours would be 105 kilometers south of the starting point. Andy, after traveling at 11 km/hr for 5 hours would be 55 kilometers north of the starting point. Therefore, they would be 160 kilometers apart. Our answer checks. √

Example 3

A plane travels round trip from Phoenix to San Francisco and back again in 510 minutes. The average rate going to San Francisco was 200 mph. On the return trip, because of a tail wind, the plane averaged 225 mph. How far is Phoenix from San Francisco?

Solution

Step 1   Problem translation

Before we can proceed, all the units must be in agreement. Since the rate is measured in miles per hour, we convert 510 minutes to $8\frac{1}{2}$, or 8.5, hours. Now, if we let t represent the time to go from San Francisco to Phoenix, $8.5 - t$ represents the time to go from Phoenix to San Francisco. Since the distance between cities is the same both ways, we obtain our equation by setting the distances equal.

$$225t = 200(8.5 - t) \leftarrow \text{This is our algebraic model.}$$

Step 2   Equation solution

$$225t = 200(8.5 - t)$$
$$225t = 1700 - 200t$$
$$425t = 1700$$
$$t = 4$$

So the time to go from San Francisco to Phoenix is 4 hours, and since the rate of the plane on this leg of the trip was 225 mph, the distance must be $225 \cdot 4 = 900$ miles.

Step 3   Check

From Phoenix to San Francisco the plane traveled 900 miles at the rate of 200 mph—that took $4\frac{1}{2}$ hours, or 270 minutes. From San Francisco to Phoenix the plane traveled 900 miles at the rate of 225 mph—that took 4 hours, or 240 minutes. $240 + 270 = 510$ minutes. Our answer checks. √

Section 4.2 Exercises

1. John travels at 55 mph. How long will it take to travel 330 miles?

2. Kristen began her trip at 7:30 A.M. and arrived at 1:00 P.M. If the train traveled at 70 mph, how far did she travel?

3. Rick drove 4 hours at a constant rate. He traveled 240 miles. How fast did he drive?

4. Joe left at 6:00 A.M. and arrived at 3:00 P.M. If he traveled 81 miles, how fast did he travel?

5. Mike jogs at a rate of 8 miles per hour. If he jogs for $2\frac{1}{2}$ hours, how far will he go?

6. Chris jogs for 2 miles and then bikes for 6 miles. He bikes 3 times faster than he jogs. If it took him $\frac{1}{2}$ hour to bike, how long did it take him to jog?

7. Danielle walked for $\frac{1}{2}$ hour at a pace of 4 mph. How far did she walk?

8. Carl walks at a rate of 4 mph for $\frac{1}{2}$ hour and then runs for $\frac{1}{2}$ hour. If he travels 6 miles, how fast does he run?

9. Katie left home at 9:00 A.M. and traveled at 55 mph toward a site 275 miles away. At what time did she arrive?

10. Jay and Bob began biking in opposite directions. One traveled at 6 mph and the other at 7 mph. How far apart were they after 3 hours?

11. Two trains leave a station at the same time and travel in opposite directions. After 6 hours they are 360 miles apart. The faster train travels at an average rate that is 20 mph faster than the slower train. Find the average rate of each train.

12. Bert and Ernie leave their apartment and walk at uniform rates in opposite directions. Bert walks twice as fast as Ernie. After 2 hours, they are 15 miles apart. How fast is each walking?

13. Sam and Nelson travel at uniform rates in opposite directions at 40 mph and 20 mph, respectively, until they are 325 miles apart. How long (in minutes) did they travel?

14. A dirt-bike rider, Sue Zooki, covered a certain motorcycle course at an average rate of 55 mph. She figured that, if she could average 5 mph faster, it would take 15 minutes *less* to finish the race. If she was able to do this, how long would it take her to finish the race?

15. Sandy is driving on the Coastal Highway at a rate of 50 mph. At 1 P.M. she is passed by a state trooper traveling a constant 75 mph. At what time will they be 50 miles apart?

16. Two joggers, Jane and Dede, pass one another traveling the same route but in opposite directions. Jane is running at a rate of 7.5 mph; Dede is running at a rate of 10 mph. How far apart will they be 30 minutes after they pass one another?

17. Two trains are on the same track 385 miles apart at 12:00 noon and begin traveling toward one another at uniform speeds of 45 mph and 65 mph. At what time will they crash? (*Hint:* Find the time it takes each—they are equal.)

18. Two horseback riders, on the same straight path 38.5 miles apart, begin to travel at 12:00 noon toward one another at uniform speeds of 4.5 mph and 6.5 mph. At what time will they meet?

19. Two trains leave a station at the same time and travel in opposite directions. After 6 hours they are 420 miles apart. The faster train travels at an average rate that is 20 mph more than the slower train. Find the average rate of each train.

20. A car leaves a certain point and travels at the rate of 45 mph. A second car leaves the same point one hour later and travels the same route at 60 mph. How long will it take the second car to catch the first car?

21. Two cars start at the same spot and travel in opposite directions. One car travels 5 mph faster than the other. If, after 4 hours, they are 460 miles apart, how fast is each traveling?

22. Two buses travel in opposite directions starting from the same spot. One is traveling at 40 mph and the other is traveling at 50 mph. After how long are they 270 miles apart?

23. Two cars are 210 miles apart driving toward the same city. One is moving at 55 mph and the other is moving at 50 mph. How long will it be before they meet?

24. John left Utica at 9 A.M. traveling toward Albany along the New York Thruway at 45 mph. Chris left at 10 A.M. traveling the same route at 50 mph. How long will it take Chris to catch up to John?

25. Two cars are 45 miles apart traveling toward one another, one at 70 mph and the other at 65 mph. How long will it take them to meet?

Supplementary Exercises

1. Renée rode her bike for 2 hours and went 6 miles. How fast did she pedal?

2. Joan traveled 75 miles at a rate of 50 mph. How long did her trip take?

3. Joe walked at a constant rate of 6 miles per hour for 2 hours. How far did he walk?

4. Marsha left home at 8 A.M. and arrived at her destination at 11 A.M. If her van was traveling at a rate of 50 mph, how far away from home did she travel?

5. Joe lives 5 miles from work. He walks at a rate of 3 mph. How long does it take to get to work?

6. Mary needs to travel from Utica to New York City. She knows the distance is about 300 miles. If she travels at a constant rate of 50 mph, how long will the trip take?

7. Jennifer leaves her home at 6:00 P.M. and travels north at 45 mph. Her brother leaves 1 hour later, traveling in the same direction at 55 mph. How long will it take him to catch up to her?

8. Two cars leave point A traveling in opposite directions. One is traveling at 60 mph and the other is traveling at 55 mph. After 3 hours, how far apart are the cars?

9. Jim and Joe are traveling in opposite directions from Smith City. Jim is driving at 55 mph and Joe is driving at 65 mph. If they left at 10:30 A.M., how far apart are they at 2:00 P.M.?

10. Mike ran for 10 minutes and then walked for 30 minutes. His walking rate is 5 mph and his running rate is 9 mph. How far did he travel?

11. Spike drove from his home to college in 2 hours at a rate of 55 mph. His sister made the same trip in $2\frac{1}{2}$ hours. How fast did she drive?

12. Kevin biked for 1 hour at a rate of 10 mph. His bike had a flat tire, and he had to walk home. The trip home took 150 minutes. How fast did he walk?

4.3   Mixture problems

Mixture problems deal with mixing or blending two types of products to produce a third product of intermediate value. To solve them, the following equation is basically all that will be used:

$$\text{Value of one unit} \cdot \text{number of units} = \text{total value}$$
$$v \quad \cdot \quad n \quad = \quad T$$

Example 1

A grocer wants to blend two types of coffee. The more expensive brand sells for $2.50 per pound, and the cheaper brand sells for $1.00 per pound. If she wants a mixture of 15 pounds to sell for $1.50 per pound, how many pounds of each type of coffee must she use?

Solution

Step 1   Problem translation

Let x equal the number of pounds of the cheaper coffee. For a better overall view of the stated problem, we convert it to the following table.

	Value of 1 pound	Number of pounds	Total value
Cheaper	1.00	x	
More expensive	2.50		
Mixture	1.50	15	

Now, since she wants a total of 15 pounds and we are letting x represent the number of pounds of cheaper coffee, $15 - x$ represents the number of pounds of more expensive coffee. We get the third column by multiplying the first two columns.

	v	$\cdot$	n	$=$	T
Cheaper	1.00		x		x
More expensive	2.50		$15 - x$		$2.5(15 - x)$
Mixture	1.50		15		22.5

Finally, we obtain the equation below by reasoning that the cost of the cheaper coffee plus the cost of the more expensive coffee equals the cost of the mixture.

$$x + 2.5(15 - x) = 22.5 \leftarrow \text{This is our algebraic model.}$$

Step 2   Equation solution

$$x + 2.5(15 - x) = 22.5$$
$$x + 37.5 - 2.5x = 22.5$$
$$-1.5x = -15$$
$$x = 10$$

So she must use 10 pounds of the cheaper coffee and $15 - 10 = 5$ pounds of the more expensive coffee to create the desired mixture.

Step 3   Check

Ten pounds of the cheaper brand sells for $10.00. Five pounds of the more expensive brand sells for $12.50. This adds up to $22.50, precisely what 15 pounds of $1.50 mixture sells for. $\checkmark$

Example 2

A chemistry student has 100 milliliters (ml) of a solution that is 12% sulfuric acid. If he wants a 20% sulfuric acid solution, how much pure sulfuric acid must be combined with the 12% solution?

Solution

Step 1   Problem translation

Let x equal the number of ml of pure sulfuric acid to be added. We complete the following table, keeping in mind that here "value" means % sulfuric acid in each solution.

	v	$\cdot$	n	$=$	T
	% of acid (in decimal)		Number of ml		Total amount of sulfuric acid
Original solution	0.12		100		12
Pure sulfuric acid	1.00		x		x
New solution	0.20		$100 + x$		$0.20(100 + x)$

The original solution combined with the pure sulfuric acid gives us the new solution. We get the following equation:

$$12 + x = 0.20(100 + x) \leftarrow \text{This is our algebraic model.}$$

Step 2   Equation solution

$$12 + x = 0.20(100 + x)$$
$$12 + x = 20 + 0.2x$$
$$0.8x = 8$$
$$x = 10$$

So 10 ml of pure sulfuric acid must be combined with the 12% solution.

Step 3   Check

Ten ml of sulfuric acid added to the 12 ml already present in the 100-ml solution would yield a solution of 110 ml, 22 ml of which are sulfuric acid. The % of sulfuric acid is $\frac{22}{110} = 20\%$. Thus our solution checks.

Example 3

How much water must be added to 4 gallons of an 80% antifreeze solution in order to make a 60% solution?

Solution

Step 1   Problem translation

Let x equal the amount of water needed. In this problem "value" means percent of antifreeze.

	% of antifreeze	Amount of fluid	Amount of antifreeze
Original	0.80	4	3.2
Water	0	x	0
New solution	0.60	$4 + x$	$0.60(4 + x)$

$3.2 + 0 = 0.60(4 + x) \leftarrow$ This is our algebraic model.

Step 2   Equation solution

$$3.2 = 0.60(4 + x)$$
$$3.2 = 2.4 + 0.6x$$
$$0.8 = 0.6x$$
$$\frac{0.8}{0.6} = x$$
$$1\frac{1}{3} = x$$

So $1\frac{1}{3}$ gallons of water must be added to the 4 gallons of 80% antifreeze solution in order to obtain the desired solution.

Step 3   Check

$$1\frac{1}{3} \text{ gallons } + 4 \text{ gallons } = 5\frac{1}{3} \text{ gallons total solution}$$

60% of $5\frac{1}{3}$ gallons is 3.2 gallons, the amount of antifreeze $\sqrt{}$

Section 4.3 Exercises

1. Mary mixes 15 pounds of peanuts and cashews. She mixes twice as much peanuts as cashews. How much of each does she mix?

2. Joe's lawn mower uses a mixture of gas and oil. He uses 5 times as much gas as oil. If he needs a total of 36 oz of mixture, how much of each does he need?

3. Andrea mixes soda and fruit juice to make punch. She uses 3 times as much soda as juice. If she needs 12 gallons of punch, how much soda does she need?

4. John's Hardware mixes two types of grass seed to make a deluxe blend. The deluxe blend has 4 times as much of Brand *A* as it does of Brand *B*. If John wishes to make 15 bushels of the mix, how much of each brand does he need to blend?

5. Jim mixes three types of nuts. He mixes one more pound of pecans than peanuts and half as many walnuts as peanuts. If he needs 16 pounds, how much of each does he mix?

6. Gloria mixes 2 parts water with 5 parts flour to make dough. If she needs 21 cups of dough, how much flour does she need?

7. Ann mixes 3 parts white flour to 2 parts wheat flour to make her bread. If she needs 20 cups of flour, how much of each type does she need?

8. Joanne mixes candied pineapple, cherries, and citron to make fruit cakes. She uses twice as much pineapple as cherries and two more cups of citron than cherries. How many cups of each does she need if the recipe calls for a total of 14 cups of fruit?

9. Andrea buys 17 cans of cat food at 39¢ per can. How much is her total bill?

10. Carol buys 20 candy bars, which cost 35¢ each, and 20 pencils, which cost 15¢ each, as party favors. What is her total cost?

11. Larry needs to mix two types of cheese to make 10 pounds of a new blend that will sell for $2.82 per pound. If he uses blue cheese and cheddar, and the blue cheese costs $2.19 per pound and the cheddar costs $3.09 per pound, how much of each should he blend?

12. How much water must be added to 2 quarts of a 13% solution to dilute it to a 10% solution?

13. How much water must be added to 5 gallons of a 25% solution to dilute it to a 15% solution?

14. How much water must be added to 9 gallons of a 75% solution to dilute it to a 60% solution?

15. How much pure alcohol must be added to 10 gallons of a 60% solution to make a 90% solution?

16. How much pure alcohol must be added to 15 gallons of a 25% solution to make a 40% solution?

17. How much salt must be added to 6 gallons of a 15% brine solution to make it a 20% brine solution?

18. How much salt must be added to 12 quarts of a 55% brine solution to make it a 65% brine solution?

19. Mark mixes cheddar cheese costing $2.10 per pound with a swiss cheese costing $1.40. If the new mix should sell for $1.60 per pound, how many pounds of each should he use to make 7 pounds of the mix?

20. Luke makes 9 pounds of a fruit–nut mix by mixing nuts costing $1.50 per pound and fruit costing $1.80 per pound. How many pounds of each should he use if the mix costs $1.60 per pound?

21. Joe mixes 2 pounds of nuts costing $1.60 per pound with 2 pounds of fruit costing $2.00 per pound. How much should the mix sell for?

22. A nut company wants to prepare a 5-pound mixture of peanuts and cashews to sell for 45¢ per pound. If peanuts sell for 30¢ per pound and cashews sell for 65¢ per pound, how many pounds of each is needed to make the mixture?

23. If the nut company wants to make a 10-pound mixture to sell for $4.05, how many pounds of each is needed? Use the same prices per pound as in Exercise 22.

24. How much pure gold must be added to a 30-kilogram alloy that is 15% gold to raise the percentage of gold to 20%?

25. A 16% acid solution is to be diluted by combining it with water. How much water must be combined with 100 ml of it to make it a 12% acid solution?

Supplementary Exercises

1. How much pure alcohol should be added to 15 quarts of a 80% solution to make it a 90% solution?

2. How much acid should be added to 7 gallons of a 75% solution to make it a 90% solution?

3. If Vinnie has 4 gallons of a 35% gas solution, how much gasoline should he add to make it a 45% gasoline solution?

4. How much water should be added to 9 cups of a 65% solution to dilute it to a 45% solution?

5. How much water should be added to 5 ml of a 40% solution to make it a 35% solution?

6. Anthony mixes water with a 35% solution to dilute it. If he has 6 liters of solution and he needs a 25% solution, how much water should he add?

7. Sue needs to mix 4 pounds of raisins, which cost $1.20 per pound, with 2 pounds of nuts, which cost $1.80 per pound. How much should the new mix cost?

8. Joe mixes dried apples costing $1.80 per pound with dried currants costing $0.90 per pound. He then has 9 pounds of mix costing $1.10 per pound. How much of each did he mix?

9. Eileen sells a tea blend for $1.06 per pound. If she wants to sell 6 pounds of the blend, and one tea costs $1.08 per pound and the other costs $1.02 per pound, how many pounds of each does she need?

10. The health food store makes a blend of powdered milk and yeast. For 10 pounds of the mix, they use 7 pounds of milk costing 79¢ per pound. The yeast costs $1.09 per pound. How much should the mix sell for?

11. Jim sells a brand of tea for $1.32 per pound. He mixes a blend selling for $2.20 a pound with a tea that sells for $0.99 a pound to make his brand. How much of each should he use to make 11 pounds of mix?

12. Nancy mixes nuts that cost $3.00 per pound with raisins that cost $2.25 per pound to make 15 pounds of a blend that sells for $2.65 per pound. How many pounds of each does she need for the mix?

4.4   Ratio and proportion

We will begin this section by examining the notion of the ratio of two numbers. Symbolically, the ratio of a to b is just the quotient $\frac{a}{b}$.

Ratio of a to b

The ratio of a to b, written

$$a:b \qquad \text{or} \qquad \frac{a}{b},$$

is the quotient $a \div b$.

Since a ratio is a fraction, it may be possible to reduce it. For example, the ratio of 14 to 4 is 14:4 or $\frac{14}{4}$ or $\frac{7}{2}$. The next example examines how we use ratios in real life.

Example 1

A classroom has a total of 45 people in it. There are 36 females and 9 males.
   (a) What is the ratio of females to males?
   (b) What is the ratio of males to females?
   (c) What is the ratio of females to total number of people present?

Solution

   (a) The ratio of females to males means the number of females *divided by* the number of males, so we have:

$$\frac{36}{9} = \frac{4}{1}$$

So the ratio of females to males is 4:1.

(b) This is just the "reverse" of part (a). So the ratio of males to females is

$$1:4$$

(c) Here we have a ratio of

$$\frac{36}{45} = \frac{4}{5}$$

So the ratio of females to total number of people is $4:5$.

A **proportion** is a statement of equality between two ratios. Symbolically it is written

$$\frac{a}{b} = \frac{c}{d}$$

and is read as "a is to b as c is to d." We refer to a and d as the **extremes** and b and c as the **means** of the proportion.

A simple example of a proportion is

$$\frac{3}{4} = \frac{9}{12}$$

Notice that the *product of the means* ($4 \times 9 = 36$) is equal to the *product of the extremes* ($3 \times 12 = 36$).

If

$$\frac{a}{b} = \frac{c}{d}$$

then

$$bc = ad$$

We use this idea, sometimes called **cross-multiplying,** in the examples that follow.

Example 2

Suppose that in a certain hospital the ratio of nurses to doctors is $10:3$. Furthermore, if there are 70 nurses, how many doctors are there?

Solution

Step 1 Problem translation

We are asked to find the number of doctors, so let x equal the number of doctors. Then we can form the *proportion:*

$$\frac{70}{x} = \frac{10}{3}$$

This is the
ratio of nurses
to doctors.

This is also the
ratio of nurses
to doctors.

Step 2 Equation solution

The proportion (our algebraic model for the problem) will now be solved by cross-multiplying.

$$\frac{70}{x} = \frac{10}{3}$$

$$\frac{70}{x} \diagdown \frac{10}{3}$$

$$10x = 210$$

$$x = 21$$

So there are 21 doctors.

Step 3   Check

The ratio of nurses to doctors is $\frac{70}{21} = \frac{10}{3}$. Hence our answer checks. ✓

Notice that, once we symbolized the unknown quantity by x, we set up a proportion and solved the problem. This idea is practiced again in the next four examples.

conver table
p. 402.

Example 3

An inspection of blue denim jeans coming off an assembly line shows that 7 out of 150 pairs have minor defects. In an order of 1650 pairs of jeans, how many pairs can we expect to have minor defects?

Solution

Step 1   Problem translation

Let x equal the number of pairs of defective jeans in the order. We form the proportion:

$$\underset{\substack{\uparrow \\ \text{This is the ratio} \\ \text{of ``defectives'' to} \\ \text{``total number.''}}}{\frac{x}{1650}} = \underset{\substack{\uparrow \\ \text{This is also the ratio} \\ \text{of ``defectives'' to} \\ \text{``total number.''}}}{\frac{7}{150}}$$

Step 2   Equation solution

$$\frac{x}{1650} = \frac{7}{150}$$

$$\frac{x}{1650} \diagdown \frac{7}{150}$$

$$150x = 11{,}550$$

$$x = 77$$

So we can expect 77 pairs of jeans to have minor defects.

Step 3   Check

The ratio of "defectives" to "total number" is $\frac{77}{1650}$. Since this reduces to $\frac{7}{150}$, our answer checks. ✓

Example 4

In a supermarket, 2 cashiers can check out 70 customers in an hour. At this rate, how many customers can be checked out by 7 cashiers in an hour?

Solution

Step 1   Problem translation

Let x equal the number of customers. Then we have

$$\underbrace{\frac{7}{x}}_{} = \underbrace{\frac{2}{70}}_{}$$

$\uparrow$ This is the ratio of cashiers to customers. $\uparrow$ This is also the ratio of cashiers to customers.

Step 2   Equation solution

$$\frac{7}{x} = \frac{2}{70}$$

$$\frac{7}{x} \times \frac{2}{70}$$

$$2x = 490$$

$$x = 245$$

So 245 customers can be checked out in an hour.

Step 3   Check

The ratio of cashiers to customers is $\frac{7}{245}$. Since this reduces to $\frac{1}{35}$, and since $\frac{2}{70}$ also reduces to $\frac{1}{35}$, our answer checks.

Example 5

An automobile travels 160 miles on 9 gallons of gasoline. At this rate, how many gallons of gasoline will be consumed on a 1000-mile trip from Jasper, Alberta to Winnipeg, Manitoba?

Solution

Step 1   Problem translation

Let g equal the number of gallons used. Then we have

$$\underbrace{\frac{1000}{g}}_{} = \underbrace{\frac{160}{9}}_{}$$

$\uparrow$ This is the ratio of miles to gallons. $\uparrow$ This is also the ratio of miles to gallons.

Step 2   Equation solution

$$\frac{1000}{g} = \frac{160}{9}$$

$$\frac{1000}{g} \times \frac{160}{9}$$

$$160g = 9000$$

$$g = 56\frac{1}{4} \text{ (or 56.25)}$$

So $56\frac{1}{4}$ gallons will be used.

Step 3   Check

We must check to see whether or not $\frac{1000}{56.25}$ has the same value as $\frac{160}{9}$. Does $\frac{1000}{56.25} = \frac{160}{9}$?

Yes, because $1000 \cdot 9 = 56.25 \cdot 160$. Our answer checks. $\checkmark$

Example 6

In the United States, the ratio of people with negative-type blood to people with positive-type blood is 1:6. In a group of 140 people how many would we expect to have negative-type blood?

Solution

Step 1   Problem translation

Let x equal the number of people with negative-type blood. Since there are 140 people, the number of people with positive-type blood can be represented by $140 - x$. So we have

$$\underbrace{\frac{x}{140 - x}}_{\uparrow} = \underbrace{\frac{1}{6}}_{\uparrow}$$

This is the ratio of people with negative-type blood to people with positive-type blood.

This is also the ratio of people with negative-type blood to people with positive-type blood.

Step 2   Equation solution

$$\frac{x}{140 - x} = \frac{1}{6}$$

$$\frac{x}{140 - x} \diagdown \frac{1}{6}$$

$$6x = 140 - x$$

$$7x = 140$$

$$x = 20$$

So we would expect 20 people to have negative-type blood.

Step 3   Check

In a group of 140 people, if 20 have negative-type blood, then 120 will have positive-type blood. The ratio of 20:120 is 1:6. Our answer checks. $\checkmark$

One use of proportions is in converting to and from the metric system of measurement. We will make use of the Table of Units in Appendix C in the next five examples. Also, we have dispensed with our usual problem translation—equation solution—check format, because many of our answers will be rounded off to agree with the accepted standard values in the metric system. Thus an exact check on our answers would not be possible.

Example 7

Determine the number of centimeters in 30 inches.

Solution

Let x equal the number of centimeters. Now, we set up a proportion.

$$\underbrace{\frac{x}{30}} = \underbrace{\frac{2.54}{1}}$$

This is the ratio of   This is the ratio of
centimeters to inches.   centimeters to 1 inch
     from the table in
     Appendix C.

$$\frac{x}{30} = \frac{2.54}{1}$$

$$\frac{x}{30} \diagdown \frac{2.54}{1}$$

$$x = 76.2$$

So there are 76.2 cm in 30 in.

Example 8

Determine the number of inches in 100 centimeters.

Solution

Let x equal the number of inches. Now set up a proportion.

$$\underbrace{\frac{100}{x}} = \underbrace{\frac{2.54}{1}}$$

This is the ratio of   This is the ratio of
centimeters to inches.   centimeters to 1 inch
     from the table in
     Appendix C.

$$\frac{100}{x} = \frac{2.54}{1}$$

$$\frac{100}{x} \diagdown \frac{2.54}{1}$$

$$2.54x = 100$$

$$x = 39.37 \text{ (to the nearest hundredth)}$$

So there are 39.37 inches in 100 cm. Since 100 cm = 1 m, there are also 39.37 inches in a meter.

Example 9

Determine the number of kilograms in 150 pounds.

Solution

Let x equal the number of kilograms. Set up the proportion:

$$\underbrace{\frac{x}{150}} = \underbrace{\frac{0.45}{1}}$$

This is the ratio of   This is the ratio of
kilograms to pounds.   kilograms to 1 pound
     from the table in
     Appendix C.

$$\frac{x}{150} = \frac{0.45}{1}$$

$$\frac{x}{150} \diagdown \frac{0.45}{1}$$

$$x = 67.5$$

So there are 67.5 kg in 150 pounds.

Example 10

Determine the number of grams in 1.5 ounces.

Solution

Let x equal the number of grams.

$$\underbrace{\frac{x}{1.5}}_{} = \underbrace{\frac{28.3}{1}}_{}$$

↑ This is the ratio of grams to ounces. ↑ This is the ratio of grams to 1 ounce from the table in Appendix C.

$$\frac{x}{1.5} = \frac{28.3}{1}$$

$$\frac{x}{1.5} \diagdown \frac{28.3}{1}$$

$$x = 42.5 \text{ (to the nearest tenth)}$$

So there are 42.5 g in 1.5 ounces.

Example 11

Determine the number of milliliters in 2 quarts.

Solution

Let x equal the number of milliliters. The table in Appendix C shows us that 1 liter = 1.06 quarts. Also, 1 liter = 1000 milliliters. So, 1000 ml = 1.06 quarts. We obtain the following proportion.

$$\underbrace{\frac{x}{2}}_{} = \underbrace{\frac{1000}{1.06}}_{}$$

↑ This is the ratio of milliliters to quarts. ↑ This is also the ratio of milliliters to quarts.

$$\frac{x}{2} = \frac{1000}{1.06}$$

$$\frac{x}{2} \diagdown \frac{1000}{1.06}$$

$$1.06x = 2000$$

$$x = \frac{2000}{1.06}$$

$$x = 1,886.8 \text{ (to the nearest tenth)}$$

So there are 1886.8 ml in 2 quarts.

Section 4.4 Exercises

1. A census taker has determined that, in a certain community, there are 8000 residents. There are 5000 adults and 3000 children.
   (a) What is the ratio of adults to children?
   (b) What is the ratio of children to adults?
   (c) What is the ratio of adults to total number of residents?
   (d) If, out of the 3000 children, there are 1600 girls, what is the ratio of boys to girls?
   (e) If, out of the 5000 adults, there are 3000 men, what is the ratio of men to women?

2. A soccer team played 50 games in a season. They won 25 games, lost 20 games, and tied 5 games.
   (a) What is the ratio of wins to losses?
   (b) What is the ratio of wins to the total number of games played?
   (c) What is the ratio of ties to the total number of games played?
   (d) What is the ratio of games either won or lost to the number of games tied?

3. An automobile dealer has 40 new cars and 60 used cars in stock.
   (a) What is the ratio of new cars to used cars?
   (b) What is the ratio of used cars to new cars?
   (c) What is the ratio of new cars to total number of cars?

In Exercises 4 through 11, solve each problem and check your solutions.

4. In the membership for NOW (National Organization for Women) the ratio of females to males is $9:1$. If there are 45,000 female members, how many male members are there?

5. The ratio of an object's weight on the surface of the moon to its weight on the surface of the earth is $1:6$. If an object weighs 50 pounds on the moon, how much does it weigh on earth?

6. A cafeteria director knows that 3 pounds of coffee will make 200 cups of coffee. In a week 5000 cups of coffee will be consumed. How many pounds of coffee will be needed?

7. In crop dusting with a certain chemical 100 gallons will cover 1.5 acres. How many gallons will be needed for a 55.2-acre plot?

8. In Pueblo, Colorado, the ratio of female registered voters to the total number of registered voters is $2:3$. If there are 13,500 registered voters in Pueblo, how many of them are females?

9. In Mitchell, South Dakota, the ratio of female registered voters to male registered voters is $4:5$. If the total number of registered voters is 11,106, how many of them are females?

10. In Exercise 9, how many of the registered voters are males?

11. It has been estimated that a city of 50,000 people would need 40 square miles of solar energy collectors to supply enough power for their needs. How many square miles of solar collectors would be needed for a city of 350,000 people?

In Exercises 12 through 38, set up a proportion and solve each problem.

12. A car travels 110 kilometers on 31 liters of gasoline. How many liters would be required to drive 400 kilometers?

13. A certain copier can make 50 copies in a minute. How many copies could be made in $2\frac{1}{2}$ minutes?

14. In northern Spain where the Basques live, the ratio of people with negative-type blood to people with positive-type blood is $1:3$. In a group of 360 Basques, how many would we expect to have negative-type blood?

15. A map of Texas is drawn with a scale of 3 inches to 250 miles. The distance between Dallas and El Paso is 654 miles. How far apart are these cities on the map?

16. A photograph 3 inches by 5 inches in size is to be enlarged so that its longer side is 12 inches. What will be the length of the shorter side of the enlarged photo?

17. On Mount Rushmore, in South Dakota, the faces of presidents Washington, Jefferson, Theodore Roosevelt, and Lincoln are "carved." During construction, the models of the faces guided the workers

on the mountain. These models were 5 feet from chin to top of head. The ratio of the model height to the actual height was 1:12. What is the distance from the chin to the top of the head for the faces on the mountain?

18. The cost of constructing a certain type of house has been estimated at $25 per square foot of floor space. If the house is to have 1800 square feet of floor space, what will be the construction cost?

19. Electrical energy is sold by the kilowatt hour (kwh). If 100 kwh cost $5, at this same rate, how much will 455 kwh cost?

20. A recipe designed to serve 4 people requires $1\frac{1}{2}$ cups of sugar. How many cups of sugar are needed to serve 10 people?

21. The ratio of the average daily river flow of the Amazon River to the average daily flow of the Mississippi River is 12:1. Knowing that the average daily flow of the Mississippi is 400,000,000,000 gallons, what is the average daily river flow of the Amazon?

22. It has been estimated that a city of 100,000 people would need 3000 windmills (each over 100 feet high and operating constantly) to generate enough power for their needs. How many windmills would be needed for a city of 1,500,000 people?

23. The ratio of white attorneys to black attorneys in the United States is 98:2. If there are 300,000 attorneys in the United States, how many are black?

24. Determine the number of centimeters in 25 inches.

25. Determine the number of millimeters in 25 inches.

26. Determine the number of centimeters in 1 foot.

27. Determine the number of inches in 200 centimeters.

28. Determine the number of kilograms in 120 pounds.

29. Determine the number of grams in 1 pound.

30. Determine the number of kilograms in 90 pounds.

31. Determine the number of pounds in 150 kilograms.

32. Determine the number of grams in 5 pounds.

33. Determine the number of liters in a gallon.

34. Determine the number of cubic meters in 5 cubic yards.

35. Determine the number of milliliters in 3 quarts.

36. Determine the number of fluid ounces in 100 milliliters.

37. Determine the number of kilometers in 55 miles.

38. Determine the number of miles in 55 kilometers.

39. The formula $F = \frac{9}{5}C + 32$ can be used to convert Celsius temperatures to Fahrenheit temperatures.

The formula $C = \frac{5}{9}(F - 32)$ can be used to convert Fahrenheit temperatures to Celsius temperatures.

Using the appropriate formula, convert each Celsius temperature to the corresponding Fahrenheit temperature.

(a) 100°C   (b) 0°C   (c) 20°C   (d) 30°C   (e) −20°C

40. Using the appropriate formula from Exercise 39, convert each Fahrenheit temperature to the corresponding Celsius temperature.

(a) 90°F   (b) 32°F   (c) −14°F   (d) 58°F   (e) −40°F

Supplementary Exercises

1. In a certain city the ratio of adults to children is 3 to 1. If there are 12,000 people in the city,

(a) How many children are there?

(b) How many adults are there?

2. The ratio of nonsmokers to smokers in a certain study is 15 to 2. If there are 68 people in the study, how many are smokers?

3. The ratio of adults to children in a certain movie theater is 1 to 7. If there are 84 children present, how many adults are present?

4. Chris knows the ratio of cement to water is always 5 pounds to 1 gallon. If he mixes 125 pounds of cement, how much water should be added?

5. Josephine knows that 15 dozen rolls cost $16.50. How much would 7 dozen cost?

6. The ratio of dogs to cats in an apartment complex is 5 to 3. If there are 15 cats, how many dogs are there?

7. A football team lost 7 games, won 15 games, and tied 3 games.

   (a) Find the ratio of the wins to losses.

   (b) Find the ratio of the wins to the total number of games played.

8. John knows that 5 pounds of seed will plant one-half an acre of field. How much seed does he need to plant 3 acres?

9. The ratio of female teachers to male teachers in a certain high school is 7:2. If there are 27 teachers, how many are women?

10. A private school likes to keep the ratio of the students to faculty at 12 to 1. If the school has 900 students, how many faculty does the school have?

11. Mary's soccer team lost $\frac{1}{3}$ of its games and tied none. If they played 27 games, what is the ratio of the wins to the losses?

12. Determine the number of centimeters in 17 inches.

13. Determine the number of centimeters in 2 yards.

14. Determine the number of inches in 300 centimeters.

15. Determine the number of pounds in 82 grams.

16. Determine the number of yards in 3 meters.

17. Determine the number of inches in 5 centimeters.

18. Determine the number of kilometers in 60 miles.

19. Determine the number of centimeters in 5 inches.

20. Determine the number of liters in 6 quarts.

4.5 Percent problems

In this section we will solve word problems that involve **percents.** Percent means "hundredths." When we write 45%, we mean 45 hundredths, $\frac{45}{100}$, or 0.45. Before we begin solving word problems, we will present several examples that involve conversions of numbers with the percent symbol.

Example 1

Convert each of the following to a fraction and its decimal equivalent.

   (a) 42%   (b) 20%   (c) 125%   (d) 250%

Solution

   (a) $42\% = \dfrac{42}{100} = \dfrac{21}{50}$ Reduced fraction

   $42\% = .42$ Decimal

(b) $20\% = \dfrac{20}{100} = \dfrac{1}{5}$ Reduced fraction

$20\% = .20$ Decimal

(c) $125\% = \dfrac{125}{100} = 1\dfrac{25}{100} = 1\dfrac{1}{4}$ Reduced fraction

$125\% = 1.25$ Decimal

(d) $250\% = \dfrac{250}{100} = 2\dfrac{50}{100} = 2\dfrac{1}{2}$ Reduced fraction

$250\% = 2.50$ Decimal

Example 2

Convert each of the following to percents.

     (a) 0.46   (b) 0.003   (c) 1.57

Solution

To convert *from* a decimal, move the decimal point two places to the right and then "attach" the % symbol:

     (a) $0.46 = 46\%$

     (b) $0.003 = 0.3\%$

     (c) $1.57 = 157\%$

To see how percents can be applied to real situations, consider this problem:

Judy got 16 questions correct on a 20-question test. What percent of the questions did she get correct?

This picture represents the situation: 16 right (R) of 20 questions [4 wrong (W)].

R	R	R	R	R
R	R	R	W	R
R	W	R	R	R
R	R	W	R	W

Thus,

$$\frac{16}{20} = \frac{80}{100} = 0.80 = 80\%$$

represents the portion of the test done correctly.

We can say

"80% of 20 is 16"

or

"16 is 80% of 20"

In general, the *total* number (20) is called the **base** (B), and the part of the base (16) is called the amount (A). The 80% is referred to as the **rate** (R). We can write the basic percent equation three ways.

(1) $A = R \cdot B$

$16 = 80\% \cdot 20$

(2) $R = \dfrac{A}{B}$

$80\% = \dfrac{16}{20}$

(3) $B = A \div R$

$20 = 16 \div 80\%$ $\left(16 \div \dfrac{80}{100} = \dfrac{16}{1} \cdot \dfrac{100}{80} = \dfrac{100}{5} = 20 \right)$

Example 3

If $R = 30\%$ and $B = 60$, find A.

Solution

$$A = R \cdot B$$

We always express R as a decimal for computational purposes.

$$A = 0.30 \cdot 60$$
$$A = 18$$

Thus, 30% of 60 is 18.

Example 4

If $A = 12$ and $B = 48$, find R.

Solution

$$R = \dfrac{A}{B}$$

$$R = \dfrac{12}{48} = \dfrac{1}{4} = 25\%$$

Thus, 12 is 25% of 48.

Example 5

If $A = 6$ and $R = 20\%$, find B.

Solution

$$B = A \div R$$
$$B = 6 \div (0.20)$$
$$B = 30$$

So, 20% of 30 is 6.

The three examples that follow illustrate how to find A, B, and R when presented with a verbal percent problem. Basically, R is always the rate expressed as a percent, B is always the number following the word "of," and A is the remaining number.

Example 6

What number is 125% of 40?

Solution

$$R = 125\% = 1.25$$
$$B = 40$$
$$A = \text{unknown}$$

So

$$A = R \cdot B$$
$$A = 1.25 \quad 40$$
$$A = 50$$

Example 7

75 is 60% of what number?

Solution

$$R = 60\% = 0.60$$
$$B = \text{unknown (follows the word ``of'')}$$
$$A = 75$$

So

$$B = A \div R$$
$$B = 75 \div 0.60$$
$$B = 125$$

Example 8

45 is what percent of 50?

Solution

$$B = 50$$
$$R = \text{unknown}$$
$$A = 45$$

So

$$R = \frac{A}{B}$$

$$R = \frac{45}{50} = 0.90 = 90\%$$

The remaining examples in this section involve percent problems in real life settings. The business concepts of commission, discount, markup, and markdown are examined.

Example 9

Andrea sells computers for the Byteso Company. Her weekly salary is $220 plus a 10% commission on her total weekly sales. If she sells $1500 in one week, what is her total salary for the week?

Solution

First, we find her commission.

$$\begin{array}{ccccc} \text{The commission} & \text{is} & 10\% & \text{of} & \$1500 \\ \downarrow & & \downarrow & \downarrow & \downarrow \\ A & = & R & \cdot & B \end{array}$$

$$A = R \cdot B$$
$$A = 0.10 \cdot 1500$$
$$A = \$150$$

Her total salary for that week is $220 + $150 = $370.

Example 10

A store is offering a 15% discount on all lamps. If the original price on one lamp was $85, how much will it sell for on sale?

Solution

$$\underbrace{\text{The discounted amount}}_{A} \underset{\downarrow}{\text{is}} \underset{\downarrow}{15\%} \underset{\downarrow}{\text{of}} \underset{\downarrow}{\$85.}$$

$$A = R \cdot B$$

$$A = 0.15 \cdot \$85$$
$$A = \$12.75$$

Final selling price = original price less discount
$$= \$85 - 12.75$$
$$= \$72.25$$

Example 11

A merchant's cost for a particular stereo amplifier is $280. He then applies a 22% markup. At what price does he sell the amplifier?

Solution

$$\underset{\downarrow}{\text{Markup}} \underset{\downarrow}{\text{is}} \underset{\downarrow}{22\%} \underset{\downarrow}{\text{of}} \underset{\downarrow}{\$280.}$$

$$A = R \cdot B$$

$$A = 0.22 \cdot \$280$$
$$A = \$61.60$$

Selling price = dealer's cost plus markup
$$= \$280 + \$61.60$$
$$= \$341.60$$

Example 12

A clothing store marked down all its merchandise 30%. If Bob wishes to purchase a coat that sold for $150 (prior to markdown), how much will it cost?

Solution

$$\underset{\downarrow}{\text{Markdown}} \underset{\downarrow}{\text{is}} \underset{\downarrow}{30\%} \underset{\downarrow}{\text{of}} \underset{\downarrow}{\$150.}$$

$$A = R \qquad B$$

$$A = .30 \cdot \$150$$
$$A = \$45$$

Selling price = original cost less markdown
$$= \$150 - \$45$$
$$= \$105$$

Section 4.5 Exercises

1. Find 15% of 20.
2. Find 6.3% of 72.
3. Find $4\frac{1}{2}$% of 90.
4. Find $6\frac{3}{4}$% of 17.
5. Find 36% of 1200.
6. Find 83% of 170.
7. What percent of 120 is 18?
8. What percent of 15,000 is 375?
9. What percent of 180 is 36?
10. What percent of 24 is 6?
11. What percent of 300 is 60?
12. What percent of 800 is 100?
13. 18% of what number is 30.6?
14. $6\frac{1}{4}$% of what number is 125?
15. 7% of what number is 147?
16. 1% of what number is 18?
17. 288 is 12% of what number?
18. 525 is 15% of what number?
19. A store offers a 15% discount off on the ticketed price of an appliance. If the ticketed price is $320, what is the discount and the new price?
20. Tony bought his new car from a credit union that offered a 10% discount on his purchase. If the car originally cost $8950, how much did Tony pay?
21. People in the $31,000–32,999 tax bracket paid 32% of their income in taxes. If the Smiths' income was $31,496 what was their tax bill?
22. Mary purchased a dress that cost $59. The sales tax rate in her city is 7%. What was the sales tax and the total cost of the dress?
23. Gerry receives a 10% commission on each house he sells. If he sells a house for $65,400, what is his commission?
24. Joe received a commission of 12% on his total sales in addition to his base pay of $200 per week. If his total sales last week were $4500, what was his salary?
25. Luke sells cars. He receives a 10% commission on each car plus a base salary of $120 per week. If he sold $21,000 worth of cars in one week, what was his salary for the week?

Supplementary Exercises

1. Find $12\frac{1}{2}$% of 120.
2. Find $7\frac{3}{4}$% of 60.
3. Find 0.2% of 75.
4. What percent of 80 is 20?

5. What percent of 200 is 30?

6. What percent of 1000 is 65?

7. 19% of what number is 133?

8. 12% of what number is 48?

9. 7% of what number is 84?

10. Sam sells computers for a small company. His weekly salary is $200 plus 15% of his total weekly sales. If he sells $1200 in one week, what was his salary for the week?

11. Frank bought a dress originally marked $60 for $48. What was his percent of discount?

12. A store marked down all its merchandise 15%. If Amy purchased a dress for $65, a skirt for $29, and a blouse for $17, what was the final cost of her purchases?

4.6 Chapter review

Summary

This chapter was about problem-solving. That is, given a "word problem" (or a problem in written form), we learned to translate it into an algebraic model, solve the resulting equation, and check the solution in the originally worded problem.

We used this step-by-step approach to solve number problems, money problems, motion problems, mixture problems, problems involving ratio and proportion, and percent problems. In Section 4.4 we also saw that we can use proportions to convert from one system (English system) of measurement to another system (metric system), and vice versa.

Vocabulary Quiz

Match the expression in Column I with the phrase in Column II that best describes it.

Column I

1. The simple annual interest formula is

2. The distance formula is

3. The ratio of a to b is

4. In the proportion $\dfrac{x_1}{x_2} = \dfrac{x_3}{x_4}$, the extremes are

5. In the proportion $\dfrac{x_1}{x_2} = \dfrac{x_3}{x_4}$, the means are

6. If $\dfrac{x_1}{x_2} = \dfrac{x_3}{x_4}$, then

7. Percent means

Column II

a. $d = rt$.

b. $\dfrac{a}{b}$.

c. x_2 and x_3.

d. x_1 and x_4.

e. $I = PR$.

f. hundredths.

g. $x_1 x_4 = x_2 x_3$.

Chapter 4 Review Exercises

1. When the sum of two consecutive integers is divided by 7, the result is 11. Find the numbers.

2. The sum of three consecutive even integers is 936. Find the integers.

3. Gail has $2.25 in quarters and dimes. If she has twice as many dimes as quarters, how many coins of each type does she have?

4. Dom invests $1000, part at 6% and part at 8%. If the annual interest from the 8% investment was $52 more than the interest from the 6% investment, how much did he invest at each rate?

In Exercises 5 and 6, there is no answer. Explain why.

5. (a) Find two consecutive odd integers whose sum is 10.

   (b) Find two consecutive even integers whose sum is 12.

6. How much water should be combined with a 12% acid solution to obtain 1000 ml of a 15% acid solution?

7. Julius spent $6.44 for 22 stamps. He bought only 20-cent stamps and 37-cent stamps. How many 20-cent stamps did he buy?

8. Cal and Joyce leave a certain place at the same time and walk in opposite directions. Cal walks 1 km/hr faster than Joyce. After 3 hours they are 33 kilometers apart. Find the rate for each.

9. Maria left San Antonio bound for New Orleans at 1 P.M. Bob left at 2 P.M. and traveled the same route 10 mph faster. Bob passed Maria at 7 P.M.

   (a) What was Bob's average rate?

   (b) What was Maria's average rate?

   (c) How far did each travel?

10. A grocer wants to blend two types of nuts: peanuts and cashews. If peanuts sell for $1 per pound and cashews sell for $1.50 per pound, how many pounds of each does the grocer need to make a mixture of 10 pounds that sell for $1.10 per pound?

11. A chemist has 100 ml of a solution that is 15% hydrochloric acid. How much pure hydrochloric acid must be combined with the 15% solution to make a 25% solution?

12. At Mattatuck Community College the ratio of part-time students to full-time students is $2:5$. If there are 4900 students in all, how many are part-time?

13. An environmentalist has been tagging striped bass. In a sample of 200 bass selected last week, the ratio of tagged bass to untagged bass is $3:7$. How many were tagged?

In Exercises 14 through 18, fill in the blanks.

14. 2.1 meters = _____ inches

15. 4 quarts = _____ liters

16. 105 grams = _____ pounds

17. 1 ton = 2000 pounds = _____ kilograms

18. 1000 meters = _____ yards

19. Mary traveled for 3 hours at a constant rate of 55 mph. How far did she travel?

20. Twice the sum of a number and -7 is -20. Find the number.

21. The sum of twice a number and 6 is the same as the number added to 10. Find the number.

22. Dick biked at a constant rate of 12 mph. If he traveled 15 miles, how long did he bike?

23. Two trains approach the station from opposite directions. Train A is traveling at 60 mph and Train B is traveling at 70 mph. If the trains arrive at the station in 2 hours, how far apart were they?

24. Paul mixes water with a 95% solution to make an 85% solution. If he started with 10 liters of solution, how much water should he add?

25. Jennifer's bank contains only dimes and nickels. She has two more nickels than dimes and has $1.00 in her bank. How many dimes and nickels does she have?

26. The ratio of smokers to nonsmokers in a certain city is 7 to 5. If there are 120,000 people in the city, how many are smokers?

27. Fifteen percent of all men in a certain study were overweight. If the survey studied 120 men, how many were overweight?

28. Mike's bank contains $3.25 in dimes, nickels, and quarters. If he has twice as many dimes as nickels and three more quarters than nickels, how many of each type of coin does he have?

29. Alice takes the bus to work in the morning and jogs home. With the bus traveling at 30 mph, the trip to work takes 10 minutes. If the trip home takes 40 minutes, how fast does Alice jog?

30. The ratio of matriculated to nonmatriculated students at a certain college was 3 to 7. If there are 4000 students, how many are matriculated?

31. Mary purchased a new television on sale at Jay's Home Appliances. The sale offered 20% off all merchandise. If Mary paid $360 for the television, what was the original price?

32. Willie earns $175 per week in base salary. In addition he receives $7\frac{1}{2}\%$ of all his sales for the week. If he sold $12,000 during the week, what was his salary for the week?

33. Ann mixes pure alcohol with 11 liters of a 75% solution to create a 80% solution. How much alcohol should she add?

34. Chris must dilute a saline solution by adding sterile water. If she has 2 liters of a solution that is presently 60% and she wishes to have a 50% solution, how much water should she add?

35. Jake mixes 2 pounds of raisins that cost $1.20 per pound with 4 pounds of peanuts that cost $1.80 per pound. How much should the mix sell for per pound?

Cumulative review exercises

1. Evaluate $(-19) - (-24)$.   (*Section 1.3*)

2. Evaluate $\sqrt{100 - 64} - \dfrac{8 - 17}{\sqrt{9}}$.   (*Section 1.8*)

3. Suppose $x = 2$, $y = 3$, and $z = -4$. Evaluate $|x + z| - y$ (*Section 2.1*)

4. Simplify $4(3x - 1) - 7x + 5(x - 2)$.   (*Section 2.2*)

5. Let $x = 1$, $y = 2$, and $z = -3$. Evaluate $(x - y)^3 + z^3$.   (*Section 2.3*)

6. Multiply and simplify $-5x(x^2 - 3x) + 2x^2(4x - 1)$.   (*Section 2.4*)

7. Solve for w: $w + 57 = 81$.   (*Section 3.1*)

8. Solve for y: $-2y = 38$.   (*Section 3.2*)

9. Solve for x: $-2(x + 2) = 5 - 3(x + 6)$.   (*Section 3.4*)

10. Solve for T: $A = P + PRT$.   (*Section 3.6*)

Chapter 4 test

Take this test to determine how well you have mastered word problems; check your answers with those found at the end of the book.

1. When two consecutive even integers are added and that sum is multiplied by 4, the result is 168. Find the numbers.   (*Section 4.1*)

2. At the university basketball game, student admission was 75¢ and nonstudent admission was $1.50. There were four times as many students in attendance as nonstudents, and the total income at the gate was $1125. How many students attended?   (*Section 4.1*)

3. Tony invested a total of $20,000 in two accounts, a 7% account and a 9% account. His total annual interest was $1540. How much did he invest in each account?   (*Section 4.1*)

4. George and Leo leave their apartment and walk at a uniform rate in opposite directions. George walks twice as fast as Leo. After two hours they are 9 miles apart. How fast is each walking?   (*Section 4.2*)

5. Paul hiked from a hut on the Appalachian Trail to the summit of Mt. Washington at a rate of 3 mph. He made the trip back (downhill) at the rate of 4 mph. The trip up took 18 minutes more than the trip down. How far is the hut from the summit?   (*Section 4.2*)

6. How much pure gold must be added to a 30-kilogram alloy that is 15% gold in order to raise the percentage of gold to 20%? (*Section 4.3*)

7. A grocer wants to sell a raisin and nut mixture. Raisins sell for $2.50 per pound and nuts sell for $2.25 per pound. How many pounds of each are necessary for a 30-pound mixture that will sell for $2.35 per pound?   (*Section 4.3*)

8. The ratio of science majors to nonscience majors at a college is 1:13. If there are 700 students, how many are science majors? (*Section 4.4*)

9. How many meters long is a 100-yd football field?   (*Section 4.4*)

10. An appliance store is offering 30% off all television sets. If one sold for $430 last week,

    (a) How much was it marked down for the sale?

    (b) What is the new selling price?   (*Section 4.5*)

1. _____

2. _____

3. _____

4. _____

5. _____

6. _____

7. _____

8. _____

9. _____

10. (a) _____

    (b) _____

chapter
five

Exponents and polynomials

5.1 Rules of exponents

We begin this section by reviewing the first two rules of exponents presented in Section 2.3.

Exponent Rule 1

$$x^a \cdot x^b = x^{a+b}$$

That is, when multiplying two expressions with the *same* base, add the exponents.

Example 1

Use Exponent Rule 1 to rewrite each of the following.

(a) $x^7 \cdot x^8$ (b) $y \cdot y^4 \cdot y^3$ (c) $(3x^4)(5x^7)$

Solution

(a) $x^7 \cdot x^8 = x^{7+8} = x^{15}$

(b) $y \cdot y^4 \cdot y^3 = y^{1+4+3} = y^8$

(c) $(3x^4)(5x^7) = 3 \cdot 5 \cdot x^{4+7} = 15x^{11}$

Exponent Rule 2

$$(x^a)^b = x^{ab}$$

That is, when raising an expression involving a base with an exponent to another power, multiply the exponents.

Example 2

Use Exponent Rule 2 to rewrite each of the following.

(a) $(x^4)^3$ (b) $(y^2)^{100}$

Solution

(a) $(x^4)^3 = x^{4 \cdot 3} = x^{12}$

(b) $(y^2)^{100} = y^{2 \cdot 100} = y^{200}$

We can extend Exponent Rule 2 to rewrite an expression like $(xy^2)^3$. Since the exponent "3" means to use xy^2 as a factor three times, we can write

$$(xy^2)^3 = xy^2 \cdot xy^2 \cdot xy^2$$
$$= x \cdot x \cdot x \cdot y^2 \cdot y^2 \cdot y^2$$
$$= x^3 y^6$$

Thus, $(xy^2)^3 = x^3 y^6$. In general we have:

Exponent Rule 3

$$(x^a y^b)^c = x^{ac} y^{bc}$$

That is, when a product of two factors is raised to a power, the result is the same as raising each factor to that power and then multiplying.

Example 3

Use Exponent Rule 3 to rewrite each of the following.

(a) $(2x^2)^4$ (b) $(3x^2y^4)^3$

Solution

(a) $(2x^2)^4 = 2^4 x^{2\cdot4}$

$\qquad\qquad = 16x^8$

(b) $(3x^2y^4)^3 = 3^3 x^{2\cdot3} y^{4\cdot3}$

$\qquad\qquad\quad = 27x^6 y^{12}$

We are now ready for the introduction of two new exponent rules. Consider the expression $\dfrac{x^8}{x^3}$. Assuming $x \neq 0$, it can be rewritten as

$$\frac{x^8}{x^3} = \frac{x \cdot x \cdot x \cdot x \cdot x \cdot x \cdot x \cdot x}{x \cdot x \cdot x} = \frac{x \cdot x \cdot x}{x \cdot x \cdot x} \cdot \frac{x \cdot x \cdot x \cdot x \cdot x}{1}$$

But

$$\frac{x \cdot x \cdot x}{x \cdot x \cdot x} = 1$$

Any number, except zero, divided by itself is 1. So,

$$\frac{x^8}{x^3} = 1 \cdot \frac{x \cdot x \cdot x \cdot x \cdot x}{1} = x^5$$

$$\frac{x^8}{x^3} = x^{8-3} = x^5$$

Exponent Rule 4

$$\frac{x^a}{x^b} = x^{a-b} \qquad (x \neq 0)$$

That is, when dividing two expressions with the same base, the exponents are subtracted.

Example 4

Use Exponent Rule 4 to simplify each of the following, if possible.

(a) $\dfrac{y^{10}}{y^2}$ (b) $\dfrac{x^5}{y^2}$ (c) $\dfrac{(x+3y)^7}{(x+3y)^4}$ (d) $\dfrac{25x^3}{5x}$

Solution

(a) $\dfrac{y^{10}}{y^2} = y^{10-2} = y^8$

(b) $\dfrac{x^5}{y^2}$ cannot be simplified using Exponent rule 4 because in the numerator the base is x and in the denominator the base is different, y.

(c) $\dfrac{(x+3y)^7}{(x+3y)^4} = (x+3y)^{7-4} = (x+3y)^3$

(d) $\dfrac{25x^3}{5x} = \dfrac{25}{5}x^{3-1} = 5x^2$

Example 5

Use Exponent Rule 4 to simplify each of the following, if possible.

(a) $\dfrac{(2x)^7}{(2x)^6}$ (b) $\dfrac{(3y)^5}{(3y)^3}$ (c) $\dfrac{\left(\dfrac{t}{2}\right)^8}{\left(\dfrac{t}{2}\right)^6}$

Solution

(a) $\dfrac{(2x)^7}{(2x)^6} = (2x)^{7-6} = (2x)^1 = 2x$

(b) $\dfrac{(3y)^5}{(3y)^3} = (3y)^{5-3} = (3y)^2 = 9y^2$

(c) $\dfrac{\left(\dfrac{t}{2}\right)^8}{\left(\dfrac{t}{2}\right)^6} = \left(\dfrac{t}{2}\right)^{8-6} = \left(\dfrac{t}{2}\right)^2 = \dfrac{t}{2}\cdot\dfrac{t}{2} = \dfrac{t^2}{4}$

Much of the discussion regarding exponents so far can be summarized by rewriting a rather complex expression such as the one that follows.

$$\left(\dfrac{x^2 y^4}{z^5}\right)^3 = \dfrac{x^2 y^4}{z^5}\cdot\dfrac{x^2 y^4}{z^5}\cdot\dfrac{x^2 y^4}{z^5}$$

Use the definition of exponents as repeated multiplication.

$$= \dfrac{x^2\cdot x^2\cdot x^2\cdot y^4\cdot y^4\cdot y^4}{z^5\cdot z^5\cdot z^5}$$

$$= \dfrac{x^6 y^{12}}{z^{15}}$$

Thus

$$\left(\dfrac{x^2 y^4}{z^5}\right)^3 = \dfrac{x^6 y^{12}}{z^{15}}$$

This leads to the *generalized* exponent rule:

Exponent Rule 5

$$\left(\dfrac{x^a y^b}{z^c}\right)^d = \dfrac{x^{ad} y^{bd}}{z^{cd}} \qquad (z \neq 0)$$

Section 5.1 Exercises

Use the rules of exponents to simplify each of the following expressions, if possible.

1. $x^2 \cdot x^5$
2. $y^4 \cdot y^7$
3. $q^{12} \cdot q^{20}$
4. $5x^2 \cdot x^2$
5. $3y^2 \cdot y^4$
6. $p \cdot p^4 \cdot p^5$
7. $(6a^2)(5a^4)$ *add exp.*
8. $(5a^5b^6)(7a^2b^8)$
9. $(xyz^2)(5x^2z)$
10. $(y^3)^4$
11. $(x^7)^8$
12. $(q^{12})^{20}$
13. $(3x^4)^4$ $= 81x^{16}$
14. $3(x^4)^4 = 3x^{16}$
15. $(2y^2)^3$
16. $2(y^2)^3$
17. $(3z^2)^4$
18. $(4xy^2)^3$
19. $(10abc^2)^3$
20. $\dfrac{x^{12}}{x^{10}}$
21. $\dfrac{y^7}{y^6}$
22. $\dfrac{x^4}{x^3}$
23. $\dfrac{z^2}{z}$
24. $\dfrac{(5x)^{12}}{(5x)^{11}}$
25. $\dfrac{18a^4}{3a^2}$
26. $\left(\dfrac{18a^4}{3a^2}\right)^2$
27. $\dfrac{(4y^2)^5}{(4y^2)^3}$
28. $\left(\dfrac{x^2y^{10}}{z}\right)^2$
29. $\left(\dfrac{ab^2}{c^4}\right)^3$
30. $\left(\dfrac{2ab^2}{c^4}\right)^3$

Supplementary Exercises

Use the rules of exponents to simplify each of the following expressions, if possible.

1. $a^3 \cdot a \cdot a^5$
2. b^7b^{26}
3. $4x^3 \cdot x^7$
4. $(9x^2y^3)(-2x^5y^7)$
5. $(-3x^2y)(4x^3y)$ *$-12x^5y^2$*
6. $(5x^2y)(-2xy^3z)$
7. $(-2x^3y)^3$ *$\sim 2x^6$*
8. $(4x^3y^2)^4$ *$256\,x^{12}y^8$*
9. $\dfrac{14x^{13}}{-7x^7}$ *$= -2x^6$*
10. $\dfrac{120y^8z^9}{48y^7z^6}$
11. $\dfrac{-9x^3y^2z^7}{3xy^2z^3}$
12. $\left(\dfrac{3x^2y}{z}\right)^3$ *$\dfrac{27x^6y^3}{z^3}$*
13. $\left(\dfrac{4x^2y^3}{12xy}\right)^2$
14. $\dfrac{9x^3y^3z}{12x^2yz}$
15. $\dfrac{(5x^3)^9}{(5x^3)^6}$ *$5x^3$*

16. Can $(x^2 + y^2)^3$ be rewritten using any of the exponent rules in this chapter? Explain.

5.2 Negative and zero exponents

If we attempt to rewrite $\dfrac{x^6}{x^8}$, we might proceed as follows:

$$\frac{x^6}{x^8} = \underbrace{\frac{x \cdot x \cdot x \cdot x \cdot x \cdot x}{x \cdot x \cdot x \cdot x \cdot x \cdot x}}_{\text{equals 1, provided } x \neq 0} \cdot \frac{1}{x \cdot x} = \frac{1}{x^2} \qquad (\text{if } x \neq 0)$$

By Exponent Rule 4, however, we have

$$\frac{x^6}{x^8} = x^{6-8} = x^{-2} \qquad (\text{if } x \neq 0)$$

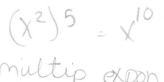

Rule #1
$x^2 \cdot x^5 = x^7$
multi - add exp.

Rule #2
$(x^2)^5 = x^{10}$
multip expon.

Rule #3
$(x^2y^3)^3 = x^6y^9$

Rule #4
Division (subtract)
$\dfrac{x^6}{x^2} = x^4$

$\dfrac{x^{16}y^3}{xy} = x^{15}y^2$

Note:

$\dfrac{(3y^2)^{10}}{(3y^2)^4} = (3y^2)^6$
exact same $= 3^6y^{12}$

$\dfrac{(2x-3y)^{10}}{(2x-3y)^4} = (2x-3y)^6$
leave it

We can conclude that $x^{-2} = \dfrac{1}{x^2}$, provided that $x \neq 0$. The concept of a negative exponent will be very useful. The general definition is:

Definition of negative exponent

$$x^{-a} = \frac{1}{x^a} \qquad (x \neq 0)$$

Example 1

Use the definition of negative exponent to evaluate each expression.

(a) 5^{-2} (b) $(-3)^{-2}$ (c) $(-3)^{-3}$

Solution

(a) $5^{-2} = \dfrac{1}{5^2} = \dfrac{1}{25}$

(b) $(-3)^{-2} = \dfrac{1}{(-3)^2} = \dfrac{1}{9}$

(c) $(-3)^{-3} = \dfrac{1}{(-3)^3} = \dfrac{1}{-27} = -\dfrac{1}{27}$

Example 2

Using the definition of negative exponent, rewrite each of the following with only positive exponents.

(a) z^{-5} (b) $x^4 y^{-2} z^{-3}$ (c) $(x^{-4}y)(x^2y^2)$

Solution

(a) $z^{-5} = \dfrac{1}{z^5}$

(b) $x^4 y^{-2} z^{-3} = x^4 \cdot \dfrac{1}{y^2} \cdot \dfrac{1}{z^3} = \dfrac{x^4}{1} \cdot \dfrac{1}{y^2} \cdot \dfrac{1}{z^3}$

$\qquad\qquad = \dfrac{x^4}{y^2 z^3}$

(c) $(x^{-4}y)(x^2y^2) = x^{-2}y^3 \qquad$ Use Exponent Rule 1.

$\qquad\qquad = \dfrac{1}{x^2} \cdot \dfrac{y^3}{1} \qquad$ Definition of Negative Exponent

$\qquad\qquad = \dfrac{y^3}{x^2}$

Two important things to note about the use of negative exponents are:

1. The negative exponent has nothing to do with the sign of an expression. For example, 5^{-3} is a positive number.
2. The negative exponent can be thought of as a way of "moving" a **factor** from the numerator to the denominator by changing the exponent's sign. For example,

$$\frac{x^{-2}y^3}{z} = \frac{y^3}{x^2 z}$$

Similarly, a negative exponent on a factor in the denominator can be moved to the numerator by changing the exponent's sign.

$$\frac{a^3}{b^{-4}} = \frac{a^3}{1} \cdot \frac{1}{b^{-4}} = a^3 \cdot b^4$$

Example 3

Rewrite each expression using only positive exponents.

(a) $\dfrac{8x^{-3}y^7}{4x^4y^{-2}}$ (b) $\dfrac{7a^2b^{-2}c^3}{21a^{-3}b^4c^{-1}}$ (c) $\dfrac{p^{-2}q^3}{r^2t^{-4}}$

Solution

(a) $\dfrac{8x^{-3}y^7}{4x^4y^{-2}} = \dfrac{8}{4} \cdot \dfrac{x^{-3}}{x^4} \cdot \dfrac{y^7}{y^{-2}}$

$= 2 \cdot x^{-3-4} \cdot y^{7-(-2)}$ Use Exponent Rule 4.

$= 2 \cdot x^{-7} \cdot y^9$

$= \dfrac{2y^9}{x^7}$ Definition of Negative Exponents

(b) $\dfrac{7a^2b^{-2}c^3}{21a^{-3}b^4c^{-1}} = \dfrac{7}{21} \cdot \dfrac{a^2}{a^{-3}} \cdot \dfrac{b^{-2}}{b^4} \cdot \dfrac{c^3}{c^{-1}}$

$= \dfrac{1}{3} \cdot a^{2-(-3)} \cdot b^{-2-4} \cdot c^{3-(-1)}$ Use Exponent Rule 4.

$= \dfrac{1}{3} \cdot a^5b^{-6}c^4$

$= \dfrac{a^5c^4}{3b^6}$ Definition of Negative Exponents

(c) $\dfrac{p^{-2}q^3}{r^2t^{-4}} = p^{-2} \cdot q^3 \cdot \dfrac{1}{r^2} \cdot \dfrac{1}{t^{-4}}$

$= \dfrac{1}{p^2} \cdot \dfrac{q^3}{1} \cdot \dfrac{1}{r^2} \cdot \dfrac{t^4}{1}$ Definition of Negative Exponents

$= \dfrac{q^3t^4}{p^2r^2}$

The expresion $\dfrac{y^3}{y^3}$ ($y \neq 0$) is equal to 1 because any nonzero number divided by itself is 1. By Exponent Rule 4,

$$\frac{y^3}{y^3} = y^{3-3} = y^0$$

When a nonzero quantity is raised to the zero exponent, the result, then, is 1.

Zero exponent

$$x^0 = 1 \quad (x \neq 0)$$

Example 4

Evaluate each of the following.

(a) z^0 (b) $(2x)^0$ (c) $2x^0$ (d) $(341)^0$ (e) $(-20)^0$

Solution

(a) $z^0 = 1,\ z \neq 0$

(b) $(2x)^0 = 1,\ x \neq 0$

(c) $2x^0 = 2 \cdot 1 = 2,\ x \neq 0$

(d) $(341)^0 = 1$

(e) $(-20)^0 = 1$

Example 5

Simplify using only positive exponents:

$$\frac{15x^2y^{-3}z^4}{20x^5y^{-3}z^4}$$

Solution

$$\frac{15x^2y^{-3}z^4}{20x^5y^{-3}z^4} = \frac{15}{20}x^{2-5}y^{-3-(-3)}z^{4-4} \qquad \text{Use Exponent Rule 4.}$$

$$= \frac{15}{20}x^{-3}y^0z^0$$

$$= \frac{15}{20} \cdot \frac{1}{x^3} \qquad\qquad x^{-3} = \frac{1}{x^3};\ y^0 = z^0 = 1$$

$$= \frac{3}{4x^3} \qquad\qquad \frac{15}{20} = \frac{5 \cdot 3}{5 \cdot 4} = \frac{3}{4}$$

Section 5.2 Exercises

In Exercises 1 through 20, evaluate each expression.

1. 2^{-1} $\frac{1}{2}$

2. 2^{-2} $\frac{1}{4}$

3. 2^{-3}

4. 3^{-1}

5. 3^{-2}

6. 3^{-3}

7. $(-4)^2$

8. $(4)^{-2}$ $\frac{1}{16}$

9. $(-4)^{-2}$ $\frac{1}{16}$

10. 5^3 125

11. 5^{-3} $\frac{1}{125}$

12. $(-5)^3$ -125

13. $(-5)^{-3}$

14. $(-7)^{-2}$

15. 7^0

16. $(-7)^0$ $+1$

17. -7^0 -1

18. $(5^{-1})(2^{-2})(16)^0$

19. $[(5^{-1})(2^{-2})16]^0$

20. $\dfrac{(10^{-2})(10^{-1})}{10^3}$ $= \dfrac{10^{-3}}{10^3} = 10^{-3-3} = 10^{-6} = \dfrac{1}{10^6}$

In Exercises 21 through 42 simplify by rewriting the expression using only positive exponents. Assume that all variables cannot be zero.

21. x^{-7}

22. y^{-2}

23. x^3y^{-5}

24. a^7b^{-4}

25. $p^{-5}q^6$

26. $r^{-2}s^{-1}t^{-8}$

27. $x^8y^{-3}z^4$

28. $x^{-8}y^{-3}z^4$

29. $x^8y^3z^{-4}$

30. $a^2b^{-3}c^0$

31. $\dfrac{1}{y^{-4}}$

32. $\dfrac{1}{x^{-2}}$

33. $\dfrac{x^{-2}}{y^{-4}}$

34. $\dfrac{x^2}{y^{-4}}$

35. $\dfrac{x^{-2}}{y^4}$

36. $\dfrac{20x^2y^{-4}}{28xyz}$

37. $\dfrac{40x^7y^{-8}}{16x^4y^{-1}z^2}$

38. $(3x^{-4}y^{-2})(5x^{-2}y^6)$

39. $\dfrac{12a^2b^{-4}c^3}{16a^5b^{-4}c^3}$

40. $\dfrac{16abc^{-2}}{12ab^2c^{-2}}$

41. $\dfrac{14x^{-2}y^{-3}z^{-4}}{21xy^{-5}z^{-3}}$

42. $\dfrac{25x^{-3}y^{-2}z^{-5}}{10x^2y^{-6}z^{-2}}$

Supplementary Exercises

In Exercises 1 through 10, evaluate each of the following.

1. 5^{-2}

2. 4^{-3}

3. $(-6)^{-2}$

4. -5^{-2}

5. -8^0

6. 9^0

7. $(-6)^0$

8. $(2^{-2})(-3^{-2})$

9. $(2)^0(3)^{-2}$

10. $(3^{-2})(4)^{-1}(5)^0$

In Exercises 11 through 25, simplify by rewriting the expression using only positive exponents. Assume that all variables cannot be zero.

11. z^{-8}

12. p^{-9}

13. x^5y^{-2}

14. x^6y^{-3}

15. $x^{-2}y^{-7}$

16. $a^{-4}b^{-9}$

17. $a^{-2}b^3c^{-5}$

18. $d^3c^{-4}f^{-5}$

19. $\dfrac{1}{x^{-7}}$

20. $\dfrac{3}{y^{-4}}$

21. $\dfrac{5x^{-3}y^3z}{4xy^5z^{-2}}$

22. $\dfrac{a^3b^{-3}c^2}{a^{-2}b^{-3}c^4}$

23. $\dfrac{-5x^3y^{-5}z}{15xy^{-7}}$

24. $\dfrac{(2x^{-3}y^3)}{10x^3y^{-5}}$

25. $\left(\dfrac{15x^{-3}y^2z}{45x^2y^{-4}}\right)^2$

5.3 Scientific notation

Because a scientist's work often involves numbers that are extremely large or extremely small, exponents (positive, negative, and zero) have applications in many branches of science. For example, a physicist deals with the velocity of light in a vacuum: 29,979,250,000 cm/sec. A chemist may have to know the mass of a hydrogen atom: 0.0000000000000000000000016734 g.

Obviously, such numbers are awkward to write. They are also difficult to read and use in calculations. To alleviate these problems, such numbers are usually written in **scientific notation,** a form that is especially useful for writing exceptionally large or extremely small numbers. To write a number in scientific notation, we write it as the product of two quantities: a number between 1 and 10 times a power of 10.

Scientific notation

1. To convert a number *greater than 10* to scientific notation:

    (a) Place a "new" decimal point to the right of the first digit.

    (b) Count the number of digits between the old and new decimal points. Since the new decimal point is located to the *left* of the old decimal point, this number is the necessary *positive* power of 10.

    Example: $563.7 \longrightarrow 5.637 \cdot 10^{②}$

    There are two digits (6 and 3) between original and new decimal points.

2. To convert a number *between 0 and 1* to scientific notation:

    (a) Place a "new" decimal point to the right of the first nonzero digit.

    (b) Count the number of digits between the old and new decimal points. Since the new decimal point is located to the *right* of the old decimal point, this number is the necessary *negative* power of 10.

    Example: $0.000357 \longrightarrow 3.57 \cdot 10^{④}$

    There are 4 digits (three 0s and 3) between original and new decimal points.

Example 1

Write each of the following in scientific notation.

(a) 8290   (b) 93,000,000   (c) 0.00091

Solution

(a) Writing 8290 as the product of a number between 1 and 10 and a power of ten is a matter of moving the decimal point. (Note that in the case of integers the decimal point is implied.) To write 8290 as a number between 1 and 10, move the decimal point three places to the left. We get 8.290. Now $8290 = 8.290 \cdot 10^3$. (Since we "moved" the decimal point *three* places to the left, multiplying by 10^3 "restores" the value.) So

$$8290 = 8.290 \cdot 10^3$$

(b) $93{,}000{,}000 = \underbrace{9.3}_{\substack{\text{Notice, this number is} \\ \text{between 1 and 10.}}} \cdot \underbrace{10^7}$

If 9.3 were to be multiplied by this number the decimal point would be moved seven places to the right.

(c) To write 0.00091 as a number between 1 and 10, move the decimal point four places *to the right,* yielding 9.1. Now $0.00091 = 9.1 \cdot 10^{-4}$. (Since we "moved" the decimal point *four* places to the right, multiplying by 10^{-4} "restores" the value.) So

$$0.00091 = 9.1 \cdot 10^{-4}$$

Example 2

Rewrite each of the following numbers in scientific notation.

(a) 30,993,600,000,000,000 (the number of feet a "particle" of light travels in 1 year)

(b) 0.0000000000000000000000395 (the mass, in grams, of an atom of the heaviest natural element, uranium)

(c) 6357 (the polar radius of the earth, in kilometers)

Solution

(a) $3.09936 \cdot 10^{16}$ (b) $3.95 \cdot 10^{-22}$ (c) $6.357 \cdot 10^{3}$

We mentioned earlier that scientific notation can be an aid when we do calculations with large or small numbers. Suppose we must multiply 3,810,000 by 24,300,000,000. We write each number in scientific notation and then multiply.

$$3{,}810{,}000 \cdot 24{,}300{,}000{,}000 = (3.81 \cdot 10^{6})(2.43 \cdot 10^{10}) \qquad \text{Rewrite in scientific notation.}$$

$$= 3.81 \cdot 2.43 \cdot 10^{6} \cdot 10^{10} \qquad \text{Rearrange factors.}$$

$$= 9.2583 \cdot 10^{16} \qquad \text{Use Exponent Rule 1.}$$

In the next example we practice this idea more.

Example 3

For each of the following, first write the numbers in scientific notation and then perform the calculation. Write the answer in scientific notation.

(a) $2{,}450{,}000 \cdot 4{,}567{,}000{,}000$

(b) $2{,}450{,}000 \cdot 0.0000004567$

(c) $(2{,}450{,}000)^{2}$

(d) $\dfrac{2{,}450{,}000}{500{,}000{,}000}$

Solution

(a) $2{,}450{,}000 \cdot 4{,}567{,}000{,}000 = 2.45 \cdot 10^{6} \cdot 4.567 \cdot 10^{9}$ Rewrite in scientific notation.

$$= 2.45 \cdot 4.567 \cdot 10^{6} \cdot 10^{9} \qquad \text{Rearrange factors.}$$

$$= 11.18915 \cdot 10^{15} \qquad \text{Use Exponent Rule 1.}$$

$$= 1.118915 \cdot 10 \cdot 10^{15} \qquad \text{Rewrite in scientific notation.}$$

$$= 1.118915 \cdot 10^{16} \qquad \text{Use Exponent Rule 1.}$$

(b) $2{,}450{,}000 \cdot 0.0000004567 = 2.45 \cdot 10^{6} \cdot 4.567 \cdot 10^{-7}$ Rewrite in scientific notation.

$$= 2.45 \cdot 4.567 \cdot 10^{6} \cdot 10^{-7} \qquad \text{Rearrange factors.}$$

$$= 11.18915 \cdot 10^{-1} \qquad \text{Use Exponent Rule 1.}$$

$$= 1.118915 \cdot 10 \cdot 10^{-1} \qquad \text{Rewrite in scientific notation.}$$

$$= 1.118915 \cdot 10^{0} \qquad \text{Use Exponent Rule 1.}$$

$$\text{or } 1.118915$$

(c) $(2{,}450{,}000)^{2} = (2.45 \cdot 10^{6})^{2}$ Rewrite in scientific notation.

$$= (2.45)^{2} \cdot (10^{6})^{2} \qquad \text{Use Exponent Rule 3.}$$

$$= 6.0025 \cdot 10^{12} \qquad \text{Use Exponent Rule 2.}$$

(d) $\dfrac{2,450,000}{500,000,000} = \dfrac{2.45 \cdot 10^6}{5 \cdot 10^8}$ Rewrite in scientific notation.

$= \dfrac{2.45}{5} \cdot \dfrac{10^6}{10^8}$

$= .49 \cdot 10^{-2}$ Use Exponent Rule 4.

$= 4.9 \cdot 10^{-1} \cdot 10^{-2}$ Rewrite in scientific notation.

$= 4.9 \cdot 10^{-3}$ Use Exponent Rule 1.

Section 5.3 Exercises

In Exercises 1 through 15, write each number in scientific notation.

1. 75.8

2. 369.04

3. 8570

4. 57

5. 5280

6. 75,800,000

7. 0.0056

8. 0.0000000123

9. 0.000550

10. 550,000

11. 205,000,000

12. 0.205

13. 0.000205

14. 29,979,250,000

15. 0.0000000000000000000000016734

In Exercises 16 through 25, perform each calculation by first writing the numbers in scientific notation.

16. $10,430 \cdot 20,600$

17. $104,300,000 \cdot 206,000,000$

18. $0.000026 \cdot 10,400,000$

19. $0.000026 \cdot 0.00000012$

20. $(10,600)^2$

21. $(70,400,000)^2$

22. $\dfrac{0.00000024}{48,000,000}$

23. $(71,000)^3$

24. $\dfrac{126,000,000}{0.000018}$

25. $\dfrac{1,024}{25,000,000,000}$

Scientific calculators are usually equipped to handle extremely large and small numbers. When 45,598 is multiplied by 2,999,239, for example, the result is displayed on many scientific calculators as:

$$1.367593 \qquad 11$$

This is merely the calculator's way of displaying 1.367593×10^{11}. Similarly, 0.00000345 times 0.000000391 is displayed as

$$1.34895 \qquad -12$$

which means 1.34895×10^{-12}.

In Exercises 26 through 30, use a calculator with scientific notation to perform the indicated operations.

26. $351,000 \cdot 467,999,761$

27. $0.000000192 \cdot 0.00000000013$

28. $0.0000000194 \cdot 0.000352$

29. $981,123,333 \cdot 9,142,200$

30. $123,000 \cdot 909,123,000 \cdot 500,000$

Supplementary Exercises

In Exercises 1 through 12, write each number in scientific notation.

1. 527

2. 6983.725

3. 4,090,103.2

4. $-62,953$ 5. $-12,325,795$ 6. $1,000,000,000$

7. 0.0023 8. 0.000000021 9. -0.002963

10. -0.000319 11. 0.52196 12. 0.000000003

In Exercises 13 through 15, perform each calculation by first writing the numbers in scientific notation.

13. $(205,000)(1,200,000)$ 14. $\dfrac{0.000026}{1,300,000}$ 15. $(0.000052)(210,000)$

5.4   Polynomials: addition and subtraction

Many of the algebraic expressions we simplified in Chapter 2 were of a particular type known as **polynomials.** In this section we will formally define a polynomial and then extend the concepts we began using in Chapter 2.

A polynomial in x

An algebraic expression that is the sum of terms of the form ax^n, where a is any real number and n is a nonnegative integer, is called a **polynomial in x.**

A one-term polynomial in x is called a **monomial** in x, a two-term polynomial in x is called a **binomial** in x, and a three-term polynomial in x is called a **trinomial** in x.

One example of a monomial in one variable is $3x$. Examples of polynomials in one variable include

$$5x + 17 \qquad \text{(binomial in } x\text{)}$$
$$x^3 - 7x^2 + 8x - 4 \qquad \text{(polynomial in } x\text{)}$$
$$12y^4 - \frac{1}{2}y + 6 \qquad \text{(trinomial in } y\text{)}$$

Notice that we write polynomials in descending powers of the variable.

but the following are *not* polynomials:

$$3x^{-2} + 5$$
$$\frac{16}{x}$$
$$\frac{1}{3x + 1}$$

None of these have terms only of the form ax^n, where $n \geq 0$.

The **degree** of a polynomial in x is the highest value of the x exponents. For example,

$$x^{③} - 7x^2 + 11$$

is a **third-degree polynomial,** or a **polynomial of degree 3.**

Example 1

Find the degree of each polynomial.

(a) $5x^6 - 11x^2 + 4x - 2$ (b) $\frac{1}{2}x^2 - 2x + 1$ (c) 19

Solution

(a) $5x^{⑥} - 11x^2 + 4x - 2$ is a 6th degree polynomial.

(b) $\frac{1}{2}x^{②} - 2x + 1$ is a 2nd degree polynomial.

(c) Because $19 = 19x^{⓪}$, we say it is a 0-degree polynomial.

A polynomial can also involve more than one variable.

> ### A polynomial in x and y
> An algebraic expression that is the sum of terms of the form $ax^m y^n$, where a is any real number and m and n are nonnegative integers, is called a **polynomial in x and y**.

Examples of polynomials in x and y include

$$12x^2 y - 5xy^2 + 14x + 6$$
$$-41x^5 y^6 + 12xy^4 - 12x^2 y - 1$$

Most of the work in this text will involve polynomials in a single variable.

Addition of Polynomials

To add polynomials, we use the notion of **collecting like terms.** We will present a series of examples to illustrate how to add polynomials horizontally (Examples 2, 3, and 4) and then vertically (Examples 5, 6, and 7).

Example 2

Add $(3x^4 - 2x^2 + 4) + (5x^4 - x^3 - 3x^2 + x - 1)$.

Solution

First, we rewrite the stated sum with like terms, being careful to include the sign preceding a term as part of the term.

$(3x^4 - 2x^2 + 4) + (5x^4 - x^3 - 3x^2 + x - 1)$

$\qquad = 3x^4 + 5x^4 - x^3 - 2x^2 - 3x^2 + x + 4 - 1$

$\qquad = 8x^4 - x^3 - 5x^2 + x + 3$ Combine like terms by adding their coefficients.

Example 3

Add $(2x^3 - 1) + (3x^3 + 7x^2 + x - 4) + (2x^2 - x - 4)$.

Solution

$(2x^3 - 1) + (3x^3 + 7x^2 + x - 4) + (2x^2 - x - 4)$

$\qquad = 2x^3 + 3x^3 + 7x^2 + 2x^2 + x - x - 1 - 4 - 4$

$\qquad = 5x^3 + 9x^2 - 9$

Example 4

Add $(5y^3 - y^2 + 7) + (12y^3 - 7y^2 + y + 1) + (-18y^3 + y + 7)$.

Solution

$$(5y^3 - y^2 + 7) + (12y^3 - 7y^2 + y + 1) + (-18y^3 + y + 7)$$
$$= 5y^3 + 12y^3 - 18y^3 - y^2 - 7y^2 + y + y + 7 + 1 + 7$$
$$= -1y^3 - 8y^2 + 2y + 15$$
$$= -y^3 - 8y^2 + 2y + 15 \qquad \text{We write } -1y^3 \text{ as } -y^3.$$

A vertical display can sometimes help to ensure that only *like* terms are added. We'll show this approach in the next few examples.

Example 5

Add $(5x^3 - 7x^2 + 2x + 1) + (x^3 + 7x - 5)$.

Solution

We line up terms with identical exponents vertically.

$$
\begin{array}{r}
5x^3 - 7x^2 + 2x + 1 \\
+ \quad x^3 \qquad\quad + 7x - 5 \\
\hline
6x^3 - 7x^2 + 9x - 4
\end{array}
$$

Notice that we leave space for "missing" terms.

Example 6

Add $(3z - 5) + (-2z^3 - 5z + 6) + (4z^3 + z^2 + z - 1)$.

Solution

Again, note that we leave appropriate spaces for "missing" terms.

$$
\begin{array}{r}
3z - 5 \\
-2z^3 \qquad\quad -5z + 6 \\
+ \quad 4z^3 + z^2 + z - 1 \\
\hline
2z^3 + z^2 - z
\end{array}
$$

Now we'll rework Example 4 vertically as Example 7.

Example 7

Add $(5y^3 - y^2 + 7) + (12y^3 - 7y^2 + y + 1) + (-18y^3 + y + 7)$.

Solution

$$
\begin{array}{r}
5y^3 - \quad y^2 \qquad\quad + 7 \\
12y^3 - 7y^2 + \quad y + 1 \\
+ \quad -18y^3 \qquad\qquad + \quad y + 7 \\
\hline
-y^3 - 8y^2 + 2y + 15
\end{array}
$$

Subtraction of Polynomials

To subtract $(3x^2 + 7x - 5)$ *from* $(8x^2 + 9x - 5)$ means to write

$$(8x^2 + 9x - 5) - (3x^2 + 7x - 5)$$

Think of the minus sign as a -1 multiplier of the second polynomial. Then,

$$(8x^2 + 9x - 5) - (3x^2 + 7x - 5)$$
$$= (8x^2 + 9x - 5) - 1(3x^2 + 7x - 5)$$
$$= 8x^2 + 9x - 5 - 3x^2 - 7x + 5$$

Using the distributive property, multiply the three terms in the second set of parentheses by -1.

$$= 8x^2 - 3x^2 + 9x - 7x - 5 + 5$$
$$= 5x^2 + 2x$$

Combine like terms.

Example 8

Subtract $3x^2 - 7x + 1$ from $x^2 - 7$.

Solution

$$(x^2 - 7) - (3x^2 - 7x + 1) = x^2 - 7 - 3x^2 + 7x - 1$$
$$= x^2 - 3x^2 + 7x - 7 - 1$$
$$= -2x^2 + 7x - 8$$

Notice the effect of the minus sign is to change the sign of each term in the second polynomial. As a final example, subtraction of polynomials will be shown using a vertical display.

Example 9

Subtract $(y^3 - 7y^2 + 8y - 4) - (y^2 + y - 5)$.

Solution

As in addition, we line up like terms under one another.

$$y^3 - 7y^2 + 8y - 4$$
$$-\ \underline{y^2 + \ y - 5}$$

Now, since the *effect* of the subtraction symbol is to change the signs of the second polynomial and then to add the terms together, we do just that.

$$y^3 - 7y^2 + 8y - 4$$
$$+\ \underline{-\ y^2 - \ y + 5}$$
$$y^3 - 8y^2 + 7y + 1$$

change bottom signs then add [handwritten annotation with asterisk]

Section 5.4 Exercises

In Exercises 1 through 10, (a) write the polynomial in descending powers of the variable and (b) determine the degree of the polynomial.

1. $3x - 4x^2 + 7$ [handwritten: $4x^2 + 3x + 7$]

2. $7y - y^2$ [handwritten: $-y^2 + 7y$]

3. $8 - 3x$ [handwritten: $-3x + 8$]

4. 12

5. 4

6. $10x^3 + 5x^5 + x^3 - x + 11$

7. $12y^{12} + 10y - 5$

8. $16x^3 + 2x^2 - x + 12$

9. $16x^2 + 2x^3 - x + 12$ [handwritten: $2x + 16x^2 - x + 12$]

10. $x^9 - x^{11} + 2$

In Exercises 11 through 30, perform the indicated operations.

11. $(2x^2 - 7x + 12) + (3x^2 + 10x + 20)$

12. $(5x^2 - 8x + 3) + (x^2 + 2x - 5)$

13. $(8x^3 - 2x^2 + x - 6) + (8x^3 - x^2 - 5x - 11)$

14. $(y^3 - 5y^2 - y - 10) + (y^3 - 3y^2 + 7y + 2)$

15. $(x^2 - 7x + 12) + (x^3 - 5x^2 + 10)$

16. $(y^3 - 7y - 5) + (y^2 - 2y - 10)$

17. $(5z^2 - 10z + 4) + (3z^3 - 5z^2 + 2z - 8) + (10z^3 - 5z + 11)$

18. $(21x^4 - 5x + 10) + (-20x^4 + x^2 - 10) + (6x - x^2 - x^4)$

19. $(21x^4 - 5x + 10) + (-20x^4 + x^2 - 10) + (5x - x^2 - x^4)$

20. $(x^3 - 5x^2 + 10x - 6) - (x^2 + 7x + 10)$

21. $(2y^2 - 5y + 7) - (y^2 + 5y - 8)$

22. Subtract $3t^2 - 8t + 17$ from $t^2 - 9$.

23. Subtract $t^2 - 9$ from $3t^2 - 8t + 17$.

24. Subtract $x^2 - x - 9$ from $3x^3 - 2x^2 + 7x + 3$.

25. Subtract $3x^3 - 2x^2 + 7x + 3$ from $x^2 - x - 9$.

26. $(5y^7 - 6y^6 + 4y^4 - y + 11) - (y^5 - 7y^4 - 8y^3 + y + 6)$

27. $(7x^8 - 5x^4 - 12) - (9x^8 + 4x^4 + 3x - 10)$

28. $(5x - 1) - (3x^6 - 7x^2 - 5x + 10)$

29. $(3y^3 - 5y + 6) + (7y^2 + 6y - 10) + (8y - 12) - (5y^2 + 12y - 4)$

30. $(15y^4 - 5y^3 + 10y^2 - 8y + 1) + (3y^4 - 5y^2 - 6y + 11) - (8y^3 - 5y^2 + 21y - 4)$

Exercises 31 through 33 involve addition and subtraction of polynomials in two variables (x and y). Be sure to combine only like terms. (Recall that like terms are those whose letters and exponents are identical; only coefficients may differ.)

31. $(x^2y + 5xy^2 - 11x + 10y) + (8xy^2 - 10y + 6)$

32. $(21x^2y^2 + 5xy - 1) + (5xy + 6x^2y - 18)$

33. $(10xy^2 - 5x^2y - 10xy) - (6xy^2 + 10xy + 6) - (5xy^2 - 5x + 6)$

34. Add $(5.123x^2 + 8.001x - 5.073) + (12.761x^2 - 5.111x + 11.51)$.

35. Subtract $(81.571x^2 - 5.072x + 14.123) - (75.199x^2 - 5.071x + 15.937)$.

Supplementary Exercises

1. What is the degree of the following polynomial?

$$5x^3 - 2x^6 + 3x^2 - 4x^7 + 2x$$

2. What is the degree of the following polynomial?

$$2 - 3x^2 + 4x - 5x^4$$

In Exercises 3 through 20, perform the indicated operations.

3. $(x^2 + 3x - 4) + (2x^2 - 5x + 1)$

4. $(5x^2 - 9x + 5) + (x^3 + 6x - 9)$

5. $(-a^2 + 2a + 5) + (2a - 3a^2 + 7)$

6. $(4x^4 - 3x^2 + 2x - 1) + (3x^4 + x^3 - 5x - 9)$

7. $(2a^2 - 9) + (5a^2 - 3a - 4) + (-4a^2 - a + 7)$

8. $(5x^2 + 3x + 2) + (-4x^2 - 4x + 3) + (-9 - 5x - x^2)$

9. $(2x^2 - 7x + 3) - (x^2 - 5x - 9)$

10. $(-3x^2 + 4x + 4) - (-6x^2 - 9x + 5)$

11. $(x^3 - 3x^2 + 2x - 1) - (x^3 + x^2 - 5) + (2x^3 - 4x^2 - 3x + 5)$

12. $(2y^2 - 3y + 5) - (y - y^2 + 1) - (3y^2 - 4y + 4)$

13. $-(a^3 - 2a^2 + 3a - 1) - (5a^3 + 4a^2 - a + 4)$

14. Find the sum of $5x^2 - 2x + 3$, $3x^2 + 3x - 9$, and $4x - 2$.

15. Subtract $3x^2 - 9x + 4$ from $x^2 + 10x - 5$.

16. Subtract $-5x + 3x^2 - 5$ from $4x^2 - 2x - 9$.

17. $(x^2y + 3xy^2 - 4x^2y^2) + (5xy^2 - 4x^2y + 6x^2y^2)$

18. $(2a^2 - 3ab + 5b^2) - (4a^2 - 4ab - 5b^2)$

19. $(a - 2b) - (2b + a) + (3a - 4b)$

20. $(5x^2 - 2xy + y^2) - (3x^2 - 4xy + 3y^2)$

5.5   Multiplying polynomials

The Distributive Property for Multiplication over Addition was introduced in Chapter 1 and used in Chapter 2. It is the basis for the work in this section, polynomial multiplication.

We begin by showing how to multiply a polynomial by a *monomial*.

Multiplication of a polynomial by a monomial

To multiply a polynomial by a monomial, apply the distributive property and multiply *each term* of the polynomial by the monomial. Then add the terms.

Example 1

Multiply $x^2(3x - 4)$.

Solution

Using the distributive property, we have

$$x^2(\,3x - 4\,) = x^2 \cdot 3x - x^2 \cdot 4$$

Thus
$$x^2(3x - 4) = 3x^3 - 4x^2$$

Example 2

Multiply $4a(3a - 2b)$.

Solution

$$4a(3a - 2b) = 4a \cdot 3a - 4a \cdot 2b$$
$$= 12a^2 - 8ab$$

Example 3

Rewrite $7 + 2x(x - 10)$ without parentheses.

Solution

To eliminate parentheses, we multiply.

$$7 + 2x(x - 10) = 7 + 2x \cdot x - 2x \cdot 10$$
$$= 7 + 2x^2 - 20x$$
$$= 2x^2 - 20x + 7$$

Either of these is an acceptable answer. However, as we said previously, it is traditional to write a polynomial in one variable in descending powers of the variable.

Example 4

Multiply $4x(x^2 - 7x + 12)$.

Solution

In $4x(x^2 - 7x + 12)$, multiply each of the three terms in parentheses by $4x$.

$$4x(x^2 - 7x + 12) = 4x \cdot x^2 - 4x \cdot 7x + 4x \cdot 12$$
$$= 4x^3 - 28x^2 + 48x$$

Example 5

Multiply $-2y^2(y^3 - 3y - 11)$.

Solution

$$-2y^2(y^3 - 3y - 11) = -2y^2 \cdot y^3 - 2y^2 \cdot -3y - 2y^2 \cdot -11$$
$$= -2y^5 + 6y^3 + 22y^2$$

The rest of this section involves the multiplication of two polynomials. We will begin with an example, the multiplication of $x + 2$ by $x + 3$.

First recall the distributive property. It tells us that, when $x + 2$ is multiplied by another quantity, the other quantity is multiplied by both the x *and* the 2. In $(x + 2)(x + 3)$, the quantity $x + 3$ must be multiplied by both the x *and* the 2. We display this idea as follows:

$$(x + 2)(x + 3) = x(x + 3) + 2(x + 3) \qquad \text{$x + 3$ is multiplied by both x and 2.}$$

$$= x^2 + 3x + 2x + 6 \qquad \text{Multiply each expression.}$$

$$= x^2 + 5x + 6 \qquad \text{Collect like terms.}$$

Some examples follow.

Example 6

Multiply $(x - 4)(x + 7)$.

Solution

$$(x - 4)(x + 7) = x(x + 7) - 4(x + 7) \qquad \text{$x + 7$ is multiplied by both x and -4.}$$

$$= x^2 + 7x - 4x - 28 \qquad \text{Multiply each expression.}$$

$$= x^2 + 3x - 28 \qquad \text{Collect like terms.}$$

Example 7

Multiply $(3y - 4)(y - 2)$.

Solution

$$(3y - 4)(y - 2) = 3y(y - 2) - 4(y - 2)$$
$$= 3y^2 - 6y - 4y + 8$$
$$= 3y^2 - 10y + 8$$

In each of the three preceding examples, the product of two binomials resulted in a four-term expression, but that, in turn, was simplified to three terms (a **trinomial**) in our final answer. The next two examples illustrate that, when two binomials are multiplied, the final answer can also contain two terms or four terms.

Example 8

Multiply $(x + 5)(x - 5)$.

Solution

$$(x + 5)(x - 5) = x(x - 5) + 5(x - 5)$$
$$= x^2 - 5x + 5x - 25 \qquad \text{Notice that}$$
$$\qquad\qquad\qquad\qquad\qquad -5x + 5x \text{ is } 0.$$
$$= x^2 - 25$$

Example 9

Find the product of $(a + b)$ and $(c + d)$.

Solution

$$(a + b)(c + d) = a(c + d) + b(c + d)$$
$$= ac + ad + bc + bd \qquad \text{Note that no sim-}$$
$$\qquad\qquad\qquad\qquad\qquad\text{plification can be}$$
$$\qquad\qquad\qquad\qquad\qquad\text{performed; there}$$
$$\qquad\qquad\qquad\qquad\qquad\text{are no like terms.}$$

Notice in the last example (as well as in all the examples preceding it) that *each* term of the first binomial is multiplied by *each* term of the second binomial. In Example 9, a and c (the first terms of each) are multiplied to produce ac; a and d (the outermost terms) are multiplied to produce ad; b and c (the innermost terms) are multiplied to produce bc; and finally b and d (the last terms of each) are multiplied to produce bd. This pattern is called **foil** and is useful when multiplying two binomials.

To multiply $(x + 2)(x + 3)$:

f	Multiply first terms	$\longrightarrow$ $x \cdot x$ or x^2
o	Multiply outermost terms	$\longrightarrow$ $x \cdot 3$ or $3x$
i	Multiply innermost terms	$\longrightarrow$ $2 \cdot x$ or $2x$
l	Multiply last terms	$\longrightarrow$ $2 \cdot 3$ or 6

$$x^2 + 3x + 2x + 6$$
$$\text{or}$$
$$x^2 + 5x + 6$$

The next two examples show products found by the foil method.

Example 10

Multiply $(2x - 5)(3x + 2)$.

Solution

$$(2x - 5)(3x + 2) \overset{\text{f o i l}}{=} 6x^2 + 4x - 15x - 10$$
$$= 6x^2 - 11x - 10$$

Example 11

Find the product of $3x + 1$ and $x - 8$.

Solution

$$(3x + 1)(x - 8) \overset{\text{f o i l}}{=} 3x^2 - 24x + x - 8$$
$$= 3x^2 - 23x - 8$$

Keep in mind that the foil method is useful only when multiplying two binomials. There is another way of obtaining the product of two binomials. It is similar to ordinary multiplication. For example, to find the product $(x + 2)(x + 3)$, we first set up the multiplication as follows:

$$\begin{array}{r} x + 3 \\ \underline{x + 2} \end{array}$$

Next, we proceed just as we would if we were multiplying numbers.

Step 1

$2 \cdot 3$ is 6.

$$\begin{array}{r} x + 3 \\ \uparrow \\ \underline{x + 2} \\ 6 \end{array}$$

Step 2

$2 \cdot x$ is $2x$.

$$\begin{array}{r} x + 3 \\ \nwarrow \\ \underline{x + 2} \\ 2x + 6 \end{array}$$

Step 3

$x \cdot 3$ is $3x$.

$$\begin{array}{r} x + 3 \\ \nearrow \\ \underline{x + 2} \\ 2x + 6 \\ 3x \end{array}$$

Step 4

$x \cdot x$ is x^2.

$$\begin{array}{r} x + 3 \\ \uparrow \\ \underline{x + 2} \\ 2x + 6 \\ \underline{x^2 + 3x} \end{array}$$

Step 5

Add.

$$\begin{array}{r} x + 3 \\ \underline{x + 2} \\ 2x + 6 \\ \underline{x^2 + 3x} \quad \text{Add.} \\ x^2 + 5x + 6 \end{array}$$

We displayed the above method here for two reasons. First, it reaffirms the multiplying idea of foil. Second, we can use this method to multiply polynomials. The next two examples further illustrate this method.

Example 12

Multiply the binomial $x + 5$ by the trinomial $x^2 + 7x - 12$.

Solution

We set up the multiplication vertically.

$$
\begin{array}{r}
x^2 + 7x - 12 \\
x + 5 \\
\hline
\end{array}
$$

Then we multiply each term of $x^2 + 7x - 12$ by 5.

$$
\begin{array}{r}
x^2 + 7x - 12 \\
x + 5 \\
\hline
5x^2 + 35x - 60
\end{array}
$$

Next we multiply each term of $x^2 + 7x - 12$ by x.

$$
\begin{array}{r}
x^2 + 7x - 12 \\
x + 5 \\
\hline
5x^2 + 35x - 60 \\
x^3 + 7x^2 - 12x \\
\hline
x^3 + 12x^2 + 23x - 60
\end{array}
$$

Example 13

Find the product of the trinomial $x^2 - x + 2$ with the trinomial $x^2 - 3x - 1$.

Solution

$$
\begin{array}{r}
x^2 - 3x - 1 \\
x^2 - x + 2 \\
\hline
2x^2 - 6x - 2 \\
-x^3 + 3x^2 + x \\
x^4 - 3x^3 - x^2 \\
\hline
x^4 - 4x^3 + 4x^2 - 5x - 2
\end{array}
$$

Section 5.5 Exercises

In Exercises 1 through 40, multiply each expression and simplify by collecting like terms.

1. $2(4x + 3)$
2. $4(2x + 3)$
3. $x(4x + 3)$
4. $x^2(4x + 3)$
5. $5y^2(y - 1)$
6. $-3y^3(y^2 - 2y + 10)$
7. $2 + x(x + 1)$
8. $8 - 3x(x - 4)$
9. $4x - x(2x - 1) + 10$
10. $5y^2 - y(3y + 7)$
11. $5x(x^2 - 7x + 1)$
12. $3y^2(2y^2 - 5y + 2)$
13. $4x(x^2 + x - 1)$
14. $3x^2(x^2 - 2x + 7)$

15. $-11y(y^5 - 2y^4 + 6)$
16. $-a^2b^2c^2(a^2 + 7ab - 4c)$.
17. $20xy^2(3x^2y - x^2 + y - 1)$
18. $x(x^5 + x^4 + x^3 + x^2 + x + 1)$
19. $(x + 5)(x + 2)$
20. $(y - 4)(y - 3)$
21. $(x - 6)(x + 3)$
22. $(x - 3)(x + 6)$
23. $(2x + 3)(x + 1)$
24. $(x + 1)(2x + 3)$
25. $(3a - 4)(a + 1)$
26. $(2x - 1)(x - 6)$
27. $(7t + 5)(2t + 1)$
28. $(7t + 5)(2t - 1)$
29. $(7t - 5)(2t + 1)$
30. $(7t - 5)(2t - 1)$
31. $(x - 2)(x + 2)$
32. $(y + 10)(y - 10)$
33. $(2y - 3)(2y + 3)$
34. $(a + b)(a - b)$
35. $(x^2 + 7)(x^2 - 2)$
36. $(x + 1)(x^2 + x + 1)$
37. $(5x - 2)(x^3 - x - 10)$
38. $(x^2 - 2)(3x^2 - 7x + 12)$
39. $(x^2 + 2x + 1)(x^2 - 5x + 6)$
40. $(x^2 + 2x - 3)^2$

41. (a) Multiply $(x + 8)^2$. [*Hint*: $(x + 8)^2 = (x + 8)(x + 8)$]
    (b) Is $(x + 8)^2$ the same as $x^2 + 64$?

42. (a) Multiply $(a + b)^2$.
    (b) Is $(a + b)^2$ the same as $a^2 + b^2$?

43. (a) Multiply $(x + 1)^3$. [*Hint*: $(x + 1)^3 = (x + 1)(x + 1)^2$ so first find $(x + 1)^2$.]
    (b) Is $(x + 1)^3$ the same as $x^3 + 1$?

44. A farmer has a fence around his square-shaped potato field to keep out unwanted critters. Let x denote the length of each side of the square (in feet).

    (a) What is the area of the field?

    (b) Suppose the farmer increases one pair of opposite sides by 6 feet and decreases the other two sides by 6 feet. Represent the area now, algebraically.

    (c) Has the area of the field increased, decreased, or remained the same? How much has it changed? (Notice that the answer does not depend on the original dimensions of the field.)

45. A rectangular box has three dimensions: length, width, and depth. For a certain box, the width is 3 inches more than twice the length and the depth is 4 inches more than the length. Let ℓ represent the length.

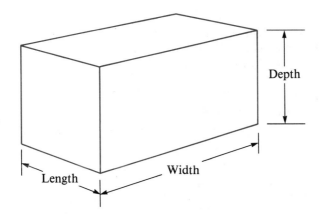

    (a) Represent the width algebraically.

    (b) Represent the depth algebraically.

    (c) Find the algebraic expression for the volume of this box. (*Note*: The volume of a rectangular box is found by multiplying the length by the width by the depth.)

    (d) Use your answer to (c) to find the volume (in cubic inches) when the length of the box is 10 inches.

Supplementary Exercises

In Exercises 1 through 23, multiply each expression and simplify by collecting like terms.

1. $5(3x - 2)$

2. $-7(5 - 2x)$

3. $y(3y - 7)$

4. $-t(5t^2 - 4)$

5. $2x^3(3x^3 - 2x + 1)$

6. $-4ab(2a - 3b)$

7. $(2x + 1)(x - 5)$

8. $(5 - 2y)(3y + 1)$

9. $(4x - 1)(x^2 + 1)$

10. $-(x + 2)(3x - 4)$

11. $5x - x(3x + 4) + 2x^2$

12. $-4y + y(y^2 - 2) - 5y^3$

13. $(2x + y)(3x + 2y) - (x^2 + 5xy - y^2)$

14. $(x - 2)^2$

15. $(a - 2)(a^2 - 3a + 4)$

16. $(2 - y)(4 + 2y - y^2)$

17. $2 - x(3 - 4x)$

18. $(5 - 2y)(3 + 5y + y^2)$

19. $(1 - x)^2$

20. $5x^2y(2x + y) - 4xy(x^3 + xy)$

21. $(a + b + c)(a - b - c)$

22. $(a - 1)^3$

23. $(x^2 + x - 1)(x^2 - 2x + 1)$

24. A rancher would like to construct a barn that is 10 feet longer than it is wide. If x represents the width, write a simplified expression for the area of the barn.

25. A rectangular box has x by $x - 2$ by $x + 4$ as its dimensions. Find a simplified expression for the volume of the box.

5.6   Dividing polynomials

Dividing a Polynomial by a Monomial

Consider the division

$$\frac{30x^4 + 6x^3}{2x^2}$$

The fraction bar acts as a grouping symbol. It means that the entire polynomial numerator, term by term, must be divided by the monomial denominator $2x^2$. So we can rewrite the expression as follows:

$$\frac{30x^4 + 6x^3}{2x^2} = \frac{30x^4}{2x^2} + \frac{6x^3}{2x^2}$$

$$= \frac{2 \cdot 15 \cdot x^2 \cdot x^2}{2x^2} + \frac{2 \cdot 3 \cdot x \cdot x^2}{2x^2}$$

$$= 15x^2 + 3x$$

In general:

Dividing a polynomial by a monomial

Divide *each term* of the polynomial by the monomial. Then add the results.

Some examples follow.

Example 1

Divide $\dfrac{8x + 2}{2}$.

Solution

$$\frac{8x + 2}{2} = \frac{8x}{2} + \frac{2}{2} \qquad \text{Divide each term of the polynomial by the monomial.}$$

$$= 4x + 1$$

Example 2

Divide $\dfrac{15x^3 - 20x^2 + 10x}{5x}$.

Solution

$$\frac{15x^3 - 20x^2 + 10x}{5x} = \frac{15x^3}{5x} - \frac{20x^2}{5x} + \frac{10x}{5x} \qquad \begin{array}{l}\text{Divide each term} \\ \text{of the polynomial} \\ \text{by the monomial.} \\ \text{Notice how the} \\ \text{sign preceding} \\ \text{each term is} \\ \text{treated.}\end{array}$$

$$= 3x^2 - 4x + 2 \qquad \begin{array}{l}\text{Simplify using} \\ \text{Exponent Rule 4.}\end{array}$$

Before proceeding to the next example, let's pause to examine the signs of a fraction. A fraction may be thought of as having three signs: the sign of the numerator, the sign of the denominator, and the sign that precedes the entire term. For example,

$$-\frac{5x}{3y}$$

will represent a negative quantity if both x and y are positive; $-\dfrac{5x}{3y}$ will be positive if x is positive and y is negative. Of course, there are other possible combinations. *Any two signs of a fraction can be changed without changing its value.* (This result is verified in Exercise 31 at the end of the section.) So, $-\dfrac{2a}{-3b}$ can be rewritten as $\dfrac{2a}{3b}$ and $-\dfrac{-9x}{-2y}$ can be rewritten as $-\dfrac{9x}{2y}$. We use this property in the next example.

Example 3

Divide $\dfrac{14x^2y^2 - 21xy^2 + 70xy}{-7xy}$.

Solution

$$\frac{14x^2y^2 - 21xy^2 + 70xy}{-7xy} = \frac{14x^2y^2}{-7xy} - \frac{21xy^2}{-7xy} + \frac{70xy}{-7xy}$$

$$= -2xy + 3y - 10 \qquad \begin{array}{l}\text{Notice that } -\dfrac{21xy^2}{-7xy} \\[2mm] \text{becomes } -\dfrac{3y}{-1} \text{ or} \\[2mm] +3y.\end{array}$$

The next example illustrates that we may be left with variables in denominators after the division.

Example 4

Divide $\dfrac{18x^2yz^2 + 60xy^2z - 120x^3y}{-6x^2y^2}$.

Solution

$$\dfrac{18x^2yz^2 + 60xy^2z - 120x^3y}{-6x^2y^2}$$

$$= \dfrac{18x^2yz^2}{-6x^2y^2} + \dfrac{60xy^2z}{-6x^2y^2} - \dfrac{120x^3y}{-6x^2y^2}$$

$$= \dfrac{6 \cdot 3 \cdot x^2 \cdot y \cdot z^2}{-1 \cdot 6 \cdot x^2 \cdot y \cdot y} + \dfrac{6 \cdot 10 \cdot x \cdot y^2 \cdot z}{-1 \cdot 6 \cdot x \cdot x \cdot y^2} - \dfrac{6 \cdot 20 \cdot x^2 \cdot x \cdot y}{-1 \cdot 6 \cdot x^2 \cdot y \cdot y}$$

$$= \dfrac{6 \cdot 3 \cdot \cancel{x^2} \cdot \cancel{y} \cdot z^2}{-1 \cdot \cancel{6} \cdot \cancel{x^2} \cdot \cancel{y} \cdot y} + \dfrac{\cancel{6} \cdot 10 \cdot \cancel{x} \cdot \cancel{y^2} \cdot z}{-1 \cdot \cancel{6} \cdot \cancel{x} \cdot x \cdot \cancel{y^2}} - \dfrac{\cancel{6} \cdot 20 \cdot \cancel{x^2} \cdot x \cdot \cancel{y}}{-1 \cdot \cancel{6} \cdot \cancel{x^2} \cdot \cancel{y} \cdot y}$$

$$= \dfrac{-3z^2}{y} - \dfrac{10z}{x} + \dfrac{20x}{y}$$

Dividing a Polynomial by a Polynomial

The method of dividing a polynomial by a polynomial resembles the long division we learned in arithmetic. Let's review that method and some associated terminology in the division $488 \div 21$.

$$\overset{\longleftarrow\text{ Quotient (to be determined)}}{21\overline{)488}}$$

Divisor   Dividend

$$\begin{array}{r} 2 \\ 21\overline{)4\,8\,8} \end{array}$$

Step 1: Determine a (trial) quotient. In this case, it is 2 because $\dfrac{48}{21}$ is $2+$.

$$\begin{array}{r} 2 \\ 21\overline{)4\,8\,8} \\ 4\,2 \end{array}$$

Step 2: Multiply divisor by (partial) quotient.

$$\begin{array}{r} 2 \\ 21\overline{)4\,8\,8} \\ \underline{4\,2} \\ 6 \end{array}$$

Step 3: Subtract.

$$\begin{array}{r} 2 \\ 21\overline{)4\,8\,8} \\ \underline{4\,2\downarrow} \\ 6\,8 \end{array}$$

Step 4: "Bring down" the next numeral.

$$
\begin{array}{r}
2\,3 \leftarrow \text{Quotient} \\
\text{Divisor} \rightarrow 21\overline{)4\,8\,8} \leftarrow \text{Dividend} \\
4\,2 \downarrow \\
\hline
6\,8 \\
6\,3 \\
\hline
5 \leftarrow \text{Remainder}
\end{array}
$$

Step 5: The ''new'' problem is to divide 21 into 68. Steps 1–4 are repeated.

We write $488 \div 21 = 23 \text{ R } 5$ or $488 \div 21 = 23\dfrac{5}{21}$.

We will follow a similar procedure for dividing polynomials. Consider the division of $2x^2 - 11x + 12$ by $x - 3$:

$$x - 3\overline{)2x^2 - 11x + 12}$$

$$
\begin{array}{r}
2x \\
x - 3\overline{)2x^2 - 11x + 12}
\end{array}
$$

Step 1: Determine the first term in the quotient by calculating $\dfrac{2x^2}{x}$. Since $\dfrac{2x^2}{x} = 2x$, we write $2x$ as the (partial) quotient and place it above its like term in the dividend.

$$
\begin{array}{r}
2x \\
x - 3\overline{)2x^2 - 11x + 12} \\
2x^2 - 6x
\end{array}
$$

Step 2: Multiply the (partial) quotient by the entire divisor.

$$
\begin{array}{r}
2x \\
x - 3\overline{)2x^2 - 11x + 12} \\
2x^2 \pm 6x \\
\hline
- 5x
\end{array}
$$

Step 3: Subtract. (Remember that subtracting is the same as adding the opposite.)

$$
\begin{array}{r}
2x \\
x - 3\overline{)2x^2 - 11x + 12} \\
2x^2 - 6x \\
\hline
- 5x + 12
\end{array}
$$

Step 4: ''Bring down'' the next term.

$$
\begin{array}{r}
2x - 5 \\
x - 3\overline{)2x^2 - 11x + 12} \\
2x^2 - 6x \\
\hline
- 5x + 12 \\
- 5x \mp 15 \\
\hline
- 3
\end{array}
$$

Step 5: Repeat steps 1–4 until the degree of the expression under the dividend is less than the degree of the divisor. (The -5 appears in the quotient because $\dfrac{-5x}{x} = -5$.)

Thus $\dfrac{2x^2 - 11x + 12}{x - 3} = 2x - 5 \text{ R } -3$ or $2x - 5 + \dfrac{-3}{x - 3}$

Some examples follow.

___

Example 5

$(6x^2 - 17x + 5) \div (3x - 1)$

Solution

$$
\begin{array}{r}
2x - 5 \\
3x - 1\overline{)6x^2 - 17x + 5} \\
\underline{6x^2 - 2x} \\
-15x + 5 \\
\underline{-15x + 5} \\
0
\end{array}
$$

Hence

$$\frac{6x^2 - 17x + 5}{3x - 1} = 2x - 5$$

Example 6

Perform the division $\dfrac{37y - 31y^2 + 6y^3 + 10}{2y - 7}$.

Solution

The procedure only works if both divisor and dividend are written in descending powers of the variable, so first we must rewrite the numerator as $6y^3 - 31y^2 + 37y + 10$.

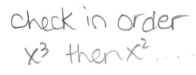

$$
\begin{array}{r}
3y^2 - 5y + 1 \\
2y - 7\overline{)6y^3 - 31y^2 + 37y + 10} \\
\underline{6y^3 - 21y^2} \\
-10y^2 + 37y \\
\underline{+10y^2 \mp 35y} \\
2y + 10 \\
\underline{-2y \mp 7} \\
17
\end{array}
$$

So $(6y^3 - 31y^2 + 37y + 10) \div (2y - 7) = 3y^2 - 5y + 1 + \dfrac{17}{2y - 7}$

answ $3y^2 - 5y + 1$ Rem $17\big/_{2y-7}$

Example 7

Divide $(x^4 - 4x^3 + 4x^2 - 6x + 8) \div (x^2 - 3x - 1)$.

Solution

$$
\begin{array}{r}
x^2 - x + 2 \\
x^2 - 3x - 1\overline{)x^4 - 4x^3 + 4x^2 - 6x + 8} \\
\underline{x^4 - 3x^3 - x^2} \\
-x^3 + 5x^2 - 6x \\
\underline{-x^3 + 3x^2 + x} \\
2x^2 - 7x + 8 \\
\underline{2x^2 - 6x - 2} \\
-x + 10
\end{array}
$$

Thus $\dfrac{x^4 - 4x^3 + 4x^2 - 6x + 8}{x^2 - 3x - 1} = x^2 - x + 2 + \dfrac{-x + 10}{x^2 - 3x - 1}$

Example 8

Divide $x^4 - 3x + 1 \div x^2 + 2$.

Solution

First, notice the "missing" terms in the dividend and divisor. Because we may have to be aligning like terms, we place the missing terms in both the dividend and divisor with 0 coefficients.

check in order
x^3 then x^2 ....

$$
\begin{array}{r}
x^2 \qquad\qquad - 2 \\
x^2 + 0x + 2\overline{)x^4 + 0x^3 + 0x^2 - 3x + 1} \\
\underline{x^4 \qquad\quad + 2x^2} \\
- 2x^2 - 3x + 1 \\
\underline{- 2x^2 \qquad - 4} \\
- 3x + 5
\end{array}
$$

Hence

$$
\frac{x^4 - 3x + 1}{x^2 + 2} = x^2 - 2 + \frac{-3x + 5}{x^2 + 2}
$$

Section 5.6 Exercises

In Exercises 1 through 14, perform the indicated division.

1. $\dfrac{4x + 12}{4}$

2. $\dfrac{8x + 12}{4}$

3. $\dfrac{15x - 5}{5}$

4. $\dfrac{15x + 5}{5}$

5. $\dfrac{14y - 21}{7}$

6. $\dfrac{14y - 21}{-7}$

7. $\dfrac{24x - 32y}{4}$

8. $\dfrac{24x - 32y}{-4}$

9. $\dfrac{24x - 32y}{4x}$

10. $\dfrac{24x - 32y}{8y}$

11. $\dfrac{8x^3 + 12x^2 + 4x}{4x}$

12. $\dfrac{-9y^5 + 15y^3 - 21y^2}{-3y^2}$

13. $\dfrac{16x^2y^2 - 8xy^2 + 32x^2y}{8xy}$

14. $\dfrac{16x^2y^2 - 8xy^2 + 32x^2y}{8x^2y^2}$

In Exercises 15 through 30, perform the indicated division. Write your result in the form

$$
\text{quotient} + \frac{\text{remainder}}{\text{divisor}}
$$

15. $(x^2 + 7x + 10) \div (x + 2)$

16. $(x^2 + 7x + 11) \div (x + 2)$

17. $(x^2 + 5x + 6) \div (x + 3)$

18. $(x^2 - 8x - 21) \div (x + 2)$

19. $(x^2 - 8x - 20) \div (x + 2)$

20. $(6x^2 + 13x - 9) \div (3x - 1)$

21. $(3x^3 + 7x^2 - 5) \div (x^2 - 2)$

22. $(z^3 + 7z^2 + 12z) \div (z^2 + 4)$

23. $(z^3 + 7z^2 + 12z) \div (z^2 + 4z)$

24. $(6x^4 - 9x^3 - 8x^2 + 15x - 3) \div (3x^2 - 6x + 2)$

25. $(x^4 - 4x^3 + 4x^2 - 5x - 2) \div (x^2 - 3x - 1)$

26. $(x^4 - 4x^3 + 4x^2 - 5x - 2) \div (x^2 - x + 2)$

27. $(2y^4 + y^3 - 8y^2 - 5y - 2) \div (y^2 - y - 3)$

28. $(8y^3 - 27) \div (2y - 3)$

29. $(x^4 - 1) \div (x - 1)$

30. $(x^4 - 1) \div (x^2 - 1)$

31. (a) Show that $-\dfrac{-12}{-5}$ is equivalent to $-\dfrac{12}{5}$.

    (b) Show that $-\dfrac{-19}{8}$ is equivalent to $\dfrac{19}{8}$.

    (c) Show that $-\dfrac{37}{-6}$ is equivalent to $\dfrac{37}{6}$.

32. Divide $\dfrac{57.129x^3 - 41.98x^2 + 7.158x}{12.005x}$.

33. Divide $\dfrac{4.058y^3 - 16.074y^2 - 91.870y}{0.003y}$.

34. Divide $\dfrac{12.098x^4 - 5.092x^3 - 18.991x^2}{71.051x^2}$.

Supplementary Exercises

In Exercises 1 through 12, perform the indicated division.

1. $\dfrac{8y - 12x}{4}$

2. $\dfrac{3a + 6b - 9c}{3}$

3. $\dfrac{-50x + 30y - 60z}{-10}$

4. $\dfrac{5x^2 - 12x^3}{x^2}$

5. $\dfrac{24x^3 - 12x^2 + 3x}{3x}$

6. $\dfrac{15x^2y - 30xy^2 + 6xy}{3xy}$

7. $\dfrac{-14x^2y^3 - 21xy^2 + 56x^2y^2}{-7xy^2}$

8. $\dfrac{9a^2b - 18ab^2 + 12a^2b^3}{a^3b^2}$

9. $\dfrac{3x^2y^2z - 6xy^3z^2 + 9xyz^2}{3x^2y^2z}$

10. $\dfrac{5a^2b - 20}{5ab}$

11. $\dfrac{-120x^4 + 60x^3 - 30x^2 + 90x}{30x}$

12. $\dfrac{9a^2b^2 - 12ab^2}{ba}$

In Exercises 13 through 25, perform the indicated division. Write your results in the form

$$\text{quotient} + \frac{\text{remainder}}{\text{divisor}}$$

13. $(x^2 - 2x + 3) \div (x + 1)$

14. $(x^2 + 4x - 1) \div (x - 2)$

15. $(y^2 - 6y + 4) \div (y + 3)$

16. $(2x^2 - 3x - 2) \div (2x + 1)$

17. $(5x^3 + 13x^2 - 5x + 3) \div (x + 3)$

18. $(2x^2 - x) \div (x + 1)$

19. $(x^4 - 2x^3 + x - 1) \div (x - 1)$

20. $(x^2 - 2x + 3) \div (x^2 + x - 1)$

21. $(4x^3 - 3x^2 + 2x - 5) \div (x^3 - 5)$

22. $(x^3 - 2x^2 + 3x - 1) \div (x^2 + x + 1)$

23. $(2x - x^2 + 3) \div (x - 1)$

24. $(5 + 3x + x^2) \div (x + 1)$

25. $(x^2 + 4x - 5) \div (5 + x)$

5.7   Chapter review

Summary

We began this chapter with the five rules of exponents to help simplify expressions. The rules are:

Exponent Rule 1: $x^a \cdot x^b = x^{a+b}$

Exponent Rule 2: $(x^a)^b = x^{ab}$

Exponent Rule 3: $(x^a y^b)^c = x^{ac} y^{bc}$

Exponent Rule 4: $\dfrac{x^a}{x^b} = x^{a-b}$ $(x \neq 0)$

Exponent Rule 5: $\left(\dfrac{x^a y^b}{z^c}\right)^d = \dfrac{x^{ad} y^{bd}}{z^{cd}}$ $(z \neq 0)$

Negative integers and zero were introduced in Section 5.2.

$$x^{-a} = \dfrac{1}{x^a}   (x \neq 0)$$

$$x^0 = 1   (x \neq 0)$$

One application of exponents, scientific notation, was studied in Section 5.3. Basically, a positive number is in scientific notation if it is the product of a number between 1 and 10 and a power of 10. For instance, 576.3 is $5.763 \cdot 10^2$ in scientific notation.

A polynomial in one variable is an algebraic expression composed of usually many terms of the form ax^n, where a is any real number and n is a positive integer.

Addition, subtraction, multiplication, division, and simplification of polynomials were examined in Sections 5.4, 5.5, and 5.6.

Vocabulary Quiz

Match the expression in Column I with the phrase in Column II that best describes it.

Column I

1. By Exponent Rule 1, $x^3 \cdot x^5 =$ X^8
2. By Exponent Rule 2, $(x^5)^3 =$ X^5
3. By Exponent Rule 3, $(x^3 y^3)^5 =$ $X^8 \, Y^8$
4. By Exponent Rule 4, $\dfrac{x^9 y^{11}}{xy^3} =$ $X^6 Y^8$

5. $\dfrac{1}{x^2} =$ X^{-2}

6. The form of each term of a polynomial in one variable is

7. A polynomial with one term is called a(n)

8. In $\dfrac{x^2 - 7x + 12}{x - 3} = x - 4$, $x^2 - 7x + 12$ is called the

9. In $\dfrac{x^2 - 7x + 12}{x - 3} = x - 4$, $x - 3$ is called the

10. In $\dfrac{x^2 - 7x + 12}{x - 3} = x - 4$, $x - 4$ is called the

Column II

a. x^8.

b. $x^8 y^8$.

c. monomial.

d. dividend.

e. quotient.

f. x^{-2}.

g. divisor.

h. x^{15}

i. ax^n.

j. $x^{15} y^{15}$

Chapter 5 Review Exercises

In Exercises 1 through 6, use the rules of exponents to simplify each expression.

1. $8x^2 \cdot x =$ $8 x^3$

2. $3y^2 \cdot y^4 \cdot y$ $3y^7$

3. $(2x^2)^3 =$ $8 X^6$

4. $(10ab^2 c^3)^2 =$ $100 \, a^2 b^4 c$

5. $\left(\dfrac{27z^4}{9z^2}\right)^3 = (3z^2)^3$
   $= 27 z^6$

6. $\left(\dfrac{2x^2 y}{z^4}\right)^3$

$\dfrac{8x^6 y^3}{z^{12}}$

In Exercises 7 through 12, evaluate each expression.

7. $(-5)^{-2}$

8. $(-5)^{-3}$

9. $\left(\dfrac{3}{8}\right)^{-2}$

10. $(3 + 5)^0$

11. $(4^{-1})(2^{-3})(8)^0$

12. 0^{-2}

In Exercises 13 through 18, rewrite each expression using only positive exponents.

13. z^{-4}

14. a^6b^{-3}

15. $\dfrac{x^{-2}y^3}{z^4}$

16. $\dfrac{x^{-2}y^{-3}}{z^{-4}}$

17. $\dfrac{12a^2b^{-3}c^4}{15ab^{-1}c^{-5}}$

18. $(3x^{-3}y^{-2})(5x^{-1}y^7)$

In Exercises 19 through 24, write each number in scientific notation.

19. 89.6

20. 578

21. 0.0571

22. 0.000509

23. 208,000,000

24. 550,000,000,000

In Exercises 25 and 26, perform each calculation by first writing the numbers in scientific notation. Write answers in scientific notation.

25. $25,700 \cdot 320,000$

26. $\dfrac{0.00058}{0.0000029}$

In Exercises 27 through 30, determine the degree of the polynomial.

27. $5x^2 - 7x - 1$

28. $5x^2 - 7x^3 - 1$

29. $3y - 7$

30. $16x^4 - 5x^3 + x - 109$

In Exercises 31 through 50, perform the indicated operations and simplify, if possible.

31. $(3x^2 - 7x + 10) + (2x^2 - 2x + 7)$

32. $(3x^2 + 9x - 10) + (8x + 1)$

33. $(14y^3 - y + 1) + (3y^2 - 4 + y)$

34. $(x^3 + x + 17) + (x^2 - 5x - 1) + (3x^3 - 19x + 21)$

35. $(x^3 - 3x^2 - 7x + 11) - (x^3 - x^2 + x + 6)$

36. $(y^4 - y^3 + 6y - 11) - (10y^3 - 8y - 5)$

37. Subtract $t - 8$ from $3t^3 - 7t^2 + 5t - 1$.

38. $(2x - 6) - (3x^3 - 4x^2 + 9x - 8)$

39. $(3x^2 - 5x + 1) + (2x^2 - 6x + 10) - (x - 1)$

40. $2x^2(3x - 5)$

41. $6y(y^2 - 7y + 3)$

42. $4z - z(2z - 1) + 10$

43. $(3x + 7)(2x - 5)$

44. $(4t - 1)(2t - 1)$

45. $(3x + 2y)(3x - 2y)$

46. $(2x - 5)(3x^2 - 7x + 1)$

47. $\dfrac{5x^2 - 10x}{5x}$

48. $\dfrac{24z - 32z^2}{8z^3}$

49. $(2x^2 - 16x - 41) \div (x + 2)$

50. $(3t^3 - 5t^2 + 11t + 1) \div (t^2 - 1)$

Cumulative review exercises

1. Evaluate $\sqrt{9} - \sqrt[3]{-64}$. *(Section 1.7)*

2. Evaluate $\dfrac{7(-8) + 12 \cdot 3}{3 - [4(-3) + 10]}$. *(Section 1.8)*

3. Diana earns $1000 more than twice Pedro's salary. If P represents Pedro's salary, what algebraic expression represents Diana's salary? *(Section 2.1)*

4. Simplify $-7(2x - 3) - 4(x + 9)$. *(Section 2.2)*

5. Simplify $(-2y)(7y^2)$. *(Section 2.3)*

6. Find the circumference of a circle of radius 30 cm. Use $\pi \approx 3.14$. *(Section 2.6)*

7. Solve for x: $-4x + 5 = 21 - 5[16 - 2(x + 5)]$. *(Section 3.4)*

8. The sum of two consecutive even integers is 206. Find the integers. *(Section 3.5)*

9. Solve and graph the solution on a real number line: *(Section 3.7)*

$$2 - 5x \le 12$$

10. Andrea has $4.50 in quarters and dimes. If there are 30 coins in all, how many of each type does she have? *(Section 4.1)*

Chapter 5 test

Take this test to see how well you have mastered exponents and polynomials. Check your answers using those found at the end of the book.

In Problems 1 through 3, use the exponent rules to simplify each expression, if possible.   *(Section 5.1)*

1. $(3t^4)(2t^3)$

2. $(4xy^2)^3$

1. _____

2. _____

3. $\left(\dfrac{30s^4}{5s^2}\right)^2$

3. _____

In Problems 4 through 6, simplify each expression by rewriting it using only positive exponents.   *(Section 5.2)*

4. $x^3 y^{-7} z^0$

5. $\dfrac{20a^2 b^{-5}}{32ab^{-2}}$

4. _____

5. _____

6. $\dfrac{16xyz^{-2}}{12xy^2 z^{-2}}$

6. _____

7. Rewrite each of the following using scientific notation.   *(Section 5.3)*
   (a) 809,000   (b) 0.00305

7. (a) _____
   (b) _____

8. What is the degree of the polynomial $17x^6 - 5x^2 - 3x + 10$?   *(Section 5.4)*

8. _____

In Problems 9 through 20, perform the indicated operation and simplify, if possible.

9. $(3x^3 - 5x + 7) + (2x^3 + 8x + 11)$ *(Section 5.4)*

9. _____

10. $(5y^3 - 7y + 1) + (y^2 - 8y - 12) - (3y^2 + 2y - 11)$ *(Section 5.4)*   10. _____

11. $(8t^3 - 5t^2 + 19) - (7t^3 + 3t^2 - 5t - 9)$ *(Section 5.4)*   11. _____

12. $2x(3x - 5)$ *(Section 5.5)*   13. $-3v^3(-5v^2 + 11v - 6)$ *(Section 5.5)*   12. _____

13. _____

14. $(7x - 2)(3x - 8)$ *(Section 5.5)*   15. $(5y - 3)(2y + 7)$ *(Section 5.5)*   14. _____

15. _____

16. $(2x + 3y)^2$ *(Section 5.5)*   17. $\dfrac{48x^2 - 40x}{8x}$ *(Section 5.6)*   16. _____

17. _____

18. $\dfrac{-8x^2y^3 + 16x^2y^2 - 24xy}{-8x^2y^2}$ *(Section 5.6)*.   18. _____

19. $(2x^2 + 9x + 13) \div (2x + 3)$ *(Section 5.6)*   19. _____

20. $(z^3 + 5z^2 + 19z - 4) \div (z^2 - z + 8)$ *(Section 5.6)*   20. _____

chapter

SIX

Factoring

6.1 Prime factorization and the GCF of integers

Most of this chapter will deal with writing *algebraic expressions* as a product of factors, an operation known as **factoring.** However, we will first begin with a discussion of factoring *positive integers.*

By **prime number** we mean any natural number greater than 1 that is only divisible by 1 and itself. For instance, since 1 and 19 are the only numbers that divide exactly into 19, we call 19 a prime number. The number 6 is not a prime number because other numbers, namely, 2 and 3, divide into it. By convention, 1 is not a prime number. The first 20 prime numbers are

$$2, 3, 5, 7, 11, 13, 17, 19, 23, 29, 31, 37, 41, 43, 47, 53, 59, 61, 67, \text{ and } 71$$

Since $6 \cdot 3 = 18$, we say that 18 is the product of the factors of 6 and 3, or that 6 and 3 are factors of 18. To factor 18 *completely* means to express it as a product of prime factors. To do this, we write 18 as $6 \cdot 3$ and then rewrite 6 as $2 \cdot 3$.

$$18 = 6 \cdot 3$$
$$18 = 2 \cdot 3 \cdot 3$$

Now, since 2 and 3 are prime numbers, we have factored 18 completely. Finally, we can write

$$18 = 2 \cdot 3 \cdot 3 \quad \text{or} \quad 18 = 2 \cdot 3^2$$

Example 1

Write the *prime factorization* of each number. (That is, factor each number completely.)

   (a) 28   (b) 36   (c)   230

Solution

(a) We can write 28 as $4 \cdot 7$. But, 4 can be "broken up" (that is, factored) as $2 \cdot 2$. So

$$28 = 4 \cdot 7$$
$$28 = 2 \cdot 2 \cdot 7 \quad \text{or} \quad 2^2 \cdot 7$$

(b) Writing 36 as $6 \cdot 6$, we see that each 6 can be factored as $2 \cdot 3$. So we have

$$36 = 6 \cdot 6$$
$$36 = 2 \cdot 3 \cdot 2 \cdot 3$$
$$36 = 2 \cdot 2 \cdot 3 \cdot 3 \quad \text{or} \quad 2^2 \cdot 3^2$$

(c) $230 = 10 \cdot 23$
     $230 = 2 \cdot 5 \cdot 23$

Since we will be concerned with complete factoring, a list of the prime factorizations of numbers between 1 and 100 is included in Table 6.1 (opposite page).

Greatest Common Factor

The **greatest common factor** (GCF) of two given numbers is the largest number that divides exactly into the given numbers. Thus 6 is the GCF of 30 and 42 because it is the

Table 6.1 Prime Factorizations

n	Prime Factorization	n	Prime Factorization
1	—	51	$3 \cdot 17$
2	prime	52	$2 \cdot 2 \cdot 13$
3	prime	53	prime
4	$2 \cdot 2$	54	$2 \cdot 3 \cdot 3 \cdot 3$
5	prime	55	$5 \cdot 11$
6	$2 \cdot 3$	56	$2 \cdot 2 \cdot 2 \cdot 7$
7	prime	57	$3 \cdot 19$
8	$2 \cdot 2 \cdot 2$	58	$2 \cdot 29$
9	$3 \cdot 3$	59	prime
10	$2 \cdot 5$	60	$2 \cdot 2 \cdot 3 \cdot 5$
11	prime	61	prime
12	$2 \cdot 2 \cdot 3$	62	$2 \cdot 31$
13	prime	63	$3 \cdot 3 \cdot 7$
14	$2 \cdot 7$	64	$2 \cdot 2 \cdot 2 \cdot 2 \cdot 2 \cdot 2$
15	$3 \cdot 5$	65	$5 \cdot 13$
16	$2 \cdot 2 \cdot 2 \cdot 2$	66	$2 \cdot 3 \cdot 11$
17	prime	67	prime
18	$2 \cdot 3 \cdot 3$	68	$2 \cdot 2 \cdot 17$
19	prime	69	$3 \cdot 23$
20	$2 \cdot 2 \cdot 5$	70	$2 \cdot 5 \cdot 7$
21	$3 \cdot 7$	71	prime
22	$2 \cdot 11$	72	$2 \cdot 2 \cdot 2 \cdot 3 \cdot 3$
23	prime	73	prime
24	$2 \cdot 2 \cdot 2 \cdot 3$	74	$2 \cdot 37$
25	$5 \cdot 5$	75	$3 \cdot 5 \cdot 5$
26	$2 \cdot 13$	76	$2 \cdot 2 \cdot 19$
27	$3 \cdot 3 \cdot 3$	77	$7 \cdot 11$
28	$2 \cdot 2 \cdot 7$	78	$2 \cdot 3 \cdot 13$
29	prime	79	prime
30	$2 \cdot 3 \cdot 5$	80	$2 \cdot 2 \cdot 2 \cdot 2 \cdot 5$
31	prime	81	$3 \cdot 3 \cdot 3 \cdot 3$
32	$2 \cdot 2 \cdot 2 \cdot 2 \cdot 2$	82	$2 \cdot 41$
33	$3 \cdot 11$	83	prime
34	$2 \cdot 17$	84	$2 \cdot 2 \cdot 3 \cdot 7$
35	$5 \cdot 7$	85	$5 \cdot 17$
36	$2 \cdot 2 \cdot 3 \cdot 3$	86	$2 \cdot 43$
37	prime	87	$3 \cdot 29$
38	$2 \cdot 19$	88	$2 \cdot 2 \cdot 2 \cdot 11$
39	$3 \cdot 13$	89	prime
40	$2 \cdot 2 \cdot 2 \cdot 5$	90	$2 \cdot 3 \cdot 3 \cdot 5$
41	prime	91	$7 \cdot 13$
42	$2 \cdot 3 \cdot 7$	92	$2 \cdot 2 \cdot 23$
43	prime	93	$3 \cdot 31$
44	$2 \cdot 2 \cdot 11$	94	$2 \cdot 47$
45	$3 \cdot 3 \cdot 5$	95	$5 \cdot 19$
46	$2 \cdot 23$	96	$2 \cdot 2 \cdot 2 \cdot 2 \cdot 2 \cdot 3$
47	prime	97	prime
48	$2 \cdot 2 \cdot 2 \cdot 2 \cdot 3$	98	$2 \cdot 7 \cdot 7$
49	$7 \cdot 7$	99	$3 \cdot 3 \cdot 11$
50	$2 \cdot 5 \cdot 5$	100	$2 \cdot 2 \cdot 5 \cdot 5$

largest number that divides exactly into both 30 and 42. The rules for finding the GCF of two numbers are as follows:

1. Write the prime factorization of the two given numbers.
2. The GCF's prime factorization will consist of the factors common to both given numbers.

We illustrate this in the next three examples.

Example 2

Find the GCF of 30 and 42.

Solution

Step 1

Table 6.1 gives the prime factorizations of 30 and 42.

$$30 = 2 \cdot 3 \cdot 5$$
$$42 = 2 \cdot 3 \cdot 7$$

Common factors

Step 2

The common factors are 2 and 3. So the GCF $= 2 \cdot 3 = 6$.

Example 3

Find the GCF of 36 and 60.

Solution

Step 1

$$36 = 2 \cdot 2 \cdot 3 \cdot 3$$
$$60 = 2 \cdot 2 \cdot 3 \cdot 5$$

Common factors

Step 2

The GCF $= 2 \cdot 2 \cdot 3 = 12$. That is, the largest number that divides *exactly* into both 36 and 60 is 12.

Example 4

Find the GCF of 90, 225, and 315.

Solution

Step 1

$$90 = 2 \cdot 3 \cdot 3 \cdot 5$$
$$225 = \quad 3 \cdot 3 \cdot 5 \cdot 5$$
$$315 = \quad 3 \cdot 3 \cdot 5 \cdot 7$$

Common factors

Step 2

The GCF $= 3 \cdot 3 \cdot 5 = 45$. Thus, the largest number that divides into 90, 225, and 315 exactly is 45.

Section 6.1 Exercises

In Exercises 1 through 12, find the prime factorization of each number. Use Table 6.1.

1. 12	2. 16	3. 36
4. 32	5. 360	6. 320
7. 98	8. 196	9. 980
10. 10,000	11. 150	12. 300

In Exercises 13 through 24, find the GCF for each set of numbers.

13. 10, 15   14. 24, 40

15. 42, 56   16. 72, 180

17. 56, 140   18. 42, 70

19. 72, 88   20. 40, 72, 88

21. 50, 72, 88   22. 45, 75, 105

23. 100, 200, 250   24. 56, 140, 532

25. If a set of numbers has no common factor, we say the GCF is 1. We may also say that the numbers are **relatively prime.** Show that 1 is the GCF of each of the following sets of numbers.

   (a) 15, 28   (b) 12, 24, 35   (c) 6, 35, 121

In Exercises 26 through 30, find the GCF for each set of numbers.

26. 12, 15, 9   27. 30, 45, 75

28. 36, 90, 126   29. 60, 108, 84

30. 80, 112, 64

Supplementary Exercises

In Exercises 1 through 8, find the prime factorization of each number. Use Table 6.1.

1. 39   2. 17   3. 125   4. 65

5. 210   6. 325   7. 100   8. 31

In Exercises 9 through 15, find the GCF for each set of numbers.

9. 8, 12   10. 9, 12, 18   11. 20, 15, 30   12. 28, 70

13. 75, 125, 50   14. 314, 785   15. 85, 119, 34

6.2   An introduction to factoring: the common factor

The concept of finding the GCF for a set of integers can be easily extended to finding the GCF for algebraic expressions. The remainder of this chapter will deal with rewriting algebraic expressions that are a sum of several terms as a product of factors. The process is called **factoring.**

We have already seen this process in reverse! When $2(x + 5)$ was rewritten as $2x + 10$, we went *from* factored form *to* a form without grouping symbols by applying the distributive property.

To see how $2x + 10$ is factored, we write

$$2x + 10 = 2 \cdot x + 2 \cdot 5$$
$$= 2(x + 5) \qquad \text{The distributive property is used here.}$$

Example 1

Factor $4x^2 + 8x$.

Solution

First we look for a common factor: 2 is common to both terms, 4 is common to both terms, and x is common to both terms. When looking for the common factor, it should be understood that we are searching for the *greatest common factor* of each term. Observe:

$$\underbrace{4x^2}_{2 \cdot 2 \cdot x \cdot x} + \underbrace{8x}_{2 \cdot 2 \cdot 2 \cdot x}$$

$4x$, that is $(2 \cdot 2 \cdot x)$ is common to both terms. We factor it out:

$$4x^2 \quad + \quad 8x$$
$$= 2 \cdot 2 \cdot x \cdot x + 2 \cdot 2 \cdot 2 \cdot x$$
$$= 2 \cdot 2 \cdot x(x + 2)$$
$$= 4x(x + 2)$$

So $4x^2 + 8x$ is factored into $4x(x + 2)$.

Check: We will multiply our answer to see if we produce the original expression.

$$4x(x + 2) = 4x \cdot x + 4x \cdot 2$$
$$= 4x^2 + 8x \ \checkmark$$

Example 2

Factor $x^3 + 3x^2 + 2x$ by finding the greatest common factor.

Solution

$$x^3 \quad + \quad 3x^2 \quad + \quad 2x$$
$$= x \cdot x \cdot x + 3 \cdot x \cdot x + 2 \cdot x$$

The greatest common factor is x.

$$x^3 \quad + \quad 3x^2 \quad + \quad 2x$$
$$= x \cdot x \cdot x + 3 \cdot x \cdot x + 2 \cdot x$$
$$= x(x \cdot x + 3 \cdot x + 2)$$
$$= x(x^2 + 3x + 2)$$

Check: $x(x^2 + 3x + 2) = x \cdot x^2 + x \cdot 3x + x \cdot 2$
$$= x^3 + 3x^2 + 2x \ \checkmark$$

Example 3

Factor $3x^2 + 3x$ by finding the greatest common factor.

Solution

$$3x^2 \quad + \quad 3x$$
$$3 \cdot x \cdot x + 3 \cdot x$$

The greatest common factor is $3x$. So,

$$3x^2 \quad + \quad 3x$$
$$= 3 \cdot x \cdot x + 3 \cdot x$$
$$= 3x(x + \underline{\quad})$$

This times $3x$ must be $3x$.

We have: $$3x^2 + 3x = 3x(x + 1)$$

Check: $$3x(x + 1) = 3x \cdot x + 3x \cdot 1$$
$$= 3x^2 + 3x \ \checkmark$$

Example 4

Factor $a^2bc^3 + 3ab^2c^2$ by finding the greatest common factor.

Solution

$$a^2bc^3 \qquad + \qquad 3ab^2c^2$$
$$= a \cdot a \cdot b \cdot c \cdot c \cdot c + 3 \cdot a \cdot b \cdot b \cdot c \cdot c$$

The greatest common factor is abc^2. We have

$$a^2bc^3 \qquad + \qquad 3ab^2c^2$$
$$= a \cdot a \cdot b \cdot c \cdot c \cdot c + 3 \cdot a \cdot b \cdot b \cdot c \cdot c$$
$$= a \cdot b \cdot c \cdot c(a \cdot c + 3 \cdot b)$$
$$= abc^2(ac + 3b)$$

Check: $$abc^2(ac + 3b) = abc^2 \cdot ac + abc^2 \cdot 3b$$
$$= a^2bc^3 + 3ab^2c^2 \ \checkmark$$

Example 5

Factor $8x + 15$ by finding the greatest common factor.

Solution

$$8x \quad + \quad 15$$
$$= \underbrace{2 \cdot 2 \cdot 2 \cdot x} + \underbrace{3 \cdot 5}$$

There is no factor common to both terms. So $8x + 15$ cannot be factored by finding the greatest common factor.

Example 6

Factor $12y^3 - 54y^2$ by finding the greatest common factor.

Solution

$$12y^3 \qquad - \qquad 54y^2$$
$$2 \cdot 2 \cdot 3 \cdot y \cdot y \cdot y - 2 \cdot 3 \cdot 3 \cdot 3 \cdot y \cdot y$$

The greatest common factor is $6y^2$.

$$12y^3 \qquad - \qquad 54y^2$$
$$= 2 \cdot 2 \cdot 3 \cdot y \cdot y \cdot y - 2 \cdot 3 \cdot 3 \cdot 3 \cdot y \cdot y$$
$$= 2 \cdot 3 \cdot y \cdot y(2 \cdot y - 3 \cdot 3)$$
$$= 6y^2(2y - 9)$$

Check: $6y^2(2y - 9) = 6y^2 \cdot 2y - 6y^2 \cdot 9$
                                              $= 12y^3 - 54y^2 \ \checkmark$

An example of factoring in an applied setting follows.

Example 7

Under certain conditions the number of animals (N) present in a population is given by the formula $N = KMx^2 - 2Kx^3$.

(a) Factor the righthand side of this formula by finding the greatest common factor.

(b) Rewrite the formula using your answer from part (a).

Solution

(a) $KMx^2 \qquad - \qquad 2Kx^3$

$= K \cdot M \cdot x \cdot x - 2 \cdot K \cdot x \cdot x \cdot x$ $\begin{cases} \text{We factor the righthand side of the equation in} \\ \text{order to find the greatest common factor there.} \end{cases}$

$= K \cdot x \cdot x(M - 2x)$

$= Kx^2(M - 2x)$

Check: $Kx^2(M - 2x) = Kx^2 \cdot M - Kx^2 \cdot 2x$
                                        $= KMx^2 - 2Kx^3 \ \checkmark$

(b) The formula can be rewritten by replacing $KMx^2 - 2Kx^3$ by its equal, $Kx^2(M - 2x)$. So we have $N = Kx^2(M - 2x)$.

always look for common factors

Example 8

Factor $-12x^2y^3 + 24xy^3 - 18x^3y$.

Solution

We must be careful with signs! In this text we will follow the convention that the leading coefficient of a factored polynomial is to be positive. So

$$-12x^2y^3 + 24xy^3 - 18x^3y$$
$$= -2 \cdot 2 \cdot 3 \cdot x \cdot x \cdot y \cdot y \cdot y + 2 \cdot 2 \cdot 2 \cdot 3 \cdot x \cdot y \cdot y \cdot y - 2 \cdot 3 \cdot 3 \cdot x \cdot x \cdot x \cdot y$$

The common factors are $2 \cdot 3 \cdot x \cdot y$ but, because we want the leading coefficient to be positive, we use $-6xy$ as the GCF. Then we have

$$-6xy(2xy^2 - 4y^2 + 3x^2)$$

We leave the check to the reader.

Section 6.2 Exercises

In Exercises 1 through 25, factor each expression by finding the greatest common factor, if possible.

1. $6x + 12$	2. $6x + 3$
3. $6x + 2$	4. $2x + 6$
5. $6x^2 + 2x$	6. $3y^2 + 6$
7. $7z - 14$	8. $8y^2 - 4y$
9. $x^4 + 5x^3$	10. $x^4 + x^3$

11. $6t^2 - 3t + 12$

12. $6t^2 - 3t + 2$

13. $-30x^2 + 10x - 15$

14. $-30x^2y^2 + 10xy^2 - 15y^2$

15. $-9x^4 - 18x^3 + 27x^2$

16. $4x^2y^2z^2 - 16xy^2z^3$

17. $-x^3 + x^2 - x$

18. $-30a^{12}b^2c^5 + 10a^3b^3c^3 - 15a$

19. $-30a^{12}b^2c^5 + 10a^3b^3c^3 - 15ab^2c^2$

20. $-30a^{12}b^2c^5 + 20a^{12}b^2c^6 - 40a^{12}b^2c^7$

21. $-5x^2y^3 + 15xy^2$

22. $18xy - 7pq$

23. $4a^2bc^2 + 12ab^2c - 4abc$

24. $12xy^2 + 8x^3y - 4x^3y^2$

25. $9t^2v^2 - 12tv + 15t^2v$

26. Under certain conditions the pressure of a gas (P_t) is given by the formula $P_t = P_0 + P_0bt$.

   (a) Factor the righthand side of the equation by finding the greatest common factor.

   (b) Rewrite the formula using your answer from part (a).

27. The accompanying diagram shows a square box with no top; it has length x, width x, and height y. The total area of the sides and bottom is given by the expression

$$x^2 + 4xy$$

Factor this expression.

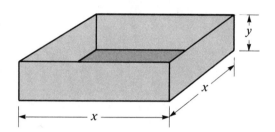

28. If a ball is thrown vertically (from the ground) with a certain velocity, the formula for the height (h) of the ball after it has traveled t seconds is

$$h = -16t^2 + 40t$$

   (a) Factor the righthand side of this equation.

   (b) Use your answer from part (a) to rewrite the formula.

29. A tennis racquet manufacturer has determined that she must spend $10,000 per month on such things as rent, insurance, salaries, and so on, whether or not she produces any racquets. For each racquet that she does make she must spend $20 on material and labor. Letting x be the number of racquets produced,

   (a) Construct a formula for the cost, C, of producing x tennis racquets.

   (b) Factor the righthand side of your answer to part (a).

   (c) Use your answer in part (b) to rewrite the cost formula.

In Example 1 of this section we saw that $4x(x + 2)$ was an alternate way of writing $4x^2 + 8x$. That is, they each represent the same quantity. So that we do not lose sight of this fact, evaluate each expression for certain values of x given in the following table.

	Value of x	Value of $4x^2 + 8x$	Value of $4x(x + 2)$
30.	3		
31.	4.5		
32.	8.0126		

Supplementary Exercises

In Exercises 1 through 15, factor each expression by finding the greatest common factor, if possible.

1. $5x + 15y$

2. $-9x + 12y$

3. $5x^2 - 7x$

4. $18x^2 + 15x$

5. $4x^2y - 8xy$

6. $-9x^2y^3 + 12xy^2 + 15xy$

7. $5x^3y - 20x^2y + 5xy$

8. $17pq - 12st$

9. $ab^2c - 4abc^2 + 8a^2bc$

10. $12a^2b^3 - 4a^4b^2 + 8a^2b^5$

11. $30m^4n^2 - 12mn^2 + 18m^2n^2$

12. $100w^2t^3 + 200w^2t^2$

13. $5x^7y^6 + 15x^3y^5 - 20x^3y^4$

14. $8p^5q^4 - 12p^3q^2 + 16p^7q^3$

15. $-2x^9y^5 + 4x^7y^7 - 8x^7y^5$

6.3   Factoring trinomials: $x^2 + bx + c$

In Chapter 5 we saw that the product of two **binomials** (two-term polynomials) was often a **trinomial** (three-term polynomial). For example, $(x + 2)(x + 3)$ was multiplied as follows using foil.

$$(x + 2)(x + 3)$$

$$\overset{\mathbf{f} \qquad \mathbf{o} \qquad \mathbf{i} \qquad \mathbf{l}}{= x^2 + 3x + 2x + 2 \cdot 3}$$

$$= x^2 + (3 + 2)x + 2 \cdot 3$$

$$= x^2 + \underbrace{5x}_{} + \underbrace{6}_{}$$

The *sum* of 2 and 3   The *product* of 2 and 3
yields the coefficient   gives us the last term.
of the middle term.

In this section we will be interested in the reverse process; that is, given a trinomial of the form $x^2 + bx + c$, we will factor it, if possible, into the product of two binomials.

To multiply the two binomials $x + 4$ and $x - 6$, notice the pattern continues regarding the sum and product of 4 and -6.

$$(x + 4)(x - 6)$$

$$= x^2 - 6x + 4x - 24$$

$$= x^2 + (-6 + 4)x - 24$$

$$= x^2 \underbrace{- 2x}_{} \underbrace{- 24}_{}$$

The *sum* of 4   The *product* of 4 and -6 gives us
and -6 gives us

Keep this pattern in mind as you read the next four examples.

Example 1

Factor the trinomial $x^2 + 5x + 4$ into the product of two binomials.

Solution

When factored, $x^2 + 5x + 4$ will look like:

$$(x \underline{\qquad})(x \underline{\qquad})$$

These two numbers must be determined. We search for two numbers whose *sum* is 5 and whose *product* is 4 because the original trinomial is $x^2 + 5x + 4$. Below we list all the pairs of integers whose product is 4. We also compute the sum in each case.

First Number	Second Number	Product	Sum
2	2	4	4
1	4	4	5
-2	-2	4	-4
-1	-4	4	-5

The pair of numbers in color satisfies both conditions (sum is 5; product is 4). So we factor $x^2 + 5x + 4$ as

$$(x + 1)(x + 4)$$

It is important to check all factoring problems, and we verify that $(x + 1)(x + 4)$ is actually $x^2 + 5x + 4$ by multiplying $(x + 1)(x + 4)$.

Check:
$$(x + 1)(x + 4)$$
$$\begin{array}{cccc} \mathbf{f} & \mathbf{o} & \mathbf{i} & \mathbf{l} \end{array}$$
$$= x^2 + 4x + x + 4$$
$$= x^2 + 5x + 4 \ \checkmark$$

Example 2

Factor $x^2 + 3x - 4$.

Solution

$x^2 + 3x - 4$ is a trinomial. When factored, it will look like

$$(x \underline{\qquad})(x \underline{\qquad})$$

The numbers that fill in the blanks must have a sum of 3 and a product of -4. Below we list all the pairs of integers whose product is -4.

First Number	Second Number	Product	Sum
1	-4	-4	-3
2	-2	-4	0
-1	4	-4	3

The pair of numbers in color satisfies both conditions (sum is 3; product is -4). So we factor $x^2 + 3x - 4$ as

$$(x - 1)(x + 4)$$

Check:
$$(x - 1)(x + 4) = x^2 + 4x - x - 4$$
$$= x^2 + 3x - 4 \ \checkmark$$

Example 3

Factor $x^2 + 7x + 12$.

Solution

Since $x^2 + 7x + 12$ is a trinomial, we must search for two numbers whose sum is 7 and whose product is 12. The only numbers that work are 3 and 4 ($3 + 4 = 7$ and $3 \cdot 4 = 12$). So we factor $x^2 + 7x + 12$ as the product of binomials

$$(x + 3)(x + 4)$$

Check:
$$(x + 3)(x + 4) = x^2 + 4x + 3x + 12$$
$$= x^2 + 7x + 12 \ \sqrt{}$$

Example 4

Factor $x^2 - 8x + 15$.

Solution

We must find two numbers whose sum is -8 and whose product is 15. (*Note:* Since their product is positive and their sum is negative, both numbers must be negative.) The two numbers are -5 and -3 ($-5 + -3 = -8$ and $-5 \cdot -3 = 15$). So $x^2 - 8x + 15$ gets factored as

$$(x - 5)(x - 3)$$

Check:
$$(x - 5)(x - 3) = x^2 - 3x - 5x + 15$$
$$= x^2 - 8x + 15 \ \sqrt{}$$

We point out here that *not* all trinomials are factorable. For example, in $x^2 + 3x + 5$, there are *not* two integers whose sum is 3 and whose product is 5. The following list illustrates this fact.

First Number	Second Number	Product	Sum
1	5	5	6
-1	-5	5	-6

We say that $x^2 + 3x + 5$ is **unfactorable.**

Now that we have examined two methods of factoring (the common factor and certain trinomials), we are ready to examine situations where *both* methods may be employed. A rule to remember when factoring is: **Always search for a common factor first.** We apply this rule in the following two examples.

Example 5

Completely factor $x^3 + 5x^2 + 6x$.

Solution

We first notice the common factor x. (To be precise, x is the greatest common factor.) So $x^3 + 5x^2 + 6x$ is factored as $x(x^2 + 5x + 6)$. The directions, however, in this example were to *completely* factor. So far we have

$$x^3 + 5x^2 + 6x = x(x^2 + 5x + 6)$$

This cannot be factored any further.

This is a trinomial. Perhaps it can be factored. We need to find two integers whose sum is 5 and whose product is 6. The integers are 2 and 3.

We find that $x^2 + 5x + 6$ factors into $(x + 2)(x + 3)$. So we have

$$x^3 + 5x^2 + 6x = x(x^2 + 5x + 6)$$
$$= x(x + 2)(x + 3)$$

The original trinomial, $x^3 + 5x^2 + 6x$, is factored as a product of three factors.

Check: $x(x + 2)(x + 3) = (x^2 + 2x)(x + 3)$
$$= x^3 + 3x^2 + 2x^2 + 6x$$
$$= x^3 + 5x^2 + 6x \checkmark$$

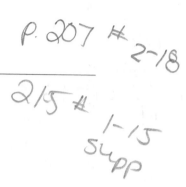

Example 6

Completely factor $x^4 - 8x^3 + 16x^2$.

Solution

In $x^4 - 8x^3 + 16x^2$, there is a common factor of x^2. So

$$x^4 - 8x^3 + 16x^2 = x^2(x^2 - 8x + 16)$$

This can be factored as $x \cdot x$, but when a factor is repeated, we usually use exponential notation.

This is a trinomial. Perhaps it can be factored. We need two integers whose sum is -8 and whose product is 16. The integers are -4 and -4.

Now, $x^2 - 8x + 16$ factors into $(x - 4)(x - 4)$ or $(x - 4)^2$. So we have

$$x^4 - 8x^3 + 16x^2 = x^2(x^2 - 8x + 16)$$
$$= x^2(x - 4)^2$$

Check: $x^2(x - 4)^2 = x^2(x - 4)(x - 4)$
$$= (x^3 - 4x^2)(x - 4)$$
$$= x^4 - 4x^3 - 4x^3 + 16x^2$$
$$= x^4 - 8x^3 + 16x^2 \checkmark$$

Example 7

Completely factor $-2x^4 + 16x^3 + 40x^2$.

Solution

First we factor out the common factor of $-2x^2$.

$$-2x^4 + 16x^3 + 40x^2 = -2x^2(x^2 - 8x - 20)$$
$$= -2x^2(x - 10)(x + 2)$$

The trinomial $x^2 - 8x - 20$ is factored as $(x - 10)(x + 2)$.

The check is left to the reader.

Section 6.3 Exercises

In Exercises 1 through 39, completely factor each expression, if possible.

1. $x^2 + 7x + 10$
2. $x^2 + 8x + 12$
3. $x^2 - 8x + 12$
4. $x^2 - 8x - 12$
5. $x^2 - 8x + 16$
6. $x^2 - 12x + 36$

7. $y^2 - 12y + 20$

8. $x^2 + 9x + 20$

9. $x^2 + 7x + 6$

10. $x^2 + 6x + 7$

11. $x^2 + 3x - 10$

12. $x^2 - 3x - 10$

13. $x^2 - 3x + 10$

14. $y^2 - 6y + 9$

15. $x^2 + 2x - 24$

16. $a^2 - 12a + 27$

17. $x^2 + 10x - 24$

18. $a^2 + a - 12$

19. $t^2 - 7t - 18$

20. $t^2 - 7t + 18$

21. $x^3 + 7x^2 + 12x$

22. $5x^2 + 10x + 5$

23. $5x^2 - 10x + 5$

24. $x^2y + 12xy - 28y$

25. $3y^2 + 18y - 48$

26. $3y^2 + 18y + 48$

27. $a^2b^2c^2 - 9ab^2c^2 + 18b^2c^2$

28. $x^3 + 4x^2 + 21x$

29. $10a^2 - 10a - 120$

30. $-5x^4y + 35x^3y - 60x^2y$

31. $-8y^4 - 24y^3 + 320y^2$

32. $-7x^5y^4 + 56x^4y^4 + 140x^3y^4$

33. $5x^3 - 10x^2 + 5x$

34. $6t^3 + 18t^2 + 12t$

35. $4m^3 - 28m^2 + 6m$

36. $5t^2 - 5$

37. $a^4 + 8a^3 + 16a^2$

38. $x^2 + 2ax + a^2$ (*Hint:* Look for two "numbers" whose sum is $2a$ and whose product is a^2.)

39. $x^2 + ax + bx + ab$ [*Hint:* Think of $x^2 + ax + bx + ab$ as $x^2 + (a + b)x + ab$.]

40. When a ball is thrown upward from a 480-ft building with an initial velocity of 112 ft/sec, its height (s) above the ground at any time (t) is given by the formula

$$s = -16t^2 + 112t + 480$$

(a) Completely factor the righthand side. (*Hint:* There is a common factor of -16.)

(b) Find s when $t = 0$ using your answer to part (a).

(c) Find s when $t = 10$ using your answer to part (a).

(d) Use part (c) to find out how long it takes the ball to hit the ground.

Supplementary Exercises

In Exercises 1 through 25, completely factor each expression, if possible.

1. $a^2 + 6a + 8$

2. $a^2 - 2a - 8$

3. $x^2 - 7x - 8$

4. $p^2 - 11p - 12$

5. $m^2 - 7m + 12$

6. $y^2 - 12y + 36$

7. $a^2 + 3a + 4$

8. $y^2 - 16y - 17$

9. $w^2 - 10w + 16$

10. $x^2 - 9x + 18$

11. $a^2 - 8a + 15$

12. $a^2 + 7a - 18$

13. $y^2 - 10y + 24$

14. $y^2 - 5y - 24$

15. $y^2 + 4y - 24$

16. $m^2 + 6m + 7$

17. $t^2 - 19t + 90$

18. $m^2 - 4m - 21$

19. $q^2 - 2q + 1$

20. $c^2 - 4c - 5$

21. $5a^2 + 15a + 10$

22. $-6x^2 - 30x - 36$

23. $4p^2 + 36p - 40$

24. $5x^2t + 55xt + 50t$

25. $7m^2x^2 - 84m^2x + 224m^2$

6.4 Factoring the difference of two perfect squares

In Section 5.5 we saw that the product of two binomials will sometimes result in another binomial. Let's look at another example of such a special situation.

$$(a + b)(a - b) = a^2 - ab + ba - b^2$$
$$= a^2 - ab + ab - b^2$$
$$= a^2 - b^2$$

If we did this process in reverse, or *factored* $a^2 - b^2$, we would write it as $(a + b)(a - b)$. Furthermore, any expression *in the form of* $a^2 - b^2$ can be factored similarly. Expressions like $a^2 - b^2$ are called the **difference of two perfect squares** because a^2 is a *perfect square* $(a \cdot a)$, b^2 is a *perfect square* $(b \cdot b)$, and $a^2 - b^2$ is the *difference* (subtraction) of two perfect squares. Following is a list of five binomials that fall into the category of the difference of two perfect squares.

1. $x^2 - 9$ x^2 is $x \cdot x$; 9 is $3 \cdot 3$.
2. $y^2 - 100$ y^2 is $y \cdot y$; 100 is $10 \cdot 10$.
3. $x^6 - 49$ x^6 is $x^3 \cdot x^3$; 49 is $7 \cdot 7$.
4. $100x^2 - 64$ $100x^2$ is $10x \cdot 10x$; 64 is $8 \cdot 8$.
5. $100x^2 - 49y^2$ $100x^2$ is $10x \cdot 10x$; $49y^2$ is $7y \cdot 7y$.

However, expressions like $x^2 + 9$, $y^2 - 99$, and $x^5 - 49$ are *not* the difference of two perfect squares. $x^2 + 9$ expresses the *sum*, not the difference between two perfect squares, and neither 99 nor x^5 are perfect squares.

Now that we can recognize expressions as the difference of two perfect squares, we see how to factor them in the four examples that follow.

Example 1

Factor $x^2 - 9$.

Solution

There is no common factor in this expression. Recall that $a^2 - b^2$ is factored as $(a + b)(a - b)$. So $x^2 - 9$ (which can be thought of as $x^2 - 3^2$) is factored as

$$(x + 3)(x - 3)$$

Check:
$$(x + 3)(x - 3) = x^2 - 3x + 3x - 9$$
$$= x^2 + 0x - 9$$
$$= x^2 - 9 \ \checkmark$$

Example 2

Factor $4x^2 - 25$.

Solution

The given expression has no common factor. However, $4x^2$ is a perfect square; it is $(2x)(2x)$, or $(2x)^2$. And 25 is a perfect square; it is $5 \cdot 5$ or 5^2. So $4x^2 - 25$ is factored as

$$(2x + 5)(2x - 5)$$

Check: $(2x + 5)(2x - 5) = 4x^2 - 10x + 10x - 25$
$$= 4x^2 + 0x - 25$$
$$= 4x^2 - 25 \ \checkmark$$

Example 3

Factor $10x^2 - 1000$.

Solution

In $10x^2 - 1000$, 10 is a common factor. We have

$$10x^2 - 1,000 = 10(x^2 - 100)$$

Now $x^2 - 100$ is the difference of two perfect squares and in factored form is $(x + 10)(x - 10)$. So we have

$$10x^2 - 1000 = 10(x^2 - 100) = 10(x + 10)(x - 10)$$

Check: $10(x + 10)(x - 10) = (10x + 100)(x - 10)$
$$= 10x^2 - 100x + 100x - 1000$$
$$= 10x^2 - 1,000 \ \checkmark$$

Example 4

Factor $t^8 - 1$.

Solution

The given expression has no common factor. However, t^8 is a perfect square; it is $(t^4)^2$. And 1 is a perfect square; it is $(1)^2$. Hence, $t^8 - 1$ is the difference of two perfect squares. So in factored form it is

$$(t^4 + 1)(t^4 - 1)$$

If we were to check $(t^4 + 1)(t^4 - 1)$, we would multiply it to obtain $t^8 - 1$. However, $(t^4 + 1)(t^4 - 1)$ is **not** the final answer because we have not factored *completely*. In particular, notice that $t^4 - 1$ is *again* factorable as the difference of two perfect squares. We put all this together below.

$t^8 - 1 = (t^4 + 1)(t^4 - 1)$
$\qquad = (t^4 + 1)(t^2 + 1)(t^2 - 1)$ $t^4 - 1$ is factored as $(t^2 + 1)(t^2 - 1)$.
$\qquad = (t^4 + 1)(t^2 + 1)(t + 1)(t - 1)$ $t^2 - 1$ is factored as $(t + 1)(t - 1)$.

So $t^8 - 1$ is factored as

$$(t^4 + 1)(t^2 + 1)(t + 1)(t - 1)$$

We stop at this point because no factor can be broken down further into other factors. Note that $t^4 + 1$ and $t^2 + 1$ are both sums of squares, neither of which can be factored further.

Check: $\underbrace{(t^4 + 1)(t^2 + 1)}\qquad \underbrace{(t + 1)(t - 1)}$
$$= (t^6 + t^4 + t^2 + 1)(t^2 - t + t - 1)$$
$$= (t^6 + t^4 + t^2 + 1)(t^2 - 1)$$

We finish this check with a vertical display.

$$
\begin{array}{r}
t^6 + t^4 + t^2 + 1 \\
t^2 - 1 \\
\hline
-\ t^6 - t^4 - t^2 - 1 \\
t^8 + t^6 + t^4 + t^2 \\
\hline
t^8 \qquad\qquad\quad - 1 \ \checkmark
\end{array}
$$

Section 6.4 Exercises

In Exercises 1 through 32, completely factor each expression, if possible.

1. $x^2 - 16$
2. $y^2 - 25$
3. $z^2 - 49$
4. $t^2 - 81$
5. $x^2 - 100$
6. $y^2 - 64$
7. $t^2 - 121$
8. $x^2 - 400$
9. $4x^2 - 4$
10. $4y^2 - 25$
11. $25 - x^2$
12. $81x^2 - 100$
13. $x^2 - 4y^2$
14. $10x^2 - 40y^2$
15. $4x^2 - 49$
16. $9x^2 - 64$
17. $2y^2 - 50$
18. $2xy^2 - 50x$
19. $z^4 - 16$
20. $8x^4 - 800$
21. $x^2 - y^2$
22. $100 - t^2$
23. $3x^6 - 27$
24. $3x^8 - 27$
25. $4x^2 + 81$
26. $9x^2 + 48$
27. $x^4 - y^2$
28. $64a^2 - 16b^2$
29. $144a^2 - 1$
30. $144a^2 - 12a$
31. $x^2y^2 - z^2$
32. $x^2y^2 - x^2$

Supplementary Exercises

In Exercises 1 through 15, completely factor each expression, if possible.

1. $a^2 - 9$
2. $49 - y^2$ $(7+y)(7-y)$
3. $m^2 + 121$
4. $144 - p^2$
5. $4y^2 - 9x^2$
6. $16x^2 - 81y^2$
7. $9x^2 - 1$
8. $3x^2 - 108$
9. $px^2 - py^2$ $p(x^2-y^2)$
10. $z^2 + 1$
11. $x^4 - y^4$
12. $5x^2 + 125$
13. $5x^2 - 125$
14. $3x^2y^2 - 12$
15. $7a^2b^4 - 28$

$5(x^2-25)(x+5)(x-5)$

$=7(a^2b^4-4)$
$=7(ab^2+2)(ab^2-2)$

6.5   Factoring trinomials: $ax^2 + bx + c$

To factor a trinomial whose x^2 coefficient is not 1, we return to the distributive property and foil to see how, for example, $2x^2 + 5x + 3$ may have been obtained.

We start by trying to factor the term $2x^2$.

$$(\underline{})(\underline{})$$

The product of first terms must be $2x^2$. So we write

$$(2x\underline{})(x\underline{})$$

The product of last terms must be 3. This gives us four possibilities:

$$(2x + 1)(x + 3) \quad \text{or} \quad (2x + 3)(x + 1) \quad \text{or} \quad (2x - 1)(x - 3) \quad \text{or} \quad (2x - 3)(x - 1)$$

Which is correct? To answer that, we must check each to see what the outermost and innermost products add up to.

Possibility 1	*Possibility 2*
$(2x + 1)(x + 3)$	$(2x + 3)(x + 1)$
$= 2x^2 + 6x + 1x + 3$	$= 2x^2 + 2x + 3x + 3$
$= 2x^2 + 7x + 3$	$= 2x^2 + 5x + 3$

Possibility 3	*Possibility 4*
$(2x - 1)(x - 3)$	$(2x - 3)(x - 1)$
$= 2x^2 - 6x - x + 3$	$= 2x^2 - 2x - 3x + 3$
$= 2x^2 - 7x + 3$	$= 2x^2 - 5x + 3$

This is the trinomial we were trying to factor. So the factors of $2x^2 + 5x + 3$ are

$$(2x + 3)(x + 1)$$

Note that possibilities 3 and 4 can be eliminated rather easily since the sum of the outermost and innermost products in both cases must be negative, whereas the trinomial to be factored had a positive middle term.

The method we used is called the **trial and error method.** We simply tried looking at all possibilities until one worked. If no possibility worked, we would conclude that the trinomial is unfactorable. We will factor more trinomials in the examples that follow.

Example 1

Factor $2x^2 + 7x + 6$

Solution

First we look to see whether there is a common factor among the terms $2x^2$, $7x$, and 6. Since there isn't, we try to factor the first term, $2x^2$.

The product of the first terms must be $2x^2$: $(2x \quad)(x \quad)$.

The product of the last terms must be 6. We list the four possible sets of positive factors. (We eliminate the four sets of negative factors since the middle term, which is the *sum* of the innermost and outermost products, must be positive.)

1. $(2x + 1)(x + 6)$
2. $(2x + 6)(x + 1)$
3. $(2x + 3)(x + 2)$
4. $(2x + 2)(x + 3)$

The product of the first terms in each case is $2x^2$.

The product of the last terms in each case is 6.

We must now check to see whether, in any of the four possible cases,

$$\left(\begin{array}{c}\text{The product of the}\\ \text{outermost terms}\end{array}\right) + \left(\begin{array}{c}\text{The product of the}\\ \text{innermost terms}\end{array}\right) = 7x$$

Outermost + Innermost

1. $(2x + 1)(x + 6)$ $12x + 1x = 13x$
2. $(2x + 6)(x + 1)$ $2x + 6x = 8x$
3. $(2x + 3)(x + 2)$ $4x + 3x = 7x$
4. $(2x + 2)(x + 3)$ $6x + 2x = 8x$

This possibility shows us that $2x^2 + 7x + 6$ is factored as

$$(2x + 3)(x + 2)$$

Check:
$$(2x + 3)(x + 2) = 2x^2 + 4x + 3x + 6$$
$$= 2x^2 + 7x + 6 \; \checkmark$$

Example 2

Factor $6x^2 + 13x - 5$.

Solution

There is no common factor among the three terms. So we start with $6x^2$. To obtain $6x^2$ as the product of first terms, there are two possibilities: $x \cdot 6x$ or $2x \cdot 3x$. Since the product of last terms is -5, they must be either -1 and 5 or 1 and -5. We list all eight possible combinations.

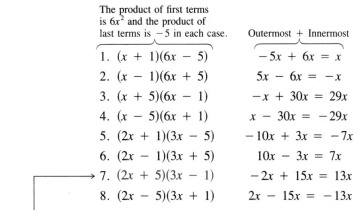

The product of first terms is $6x^2$ and the product of last terms is -5 in each case.	Outermost + Innermost
1. $(x + 1)(6x - 5)$	$-5x + 6x = x$
2. $(x - 1)(6x + 5)$	$5x - 6x = -x$
3. $(x + 5)(6x - 1)$	$-x + 30x = 29x$
4. $(x - 5)(6x + 1)$	$x - 30x = -29x$
5. $(2x + 1)(3x - 5)$	$-10x + 3x = -7x$
6. $(2x - 1)(3x + 5)$	$10x - 3x = 7x$
7. $(2x + 5)(3x - 1)$	$-2x + 15x = 13x$
8. $(2x - 5)(3x + 1)$	$2x - 15x = -13x$

This possibility shows us that $6x^2 + 13x - 5$ is factored as

$$(2x + 5)(3x - 1)$$

Check:
$$(2x + 5)(3x - 1) = 6x^2 - 2x + 15x - 5$$
$$= 6x^2 + 13x - 5 \; \checkmark$$

Example 3

Factor $3x^2 - 11x + 10$.

Solution

There is no common factor among the three terms. So we begin by factoring the first term, $3x^2$.

The product of first terms must be $3x^2$: $(3x \quad)(x \quad)$.

The product of last terms must be 10. The possibilities are

$$1 \cdot 10$$
$$2 \cdot 5$$
$$-1 \cdot -10$$
$$-2 \cdot -5$$

But, since the middle term is $-11x$ (since the sum of outermost and innermost products must be negative), we can eliminate the first two cases. Now we must pair up both remaining cases of products of 10 ($-1 \cdot -10$ and $-2 \cdot -5$) with the $3x$ and x. This yields the following four cases.

$$1. \ (3x - 1)(x - 10)$$
$$2. \ (3x - 10)(x - 1)$$
$$3. \ (3x - 2)(x - 5)$$
$$4. \ (3x - 5)(x - 2)$$

Again, we examine the middle term of each possibility.

$$1. \ (3x - 1)(x - 10) \qquad -30x - 1x = -31x$$
$$2. \ (3x - 10)(x - 1) \qquad -3x - 10x = -13x$$
$$3. \ (3x - 2)(x - 5) \qquad -15x - 2x = -17x$$
$$\longrightarrow 4. \ (3x - 5)(x - 2) \qquad -6x - 5x = -11x$$

This possibility shows us that $3x^2 - 11x + 10$ is factored as

$$(3x - 5)(x - 2)$$

Check:
$$(3x - 5)(x - 2) = 3x^2 - 6x - 5x + 10$$
$$= 3x^2 - 11x + 10 \ \checkmark$$

Example 4

Factor completely $6x^2 + 15x + 9$.

Solution

The common factor, 3, is extracted first.

$$6x^2 + 15x + 9 = 3(2x^2 + 5x + 3)$$

We have already factored $2x^2 + 5x + 3$ as $(2x + 3)(x + 1)$ at the beginning of this section. So

$$6x^2 + 15x + 9 = 3(2x^2 + 5x + 3)$$
$$= 3(2x + 3)(x + 1)$$

Check:
$$3(2x + 3)(x + 1) = (6x + 9)(x + 1)$$
$$= 6x^2 + 6x + 9x + 9$$
$$= 6x^2 + 15x + 9 \ \checkmark$$

Section 6.5 Exercises

Completely factor each expression, if possible.

1. $2x^2 + 7x + 3$
2. $2x^2 + 4x + 2$
3. $4x^2 + 4x - 1$
4. $4x^2 + 4x - 4$
5. $3x^2 + 4x + 1$
6. $3x^2 + 14x + 8$
7. $3x^2 + 11x + 8$
8. $3x^2 - 10x - 8$
9. $4x^2 - x - 4$
10. $12x^2 - 36x + 24$
11. $7x^2 - 13x - 2$
12. $18x^2 - 9x + 1$
13. $9x^2 - 1$
14. $15x^2 + 8x + 1$
15. $15x^2 + 16x + 1$
16. $6x^2 - 11x + 4$
17. $6x^2 - 12x + 4$
18. $6x^2 + 30x + 36$

19. $6x^2 - 10x + 1$ 20. $9x^2 - 6x - 1$
21. $x^2 + 2x$ 22. $2x^2 + 8x + 6$
23. $x^2 + x + 1$ 24. $x^2 + 1$
25. $x^2 - 1$ 26. $2x^2 - 5x - 3$
27. $5x^2 - 4x - 1$ 28. $4x^2 + 4x + 1$
29. $6x^2 - x - 2$ 30. $10x^2 - 3x - 1$

Supplementary Exercises

Completely factor each expression, if possible.

1. $2p^2 + 5p + 3$ 2. $3m^2 + 7m + 4$
3. $5q^2 + 4q - 1$ 4. $5w^2 + 10w - 15$
5. $6a^2 + a - 1$ 6. $8q^2 + 2q - 3$
7. $8m^2 - 15m - 2$ 8. $5t^2 + 12t + 4$
9. $7w^2 - 20w - 3$ 10. $6m^2 + 31m + 5$
11. $4q^2 - 25q - 21$ 12. $6m^2 - 5m - 6$
13. $10t^2 + 25t + 10$ 14. $12t^2 - 56t - 20$
15. $3c^2 + 10c + 7$

6.6 Using factoring to solve quadratic equations

Now that we have learned how to factor, it is time to look at some *uses* of factoring. In this section we will apply the concepts of factoring to solving equations.

In Chapters 3 and 4 we solved equations of the first degree, so called because the variable was always raised to the **first** power. In this section we will examine second-degree equations, also known as **quadratic equations.**

A quadratic equation

Any equation that has the form

$$ax^2 + bx + c = 0$$

where a, b, and c are real numbers ($a \neq 0$), is called a **quadratic equation.** We call $ax^2 + bx + c = 0$ the **general form** of a quadratic equation in the single variable, x.

Before we begin to study the method of solving a quadratic equation, we will need to recall a property of zero from Chapter 1.

Property of zero

Any number multiplied by 0 is 0. That is,

If $pq = 0$, then $p = 0$ or $q = 0$.

We see how to make use of that property in the following example.

Example 1

Solve $(x - 3)(x - 4) = 0$.

Solution

We know the product of two quantities equals zero. Then either

$$(x - 3) = 0 \quad \text{or} \quad (x - 4) = 0$$

(Think of $x - 3$ as p and $x - 4$ as q in the property statement we just made.)

$$
\begin{array}{rcl}
x - 3 = & 0 \\
+ 3 = & +3 \\
\hline
x \quad = & 3
\end{array}
\qquad
\begin{array}{rcl}
x - 4 = & 0 \\
+ 4 = & +4 \\
\hline
x \quad = & 4
\end{array}
$$

Thus, $x = 3$ and $x = 4$ are solutions to $(x - 3)(x - 4) = 0$.

Check for $x = 3$:
$(x - 3)(x - 4) = 0$
$(3 - 3)(3 - 4) = 0$
$(0)(-1) = 0 \ \checkmark$

Check for $x = 4$:
$(x - 3)(x - 4) = 0$
$(4 - 3)(4 - 4) = 0$
$(1)(0) = 0 \ \checkmark$

Example 2

Solve $(2x + 3)(x - 5) = 0$.

Solution

We set each factor equal to 0.

$$
\begin{array}{rcl}
2x + 3 = & 0 \\
- 3 = & -3 \\
\hline
2x \quad = & -3 \\
x = & -\dfrac{3}{2}
\end{array}
\qquad
\begin{array}{rcl}
x - 5 = & 0 \\
+ 5 = & +5 \\
\hline
x \quad = & 5 \\
x = & 5
\end{array}
$$

The solutions are $x = -\dfrac{3}{2}$ and $x = 5$. The checks are left to the reader.

To solve a quadratic equation, we perform the following four steps:

Solving a quadratic equation

1. Be sure 0 is on one side of the equation.
2. Factor the polynomial that appears on the other side. (If the expression is not factorable, the method of solution presented here fails.)
3. Set each factor equal to zero and solve.
4. Check solutions in the original equation.

Some examples follow.

Example 3

Solve $x^2 - 3x - 10 = 0$.

Solution

$$x^2 - 3x - 10 = 0$$

$$(x - 5)(x + 2) = 0 \qquad \text{Factor the lefthand side.}$$

Set $x - 5$ equal to zero.	$x - 5 = 0$	$x + 2 = 0$	Set $x + 2$ equal to zero.
Solve for x.	$x = 5$	$x = -2$	Solve for x.

Our two solutions are 5 and -2. We check them below:

Check for $x = 5$:

$x^2 - 3x - 10 = 0$

$5^2 - 3 \cdot 5 - 10 = 0$

$25 - 15 - 10 = 0$

$0 = 0$ ✓

Check for $x = -2$:

$x^2 - 3x - 10 = 0$

$(-2)^2 - 3(-2) - 10 = 0$

$4 + 6 - 10 = 0$

$0 = 0$ ✓

Example 4

Solve $2x^2 + 2x = 40$.

Solution

First we must rewrite the equation so that zero appears on one side of it. We do this by subtracting 40 from both sides of the equation.

$$2x^2 + 2x - 40 = 0$$

$$2(x^2 + x - 20) = 0 \qquad \text{Factor the lefthand side.}$$

$$2(x + 5)(x - 4) = 0$$

Now, as before, we have a product of factors equal to zero, and this happens when any of the factors is zero. Since 2 can never be zero, we concentrate on $(x + 5)$ and $(x - 4)$:

$$2(x + 5)(x - 4) = 0$$

Set $x + 5$ equal to zero.	$x + 5 = 0$	$x - 4 = 0$	Set $x - 4$ equal to zero.
Solve for x.	$x = -5$	$x = 4$	Solve for x.

Check for $x = -5$:

$2x^2 + 2x = 40$

$2(-5)^2 + 2(-5) = 40$

$2(25) - 10 = 40$

$50 - 10 = 40$

$40 = 40$ ✓

Check for $x = 4$:

$2x^2 + 2x = 40$

$2(4)^2 + 2 \cdot 4 = 40$

$2 \cdot 16 + 2 \cdot 4 = 40$

$32 + 8 = 40$

$40 = 40$ ✓

Example 5

Solve $y^2 = 16$.

Solution

First, we must rewrite the equation so that zero appears on the righthand side. To do this, we subtract 16 from both sides of the equation.

$$y^2 - 16 = 0$$

$$(y + 4)(y - 4) = 0 \qquad \text{Factor the lefthand side.}$$

Set $y + 4$ equal to zero.	$y + 4 = 0$	$y - 4 = 0$	Set $y - 4$ equal to zero.
Solve for y.	$y = -4$	$y = 4$	Solve for y.

$$\begin{array}{ll}
\textbf{Check } \text{for } y = -4: & \textbf{Check } \text{for } y = 4: \\
y^2 = 16 & y^2 = 16 \\
(-4)^2 = 16 & (4)^2 = 16 \\
16 = 16 \ \checkmark & 16 = 16 \ \checkmark
\end{array}$$

In Chapters 2 and 3 we saw that, once we know how to solve an equation, we can use that knowledge to help us solve problems of a verbal nature. The next three examples all lead to quadratic equations.

Example 6

Find two consecutive positive even integers whose product is 224.

Solution

Step 1   Problem translation

The two numbers are, at present, unknown. Let n be the smaller one. The other is represented by $n + 2$. Furthermore, since the product of the numbers is 224, we have:

$$n(n + 2) = 224 \leftarrow \text{This is our algebraic model.}$$

Step 2   Equation solution

We need to solve $n(n + 2) = 224$. First we multiply the lefthand side.

$$n^2 + 2n = 224$$

Next we subtract 224 from both sides.

$$n^2 + 2n - 224 = 0 \leftarrow \text{This is a } \textit{quadratic equation.}$$
$$(n + 16)(n - 14) = 0$$
$$n + 16 = 0 \qquad n - 14 = 0$$
$$n = -16 \qquad n = 14$$

These are both solutions to the quadratic equation $n^2 + 2n - 224 = 0$. Since n represents a positive even integer, -16 cannot possibly be an allowable solution; we *reject* it. The other value of n, namely 14, must be the smaller positive even integer. Thus the two consecutive positive even integers we wanted to find are 14 and 16.

Step 3   Check

Our solution checks because 14 and 16 (consecutive positive even integers) have a product of 224.

Example 7

Pam, a homeowner, is planning to build an addition to her house. Originally, she planned to have a square room (with each side x meters), but she now plans an even larger addition by adding 3 meters to one dimension and 1 meter to the other. The floor area of the addition will be 48 square meters. What are the dimensions of the new addition?

Solution

Step 1   Problem translation

Originally Pam planned a room that looked like this:

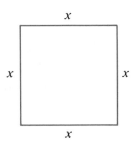

Now she has planned a room that looks like this:

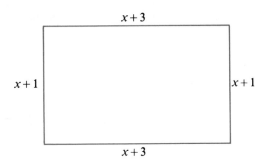

The area is 48 square meters, so we have:

$$(x + 3)(x + 1) = 48 \leftarrow \text{This is our algebraic model.}$$

Step 2   Equation solution

We need to solve $(x + 3)(x + 1) = 48$. So we first multiply the lefthand side.

$$x^2 + 4x + 3 = 48$$

Then we subtract 48 from both sides.

$$x^2 + 4x - 45 = 0 \leftarrow \text{This is a } quadratic\ equation.$$
$$(x + 9)(x - 5) = 0$$
$$x + 9 = 0 \qquad x - 5 = 0$$
$$x = -9 \qquad\quad x = 5$$

These are both solutions to the quadratic equation $x^2 + 4x - 45 = 0$. Since x represents a dimension for a room, -9 cannot possibly be an allowable solution; we *reject* it. The other value of x, namely 5, must be the side of the square room. Thus the dimensions of the rectangular addition are 6 meters by 8 meters.

Step 3   Check

Originally, Pam's square room would have been 5 meters on a side. By increasing one dimension by 3 meters and the other by 1 meter, she obtains a room that is 8 meters by 6 meters. The area of this room would be 48 square meters, so our solution checks.

Example 8

The All-Safe Home Security Company manufactures smoke detectors for the home. The profit in dollars (P) they make by selling d smoke detectors is given by the formula $P = d^2 + 5d - 50$.

(a) If they sell 100 smoke detectors, what is the profit?

(b) How many smoke detectors must they sell to realize a profit of $250?

(c) How many smoke detectors must they sell to "break even"?

Solution

(a) We must determine the profit (P) when $d = 100$

$$P = d^2 + 5d - 50$$
$$P = (100)^2 + 5 \cdot 100 - 50$$
$$P = 10,000 + 500 - 50$$
$$P = 10,450$$

So the profit is $10,450.

(b) We are asked to find d when $P = 250$. So, in the profit equation given to us, we let $P = 250$ and then solve the equation for d.

$$250 = d^2 + 5d - 50$$

or

$$d^2 + 5d - 50 = 250$$
$$d^2 + 5d - 300 = 0$$
$$(d + 20)(d - 15) = 0$$
$$d + 20 = 0 \qquad d - 15 = 0$$
$$d = -20 \qquad d = 15$$

These are both solutions to the quadratic equation, $d^2 + 5d - 50 = 250$. However, d represents a number of smoke detectors, so the value -20 must be discarded; we *reject* it. Thus the solution is 15. All-Safe must sell 15 smoke detectors to make a profit of $250.

(c) If All-Safe is to "break even," the profit must be zero dollars ($P = 0$). So, in the profit equation given to us, we let $P = 0$.

$$0 = d^2 + 5d - 50$$

or

$$d^2 + 5d - 50 = 0$$
$$(d + 10)(d - 5) = 0$$
$$d + 10 = 0 \qquad d - 5 = 0$$
$$d = -10 \qquad d = 5$$

We *reject* this solution.   This is the solution we consider.

So All-Safe must sell 5 detectors in order to "break even."

Section 6.6 Exercises

In Exercises 1 through 21, solve each equation.

1. $x^2 - 5x + 4 = 0$
2. $x^2 + 5x + 4 = 0$
3. $2x^2 + 10x + 8 = 0$
4. $200x^2 + 1000x + 800 = 0$
5. $x^2 - 3x + 2 = 0$
6. $x^2 + 4x - 12 = 0$
7. $x^2 + 8x + 12 = 0$
8. $3y^2 - 6y - 24 = 0$
9. $x^2 - 3x = 10$
10. $x^2 = 3x + 10$
11. $y^2 = 64$
12. $x^2 + 6x = -9$

13. $z^2 + 4z = 77$

14. $3x^2 + 6x - 24 = 0$

15. $t^2 = 49$

16. $y^2 = y$

17. $n^2 = 7n + 8$

18. $y^2 = 10y - 25$

19. $3x^2 + 12x - 63 = 0$

20. $(2x + 1)^2 = 9$

21. $4x^2 + 7x + 21 = 3x^2 + 9$

22. Find two numbers whose sum is 13 and whose product is 36.

23. A positive integer plus its square is equal to 56. Determine the positive integer.

24. Find two numbers whose sum is 14 and whose product is 45.

25. A negative integer plus its square is 90. Determine the negative integer.

26. The product of two consecutive positive integers is 132. What are the integers?

27. The product of two consecutive positive integers is 210. What are the integers?

28. An architect has been directed to design a room that has an area of 154 square feet. The owners have purchased certain pieces of furniture that require the length of the room to be 3 feet more than the width. What must the dimensions of the room be?

29. In a certain machine shop the total cost in dollars (C) for producing a certain type of firing pin is given by

$$C = x^2 + 11x + 24$$

where x is the number of firing pins produced.

(a) What does it cost to produce 20 firing pins?

(b) If the shop can spend $204 on this project, how many firing pins can be produced?

30. A gardener designs a rectangular flower bed 450 square feet in area. The length of the bed is twice the width. What are the dimensions?

31. The page of a book has an area of 500 square centimeters. The length of a page is 5 centimeters more than its width. Find the dimensions of the page.

32. The cost of carpeting a rectangular living room was $600. The carpeting cost $10 per square yard. The length exceeded the width by 7 feet. What were the dimensions of the room? (*Note:* There are 9 square feet in 1 square yard.)

33. A carpenter is building the rear of an A-frame house that looks like this:

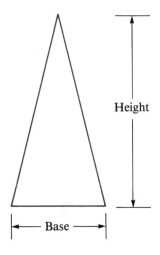

The area is 100 square meters and the height is twice the base. Determine the length of the base and height. Note that the area of a triangle is

$$\frac{\text{base} \cdot \text{height}}{2}$$

34. Along the North Slope of Alaska are thousands of discarded oil drums that have accumulated there since the 1940s. An ecology-minded scrap metal company has considered removing and recycling the drums for profit. Preliminary studies show that the profit can be calculated by the following equation:

$$P = x^2 - 2x - 4$$

where P is profit in thousands of dollars and x is thousands of barrels removed. (That is, if $P = 20$, it means $20,000 and if $x = 3$, it means 3000 barrels.)

(a) If 10,000 barrels are removed, what profit is realized?

(b) If 4000 barrels are removed, what is the profit?

(c) If the company wants to realize $20,000 in profit, how many barrels must be removed?

(d) If the company wants to realize $44,000 in profit, how many barrels must be removed?

35. If a football is thrown into the air from a quarterback's hand 8 feet above the ground with an initial speed of 48 ft/sec, its distance (y) above the ground after t seconds have elapsed is given by:

$$y = 8 + 48t - 16t^2$$

(a) How high is the football after 2 seconds have elapsed?

(b) In order to be caught by the intended wide receiver, the football should be a height of 8 feet. How long will it take the football to return to a height of 8 feet?

The method of factoring one side of an equation to find solutions (provided the other side is always zero) can be extended to equations whose variable is raised to the third, fourth, or higher powers. Use the method for the following.

36. $x^3 + 7x^2 + 12x = 0$

37. $x^4 - 6x^3 + 5x^2 = 0$

38. $(x - 1)(x - 2)(x + 3)(x + 4) = 0$

Supplementary Exercises

In Exercises 1 through 15, solve each equation.

1. $x^2 + x - 2 = 0$

2. $a^2 - 9a - 10 = 0$

3. $x^2 + 5x = 14$

4. $x^2 - 9 = 0$

5. $4 = x^2$

6. $x^2 = -5 - 6x$

7. $p^2 + p + 1 = 0$

8. $4a^2 - 24a + 20 = 0$

9. $p^2 + p - 18 = 2$

10. $m^2 + 10 = 7m$

11. $(3x + 5)^2 = 16$

12. $3x^2 + 4x + 5 = 2x^2 + 1$

13. $5a^2 + 3a - 5 = 4a^2 + a - 2$

14. $x^2 = x$

15. $4x^2 = 2x$

16. Find two consecutive integers whose product is 72.

17. Find two consecutive even integers whose product is 440.

18. A rectangular room is 5 feet longer than it is wide. If its area is 126 square feet, find the dimensions of the room.

19. A garden must be 6 feet shorter than it is wide. If its area must be 135 square feet, find the dimensions of the garden.

20. Mary decides to make her living room 5 feet longer than it presently is. If its present width is one-half the present length and the new area will be 168 square feet, find the present dimensions of the room.

6.7   Chapter review

Summary

We began this chapter by examining the prime factorization of integers and the greatest common factor of a collection of integers. The notion of factoring integers was extended to factoring algebraic expressions in the rest of the chapter. The four types of factoring we did are summarized in Table 6.2. In Section 6.6 we applied our factoring techniques to solving quadratic equations and word problems.

Table 6.2   Techniques of Factoring Polynomials

Type of Expression	Example	What to Do	Factored Results
1. *Any* expression involving a common factor	$12x^2 + 30x$	Extract the greatest common factor, $6x$.	$6x(2x + 5)$
2. Trinomial of the type $x^2 + bx + c$	$x^2 + 7x + 12$	Find two numbers whose sum is 7 and whose product is 12.	$(x + 3)(x + 4)$
3. Difference of two perfect squares	$9x^2 - 4$	Think of $9x^2 - 4$ as $(3x)^2 - (2)^2$	$(3x + 2)(3x - 2)$
4. Trinomial of the type $ax^2 + bx + c$	$2x^2 + 7x + 3$	Use trial and error to examine factors of $2x^2$ and factors of 3.	$(2x + 1)(x + 3)$

Vocabulary Quiz

Match the expression in Column I with the phrase in Column II that *best* describes it.

Column I

1. A trinomial is
2. The GCF is
3. The general form of a quadratic equation is
4. Factoring is
5. The difference of two perfect squares is
6. A rejected solution is

Column II

a. an answer to an equation derived from a word problem but is not considered because it is physically impossible.
b. $x^2 - a^2$, for example.
c. the process of writing expressions as the product of factors.
d. $ax^2 + bx + c = 0$.
e. the greatest common factor of two expressions.
f. a polynomial having three terms.

Chapter 6 Review Exercises

In Exercises 1 through 5, find the prime factorization of each number. Use Table 6.1.

1. 86   2. 84
3. 162   4. 810
5. 150

In Exercises 6 through 10, find the GCF for each set of numbers.

6. 8, 10

7. 24, 50

8. 24, 60

9. 24, 60, 72

10. 24, 72, 144

In Exercises 11 through 28, completely factor each expression, if possible.

11. $5x^2 - 20x$

12. $5x^2 - 20$

13. $5a^2b^3 - 15abc$

14. $y^2 + 6y - 27$

15. $t^2 - 11t + 30$

16. $t^4 - 11t^3 + 30t^2$

17. $x^2 - 2x - 48$

18. $x^2 + 2x - 48$

19. $x^2 - 2x + 48$

20. $x^2 - 8x - 48$

21. $x^2 - 16x + 48$

22. $2y^3 - 22y^2 + 60y$

23. $2y^3 - 29y^2 + 60y$

24. $64x^2 - 1$

25. $t^4 - 1$

26. $2x^2 - 3x - 5$

27. $2x^2 - 7x + 5$

28. $6x^2 - 20x + 6$

In Exercises 29 through 35, solve each quadratic equation.

29. $x^2 - 5x - 6 = 0$

30. $x^2 - 5x + 6 = 0$

31. $x^2 + 6x - 27 = 0$

32. $x^2 - 7x + 10 = 0$

33. $2x^2 + 5x - 3 = 0$

34. $25 = x^2$

35. $3x^2 = 2x^2 + 7x - 12$

Cumulative review exercises

1. The expression $\frac{x}{x}$ is equal to one for every possible value of x except one value. What is that value?   (*Section 1.6*)

2. Evaluate -2^4.   (*Section 2.3*)

3. Solve for x: $8(2x - 5) = x - 40$.   (*Section 3.4*)

4. Solve for x: $2x + 3y = 5$.   (*Section 3.6*)

5. Solve for x and graph the solution on a real number line:

$$12 - 4x > 2 - 3x \quad (Section\ 3.7)$$

6. Two runners, Gloria and Vinnie, pass each other at noon traveling the same route but in opposite directions. Gloria is running at a rate of 7.5 mph; Vinnie is running at a rate of 10 mph. At what time will they be 35 miles apart?   (*Section 4.2*)

7. Simplify $(3a^2bc^2)^3(4a^2b^3)$.   (*Section 5.1*)

8. Write 0.00219 in scientific notation.   (*Section 5.3*)

9. Multiply $(3y - 7)(2y + 3)$.   (*Section 5.5*)

10. Perform the indicated division: $(x^2 + 8x - 20) \div (x - 2)$.   (*Section 5.6*)

Chapter 6 test

Take this test to see how well you have mastered factoring; check your answers using those found at the end of the book.

1. Write the prime factorization of 180. *(Section 6.1)*

2. Find the GCF of 30 and 105. *(Section 6.1)*

1. _____

2. _____

In Problems 3 through 16, completely factor each expression, if possible.

3. $10x + 2$ *(Section 6.2)*

4. $x^4 - 3x^3$ *(Section 6.2)*

3. _____

4. _____

5. $-30s^2t^2 + 10st^2 - 15t^2$ *(Section 6.2)*

6. $x^2 + 9x + 20$ *(Section 6.3)*

5. _____

6. _____

7. $x^2 - 6x - 16$ *(Section 6.3)*

8. $-2x^4 + 16x^3 + 40x^2$ *(Section 6.3)*

7. _____

8. _____

9. $3y^3 + 27y^2 + 30y$ *(Section 6.3)*

10. $z^2 - 9$ *(Section 6.4)*

9. _____

10. _____

11. $8t^4 - 8$ *(Section 6.4)*

12. $4x^2 - 25y^2$ *(Section 6.4)*

11. _____

12. _____

13. $40x^2 - 250y^2$ *(Section 6.4)*

14. $3y^2 - 10y - 8$ *(Section 6.5)*

13. _____

14. _____

15. $6t^2 + 22t + 8$ (*Section 6.5*)

16. $12y^4 - 16y^3 + 5y^2$ (*Section 6.5*)

15. _____

16. _____

In Problems 17 through 20, solve the given equation.   (*Section 6.6*)

17. $x^2 - 5x - 6 = 0$

18. $y^2 - 7y - 8 = 0$

17. _____

18. _____

19. $y^2 - 6y + 8 = 0$

20. $2x^2 + 13x = 7$

19. _____

20. _____

chapter seven

Fractions

7.1 Reducing fractions

The tool of factoring developed in Chapter 6 will be used throughout Chapter 7 in dealing with fractions. Our main interest is to study *algebraic* fractions, and we do so by first examining their *arithmetic* counterparts.

To reduce an arithmetic fraction like $\dfrac{30}{42}$, for example, we follow the following rules:

Reducing arithmetic fractions

1. Find the greatest common factor (GCF) of numerator and denominator.
2. Divide the numerator and denominator by that GCF.

In the illustration $\dfrac{30}{42}$, we find the GCF first.

$$30 = 2 \cdot 3 \cdot 5$$
$$42 = 2 \cdot 3 \cdot 7$$

The GCF $= 2 \cdot 3 = 6$.

The second step is to divide numerator and denominator by 6.

$$\frac{30}{42} = \frac{6 \cdot 5}{6 \cdot 7} = \frac{5}{7} \qquad \text{Notice that 6 divided by 6 is 1.}$$

A third step is to check to see whether, in fact, the fractions are equal. We say that $\dfrac{30}{42}$ *equals* $\dfrac{5}{7}$, or that $\dfrac{30}{42}$ and $\dfrac{5}{7}$ are **equivalent fractions.** The test for equivalent fractions is as follows:

Equality of fractions principle

Two fractions are equal if the numerator of the first times the denominator of the second equals the denominator of the first times the numerator of the second. That is, $\dfrac{30}{42}$ and $\dfrac{5}{7}$ are equal because

$$\frac{30}{42} \underset{\longleftarrow}{\overset{\longrightarrow}{\times}} \frac{5}{7}$$

$$30 \cdot 7 = 42 \cdot 5 = 210$$

This process is sometimes called **cross-multiplying.**

Three examples of reducing arithmetic fractions follow.

Example 1

Simplify $\dfrac{12}{15}$.

Solution

Step 1

Find the GCF of 12 and 15.

$$12 = 2 \cdot 2 \cdot 3$$
$$15 = 3 \cdot 5$$

So the GCF $= 3$

Step 2

Divide numerator and denominator by 3.

$$\frac{12}{15} = \frac{3 \cdot 4}{3 \cdot 5} = \frac{4}{5}$$

Thus, $\frac{12}{15}$ reduces to $\frac{4}{5}$.

Step 3   Check:

$$\frac{12}{15} \times \frac{4}{5}$$

$$12 \cdot 5 = 15 \cdot 4$$
$$60 = 60 \ \checkmark$$

Example 2

Reduce $\frac{36}{60}$ to lowest terms.

Solution

Step 1

Find the GCF of 36 and 60.

$$36 = 2 \cdot 2 \cdot 3 \cdot 3$$
$$60 = 2 \cdot 2 \cdot 3 \cdot 5$$

So the GCF $= 2 \cdot 2 \cdot 3 = 12$.

Step 2

Divide 36 and 60 by 12.

$$\frac{36}{60} = \frac{12 \cdot 3}{12 \cdot 5} = \frac{3}{5}$$

Hence, $\frac{36}{60}$ is reduced to $\frac{3}{5}$.

Step 3   Check:

$$\frac{36}{60} \times \frac{3}{5}$$

$$36 \cdot 5 = 60 \cdot 3$$
$$180 = 180 \ \checkmark$$

Example 3

Reduce $\frac{28}{99}$.

Solution

Step 1

Find the GCF of 28 and 99.

$$28 = 2 \cdot 2 \cdot 7$$
$$99 = 3 \cdot 3 \cdot 11$$

The GCF is 1.

Step 2

If the GCF is 1, there is no simplification of the fraction. That is, dividing both the numerator and denominator by 1 will have no effect.

The remainder of this section deals with reducing **algebraic fractions.** An **algebraic fraction** is a fraction whose numerator or denominator involves one or more variables. The process for reducing them is very similar to the one we used for arithmetic fractions.

Reducing algebraic fractions

1. Completely factor the numerator and denominator.
2. Divide the numerator and denominator by common factors.
3. Check using the Equality of Fractions (cross-multiplying) Principle.

Example 4

Simplify the fraction $\dfrac{15a^5b}{10a^2b^3}$.

Solution

We break down the numerator and denominator into factors.

$$\frac{15a^5b}{10a^2b^3} = \frac{3 \cdot 5 \cdot a \cdot a \cdot a \cdot a \cdot a \cdot b}{2 \cdot 5 \cdot a \cdot a \cdot b \cdot b \cdot b}$$

So

$$\frac{15a^5b}{10a^2b^3} = \frac{3a^3}{2b^2}$$

Check:

$$\frac{15a^5b}{10a^2b^3} = \frac{3a^3}{2b^2}$$

$$\frac{15a^5b}{10a^2b^3} \times \frac{3a^3}{2b^2}$$

$$30a^5b^3 = 30a^5b^3 \ \checkmark$$

Remember that a fraction's denominator may never be zero. So, when we write

$$\frac{15a^5b}{10a^2b^3} = \frac{3a^3}{2b^2}$$

we assume that the variables a and b may not be zero. Throughout this chapter we will assume that any values of the variable that make the denominator zero are not permitted.

Example 5

Reduce $\dfrac{x-2}{x^2-4}$ to lowest terms.

Solution

$$\frac{x-2}{x^2-4} = \frac{x-2}{(x+2)(x-2)}$$ Factor the denominator.

$$= \frac{\cancel{x-2}}{(x+2)\cancel{(x-2)}}$$ Divide numerator and denominator by the GCF, which is $(x-2)$.

$$= \frac{1}{x+2}$$

Check: $$\frac{x-2}{x^2-4} = \frac{1}{x+2}$$

$$\frac{x-2}{x^2-4} \diagdown \frac{1}{x+2}$$

$$(x-2)(x+2) = 1(x^2-4)$$
$$x^2-4 = x^2-4 \ \checkmark$$

Example 6

Simplify the fraction $\dfrac{2y^3}{y^3+y^2}$.

Solution

$$\frac{2y^3}{y^3+y^2} = \frac{2\cdot \cancel{y}\cdot \cancel{y}\cdot y}{\cancel{y}\cdot \cancel{y}(y+1)}$$ Factor the numerator and denominator.

$$= \frac{2y}{y+1}$$

Notice that we cannot cancel the y's in $\dfrac{2y}{y+1}$. We can only cancel common factors; the y in the denominator is *not* a factor.

Check: $$\frac{2y^3}{y^3+y^2} = \frac{2y}{y+1}$$

$$\frac{2y^3}{y^3+y^2} \diagdown \frac{2y}{y+1}$$

$$2y^3(y+1) = (y^3+y^2)2y$$
$$2y^4+2y^3 = 2y^4+2y^3 \ \checkmark$$

Example 7

Simplify the fraction $\dfrac{x^2y^2-xy^3}{x^2-y^2}$.

Solution

Factoring the numerator and denominator yields

$$\frac{x^2y^2-xy^3}{x^2-y^2} = \frac{xy^2\cancel{(x-y)}}{(x+y)\cancel{(x-y)}}$$

So

$$\frac{x^2y^2-xy^3}{x^2-y^2} = \frac{xy^2}{x+y}$$

Check:
$$\frac{x^2y^2 - xy^3}{x^2 - y^2} = \frac{xy^2}{x + y}$$

$$\frac{x^2y^2 - xy^3}{x^2 - y^2} \underset{\longleftarrow}{\overset{\longrightarrow}{\times}} \frac{xy^2}{x + y}$$

$$(x^2y^2 - xy^3)(x + y) = (x^2 - y^2)(xy^2)$$
$$x^3y^2 + x^2y^3 - x^2y^3 - xy^4 = x^3y^2 - xy^4$$
$$x^3y^2 - xy^4 = x^3y^2 - xy^4 \; \checkmark$$

Recall from Section 5.6 that any two of the three signs of a fraction can be changed without changing the value of the fraction. We use this property in Examples 8 and 9 for arithmetic fractions and in Example 10 for an algebraic fraction.

Example 8

(a) Does $\frac{-4}{-3}$ have the same value as $\frac{4}{3}$?

(b) Does $-\frac{-4}{3}$ have the same value as $\frac{4}{3}$?

(c) Does $-\frac{-4}{-3}$ have the same value as $\frac{4}{3}$?

Solution

(a) Yes, $\frac{-4}{-3}$ has the same value as $\frac{4}{3}$ because, if we change the sign of the numerator and the sign of the denominator in $\frac{-4}{-3}$, we get $\frac{4}{3}$. (Remember we can change any two of the three signs.)

(b) Yes, $-\frac{-4}{3}$ has the same value as $\frac{4}{3}$ because, if we change the sign of the numerator and the sign preceding the fraction in $-\frac{-4}{3}$, we get $\frac{4}{3}$.

(c) No, $-\frac{-4}{-3}$ does not have the same value as $\frac{4}{3}$. No matter which two signs are changed in $-\frac{-4}{-3}$, we do not get $\frac{4}{3}$.

Example 9

Rewrite the following fractions so that each has at most one "$-$" sign.

(a) $-\frac{-5}{2}$ (b) $-\frac{-5}{-2}$ (c) $-\frac{5}{-2}$ (d) $\frac{5}{-2}$

Solution

(a) $-\frac{-5}{2}$ has the value $\frac{5}{2}$. This is obtained by changing the sign of the numerator and the sign preceding the fraction.

(b) $-\frac{-5}{-2}$ has the value $\frac{-5}{2}$. This is obtained by changing the sign of the denominator and the sign preceding the fraction. Alternate answers are $-\frac{5}{2}$ or $\frac{5}{-2}$.

However, when a fraction has one "$-$" sign, we prefer to have it in the numerator.

(c) $-\dfrac{5}{-2}$ has the value $\dfrac{5}{2}$ by changing the sign of the denominator and the sign preceding the fraction.

(d) $\dfrac{5}{-2}$ already has just one "$-$" sign. However, by changing the sign of the numerator and the sign of the denominator, we obtain the preferred form $\dfrac{-5}{2}$.

Example 10

Simplify $\dfrac{x^2 - 4x}{16 - x^2}$.

Solution

Factoring yields

$$\frac{x(x - 4)}{(4 + x)(4 - x)}$$

There *is* a factor common to both the numerator and denominator, although it is well hidden! Recall that in $(4 - x)$ we can factor a "-1." That is,

$$(4 - x) \quad \text{is the same as} \quad -1(x - 4)$$

So we have

$$\frac{x(x - 4)}{(4 + x)(4 - x)} = \frac{x(x - 4)}{(4 + x)(-1)(x - 4)}$$

$$= \frac{x}{(x + 4)(-1)}$$

$$= \frac{x}{-(x + 4)} \quad \text{or} \quad \frac{-x}{x + 4}$$

Check: $\qquad \dfrac{x^2 - 4x}{16 - x^2} = \dfrac{-x}{x + 4}$

$$\frac{x^2 - 4x}{16 - x^2} \diagdown\!\!\!\diagup \frac{-x}{x + 4}$$

$$(x^2 - 4x)(x + 4) = (16 - x^2)(-x)$$
$$x^3 + 4x^2 - 4x^2 - 16x = -16x + x^3$$
$$x^3 - 16x = x^3 - 16x \;\checkmark$$

Example 11

Reduce $\dfrac{t^2 - 2t - 3}{t^2 + 3t + 2}$ to lowest terms.

Solution

Factoring yields

$$\frac{t^2 - 2t - 3}{t^2 + 3t + 2} = \frac{(t - 3)(t + 1)}{(t + 2)(t + 1)}$$

$$= \frac{t - 3}{t + 2}$$

Check:

$$\frac{t^2 - 2t - 3}{t^2 + 3t + 2} = \frac{t - 3}{t + 2}$$

$$\frac{t^2 - 2t - 3}{t^2 + 3t + 2} \underset{\longleftarrow}{\overset{\longrightarrow}{\times}} \frac{t - 3}{t + 2}$$

$$\underbrace{(t^2 - 2t - 3)(t + 2)} = \underbrace{(t^2 + 3t + 2)(t - 3)}$$

$$
\begin{array}{c}
t^2 - 2t - 3 \\
t + 2 \\
\hline
2t^2 - 4t - 6 \\
t^3 - 2t^2 - 3t \\
\hline
t^3 \qquad\quad - 7t - 6 \\
t^3 \qquad\quad - 7t - 6
\end{array}
\qquad
\begin{array}{c}
t^2 + 3t + 2 \\
t - 3 \\
\hline
- 3t^2 - 9t - 6 \\
t^3 + 3t^2 + 2t \\
\hline
t^3 \qquad\quad - 7t - 6 \\
t^3 \qquad\quad - 7t - 6 \ \checkmark
\end{array}
$$

$$t^3 \quad - 7t - 6 = t^3 \quad - 7t - 6 \ \checkmark$$

Section 7.1 Exercises

In Exercises 1 through 5, rewrite each fraction so that each has at most one "$-$" and no "$-$" signs in its denominator.

1. $-\dfrac{3}{-4}$

2. $-\dfrac{-8}{17}$

3. $-\dfrac{-9}{-16}$

4. $\dfrac{5}{-8}$

5. $-\dfrac{-12}{25}$

In Exercises 6 through 44, reduce the fraction to lowest terms, if possible. Check your answers in each case.

6. $\dfrac{6}{15}$

7. $\dfrac{8}{16}$

8. $\dfrac{8}{12}$

9. $\dfrac{12}{42}$

10. $\dfrac{120}{420}$

11. $\dfrac{51}{66}$

12. $\dfrac{90}{270}$

13. $\dfrac{66}{99}$

14. $\dfrac{75}{105}$

15. $\dfrac{75x^2}{105x^3}$

16. $\dfrac{x^4}{x^7}$

17. $\dfrac{t^6}{t^4}$

18. $\dfrac{x^2 y^2}{xy^3}$

19. $\dfrac{4xy}{-2x}$

20. $\dfrac{-3a^3}{-6b^3}$

21. $\dfrac{ab^2 c^3}{a^2 bc}$

22. $\dfrac{-x^2 y^2 z^2}{x^3 y^4 z^6}$

23. $\dfrac{x - 3}{x^2 - 9}$

24. $\dfrac{3y^3}{y^3 + y^2}$

25. $\dfrac{2y^2}{y^3 + 2y^2}$

26. $\dfrac{x^3 + 7x^2}{x^2 + 9x + 14}$

27. $\dfrac{x^2 - 9}{x^2 + 6x + 9}$

28. $\dfrac{x - 4}{4 - x}$

29. $\dfrac{a^2 b^2 + ab^2}{ab^2}$

30. $\dfrac{x + y}{y^2 - x^2}$

31. $\dfrac{x^2 + 9x + 20}{x^2 + 7x + 12}$

32. $\dfrac{5a + 10b}{6a + 12b}$

33. $\dfrac{-7y}{7y^2 + 21y}$

34. $\dfrac{x - 5}{x^2 - x - 20}$

35. $\dfrac{x - 5}{25 - x^2}$

36. $\dfrac{5a - 5b}{a^2 - 2ab + b^2}$

37. $\dfrac{x^2 - 36}{6 - x}$

38. $\dfrac{x^2 - 2x}{4 - x^2}$

39. $\dfrac{y^2 - 7y + 12}{y^2 - 9}$

40. $\dfrac{y^2 - 7y + 12}{9 - y^2}$

41. $\dfrac{3x^2 - 4x + 1}{x^2 - 2x + 1}$

42. $\dfrac{ab^2c^3}{ab + c}$

43. $\dfrac{x - 5}{5}$

44. $\dfrac{(x - 1)^8}{x^8 - 1}$

45. Consider the following "proof" that $2 = 1$. Start with two *equal, nonzero* quantities, which we'll call a and b.

$$a = b$$
$$a^2 = ab \qquad \text{Multiply both sides of the equation by } a.$$

$$a^2 - b^2 = ab - b^2 \qquad \text{Subtract } b^2 \text{ from both sides of the equation.}$$

$$(a + b)(a - b) = b(a - b) \qquad \text{Factor both sides of the equation.}$$

$$\frac{(a + b)\cancel{(a - b)}}{\cancel{a - b}} = \frac{b\cancel{(a - b)}}{\cancel{a - b}} \qquad \text{Divide both sides of the equation by } a - b \text{; cancel the } a - b \text{ on both sides of the equation.}$$

$$a + b = b$$
$$b + b = b \qquad \text{Since } a \text{ and } b \text{ are equal, substitute } b \text{ for } a.$$

$$2b = b \qquad \text{Simplify the lefthand side of the equation.}$$

$$\frac{2\cancel{b}}{\cancel{b}} = \frac{\cancel{b}}{\cancel{b}} \qquad \text{Divide both sides of the equation by } b \text{; cancel the common factor, } b \text{, on both sides of the equation.}$$

$$2 = 1$$

There is something wrong with this "proof." What is it?

Supplementary Exercises

In Exercises 1 through 25 reduce the fraction to lowest terms, if possible.

1. $\dfrac{-75}{125}$

2. $\dfrac{-18}{108}$

3. $\dfrac{-7}{-9}$

4. $\dfrac{p^9}{p^4}$

5. $\dfrac{-x^7}{x^4}$

6. $\dfrac{a^{12}}{a^7}$

7. $\dfrac{-x^3y^2}{x^2y}$

8. $\dfrac{8x^3y^2}{4x^2y}$

9. $\dfrac{-12a^3b^4c^2}{-6a^2b^3c}$

10. $\dfrac{a^2 - b^2}{a + b}$

11. $\dfrac{4x^3}{12x^5 + 4x^3}$

12. $\dfrac{x^2 + 2x - 3}{x^2 + 5x + 6}$

13. $\dfrac{2x^2 - x - 1}{x^2 - 1}$

14. $\dfrac{a^2 - 9}{a^2 - 6a + 9}$

15. $\dfrac{u^2 - v^2}{u^2 - uv - 2v^2}$

16. $\dfrac{6p - 6}{p^2 - 7p + 6}$

17. $\dfrac{4 - 4x}{x^2 + 6x - 7}$

18. $\dfrac{a^2bc}{a^2x^2 + a^2}$

19. $\dfrac{5 + x}{5}$

20. $\dfrac{(x - 1)^4}{x^2 - 1}$

21. $\dfrac{x^3 + x^2}{x^3 - x}$

22. $\dfrac{y^2 + y - 6}{y^2 + 4y + 3}$

23. $\dfrac{a^2 - 144}{a^2 + 15a + 36}$

24. $\dfrac{m^2 + 4m + 4}{m^2 + 5m + 6}$

25. $\dfrac{2 - 3m + m^2}{m^2 - 1}$

7.2   Multiplying and dividing fractions

Before we examine how to multiply and divide algebraic fractions, let's review the operations of multiplication and division of arithmetic fractions.

To multiply fractions, we multiply numerators and multiply denominators. Factoring first, however, can ease the calculation. Three examples of multiplying and dividing arithmetic fractions follow.

Example 1

Multiply $\dfrac{12}{35} \cdot \dfrac{15}{16}$.

Solution

We rewrite each numerator and denominator in prime-factored form.

$$\frac{12}{35} \cdot \frac{15}{16} = \frac{2 \cdot 2 \cdot 3}{5 \cdot 7} \cdot \frac{3 \cdot 5}{2 \cdot 2 \cdot 2 \cdot 2}$$

Now wherever a factor appears in both a numerator and denominator, it can be divided out. So

$$\frac{12}{35} \cdot \frac{15}{16} = \frac{\cancel{2} \cdot \cancel{2} \cdot 3}{\cancel{5} \cdot 7} \cdot \frac{3 \cdot \cancel{5}}{\cancel{2} \cdot \cancel{2} \cdot 2 \cdot 2}$$

$$= \frac{3}{7} \cdot \frac{3}{4}$$

$$= \frac{9}{28}$$

Example 2

Divide $\dfrac{5}{27} \div \dfrac{25}{36}$.

Solution

First we "convert" this division into multiplication by rewriting $\dfrac{5}{27} \div \dfrac{25}{36}$ as $\dfrac{5}{27} \cdot \dfrac{36}{25}$. Then we proceed as in the previous example.

$$\frac{5}{27} \cdot \frac{36}{25} = \frac{5}{3 \cdot 3 \cdot 3} \cdot \frac{2 \cdot 2 \cdot 3 \cdot 3}{5 \cdot 5}$$

$$= \frac{\cancel{5}}{\cancel{3} \cdot \cancel{3} \cdot 3} \cdot \frac{2 \cdot 2 \cdot \cancel{3} \cdot \cancel{3}}{\cancel{5} \cdot 5}$$

$$= \frac{4}{15}$$

Example 3

Perform the indicated operations:

$$\left(\frac{5}{8} \cdot \frac{4}{15} \right) \div \frac{5}{12}$$

Solution

Perform the multiplication within parentheses first.

$$\left(\frac{5}{8} \cdot \frac{4}{15}\right) \div \frac{5}{12} = \frac{5}{2 \cdot 2 \cdot 2} \cdot \frac{2 \cdot 2}{3 \cdot 5} \div \frac{5}{12}$$

$$= \frac{\cancel{5}}{\cancel{2} \cdot \cancel{2} \cdot 2} \cdot \frac{\cancel{2} \cdot \cancel{2}}{3 \cdot \cancel{5}} \div \frac{5}{12}$$

$$= \frac{1}{6} \div \frac{5}{12}$$

$$= \frac{1}{6} \cdot \frac{12}{5} \qquad \text{Convert to multiplication.}$$

$$= \frac{1}{\cancel{2} \cdot \cancel{3}} \cdot \frac{\cancel{2} \cdot 2 \cdot \cancel{3}}{5}$$

$$= \frac{1}{1} \cdot \frac{2}{5}$$

$$= \frac{2}{5}$$

Now let's outline the procedure for multiplying algebraic fractions.

Multiplying algebraic fractions

1. Factor the numerators and denominators.
2. Any factor common to a numerator and denominator can be "divided out" (that is, divided by itself).
3. Multiply the remaining factors in the numerators; multiply the remaining factors in the denominators.

Some examples follow.

Example 4

Find the product of $\frac{x^2}{yz}$ and $\frac{5}{xz}$.

Solution

First, we factor in order to find any factors that may be common to both numerator and denominator.

$$\frac{x^2}{yz} \cdot \frac{5}{xz} = \frac{x \cdot x}{y \cdot z} \cdot \frac{5}{x \cdot z} \qquad \text{Factor.}$$

$$= \frac{\cancel{x} \cdot x}{y \cdot z} \cdot \frac{5}{\cancel{x} \cdot z} \qquad \text{Divide out } x.$$

$$= \frac{5x}{yz^2} \qquad \text{Multiply.}$$

Example 5

Multiply $\dfrac{-15c^2}{d^3} \cdot \dfrac{-3d^2}{10c^3}$.

Solution

$$\dfrac{-15c^2}{d^3} \cdot \dfrac{-3d^2}{10c^3} = \dfrac{-3 \cdot 5 \cdot c \cdot c}{d \cdot d \cdot d} \cdot \dfrac{-3 \cdot d \cdot d}{2 \cdot 5 \cdot c \cdot c \cdot c} \qquad \text{Factor.}$$

$$= \dfrac{-3 \cdot \cancel{5} \cdot \cancel{c} \cdot \cancel{c}}{d \cdot \cancel{d} \cdot \cancel{d}} \cdot \dfrac{-3 \cdot \cancel{d} \cdot \cancel{d}}{2 \cdot \cancel{5} \cdot \cancel{c} \cdot \cancel{c} \cdot c} \qquad \begin{array}{l}\text{Divide out } c^2, \\ 5, \text{ and } d^2.\end{array}$$

$$= \dfrac{-3}{d} \cdot \dfrac{-3}{2c}$$

$$= \dfrac{9}{2cd} \qquad \text{Multiply.}$$

Example 6

Multiply $\dfrac{5x - 10}{x + 2} \cdot \dfrac{6x + 12}{7x - 14}$.

Solution

$$\dfrac{5x - 10}{x + 2} \cdot \dfrac{6x + 12}{7x - 14} = \dfrac{5(x - 2)}{x + 2} \cdot \dfrac{6(x + 2)}{7(x - 2)} \qquad \text{Factor.}$$

$$= \dfrac{5(\cancel{x - 2})}{\cancel{x + 2}} \cdot \dfrac{6(\cancel{x + 2})}{7(\cancel{x - 2})} \qquad \begin{array}{l}\text{Divide out } x - 2 \\ \text{and } x + 2.\end{array}$$

$$= \dfrac{5}{1} \cdot \dfrac{6}{7}$$

$$= \dfrac{30}{7} \qquad \text{Multiply.}$$

Example 7

Multiply $\dfrac{y^5}{y - 1} \cdot \dfrac{y^2 + y - 2}{y^2}$.

Solution

$$\dfrac{y^5}{y - 1} \cdot \dfrac{y^2 + y - 2}{y^2} = \dfrac{y \cdot y \cdot y \cdot y \cdot y}{y - 1} \cdot \dfrac{(y + 2)(y - 1)}{y \cdot y} \qquad \text{Factor.}$$

$$= \dfrac{y \cdot y \cdot y \cdot \cancel{y} \cdot \cancel{y}}{\cancel{y - 1}} \cdot \dfrac{(y + 2)(\cancel{y - 1})}{\cancel{y} \cdot \cancel{y}} \qquad \begin{array}{l}\text{Divide out} \\ y^2 \text{ and } y - 1.\end{array}$$

$$= \dfrac{y^3}{1} \cdot \dfrac{y + 2}{1}$$

$$= y^3(y + 2) \qquad \text{Multiply.}$$

Note that the answer could be rewritten as $y^4 + 2y^3$. However, we prefer to show the answers to algebraic fraction exercises in factored form.

Example 8

Find the product:

$$\dfrac{x^2 - 9}{4x^2 - 9} \cdot \dfrac{2x^2 - 7x - 15}{2x^2 - 4x - 30}$$

Solution

$$\frac{x^2 - 9}{4x^2 - 9} \cdot \frac{2x^2 - 7x - 15}{2x^2 - 4x - 30} = \frac{(x + 3)(x - 3)}{(2x + 3)(2x - 3)} \cdot \frac{(2x + 3)(x - 5)}{2(x + 3)(x - 5)}$$ Factor.

$$= \frac{(x + 3)(x - 3)}{(2x + 3)(2x - 3)} \cdot \frac{(2x + 3)(x - 5)}{2(x + 3)(x - 5)}$$ Divide out $2x + 3$, $x + 3$, and $x - 5$.

$$= \frac{x - 3}{2x - 3} \cdot \frac{1}{2}$$

$$= \frac{x - 3}{2(2x - 3)}$$ Multiply.

To divide two algebraic fractions $\frac{a}{b} \div \frac{c}{d}$, we follow the rule we learned in arithmetic: invert the second fraction and multiply. So

$$\frac{a}{b} \div \frac{c}{d}$$

is the same as

$$\frac{a}{b} \cdot \frac{d}{c}$$

Example 9

Find the following quotient:

$$\frac{5a^2}{b^2} \div \frac{10c^2}{3}$$

Solution

"Find the quotient" means "divide." So

$$\frac{5a^2}{b^2} \div \frac{10c^2}{3} = \frac{5a^2}{b^2} \cdot \frac{3}{10c^2}$$ Convert to multiplication.

$$= \frac{5 \cdot a \cdot a}{b \cdot b} \cdot \frac{3}{2 \cdot 5 \cdot c \cdot c}$$ Factor and divide common factors.

$$= \frac{3a^2}{2b^2c^2}$$ Multiply.

Example 10

Divide $\frac{x^2 - 9}{12}$ by $\frac{3x^2 + 21x + 36}{8x}$.

Solution

$$\frac{x^2 - 9}{12} \div \frac{3x^2 + 21x + 36}{8x} = \frac{x^2 - 9}{12} \cdot \frac{8x}{3x^2 + 21x + 36}$$ Convert to multiplication.

$$= \frac{(x + 3)(x - 3)}{4 \cdot 3} \cdot \frac{4 \cdot 2 \cdot x}{3(x + 3)(x + 4)}$$ Factor and divide out $x + 3$ and 4.

$$= \frac{2x(x - 3)}{9(x + 4)}$$ Multiply.

Example 11

Find the following quotient:

$$\frac{2x^2 - 9x + 7}{(x + 2)^2} \div \frac{49 - 4x^2}{x^2 + 5x + 6}$$

Solution

$$\frac{2x^2 - 9x + 7}{(x + 2)^2} \div \frac{49 - 4x^2}{x^2 + 5x + 6}$$

$$= \frac{2x^2 - 9x + 7}{(x + 2)^2} \cdot \frac{x^2 + 5x + 6}{49 - 4x^2}$$

Convert to multiplication.
Note:

$$49 - 4x^2 = -4x^2 + 49$$
$$= -1(4x^2 - 49)$$
$$= -1(2x + 7)(2x - 7)$$

$$= \frac{(2x - 7)(x - 1)}{(x + 2)(x + 2)} \cdot \frac{(x + 2)(x + 3)}{-1(2x + 7)(2x - 7)}$$

$$= \frac{x - 1}{x + 2} \cdot \frac{x + 3}{-1(2x + 7)}$$

Factor and divide out $2x - 7$ and $x + 2$.

$$= \frac{(x - 1)(x + 3)}{-(x + 2)(2x + 7)}$$

Multiply.

$$= \frac{-(x - 1)(x + 3)}{(x + 2)(2x + 7)}$$

Change the signs of the numerator and denominator.

We conclude this section with an example that combines multiplication and division of algebraic fractions.

Example 12

Perform the indicated operations.

$$\frac{x^2 - 8x + 15}{7x^2} \cdot \frac{x}{3} \div \frac{x - 3}{2}$$

Solution

We convert the division to multiplication and find the product of the three resulting fractions.

$$\frac{x^2 - 8x + 15}{7x^2} \cdot \frac{x}{3} \div \frac{x - 3}{2} = \frac{x^2 - 8x + 15}{7x^2} \cdot \frac{x}{3} \cdot \frac{2}{x - 3}$$

Convert division to multiplication.

$$= \frac{(x - 3)(x - 5)}{7 \cdot x \cdot x} \cdot \frac{x}{3} \cdot \frac{2}{x - 3}$$

Factor and divide common factors.

$$= \frac{2(x - 5)}{21x}$$

Multiply.

Section 7.2 Exercises

In Exercises 1 through 40, perform the indicated operations and simplify your results, if possible.

1. $\dfrac{7}{15} \cdot \dfrac{3}{8}$

2. $\dfrac{11}{24} \cdot \dfrac{15}{22}$

3. $\dfrac{3}{8} \cdot \dfrac{7}{9}$

4. $\dfrac{6}{35} \cdot \dfrac{7}{2}$

5. $\dfrac{1}{4} \div \dfrac{1}{2}$

6. $\dfrac{8}{13} \div \dfrac{16}{15}$

7. $\dfrac{1}{9} \div \dfrac{5}{18}$

8. $\left(\dfrac{7}{8} \div \dfrac{21}{32}\right) \cdot \dfrac{3}{4}$

9. $\dfrac{x^2}{y} \cdot \dfrac{y^2}{2xyz}$

10. $\dfrac{5x}{-3} \cdot \dfrac{-1}{x^3}$

11. $\dfrac{-6ab}{5} \cdot \dfrac{15b^2}{7a^3}$

12. $\dfrac{-48x^2y^2z^2}{12x^3} \cdot \dfrac{-3y^3}{-9z^3}$

13. $\dfrac{3x - 12}{2x + 6} \cdot \dfrac{3x + 4}{x - 4}$

14. $\dfrac{x^2 - 7x + 12}{x^2 - 16} \cdot \dfrac{x^2 + 5x + 4}{x^2 - 2x - 3}$

15. $\dfrac{x^2 - 7x + 12}{16 - x^2} \cdot \dfrac{x^2 + 5x + 4}{x^2 - 2x - 3}$

16. $\dfrac{y}{y^2 - 9} \cdot \dfrac{y^2 - 6y + 9}{y + 3}$

17. $\dfrac{3x^2 - 7x + 2}{4 - x^2} \cdot \dfrac{3x^2 + 10x + 8}{1 - 9x^2}$

18. $\dfrac{x^2 - 9x + 20}{x - 4} \cdot \dfrac{x - 5}{3}$

19. $\dfrac{2x - 10}{3x - 9} \cdot \dfrac{x - 3}{x - 5} \cdot \dfrac{5}{x^2}$

20. $\dfrac{2a}{3b} \div \dfrac{7a}{9b}$

21. $\dfrac{2a}{3b} \div \dfrac{9b}{7a}$

22. $\dfrac{8x^2}{7y} \div (-x^2)$

23. $\dfrac{x}{x^2 - 1} \div \dfrac{1}{x + 1}$

24. $\dfrac{x^2 - 9}{x^2 - 4} \div \dfrac{x^2 + 6x + 9}{x^2 + 4x + 4}$

25. $\dfrac{x^3 - x^2}{y^2 - 1} \div \dfrac{x^3}{y - 1}$

26. $\dfrac{x^2 + 6x + 9}{4x^2 + 4x + 1} \div \dfrac{x + 3}{2x + 1}$

27. $\dfrac{3y^2 + 2y - 1}{1 - y^2} \div \dfrac{-6y^2 + 8y - 2}{y^2 - 2y + 1}$

28. $\dfrac{y^2 - y - 6}{10y - 20} \div \dfrac{3 - y}{-5}$

29. $\dfrac{x^2 - 7x + 12}{x^2 - 16} \div \dfrac{x^2 - 2x - 3}{2x + 8}$

30. $\dfrac{-4y^2}{7} \cdot \dfrac{3y + 1}{8y^5} \div \dfrac{9y^2 - 1}{y^3}$

31. $\dfrac{4t^2 - 9}{t^2 - 16} \cdot \dfrac{t^2 - t - 20}{2t^4 + 3t^3} \div \dfrac{t - 5}{t^2}$

32. $\dfrac{y^2 - 9}{9 - x^2} \cdot \dfrac{x + 3}{y + 3} \div \dfrac{7}{x - 3}$

33. $\left(\dfrac{x^3 - x}{y^4 - y^2} \div \dfrac{x^3}{y^3}\right) \cdot \dfrac{1}{x}$

34. $\dfrac{x^3 - x}{y^4 - y^2} \div \left(\dfrac{x^3}{y^3} \cdot \dfrac{1}{x}\right)$

35. $(5x^2 - 20) \div \dfrac{x - 2}{3}$

36. $\dfrac{x^2 - 9}{x^2 + 5x} \cdot \dfrac{6x}{x^2 - x - 6} \cdot \dfrac{x^2 - 25}{3x + 9}$

37. $\dfrac{5x^2y^3}{x^2 + 6x + 9} \div \dfrac{-10xy^5}{x^2 - 9}$

38. $\dfrac{3a}{a^2 + 2ab + b^2} \cdot \dfrac{a^2 - b^2}{6a^2}$

39. $\dfrac{4x + 8}{y^4} \cdot \dfrac{x^4y}{x^2 - 4x + 4} \div \dfrac{6x^5}{x^2 - 4}$

40. $\dfrac{9x^2 - 1}{y} \div (3x + 1)$

Supplementary Exercises

In Exercises 1 through 15, perform the indicated operations and simplify your results, if possible.

1. $\dfrac{4}{9} \cdot \dfrac{27}{16}$

2. $\dfrac{-2}{5} \cdot \dfrac{35}{8}$

3. $\dfrac{9}{16} \div \dfrac{1}{4}$

4. $\dfrac{-3}{5} \div \dfrac{6}{7}$

5. $\dfrac{x^3}{y} \cdot \dfrac{5y^2}{x}$

6. $\dfrac{3x^2}{y^3} \cdot \dfrac{-2y^4}{9x^4}$

7. $\dfrac{2x^2}{y} \div 4x^3$

8. $\dfrac{-4ab}{7} \cdot \dfrac{14}{b^2} \div 2a^2b$

9. $\dfrac{-8}{x^2 - 9} \cdot \dfrac{x^2 - 6x + 9}{4}$

10. $\dfrac{3x - 9}{10 - 2x} \cdot \dfrac{x - 5}{x - 3}$

11. $\dfrac{x^2 + x - 2}{x + 1} \cdot \dfrac{3x + 3}{1 - x}$

12. $\dfrac{x(x + 2)}{y(2y - 1)} \cdot \dfrac{y}{4x + 8} \div \dfrac{x^2}{y^3}$

13. $\dfrac{(x + 1)(x + 2)}{y} \cdot \dfrac{xy^2}{x^3 + 2x^2}$

14. $\dfrac{x + 3}{6y - 6} \div \dfrac{3}{y^2 - 1}$

15. $\dfrac{-2a^2b}{5xy^3} \div \dfrac{4ab^2}{15x^2y}$

7.3   The LCM and arithmetic fractions

In this section we lay the groundwork for adding and subtracting *algebraic* fractions. That groundwork consists of studying how fractions are added and subtracted in *arithmetic*.

Fractions with the same denominator are called **like fractions.** To add them, we simply add the numerators and keep the denominator. For example,

$$\frac{3}{7} + \frac{2}{7} = \frac{5}{7}$$

When two fractions do not have the same denominator, they are called **unlike fractions** and the procedure for adding them is somewhat more lengthy. To add unlike fractions $\left(\text{such as } \dfrac{3}{8} + \dfrac{3}{10}\right)$, we must first rewrite them so that they have the same denominator. To do so, we need to find the **least common multiple (LCM)** of the denominators, sometimes referred to as the **lowest common denominator (LCD).**

Finding the LCM of two numbers

Step 1 Write the prime factorizations of each number.

Step 2 Note the greatest number of times *each* factor occurs in *each* number.

Step 3 The LCM is the product of factors noted in Step 2, each raised to the power of its highest frequency.

Example 1

Find the LCM of 8 and 10.

Solution

Step 1

$$\left.\begin{array}{l} 8 = 2 \cdot 2 \cdot 2 \\ 10 = 2 \cdot 5 \end{array}\right\} \quad \text{Prime factorizations}$$

Step 2

Note that the greatest number of times the factor 2 occurs is 3 times. So the LCM will contain $2 \cdot 2 \cdot 2$ or 2^3. The greatest number of times the factor 5 occurs is once. So the LCM will contain 5^1 or 5.

Step 3

$$\text{LCM} = 2^3 \cdot 5 = 40$$

In other words, the least common multiple of 8 and 10, or, equivalently, the smallest number that both 8 and 10 will evenly divide into is 40.

Example 2

Find the LCM of 36 and 60.

Solution

Step 1

$$36 = 2 \cdot 2 \cdot 3 \cdot 3$$
$$60 = 2 \cdot 2 \cdot 3 \cdot 5$$

Step 2

The greatest number of times the factor 2 appears is 2 times.
The greatest number of times the factor 3 appears is 2 times.
The greatest number of times the factor 5 appears is 1 time.

Step 3

$$\text{LCM} = 2^2 \cdot 3^2 \cdot 5 = 180$$

The smallest number that both 36 and 60 divide into evenly is 180. This also means that both 36 and 60 can be "converted" to 180 by multiplication of a suitable number. It is precisely this property that allows us to rewrite two unlike fractions so that they have the same denominator. Once we do so, we can add them as we would add like fractions. In general, the procedure is:

Adding unlike fractions

1. Find the LCM of the denominators.
2. "Build up" each denominator to the LCM by multiplying each by an appropriate integer. Remember to multiply each numerator by the same appropriate integer.
3. Add the resulting numerators and keep the LCM denominator.
4. Reduce, if possible.

Example 3

Add $\dfrac{11}{36} + \dfrac{7}{60}$.

Solution

From Example 2 we know that the LCM of 36 and 60 is 180. Thus, 180 is the lowest common denominator. Next, we "build up" the fractions.

$$\frac{11}{36} + \frac{7}{60} = \frac{11 \cdot 5}{36 \cdot 5} + \frac{7 \cdot 3}{60 \cdot 3}$$

We change each denominator to 180. (*Note*: Multiplying numerator and denominator by the same quantity does not change the value of the fraction.)

$$= \frac{55}{180} + \frac{21}{180}$$

$$= \frac{76}{180}$$

Add numerators, keep denominator.

$$= \frac{4 \cdot 19}{4 \cdot 45}$$

$$= \frac{19}{45}$$

Reduce the fraction.

That is, $\frac{11}{36} + \frac{7}{60} = \frac{19}{45}$.

Example 4

Perform the indicated operations.

$$\frac{1}{12} + \frac{5}{42} - \frac{2}{15}$$

Solution

The lowest common denominator will be the LCM of 12, 42, and 15.

$$12 = 2 \cdot 2 \cdot 3$$
$$42 = 2 \cdot 3 \cdot 7$$
$$15 = 3 \cdot 5$$

The greatest number of times that the factor 2 occurs is twice . The factors 3, 5, and 7 each occur once. Thus, the LCM $= 2^2 \cdot 3 \cdot 5 \cdot 7 = 420$. So we build up each denominator to 420.

$$\frac{1}{12} + \frac{5}{42} - \frac{2}{15} = \frac{1 \cdot 35}{12 \cdot 35} + \frac{5 \cdot 10}{42 \cdot 10} - \frac{2 \cdot 28}{15 \cdot 28}$$

Note: $12 \cdot 35 = 420$
$42 \cdot 10 = 420$
$15 \cdot 28 = 420$

$$= \frac{35}{420} + \frac{50}{420} - \frac{56}{420}$$

$$= \frac{35 + 50 - 56}{420}$$

$$= \frac{29}{420}$$

In the next section, when we deal with algebraic fractions, we will use a very similar procedure. There, however, the factorization is algebraic factorization.

Section 7.3 Exercises

In Exercises 1 through 18, find the LCM for each set of numbers.

1. 6, 15   2. 14, 7   3. 6, 12

4. 14, 12   5. 15, 15   6. 60, 72

7. 64, 96   8. 20, 30   9. 30, 40

10. 30, 60   11. 20, 30, 60   12. 6, 12, 15

13. 2, 4, 6   14. 9, 18, 81   15. 45, 63, 315

16. 45, 63, 99   17. 10, 11, 12   18. 10, 12, 14

In Exercises 19 through 40, perform the indicated operation(s) and express each fraction in reduced form.

19. $\dfrac{3}{10} + \dfrac{1}{6}$

20. $\dfrac{3}{10} + \dfrac{5}{6}$

21. $\dfrac{5}{10} + \dfrac{3}{6}$

22. $\dfrac{7}{36} + \dfrac{11}{60}$

23. $\dfrac{11}{36} + \dfrac{13}{60}$

24. $\dfrac{5}{12} + \dfrac{1}{18}$

25. $\dfrac{5}{12} - \dfrac{1}{18}$

26. $\dfrac{7}{30} + \dfrac{2}{75}$

27. $\dfrac{7}{30} - \dfrac{2}{75}$

28. $\dfrac{2}{75} - \dfrac{7}{30}$

29. $\left(\dfrac{7}{12} + \dfrac{2}{9}\right) - \dfrac{1}{18}$

30. $\left(\dfrac{7}{12} - \dfrac{2}{9}\right) - \dfrac{1}{18}$

31. $\left(\dfrac{1}{2} - \dfrac{1}{3}\right) + \dfrac{3}{4}$

32. $\left(\dfrac{1}{2} - \dfrac{1}{3}\right) - \dfrac{3}{4}$

33. $\left(\dfrac{1}{20} - \dfrac{1}{30}\right) + \dfrac{3}{40}$

34. $\left(\dfrac{7}{12} + \dfrac{2}{9}\right) \cdot \dfrac{1}{18}$

35. $\left(\dfrac{7}{12} + \dfrac{2}{9}\right) \div \dfrac{1}{18}$

36. $\dfrac{1}{2} \cdot \dfrac{8}{9} - \dfrac{1}{3}$

37. $\dfrac{3}{4} \div \dfrac{2}{9} + \dfrac{1}{12}$

38. $\left(\dfrac{2}{3} + \dfrac{1}{2}\right) - \left(\dfrac{1}{3} + \dfrac{1}{4}\right)$

39. $\left(\dfrac{3}{5} - \dfrac{1}{3}\right) - \left(\dfrac{6}{5} - \dfrac{2}{3}\right)$

40. $\dfrac{1}{4} - \left(\dfrac{2}{3} + \dfrac{1}{5}\right)$

Supplementary Exercises

In Exercises 1 through 5, find the LCM for each set of numbers.

1. 3, 4, 5   2. 6, 9, 12   3. 8, 12, 15

4. 3, 5, 7   5. 2, 8, 9

In Exercises 6 through 20, perform the indicated operation(s) and express each fraction in reduced form.

6. $\dfrac{2}{3} + \dfrac{3}{5}$

7. $\dfrac{1}{4} - \dfrac{2}{3}$

8. $\dfrac{1}{5} + \dfrac{2}{3} + \dfrac{1}{4}$

9. $\dfrac{2}{5} - \dfrac{1}{3}$

10. $\dfrac{3}{7} - \dfrac{1}{9}$

11. $\dfrac{1}{7} - \dfrac{2}{3} - \dfrac{1}{14}$

12. $\dfrac{3}{4} + \dfrac{1}{2} - \dfrac{2}{3}$

13. $\dfrac{1}{7} + \dfrac{3}{8} - \dfrac{1}{4}$

14. $\dfrac{3}{4} + \dfrac{1}{3} - \dfrac{1}{5}$

15. $\dfrac{1}{8} - \dfrac{11}{12} + \dfrac{2}{9}$

16. $\dfrac{3}{4} - \dfrac{2}{3} + \dfrac{4}{5}$

17. $\dfrac{6}{7} - \dfrac{1}{8} + \dfrac{3}{4}$

18. $\dfrac{3}{4} \cdot \dfrac{2}{9} - \dfrac{1}{6}$

19. $\dfrac{4}{5} + \dfrac{3}{8} \div \dfrac{9}{16}$

20. $\left(\dfrac{2}{3} - \dfrac{1}{2}\right) \div \dfrac{5}{12}$

7.4 Adding and subtracting algebraic fractions

The same procedure for adding and subtracting arithmetic fractions is used for adding and subtracting algebraic fractions. If the algebraic fractions are *like* fractions, we add numerators and keep the like denominator. Two examples follow.

Example 1

Add $\dfrac{5}{x} + \dfrac{2}{x}$.

Solution

$$\frac{5}{x} + \frac{2}{x} = \frac{5 + 2}{x} \qquad \text{Add numerators; keep the denominator.}$$

$$= \frac{7}{x}$$

Example 2

Add $\dfrac{2x}{x^2 - 1} + \dfrac{2}{x^2 - 1}$.

Solution

We proceed as in the previous example. However, notice that we can reduce the fraction before we state the final answer.

$$\frac{2x}{x^2 - 1} + \frac{2}{x^2 - 1} = \frac{2x + 2}{x^2 - 1} \qquad \begin{array}{l}\text{Add the numerators;}\\ \text{keep the denominator.}\end{array}$$

$$= \frac{2(\cancel{x + 1})}{(\cancel{x + 1})(x - 1)} \qquad \begin{array}{l}\text{Factor and divide out}\\ x + 1.\end{array}$$

$$= \frac{2}{x - 1}$$

When the denominators of two fractions are not the same, we must first find the least common multiple (LCM) of the two denominators. Although we will add algebraic fractions in the same manner, finding the LCM must be mastered first. The next four examples show how to find the LCM of algebraic quantities.

Example 3

Find the LCM of $x^2 y$ and xy^3

Solution

We create the LCM by writing a certain product of factors. We obtain that product of factors as follows.

$$\left.\begin{array}{l} x^2 y = x \cdot x \cdot y \\ xy^3 = x \cdot y \cdot y \cdot y \end{array}\right\} \quad \begin{array}{l}\text{Now compare these two expres-}\\ \text{sions and make note of the great-}\\ \text{est number of times a factor}\\ \text{occurs in either.}\end{array}$$

The greatest number of times x appears as a factor is 2 times. Thus the LCM will contain $x \cdot x$, or x^2.

The greatest number of times y appears as a factor is 3 times. Thus the LCM will contain $y \cdot y \cdot y$, or y^3.

So LCM $= x^2 y^3$.

Example 4

Find the LCM of $x^3 yz$ and xz^4.

Solution

First we completely factor each expression.

$$\left. \begin{aligned} x^3 yz &= x \cdot x \cdot x \cdot y \cdot z \\ xz^4 &= x \cdot z \cdot z \cdot z \cdot z \end{aligned} \right\} \quad \begin{aligned} &\text{Now compare these two} \\ &\text{expressions.} \end{aligned}$$

The greatest number of times x appears as a factor is 3 times. Thus the LCM must contain $x \cdot x \cdot x$, or x^3.

The greatest number of times y appears as a factor is 1 time. Thus the LCM must contain y.

The greatest number of times z appears as a factor is 4 times. Thus the LCM must contain $z \cdot z \cdot z \cdot z$, or z^4.

So LCM $= x^3 yz^4$.

Example 5

Find the LCM of $x^2 - 1$ and $x^2 + 2x + 1$.

Solution

$$\left. \begin{aligned} x^2 - 1 &= (x + 1)(x - 1) \\ x^2 + 2x + 1 &= (x + 1)(x + 1) \end{aligned} \right\} \quad \begin{aligned} &\text{Completely factor} \\ &\text{each expression.} \end{aligned}$$

The LCM contains $(x + 1)(x + 1)$ (since it appears twice in $x^2 + 2x + 1$) and $x - 1$ (since it appears once in $x^2 - 1$).

So LCM $= (x + 1)^2 (x - 1)$.

The LCM of three (or more) quantities can also be found as the next example illustrates.

Example 6

Find the LCM of $5x - 15$, $x^2 - 6x + 9$, and $x^2 - 7x + 12$.

Solution

$$\left. \begin{aligned} 5x - 15 &= 5(x - 3) \\ x^2 - 6x + 9 &= (x - 3)(x - 3) \\ x^2 - 7x + 12 &= (x - 3)(x - 4) \end{aligned} \right\} \quad \text{Factor.}$$

The LCM is composed of 5, $(x - 3)(x - 3)$, and $(x - 4)$. (Every time a factor appears it must be included in the LCM *and* it must be included the greatest number of times it occurs in any one expression.)

So LCM $= 5(x - 3)^2 (x - 4)$.

We are now ready to add algebraic fractions. We restate the procedure developed in Section 7.3.

Adding unlike algebraic fractions

1. Find the LCM of the denominators.
2. "Build up" each denominator to the LCM by multiplying by appropriate factor(s). Also, the numerator must be multiplied by those same appropriate factor(s).
3. Add the resulting numerators; keep the LCM denominator.
4. Reduce, if possible.

Example 7

Add $\dfrac{2}{x^2y} + \dfrac{7}{xy^3}$.

Solution

The LCM of x^2y and xy^3 is x^2y^3 (see Example 3). Thus, we must build up both fractions so that each has the denominator x^2y^3.

$$\frac{2}{x^2y} \qquad + \qquad \frac{7}{xy^3}$$

This denominator must be multiplied by y^2 for it to become x^2y^3.

This denominator must be multiplied by x for it to become x^2y^3.

$$\frac{2 \cdot y^2}{x^2y \cdot y^2} \qquad + \qquad \frac{7 \cdot x}{xy^3 \cdot x}$$

Multiply *numerator and denominator* by y^2.

Multiply *numerator and denominator* by x.

$$\frac{2}{x^2y} + \frac{7}{xy^3} = \frac{2y^2}{x^2y^3} + \frac{7x}{x^2y^3} = \frac{2y^2 + 7x}{x^2y^3}$$

Example 8

Add $\dfrac{3}{x^2 - 3x + 2} + \dfrac{4}{x^2 - 1}$.

Solution

We leave it to the reader to verify that the LCM of $x^2 - 3x + 2$ and $x^2 - 1$ is $(x - 1)(x - 2)(x + 1)$. (Do it!) We build up and add the fractions

$$\frac{3}{x^2 - 3x + 2} + \frac{4}{x^2 - 1}$$

$$= \frac{3}{(x - 1)(x - 2)} + \frac{4}{(x + 1)(x - 1)}$$ Factor denominators.

$$= \frac{3(x + 1)}{(x - 1)(x - 2)(x + 1)} + \frac{4(x - 2)}{(x + 1)(x - 1)(x - 2)}$$ Multiply numerator and denominator by the right quantity so that both denominators become $(x - 1)(x - 2)(x + 1)$, which is the LCM of the denominators.

$$= \frac{3x + 3}{(x - 1)(x - 2)(x + 1)} + \frac{4x - 8}{(x + 1)(x - 1)(x - 2)}$$ Multiply numerators.

$$= \frac{7x - 5}{(x - 1)(x - 2)(x + 1)}$$ Add numerators; keep denominators.

Example 9

Add $\dfrac{x}{x^2 + 6x + 9} + \dfrac{2}{x^2 + 8x + 15}$.

Solution

The LCM of $x^2 + 6x + 9$ and $x^2 + 8x + 15$ is $(x + 3)^2(x + 5)$. *(Why?)*

$$\dfrac{x}{x^2 + 6x + 9} + \dfrac{2}{x^2 + 8x + 15}$$

$$= \dfrac{x}{(x + 3)^2} + \dfrac{2}{(x + 3)(x + 5)} \qquad \text{Factor denominators.}$$

$$= \dfrac{x(x + 5)}{(x + 3)^2(x + 5)} + \dfrac{2(x + 3)}{(x + 3)(x + 5)(x + 3)} \qquad \text{Build up each fraction so that each denominator is } (x + 3)^2(x + 5).$$

$$= \dfrac{x^2 + 5x}{(x + 3)^2(x + 5)} + \dfrac{2x + 6}{(x + 3)^2(x + 5)} \qquad \text{Multiply numerators.}$$

$$= \dfrac{x^2 + 7x + 6}{(x + 3)^2(x + 5)} \qquad \text{Add numerators; keep denominator.}$$

$$= \dfrac{(x + 6)(x + 1)}{(x + 3)^2(x + 5)} \qquad \text{Factor numerator.}$$

Since every subtraction problem can be converted to addition (recall, $a - b = a + -b$), subtracting two fractions is very similar to adding two fractions. We present two examples involving subtraction.

Example 10

Subtract $\dfrac{x}{x + 3} - \dfrac{7}{x - 2}$.

Solution

First we convert subtraction to addition.

$$\dfrac{x}{x + 3} - \dfrac{7}{x - 2} = \dfrac{x}{x + 3} + \dfrac{-7}{x - 2}$$

The LCM of $x + 3$ and $x - 2$ is $(x + 3)(x - 2)$.

$$\dfrac{x}{x + 3} + \dfrac{-7}{x - 2}$$

$$= \dfrac{x(x - 2)}{(x + 3)(x - 2)} + \dfrac{-7(x + 3)}{(x - 2)(x + 3)} \qquad \text{Build up each fraction so that each denominator is } (x + 3)(x - 2).$$

$$= \dfrac{x^2 - 2x}{(x + 3)(x - 2)} + \dfrac{-7x - 21}{(x + 3)(x - 2)} \qquad \text{Multiply numerators.}$$

$$= \dfrac{x^2 - 2x - 7x - 21}{(x + 3)(x - 2)} \qquad \text{Add numerators; keep denominators.}$$

$$= \dfrac{x^2 - 9x - 21}{(x + 3)(x - 2)}$$

Example 11

Subtract $\dfrac{x}{x^2 - 16} - \dfrac{6}{x^2 + 5x + 4}$.

Solution

$$\frac{x}{x^2 - 16} - \frac{6}{x^2 + 5x + 4}$$

$$= \frac{x}{x^2 - 16} + \frac{-6}{x^2 + 5x + 4} \qquad \text{Convert to addition.}$$

$$= \frac{x}{(x + 4)(x - 4)} + \frac{-6}{(x + 4)(x + 1)} \qquad \text{Factor denominators.}$$

The LCM of $x^2 - 16$ and $x^2 + 5x + 4$ is $(x + 4)(x - 4)(x + 1)$.

$$\frac{x}{(x + 4)(x - 4)} + \frac{-6}{(x + 4)(x + 1)}$$

$$= \frac{x(x + 1)}{(x + 4)(x - 4)(x + 1)} + \frac{-6(x - 4)}{(x + 4)(x - 4)(x + 1)} \qquad \text{Build up fractions.}$$

$$= \frac{x^2 + x}{(x + 4)(x - 4)(x + 1)} + \frac{-6x + 24}{(x + 4)(x - 4)(x + 1)} \qquad \text{Multiply numerators.}$$

$$= \frac{x^2 - 5x + 24}{(x + 4)(x - 4)(x + 1)} \qquad \text{Add numerators; keep denominators.}$$

Section 7.4 Exercises

In Exercises 1 through 5, add the given fractions.

1. $\dfrac{7}{a} + \dfrac{8}{a}$

2. $\dfrac{2}{x - 3} + \dfrac{7}{x - 3}$

3. $\dfrac{3x}{x + 7} + \dfrac{21}{x + 7}$

4. $\dfrac{5x}{x^2 + x - 20} + \dfrac{21}{x^2 + x - 20}$

5. $\dfrac{5x}{x^2 + x - 20} + \dfrac{25}{x^2 + x - 20}$

In Exercises 6 through 15, find the LCM of the given expressions.

6. $a^2b^3,\ a^4$

7. $x^4y,\ xy^4$

8. $4x + 8,\ x + 2$

9. $5y,\ 3y$

10. $5y,\ 3y^2$

11. $x^2 - 9,\ x^2 + 4x - 21$

12. $x^2 - 7x + 12,\ x^2 + 4x - 21$

13. $x^4 - x^2,\ x^3 + 2x^2 + x$

14. $x^4 - x^2,\ x^3 + 2x^2 + x,\ x^3$

15. $6x - 12,\ 9x^2 - 36,\ x^2 + 4x + 4$

In Exercises 16 through 40, perform the indicated operation(s).

16. $\dfrac{1}{x} + \dfrac{1}{y}$

17. $\dfrac{1}{x^2} + \dfrac{8}{xy^2}$

18. $\dfrac{7}{a^2b^3} + \dfrac{b}{a^4}$

19. $\dfrac{9}{4x + 8} + \dfrac{7}{x + 2}$

20. $\dfrac{x}{x+4} + \dfrac{1}{x+3}$

21. $\dfrac{x}{x^2-4} + \dfrac{3}{x^2+5x-14}$

22. $\dfrac{1}{x} + \dfrac{1}{y} + \dfrac{1}{z}$

23. $\dfrac{1}{x^2+x-2} + \dfrac{3}{x^2-4x+4}$

24. $\dfrac{x}{x-1} + \dfrac{x^2+1}{x^2-1}$

25. $\dfrac{3}{3x^2-4x+1} + \dfrac{x}{x^2-1}$

26. $\dfrac{10}{x^2-y^2} + \dfrac{2}{x^2+2xy+y^2}$

27. $\dfrac{1}{x^2-9} + \dfrac{1}{x^2-7x+12}$

takeout common factor

−28. $\dfrac{x}{x^4-x^2} + \dfrac{7}{x^3+2x^2+x}$

29. $\dfrac{1}{6x-12} + \dfrac{2}{9x^2-36} + \dfrac{x+1}{x^2+4x+4}$

go before page 2

30. $\dfrac{10}{2x+3} - \dfrac{1}{4x+6} + \dfrac{x}{6x+9}$

31. $\dfrac{x}{x^2-4} - \dfrac{3}{x^2+5x-14}$

32. $\dfrac{x}{x^2-4} - \dfrac{3}{x^2+9x+14}$

33. $\dfrac{3x}{x^2+5x+6} - \dfrac{2x-1}{x^2+6x+9}$

34. $\dfrac{3x}{x^2+5x+6} - \dfrac{2x-1}{x^2+6x+9} + \dfrac{1}{x+2}$

35. $\dfrac{1}{x^2-9} - \dfrac{3}{x^2-6x+9} + \dfrac{2}{x^2+6x+9}$

36. $\dfrac{1}{a^2-4} - \dfrac{2}{a-2}$

37. $\dfrac{a}{a^2-9} + \dfrac{3}{a^2+6a+9}$

38. $\dfrac{2}{x-2} + \dfrac{3}{x^2-4} - \dfrac{1}{x+2}$

39. $\dfrac{3}{x-2} - \left(\dfrac{2}{x+2} - \dfrac{3}{x^2-4} \right)$

40. $\dfrac{2x}{x^2-16} - \dfrac{2}{x+4}$

Supplementary Exercises

In Exercises 1 through 3, add the given fractions.

1. $\dfrac{3x}{x+5} + \dfrac{15}{x+5}$

2. $\dfrac{4x}{x+2} + \dfrac{8}{x+2}$

3. $\dfrac{x}{x^2+2x+1} + \dfrac{1}{x^2+2x+1}$

In Exercises 4 through 8, find the LCM of the given expressions.

4. $3x^2y^2,\ -6xy^3$

5. $a^2b,\ ab^3$

6. $x^2 - 100,\ x^2 - 11x + 10$

7. $x - 3,\ 9 - x^2$

8. $x^2 - 9,\ x^2 - x - 6$

In Exercises 9 through 20, perform the indicated operation(s) and simplify, if possible.

9. $\dfrac{5}{a^2b} + \dfrac{4}{ab^2}$

10. $\dfrac{6}{5x^2y} - \dfrac{4}{3xy^2}$

11. $\dfrac{3x}{x^2-9} + \dfrac{2}{x+3}$

12. $\dfrac{5x}{x^2-16} - \dfrac{5}{x-4}$

13. $\dfrac{-2}{9-x^2} + \dfrac{3}{x^2-6x+9}$

14. $\dfrac{5}{x+2} - \dfrac{x-1}{x^2+3x+2}$

15. $\dfrac{3x}{x^2-1} + \dfrac{2x}{x^2+2x+1}$

16. $\dfrac{3}{a^2b} - \dfrac{2}{ab^2} + \dfrac{4}{a^2b^2}$

17. $\dfrac{2}{x+1} - \dfrac{3}{x^2 - 1} + \dfrac{5}{x - 1}$ 18. $\dfrac{2x}{x^2 + 3x - 4} - \dfrac{x}{x^2 - 16}$

19. $\dfrac{3}{x + 1} - \dfrac{2}{x + 2} + \dfrac{-2}{x^2 + 3x + 2}$ 20. $\dfrac{2}{x^2 + 5x + 4} - \dfrac{3}{x^2 + 6x + 8}$

7.5 Complex fractions

When the numerator or denominator of a fraction contains a fraction, the main fraction is called a **complex fraction.**

$$\dfrac{\dfrac{3}{x^2 - 1} + \dfrac{6}{x + 1}}{\dfrac{8}{x - 1}} \quad \begin{array}{l} \text{Numerator} \\[18pt] \text{Denominator} \end{array}$$

Example of a Complex Fraction

In the example $\dfrac{3}{x^2 - 1}, \dfrac{6}{x + 1}$, and $\dfrac{8}{x - 1}$ are called **secondary fractions.** They are, in fact, fractions within fractions.

Our goal in this section is to be able to rewrite a complex fraction as a **simple fraction,** that is, a fraction whose numerator and denominator do *not* contain fractions. There are two methods for rewriting complex fractions. The first one involves multiplying the main numerator and main denominator by the LCM of secondary denominators. The procedures is as follows.

Simplifying complex fractions

Method 1

1. Find the LCM of all secondary denominators.
2. Multiply the main numerator and main denominator by the LCM found in part 1.
3. Factor and reduce, if possible.

Three examples employing Method 1 follow.

Example 1

Simplify

$$\dfrac{\dfrac{1}{4} + \dfrac{2}{3}}{\dfrac{5}{24}}$$

Solution

The secondary denominators are 4, 3, and 24.

$$4 = 2 \cdot 2$$
$$3 = 3$$
$$24 = 2 \cdot 2 \cdot 2 \cdot 3$$

The LCM of secondary denominators is $2 \cdot 2 \cdot 2 \cdot 3 = 24$.

Next, we multiply the numerator and denominator by 24.

$$\frac{\frac{1}{4}+\frac{2}{3}}{\frac{5}{24}} = \frac{\frac{1}{4}+\frac{2}{3}}{\frac{5}{24}} \cdot \frac{24}{24}$$

$$= \frac{\frac{1}{4}\cdot\frac{24}{1}+\frac{2}{3}\cdot\frac{24}{1}}{\frac{5}{24}\cdot\frac{24}{1}} \qquad \text{Use the distributive property.}$$

$$= \frac{6+16}{5}$$

$$= \frac{22}{5} = 4\frac{2}{5}$$

Example 2

Simplify

$$\frac{\frac{2}{x}+\frac{1}{y}}{\frac{1}{x}}$$

Solution

The secondary denominators are x, y, and x. Their LCM is xy.

$$\frac{\frac{2}{x}+\frac{1}{y}}{\frac{1}{x}} = \frac{\frac{2}{x}+\frac{1}{y}}{\frac{1}{x}} \cdot \frac{xy}{xy}$$

$$= \frac{\frac{2}{x}\cdot\frac{xy}{1}+\frac{1}{y}\cdot\frac{xy}{1}}{\frac{1}{x}\cdot\frac{xy}{1}}$$

$$= \frac{2y+x}{y}$$

$$= \frac{x+2y}{y}$$

Example 3

Simplify

$$\frac{\frac{1}{x^2}-9}{\frac{1}{x}+3}$$

Solution

The secondary denominators are x and x^2; their LCM is x^2.

$$\frac{\dfrac{1}{x^2} - 9}{\dfrac{1}{x} + 3} = \frac{\dfrac{1}{x^2} - 9}{\dfrac{1}{x} + 3} \cdot \frac{x^2}{x^2}$$

$$= \frac{\dfrac{1}{x^2} \cdot \dfrac{x^2}{1} - \dfrac{9}{1} \cdot \dfrac{x^2}{1}}{\dfrac{1}{x} \cdot \dfrac{x^2}{1} + \dfrac{3}{1} \cdot \dfrac{x^2}{1}} \qquad \text{Multiply by } x^2 \text{ using the distributive property.}$$

$$= \frac{1 - 9x^2}{x + 3x^2}$$

$$= \frac{(1 + 3x)(1 - 3x)}{x(1 + 3x)} \qquad \text{Factor.}$$

$$= \frac{1 - 3x}{x} \qquad \text{Reduce.}$$

We can also simplify complex fractions by first performing any necessary operations to obtain at most one fraction in the main fraction's numerator and at most one fraction in the main fraction's denominator. This second method is summarized in the following.

Simplifying complex fractions

Method 2

1. Perform any indicated operations in the numerator so that it contains at most one fraction.
2. Do the same for the denominator.
3. Divide the numerator by the denominator.

For example, consider

$$\frac{\dfrac{1}{4} + \dfrac{2}{3}}{\dfrac{5}{24}}$$

first seen in Example 1. To apply Method 2, we would first add $\dfrac{1}{4} + \dfrac{2}{3}$.

$$\frac{1}{4} + \frac{2}{3} = \frac{3}{12} + \frac{8}{12} = \frac{11}{12}$$

Now, we can replace $\dfrac{1}{4} + \dfrac{2}{3}$ with $\dfrac{11}{12}$.

$$\frac{\dfrac{1}{4} + \dfrac{2}{3}}{\dfrac{5}{24}} = \frac{\dfrac{11}{12}}{\dfrac{5}{24}} \quad \longleftarrow \text{means} \div$$

$$= \frac{11}{12} \div \frac{5}{24}$$

$$= \frac{11}{12} \cdot \frac{24}{5} \qquad \text{Convert division to multiplication.}$$

$$= \frac{22}{5} \qquad \text{Multiply.}$$

$$= 4\frac{2}{5}$$

It is important to realize that there is no "best" method. In certain cases Method 1 is easier; sometimes Method 2 is. We conclude this section with two more examples of applying Method 2.

Example 4

Simplify

$$\frac{\dfrac{6}{a} - \dfrac{3}{a^2}}{4 - \dfrac{1}{a^2}}$$

Solution

(1) Combine numerator.

$$\frac{6}{a} - \frac{3}{a^2} = \frac{6a}{a^2} - \frac{3}{a^2} \qquad \text{The LCD is } a^2.$$

$$= \frac{6a - 3}{a^2}$$

$$= \frac{3(2a - 1)}{a^2} \qquad \text{Factor.}$$

(2) Combine denominator.

$$4 - \frac{1}{a^2} = \frac{4a^2}{a^2} - \frac{1}{a^2} \qquad \text{The LCD is } a^2.$$

$$= \frac{4a^2 - 1}{a^2}$$

$$= \frac{(2a + 1)(2a - 1)}{a^2} \qquad \text{Factor.}$$

(3) Divide.

$$\frac{\dfrac{6}{a} - \dfrac{3}{a^2}}{4 - \dfrac{1}{a^2}} = \frac{\dfrac{3(2a - 1)}{a^2}}{\dfrac{(2a + 1)(2a - 1)}{a^2}}$$

$$= \frac{3(2a - 1)}{a^2} \div \frac{(2a + 1)(2a - 1)}{a^2}$$

$$= \frac{3(2a - 1)}{a^2} \cdot \frac{a^2}{(2a + 1)(2a - 1)} \qquad \begin{array}{l}\text{Invert and}\\ \text{multiply.}\end{array}$$

$$= \frac{3}{2a + 1} \qquad \begin{array}{l}\text{Multiply numerators}\\ \text{and denominators}\\ \text{and simplify.}\end{array}$$

Example 5

Simplify

$$\frac{x^2 - y^2}{\dfrac{1}{x} + \dfrac{1}{y}}$$

Solution

Combine

$$\frac{1}{x} + \frac{1}{y} = \frac{y}{xy} + \frac{x}{xy} \qquad \text{The LCD is } xy.$$

$$= \frac{y + x}{xy}$$

We have

$$\frac{x^2 - y^2}{\dfrac{1}{x} + \dfrac{1}{y}} = x^2 - y^2 \div \frac{x + y}{xy} \qquad \text{Divide.}$$

$$= \frac{x^2 - y^2}{1} \cdot \frac{xy}{x + y} \qquad \text{Multiply.}$$

$$= \frac{(\cancel{x + y})(x - y)}{1} \cdot \frac{xy}{\cancel{x + y}} \qquad \text{Factor.}$$

$$= (x - y)xy \qquad \text{Multiply numerators}$$
$$\text{or} \quad xy(x - y) \qquad \begin{array}{l}\text{and denominators and}\\ \text{simplify.}\end{array}$$

Section 7.5 Exercises

In Exercises 1 through 21, simplify each complex fraction. Use either Method 1 or Method 2.

1. $\dfrac{\dfrac{5}{6} - \dfrac{1}{3}}{\dfrac{5}{9} + \dfrac{1}{6}}$

2. $\dfrac{\dfrac{5}{6} + \dfrac{1}{3}}{\dfrac{5}{9} - \dfrac{1}{6}}$

3. $\dfrac{\dfrac{7}{12} - \dfrac{1}{2}}{\dfrac{2}{3} + \dfrac{3}{4}}$

4. $\dfrac{\dfrac{7}{12} + \dfrac{1}{2}}{\dfrac{2}{3} + \dfrac{3}{4}}$

5. $\dfrac{\dfrac{7}{12x} + \dfrac{1}{2x}}{\dfrac{2}{3x} + \dfrac{3}{4x}}$

6. $\dfrac{\dfrac{7x}{12} + \dfrac{x}{2}}{\dfrac{2x}{3} + \dfrac{3x}{4}}$

7. $\dfrac{\dfrac{5}{6x} - \dfrac{1}{3y}}{\dfrac{5}{9x} + \dfrac{1}{6y}}$

8. $\dfrac{\dfrac{3x^2}{4y}}{\dfrac{8x}{9y^2}}$

9. $\dfrac{\dfrac{6xy^2}{5z}}{\dfrac{12xy}{25z^2}}$

10. $\dfrac{\dfrac{6xy^2}{5 - z}}{\dfrac{12xy}{25 - z^2}}$

11. $\dfrac{\dfrac{1}{x} + \dfrac{1}{y}}{\dfrac{1}{x}}$

12. $\dfrac{\dfrac{1}{x} - \dfrac{1}{y}}{\dfrac{1}{x}}$

13. $\dfrac{\dfrac{1}{x} + \dfrac{1}{y}}{x}$

14. $\dfrac{\dfrac{1}{x^2} - 16}{\dfrac{1}{x} + 4}$

15. $\dfrac{\dfrac{y^2 - 9}{4}}{\dfrac{y^2 + 6y + 9}{6}}$

16. $\dfrac{\dfrac{z^2 - 1}{z}}{\dfrac{z^2 + 2z + 1}{z^2}}$

17. $\dfrac{\dfrac{6}{x} - 2}{7 - \dfrac{21}{x}}$

18. $\dfrac{a - \dfrac{1}{a}}{1 + \dfrac{1}{a}}$

19. $\dfrac{a - \dfrac{1}{a}}{a + \dfrac{1}{a}}$

20. $\dfrac{\dfrac{3}{x^2 - 1} + \dfrac{6}{x + 1}}{\dfrac{8}{x - 1}}$

21. $\dfrac{1 - \dfrac{7}{x - 2}}{3x + \dfrac{1}{x}}$

Supplementary Exercises

In Exercises 1 through 12 simplify each complex fraction.

1. $\dfrac{\frac{5}{8}}{\frac{-5}{4}}$

2. $\dfrac{1 + \frac{2}{3}}{\frac{1}{9}}$

3. $\dfrac{\frac{3}{4} \cdot \frac{8}{6}}{\frac{7}{8}}$

4. $\dfrac{\frac{a^2}{b}}{\frac{a}{b^3}}$

5. $\dfrac{\frac{ab^3}{c}}{\frac{2a}{c} \cdot \frac{b^2}{c^2}}$

6. $\dfrac{1 + \frac{1}{x}}{1 - \frac{1}{x}}$

7. $\dfrac{5 - \frac{1}{a}}{5 + \frac{1}{a}}$

8. $\dfrac{a + \frac{1}{b}}{b + \frac{1}{a}}$

9. $\dfrac{x - \frac{1}{x}}{x^3}$

10. $\dfrac{y - \frac{1}{y}}{y^2 + y}$

11. $\dfrac{\frac{x^2 - 9}{y}}{x - \frac{9}{x}}$

12. $\dfrac{\frac{5}{x} + 1}{x^2 + 6x + 5}$

7.6   Fractional equations

An equation in which a variable appears in at least one fraction's denominator is called a **fractional equation**. In this section we will introduce methods for solving fractional equations. The first three examples use the notion of cross-multiplying; the last four examples make use of the LCM principle. Remember that cross-multiplication can be used only when a **single** fraction appears on each side of the equation. If more than one fraction appears on either side of the equation, we *must* use the LCM principle.

Example 1

Solve for x: $\dfrac{1}{x} = \dfrac{2}{3}$.

Solution

To solve $\dfrac{1}{x} = \dfrac{2}{3}$, use the definition of equality of fractions, that is, cross-multiplying.

$$\frac{1}{x} = \frac{2}{3}$$

$$\frac{1}{x} \diagdown\!\!\!\!\diagup \frac{2}{3}$$

$$2x = 3$$

$$x = \frac{3}{2}$$

Check:
$$\frac{1}{x} = \frac{2}{3}$$

We return to the *original equation* for our check.

$$\frac{1}{\frac{3}{2}} = \frac{2}{3}$$

$$1 \div \frac{3}{2} = 1 \cdot \frac{2}{3} = \frac{2}{3}$$

$$\frac{2}{3} = \frac{2}{3} \checkmark$$

Example 2

Solve $\dfrac{7}{3x - 1} = \dfrac{1}{x - 3}$.

Solution

Here we have *one* fraction equal to *one* other fraction. So we use the notion of cross-multiplying.

$$\frac{7}{3x - 1} = \frac{1}{x - 3}$$

$$\frac{7}{3x - 1} \diagdown\!\!\!\!\diagup \frac{1}{x - 3}$$

$$7(x - 3) = 3x - 1$$
$$7x - 21 = 3x - 1$$
$$4x = 20$$
$$x = 5$$

Check:

$$\frac{7}{3x - 1} = \frac{1}{x - 3}$$

$$\frac{7}{3 \cdot 5 - 1} = \frac{1}{5 - 3}$$

$$\frac{7}{15 - 1} = \frac{1}{2}$$

$$\frac{7}{14} = \frac{1}{2}$$

$$\frac{1}{2} = \frac{1}{2} \checkmark$$

Example 3

Solve $\dfrac{y}{y - 4} = \dfrac{4}{y - 4}$.

Solution

$$\frac{y}{y - 4} = \frac{4}{y - 4}$$

$$\frac{y}{y - 4} \diagdown\!\!\!\!\diagup \frac{4}{y - 4}$$

$$y^2 - 4y = 4y - 16$$
$$y^2 - 8y + 16 = 0 \qquad \text{We subtract } 4y - 16 \text{ from both sides.}$$
$$(y - 4)(y - 4) = 0$$

Thus $$y = 4$$

It looks like $y = 4$ is the answer; however, let us perform the usual check.

Check:
$$\frac{y}{y - 4} = \frac{4}{y - 4}$$

$$\frac{4}{4 - 4} = \frac{4}{4 - 4}$$

Here we have 0 in the denominator. Since division by zero is not defined, our result of $y = 4$ is *not* an answer. The equation $\dfrac{y}{y - 4} = \dfrac{4}{y - 4}$ has *no solution*.

The final four examples show the method that must be used to solve fractional equations involving more than one fraction on either side of the equation. That method is multiplying both sides of the equation by the LCM of all the denominators.

Example 4

Solve $\dfrac{2}{x} + \dfrac{1}{3} = 1$.

Solution

The equation can be written as

$$\frac{2}{x} + \frac{1}{3} = \frac{1}{1}$$

We next multiply both sides by the LCM of the three denominators, $(x, 3, \text{ and } 1)$, which is $3x$.

$$\frac{3x}{1}\left(\frac{2}{x} + \frac{1}{3}\right) = \left(\frac{1}{1}\right)\frac{3x}{1}$$

$$\frac{6x}{x} + \frac{3x}{3} = \frac{3x}{1}$$

In the first two fractions simplification is possible. We obtain

$$6 + x = 3x$$
$$6 = 2x$$
$$3 = x$$
$$x = 3$$

Notice that, by multiplying both sides of the original equation by the LCM of all the denominators, we get an equation that is free of fractions.

Check:
$$\frac{2}{x} + \frac{1}{3} = 1$$

$$\frac{2}{3} + \frac{1}{3} = 1$$

$$\frac{3}{3} = 1$$

$$1 = 1 \checkmark$$

Example 5

Solve $\dfrac{2}{x} + \dfrac{8}{x^2} = 1$.

Solution

This equation can be written as

$$\frac{2}{x} + \frac{8}{x^2} = \frac{1}{1}$$

We next multiply both sides by the LCM of the three denominators, $(x,\ x^2,$ and $1)$, which is x^2.

$$\frac{x^2}{1}\left(\frac{2}{x} + \frac{8}{x^2}\right) = \left(\frac{1}{1}\right)\frac{x^2}{1}$$

$$\frac{2x^2}{x} + \frac{8x^2}{x^2} = \frac{x^2}{1}$$

In the first two fractions simplification is possible. We obtain

$$2x + 8 = x^2$$

Notice again that, by multiplying both sides of the original equation by the LCM of all the denominators, we get an equation that is free of fractions.

So $0 = x^2 - 2x - 8$

or $x^2 - 2x - 8 = 0$ This is a quadratic equation.

$$(x - 4)(x + 2) = 0$$

$$x - 4 = 0 \qquad x + 2 = 0$$

$$x = 4 \qquad x = -2$$

Check for $x = 4$:   **Check** for $x = -2$:

$$\frac{2}{x} + \frac{8}{x^2} = 1 \qquad\qquad \frac{2}{x} + \frac{8}{x^2} = 1$$

$$\frac{2}{4} + \frac{8}{4^2} = 1 \qquad\qquad \frac{2}{-2} + \frac{8}{(-2)^2} = 1$$

$$\frac{1}{2} + \frac{8}{16} = 1 \qquad\qquad -1 + \frac{8}{4} = 1$$

$$\frac{1}{2} + \frac{1}{2} = 1 \qquad\qquad -1 + 2 = 1$$

$$1 = 1 \ \checkmark \qquad\qquad 1 = 1 \ \checkmark$$

Example 6

Solve $\dfrac{3x}{x + 1} + \dfrac{9}{x^2 - 1} = \dfrac{5}{x - 1}$.

Solution

We use the method of multiplying both sides of the equation by the LCM of the three denominators $(x + 1,\ x^2 - 1,$ and $x - 1)$, which is $(x + 1)(x - 1)$.

$$\frac{3x}{x + 1} + \frac{9}{x^2 - 1} = \frac{5}{x - 1}$$

$$\frac{(x + 1)(x - 1)}{1}\left(\frac{3x}{x + 1} + \frac{9}{x^2 - 1}\right) = \left(\frac{5}{x - 1}\right)\frac{(x + 1)(x - 1)}{1}$$

$$\frac{3x(x + 1)(x - 1)}{x + 1} + \frac{9(x + 1)(x - 1)}{x^2 - 1} = \frac{5(x + 1)(x - 1)}{x - 1}$$

$$\frac{3x(x + 1)(x - 1)}{x + 1} + \frac{9(x + 1)(x - 1)}{(x + 1)(x - 1)} = \frac{5(x + 1)(x - 1)}{x - 1}$$

We now have

$$3x(x - 1) + 9 = 5(x + 1)$$

Again, notice that, by multiplying both sides of the original equation by the LCM of all the denominators, we get an equation that is free of fractions.

$$3x(x - 1) + 9 = 5(x + 1)$$

$$3x^2 - 3x + 9 = 5x + 5 \qquad \text{Subtract } 5x + 5 \text{ from both sides.}$$

$$3x^2 - 8x + 4 = 0 \longleftarrow \text{This is a quadratic equation.}$$

$$(3x - 2)(x - 2) = 0$$

$$\begin{array}{rl} 3x - 2 = 0 & \quad x - 2 = 0 \\ 3x = 2 & \quad x = 2 \\ x = \dfrac{2}{3} & \end{array}$$

We have two solutions:

$$x = \frac{2}{3} \quad \text{and} \quad x = 2$$

We check each solution below.

Check for $x = \dfrac{2}{3}$:

$$\frac{3x}{x + 1} + \frac{9}{x^2 - 1} = \frac{5}{x - 1}$$

$$\frac{\frac{2}{5}}{\frac{5}{3}} + \frac{9}{\frac{4}{9} - 1} = \frac{5}{\frac{2}{3} - 1}$$

$$\frac{6}{5} - \frac{81}{5} = -15$$

$$\frac{-75}{5} = -15$$

$$-15 = -15 \; \checkmark$$

Check for $x = 2$:

$$\frac{3x}{x + 1} + \frac{9}{x^2 - 1} = \frac{5}{x - 1}$$

$$\frac{6}{3} + \frac{9}{4 - 1} = \frac{5}{2 - 1}$$

$$2 + 3 = 5$$

$$5 = 5 \; \checkmark$$

Example 7

Solve $\dfrac{x}{x - 2} - \dfrac{2}{x + 2} = \dfrac{4x}{x^2 - 4}$.

Solution

We multiply both sides of the equation by the LCM of the three denominators ($x - 2$, $x + 2$, and $x^2 - 4$), which is $(x + 2)(x - 2)$.

$$\frac{x}{x - 2} - \frac{2}{x + 2} = \frac{4x}{x^2 - 4}$$

$$\frac{(x + 2)(x - 2)}{1}\left(\frac{x}{x - 2} - \frac{2}{x + 2}\right) = \left(\frac{4x}{x^2 - 4}\right)\frac{(x + 2)(x - 2)}{1}$$

$$\frac{x(x + 2)(x - 2)}{x - 2} - \frac{2(x + 2)(x - 2)}{x + 2} = \frac{4x(x + 2)(x - 2)}{(x + 2)(x - 2)}$$

The denominator of the third fraction has been factored.

$$x(x + 2) - 2(x - 2) = 4x$$

$$x^2 + 2x - 2x + 4 = 4x \qquad \text{Subtract } 4x \text{ from both sides.}$$

$$x^2 - 4x + 4 = 0 \quad \leftarrow \text{ This is a quadratic equation.}$$

$$(x - 2)(x - 2) = 0$$

$$x = 2$$

It appears that $x = 2$ is the answer; however, let us perform the usual check.

Check:

$$\frac{x}{x - 2} - \frac{2}{x + 2} = \frac{4x}{x^2 - 4}$$

$$\frac{2}{2 - 2} - \frac{2}{2 + 2} = \frac{4 \cdot 2}{2^2 - 4}$$

Here we have 0 in the denominator. This is impossible, since we cannot divide by zero. So $x = 2$ is *not* an answer. The equation

$$\frac{x}{x - 2} - \frac{2}{x + 2} = \frac{4x}{x^2 - 4}$$

has *no solution.*

We conclude this section with a return to literal equations. In particular, we will solve **fractional literal equations** for a specified variable. The next three examples illustrate this.

Example 8

Solve for x: $\dfrac{1}{x} = \dfrac{1}{a} + \dfrac{1}{b}$.

Solution

We multiply, just as before, by the LCM of the three denominators (a, b, and x), which is abx.

$$\frac{1}{x} = \frac{1}{a} + \frac{1}{b}$$

$$\frac{abx}{1}\left(\frac{1}{x}\right) = \left(\frac{1}{a} + \frac{1}{b}\right)\frac{abx}{1}$$

$$\frac{abx}{x} = \frac{abx}{a} + \frac{abx}{b}$$

$$ab = bx + ax$$

We want to isolate the variable x; that is, eventually we want x on one side of the equation. One way to accomplish this is by factoring the righthand side.

$$ab = bx + ax$$

$$ab = x(b + a) \qquad \text{Factor.}$$

$$\frac{ab}{b + a} = x \qquad\qquad \text{Divide both sides by } b + a.$$

or

$$x = \frac{ab}{a + b}$$

Example 9

Solve for r: $S = \dfrac{a}{1-r}$.

Solution

We can write $S = \dfrac{a}{1-r}$ as $\dfrac{S}{1} = \dfrac{a}{1-r}$ and use the cross-multiplying technique.

$$\frac{S}{1} = \frac{a}{1-r}$$

$$\frac{S}{1} \diagdown\!\!\!\!\diagup \frac{a}{1-r}$$

$$S(1-r) = a$$

$$S - Sr = a$$

$$-Sr = a - S \qquad \text{We must isolate } r\text{, so we subtract } S \text{ from both sides.}$$

$$r = \frac{a-S}{-S} \qquad \text{We do not want } r \text{ being multiplied by } -S\text{, so we divide by } -S \text{ on both sides.}$$

$$r = \frac{S-a}{S} \qquad \text{Multiply numerator and denominator by } -1.$$

Example 10

Solve for x: $\dfrac{a+b}{x+1} - \dfrac{1}{x} = \dfrac{b^2}{x}$.

Solution

We multiply by the LCM of the three denominators $(x+1,\ x,\ \text{and } x)$, which is $x(x+1)$.

$$\frac{x(x+1)}{1}\left(\frac{a+b}{x+1} - \frac{1}{x}\right) = \left(\frac{b^2}{x}\right)\frac{x(x+1)}{1}$$

$$\frac{(a+b)x(x+1)}{x+1} - \frac{x(x+1)}{x} = \frac{b^2 x(x+1)}{x}$$

$$x(a+b) - (x+1) = b^2(x+1)$$

$$ax + bx - x - 1 = b^2 x + b^2 \qquad \text{Multiply.}$$

$$ax + bx - x - b^2 x = b^2 + 1 \qquad \text{Collect terms involving } x \text{ on the lefthand side.}$$

$$x(a + b - 1 - b^2) = b^2 + 1 \qquad \text{Factor out } x.$$

$$x = \frac{b^2+1}{a+b-1-b^2} \qquad \text{Divide by } (a+b-1-b^2).$$

Section 7.6 Exercises

In Exercises 1 through 30, solve for x.

1. $\dfrac{12}{x} = \dfrac{6}{5}$

2. $\dfrac{x}{x+1} = \dfrac{3}{4}$

3. $\dfrac{4}{x+1} = \dfrac{9}{x^2}$

4. $\dfrac{x}{2} = \dfrac{2}{x}$

5. $\dfrac{1}{x} = \dfrac{1}{4}$

6. $\dfrac{5}{x} = \dfrac{1}{2}$

7. $\dfrac{2x}{x-1} = 3$

8. $\dfrac{2x}{x-1} = 2$

9. $\dfrac{2}{7} + \dfrac{5}{x} = 1$

10. $\dfrac{3}{x} + \dfrac{1}{4} = 1$

11. $\dfrac{3}{4} + \dfrac{x}{x+1} = \dfrac{3}{2}$

12. $\dfrac{x}{x+1} + \dfrac{1}{2} = 1$

13. $\dfrac{1}{x+1} - \dfrac{1}{x-1} = \dfrac{2x}{x^2-1}$

14. $\dfrac{7}{3x-1} - \dfrac{1}{x-3} = \dfrac{-4}{11}$

15. $\dfrac{x^2}{x-4} = \dfrac{16}{x-4}$

16. $1 - \dfrac{2}{x+2} = \dfrac{5}{x+2}$

17. $\dfrac{x}{x+1} - \dfrac{7}{2x+1} = \dfrac{1}{2x^2+3x+1}$

18. $\dfrac{x+3}{x} + \dfrac{1}{x-4} = \dfrac{12}{x}$

19. $\dfrac{x}{3x+5} + \dfrac{1}{x} = \dfrac{9}{3x^2+5x}$

20. $\dfrac{2-x}{10x} + \dfrac{1}{12} = \dfrac{2}{15x}$

21. $\dfrac{x}{2} = \dfrac{3x+1}{x+7}$

22. $\dfrac{2}{x+4} + \dfrac{1}{2(2x+3)} = \dfrac{1}{2}$

23. $\dfrac{1}{x} + \dfrac{1}{y} = a + \dfrac{1}{b}$

24. $\dfrac{1}{ax+b} + \dfrac{1}{c} = 2$

25. $\dfrac{-c^2}{ax} + \dfrac{11}{x-1} = \dfrac{1}{ax^2-ax}$

26. $\dfrac{5x}{3-2x} = -3$

27. $\dfrac{x+1}{x} = \dfrac{4}{3}$

28. $\dfrac{x+1}{x} + \dfrac{2}{x-1} = \dfrac{1}{x}$

29. $\dfrac{3x-1}{x+2} + \dfrac{4}{3} = \dfrac{-1}{3}$

30. $\dfrac{x-1}{x} = \dfrac{x}{x-2}$

In Exercises 31 through 35, solve for the specified variable.

31. $\dfrac{a}{x} + \dfrac{b}{x} = c$ (for x)

32. $\dfrac{1}{f} = \dfrac{1}{s_1} + \dfrac{1}{s_2}$ (for s_1)

33. $F = \dfrac{GmM}{r^2}$ (for M)

34. $l = a + (n-1)d$ (for n)

35. $l = a + (n-1)d$ (for d)

36. We have seen that the process of cross-multiplying will sometimes lead to a situation in which no solution exists. In parts (a) and (b) below, what happens when the process is used? Solve each equation for x.

(a) $\dfrac{x}{x} = \dfrac{x}{x}$

(b) $\dfrac{x-1}{x-2} = \dfrac{x-1}{x-2}$

Supplementary Exercises

In Exercises 1 through 18 solve for x.

1. $\dfrac{8}{x} = \dfrac{4}{3}$

2. $\dfrac{3}{x} = \dfrac{-6}{5}$

3. $\dfrac{x}{5} = \dfrac{x}{-6}$

4. $\dfrac{x}{x-1} = \dfrac{2}{3}$

5. $\dfrac{3}{x+3} = \dfrac{2}{x}$

6. $\dfrac{2x}{x-1} = \dfrac{1}{2}$

7. $\dfrac{5-x}{3} = \dfrac{1-x}{2}$

8. $\dfrac{1}{2} + \dfrac{2}{x} = 3$

9. $\dfrac{3}{4} - \dfrac{1}{x+1} = 1$

10. $\dfrac{3}{x+3} - \dfrac{1}{x-3} = \dfrac{6}{x^2-9}$

11. $\dfrac{x^2}{x-5} = \dfrac{25}{x-5}$

12. $\dfrac{x}{x+3} - \dfrac{1}{x} = \dfrac{x}{x^2+3x}$

13. $\dfrac{3}{x+2} - \dfrac{1}{2x+1} = \dfrac{2}{2x^2+5x+2}$

14. $\dfrac{1}{4} + \dfrac{x}{x-1} = \dfrac{1}{4}$

15. $\dfrac{2}{3} - \dfrac{x+3}{x-9} = \dfrac{2}{3}$

16. $\dfrac{x}{4} = \dfrac{5-x^2}{x}$

17. $\dfrac{x^2-12}{x} = \dfrac{-x}{3}$

18. $\dfrac{m}{x} + \dfrac{n}{x} = 2$

7.7   Chapter review

Summary

We began this chapter by discussing the use of the greatest common factor (GCF) to reduce arithmetic fractions. We saw that a similar process was used to simplify (reduce) algebraic fractions. In Section 7.1 we also discovered that cross-multiplying can be used to determine whether two fractions are equivalent.

$$\frac{5}{8} = \frac{10}{16}$$

$$\frac{5}{8} \diagdown \frac{10}{16}$$

$$5 \cdot 16 = 8 \cdot 10$$
$$80 = 80$$

In Section 7.2 we examined the operations of multiplying and dividing fractions. Again, we noted the strong analogy between what happens in arithmetic and what happens in algebra.

Fractions were added and subtracted in Sections 7.3 (arithmetic) and 7.4 (algebraic). To add or subtract unlike fractions, the least common multiple (LCM) of the denominators of fractions must be found. Then all the denominators are "built up" to that LCD (the lowest common denominator). Once that is done, the fractions are "like" and can be added by adding the numerators and keeping the denominator the same.

When a fraction appears within a fraction, the expression is called a complex fraction. Two ways to simplify an algebraic complex fraction were examined in Section 7.5.

Fractional equations were discussed in Section 7.6. We learned that multiplying both sides of a fractional equation by the LCM of all denominators will produce an equation free of fractions. We also saw that solutions to fractional equations must be checked carefully. Sometimes apparent solutions do not check out in the original equation.

Vocabulary Quiz

Match the expression in Column I with the phrase in Column II that best describes it.

Column I

1. To reduce a fraction, divide the numerator and denominator by

2. If $\dfrac{a}{b} = \dfrac{c}{d}$, then

3. If two fractions have the same denominator, they are called

4. Cross-multiplying is a technique used to find

5. The first step in adding two fractions is to examine the denominators and find their

6. Fractions whose numerators or denominators are themselves fractions, are called

Column II

a. like fractions.

b. LCM.

c. equivalent fractions.

d. GCF.

e. complex fractions.

f. $ad = bc$.

Chapter 7 Review Exercises

In Exercises 1 through 10, reduce each fraction to lowest terms, if possible.

1. $\dfrac{15}{18}$

2. $\dfrac{35}{55}$

3. $\dfrac{350}{550}$

4. $\dfrac{9x^3}{12x^5}$

5. $\dfrac{x^2 - 8x + 15}{x^2 - 25}$

6. $\dfrac{2y^2}{4y^2 - 24y}$

7. $\dfrac{2y^2 - 12y}{4y^3 - 24y^2}$

8. $\dfrac{12y - 2y^2}{4y^3 - 24y^2}$

9. $\dfrac{a + b}{a^2 - b^2}$

10. $\dfrac{a - b}{b^2 - a^2}$

11. Rewrite $-\dfrac{7}{-15}$ so that there is at most one "$-$" symbol and no "$-$" symbol in the denominator.

In Exercises 12 through 21, perform the indicated operation(s) and simplify the result.

12. $\dfrac{20}{77} \cdot \dfrac{33}{64}$

13. $\dfrac{4}{27} \div \dfrac{2}{15}$

14. $\left(\dfrac{7}{8} \div \dfrac{7}{32}\right) \cdot \dfrac{3}{4}$

15. $\dfrac{-10x^2y}{9x^3} \cdot \dfrac{21xy}{4y^2}$

16. $\dfrac{3a - 15}{6a - 42} \cdot \dfrac{a^2 + 7a + 10}{a^2 - 25}$

17. $\dfrac{3y - 15}{6y - 42} \div \dfrac{25 - y^2}{y^2 + 7y + 10}$

18. $\dfrac{x}{x^2 - 16} \cdot \dfrac{x^2 - 8x + 16}{x + 3}$

19. $\dfrac{3x - 5}{6x - 21} \cdot \dfrac{6x^2 - 11x - 35}{9x^2 - 25}$

20. $\dfrac{2x + y}{2x - y} \cdot \dfrac{4x^2 - y^2}{4x^2 + 4xy + y^2}$

21. $\dfrac{2x + y}{y - 2x} \cdot \dfrac{4x^2 - y^2}{4x^2 + 4xy + y^2}$

In Exercises 22 through 28, find the LCM.

22. 15, 21

23. 30, 42

24. 12, 18, 60

25. $xy^2,\ x^3y,\ x^2y^4$

26. $y^2 - 9y + 20, \; y^2 + y - 20$ 27. $x^2 - 1, \; x^4 - 1$

28. $2x^2 + 3x - 5, \; 6x - 15, \; 4x^2 - 25$

In Exercises 29 through 39, perform the indicated operation(s) and simplify the result.

29. $\dfrac{8}{36} + \dfrac{11}{60}$ 30. $\dfrac{8}{60} + \dfrac{11}{36}$

31. $\dfrac{19}{25} - \dfrac{7}{30}$ 32. $\dfrac{8}{x^2 y} + \dfrac{3}{x^2 y}$

33. $\dfrac{4a}{4a^2 - 1} - \dfrac{1}{4a^2 - 1}$ 34. $\dfrac{5}{x} + \dfrac{7}{x^2}$

35. $\dfrac{2y}{y^2 - 2y - 48} + \dfrac{3}{y + 6}$ 36. $\dfrac{2y}{y^2 - 2y - 48} - \dfrac{3}{y + 6}$

37. $\dfrac{x}{x^2 - 4} + \dfrac{x + 7}{x^2 + 7x + 10}$ 38. $\dfrac{2x}{2x^2 - 3x + 1} - \dfrac{2x}{x^2 - 1}$

39. $\dfrac{1}{y^2 - 9} - \dfrac{3}{y^2 - 6y + 9} + \dfrac{2}{y^2 + 6y + 9}$

In Exercises 40 through 45, simplify each complex fraction.

40. $\dfrac{\dfrac{1}{a} + \dfrac{1}{b}}{a}$ 41. $\dfrac{\dfrac{1}{a} - \dfrac{1}{b}}{\dfrac{1}{a}}$

42. $\dfrac{\dfrac{1}{x - 2} - \dfrac{3}{x + 3}}{\dfrac{2x}{x^2 + x - 6}}$ 43. $\dfrac{\dfrac{8x}{2x - 1} + \dfrac{3}{x - 1}}{\dfrac{4x}{2x^2 - 3x + 1}}$

44. $\dfrac{\dfrac{7a^2 b}{3 - a}}{\dfrac{14ab}{a^2 - 9}}$ 45. $\dfrac{\dfrac{7a^2 b}{3 - a}}{\dfrac{14ab}{9 - a^2}}$

In Exercises 46 through 50 solve each equation for x.

46. $\dfrac{2}{x} = \dfrac{4}{9}$ 47. $\dfrac{x}{4} = \dfrac{x + 1}{3}$

48. $\dfrac{1}{2x + 1} - \dfrac{2}{x + 2} = \dfrac{x - 5}{2x + 4}$ 49. $\dfrac{4}{x + 4} + \dfrac{1}{2x + 3} = 1$

50. $\dfrac{1}{x} - \dfrac{1}{a} = \dfrac{1}{y} + 2$

Cumulative review exercises

1. Evaluate $-6^2 + 2[3 - 2(\sqrt[3]{-27} - 6)]$.   (*Section 1.8*)
2. Simplify $(3xy)(-2xy^2)(4x^2 y^3)$.   (*Section 2.3*)
3. Find the area of the triangle with base = 6 m and height = 16 m.   (*Section 2.5*)

4. Solve for x: $-5 = \dfrac{3x}{2} - 1$.   (*Section 3.3*)

5. How much pure silver must be added to a 45-kilogram alloy that is 20% silver in order to raise the percentage of silver to 25%?   (*Section 4.3*)

6. Simplify by rewriting the expression using only positive exponents. Assume the variables cannot be zero.   (*Section 5.2*)

$$\frac{8x^{-3}y^2}{48x^4y^{-3}}$$

7. Perform the indicated operations.

$$(2x^3 - x + 3) - (3x^2 - 2x - 5) + (x^3 - x^2 + 6x - 3)$$ (*Section 5.4*)

8. Divide $(2x^2 - 10x + 17)$ by $(x + 3)$.   (*Section 5.6*)

9. Completely factor $2x^3 - 4x^2 - 30x$.   (*Section 6.3*)

10. Solve for x: $2x^2 - 5x = 12$.   (*Section 6.6*)

Chapter 7 test

Take this test to determine how well you have mastered fractions; check your answers with those found at the end of the book.

1. Rewrite the fraction $-\dfrac{-8}{9}$ so that it has at most one "$-$" sign and no "$-$" sign in its denominator.   (*Section 7.1*)

1. _____

In Problems 2 through 4, reduce each fraction, if possible.   (*Section 7.1*)

2. $\dfrac{15a^2b^3}{21a^3b^2}$

3. $\dfrac{x^2 - 9x + 20}{x^2 - 25}$

2. _____

3. _____

4. $\dfrac{6y^2 - 7y + 2}{1 - 4y^2}$

4. _____

In Problems 5 through 12 perform the indicated operation(s), and simplify, if possible.

5. $\dfrac{3x - 9}{2x + 6} \cdot \dfrac{3x + 9}{2x^2 - 6x}$ (*Section 7.2*)

6. $\dfrac{3x^2 - 2x - 5}{2x^2 - 2x} \cdot \dfrac{8x - 8}{3x^2 + x - 10}$
(*Section 7.2*)

5. _____

6. _____

7. $\dfrac{x^2 - 25}{x^2 - 4} \div \dfrac{x^2 + 10x + 25}{x^2 + 2x + 4}$ (*Section 7.2*)

7. _____

8. $\dfrac{4y^2 - 9}{y^2 - 16} \cdot \dfrac{y^2 - 8y + 16}{2y^4 + 3y^3} \div \dfrac{y - 4}{y}$ (*Section 7.2*)

8. _____

9. $\dfrac{7}{25} + \dfrac{4}{15}$ (*Section 7.3*)

10. $\dfrac{5}{a^2b^3} + \dfrac{7}{ab^4}$ (*Section 7.4*)

9. _____

10. _____

11. $\dfrac{x-6}{x^2-10x+16} + \dfrac{4}{x-2}$ (*Section 7.4*)

12. $\dfrac{3z}{z^2-6z-16} - \dfrac{2}{z+2}$ (*Section 7.4*)

11. _____

12. _____

13. Find the LCM of each set of expressions.

    (a) $25, 45, 75$ (*Section 7.3*)

    (b) $x^3 - x, x^2 + 2x + 1$ (*Section 7.4*)

13. (a) _____

    (b) _____

In Problems 14 and 15, simplify each complex fraction.   (*Section 7.5*)

14. $\dfrac{\dfrac{3x^2}{5y}}{\dfrac{9x}{10y^2}}$

15. $\dfrac{\dfrac{5}{x-1} - \dfrac{3}{2x+1}}{\dfrac{8x}{2x^2-x-1}}$

14. _____

15. _____

In Problems 16 through 19, solve each equation for the variable.   (*Section 7.6*)

16. $\dfrac{15}{y} = \dfrac{3}{8}$

17. $\dfrac{x-3}{x-1} = \dfrac{x-5}{3}$

16. _____

17. _____

18. $\dfrac{2x}{x^2-9} = \dfrac{1}{x-3} - \dfrac{1}{x+3}$

19. $\dfrac{3x}{x^2-9} = \dfrac{1}{x-3} - \dfrac{1}{x+3}$

18. _____

19. _____

20. The literal equation

$$\frac{1}{f} = \frac{1}{s_1} + \frac{1}{s_2}$$

occurs in the study of optics. Solve it for s_1.   (*Section 7.6*)

20. _____

chapter eight

Linear equations in two variables

8.1 An introduction to the plane and equations in two variables

Much of the discussion in this text has involved solving equations with only one variable. Some examples are

$$3x + 1 = 13$$

$$\frac{x}{5} = \frac{x + 1}{2}$$

$$x^2 + 7x + 12 = 0$$

Equations in *One* Variable

In this chapter we will extend our equation-solving capabilities to include equations involving *two* variables, such as:

$$y = 3x + 1$$

$$5t + 6s = 11$$

$$x - y = 4$$

Let us examine one of these equations with two variables more closely. First, observe that $x - y = 4$ has *more than one solution,* because $x - y = 4$ is the algebraic representation of the statement "the difference of two numbers (x and y) is 4." Many pairs of numbers have that property. That is, if $x = 5$ and $y = 1$, then $x - y = 5 - 1 = 4$, and that *pair* of numbers is a solution to $x - y = 4$. Similarly, the pair $x = 6$ and $y = 2$ is a solution to $x - y = 4$ because $6 - 2 = 4$. In general:

Solutions to equations in two variables

Each solution to an equation in the two variables x and y is a *pair of numbers*. The pair is composed of an x value and a y value that, when substituted in the original equation simultaneously, make a true statement.

Returning to the equation $x - y = 4$, note that in addition to the two pairs $x = 5$, $y = 1$ and $x = 6$, $y = 2$, there are infinitely many pairs that solve the equation. For any value of x, we can find a value of y to satisfy $x - y = 4$. If $x = 10$, $y = 6$. If $x = 107$, $y = 103$; if $x = -2$, $y = -6$. Since a listing of all solutions is impossible, we use an alternative: display the solutions *graphically*.

Points in a plane (a vast flat surface) can be represented by pairs of numbers in the **rectangular coordinate system,** also called the **Cartesian coordinate system.** To construct the system, we use two real number lines and intersect them perpendicularly at their zeros. The point of intersection is called the **origin.** The system looks like this:

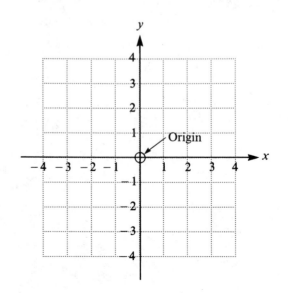

Notice the arrows in the diagram; they indicate positive direction. We label the horizontal real number line with an "x" (it represents the values x can assume) and call it the **x-axis.** The vertical real number line is the **y-axis.**

Every point in the plane can now be labeled by associating with it a pair of numbers (x, y), where x represents the distance traveled horizontally and y represents the distance traveled vertically to get to that point from the origin. For example, the point $(2, 3)$ is arrived at by starting at the origin and traveling 2 units to the right and then 3 units upward. This is shown in the following diagram:

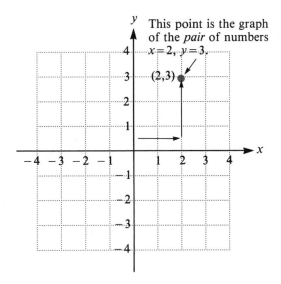

Negative values represent movement to the left (for x) and downward (for y). For example, the points labeled $(-4, 2)$, $(-3, -2)$, and $(2, -4)$ are shown in the following graph.

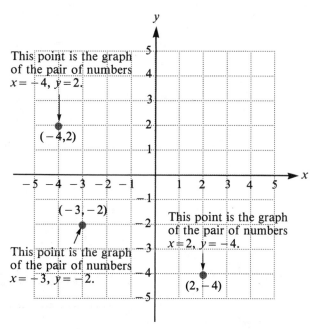

Notice from the illustration that *one pair* of numbers (an x and a y) represents *one* point in the rectangular coordinate system. Notice also that when we write a pair of numbers, the first number is the x value and the second number is the y value. Since the order is important, we refer to a pair of numbers like $(2, -4)$ as an **ordered pair.** In Example 1 we "plot" (or graph) some more points in a rectangular coordinate system.

Example 1

Graph the following pairs of numbers: $(1, 4)$, $(-2, 3)$, $(-3, 2)$, $(2, -3)$, $\left(-\frac{1}{2}, -\frac{1}{2}\right)$, $(0, 3)$, $(3, 0)$, $\left(\frac{5}{2}, -1\right)$.

Solution

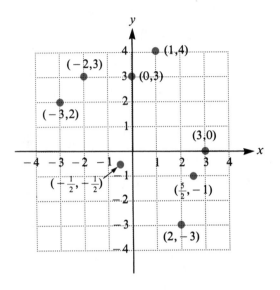

When working with ordered pairs of numbers, we call the x value in the ordered pair the **abscissa** and the y value the **ordinate.** Taken together, the abscissa and the ordinate are called the **coordinates** of a point. Some examples follow.

Example 2

(a) Graph the points $(0, -1)$, $(0, -2)$, $(0, 0)$, $(0, 1)$, and $(0, 3)$. What can be said about all the points whose abscissas are zero?

(b) Graph the points $(-2, 0)$, $(-4, 0)$, $(0, 0)$, $(3, 0)$, and $(4, 0)$. What can be said about all the points whose ordinates are zero?

Solution

(a) Graphing the points, we can see that all points whose x value is zero lie on the y-axis.

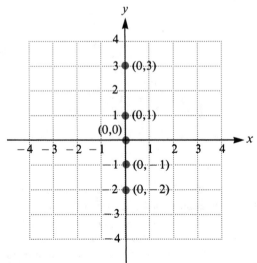

(b) Notice that all points whose *y* value is zero lie on the *x*-axis.

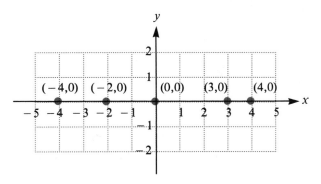

Since the two axes naturally divide a plane into four regions, we label the regions I, II, III, IV, as shown in the following diagram. Each region is called a **quadrant.**

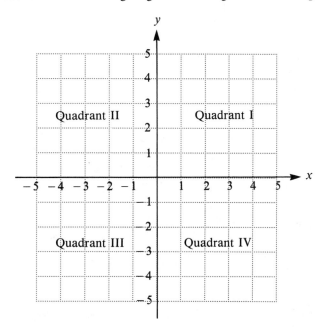

Notice that when a point is plotted anywhere in the plane, that point must lie either on one of the two axes or in one of the four regions.

Section 8.1 Exercises

1. On a sheet of graph paper, graph each of the following points: $(4, 6)$, $(3, -1)$, $(-1, 3)$, $(-3, -3)$, $\left(\frac{3}{2}, \frac{5}{2}\right)$, $(0, -1)$, $(\pi, 0)$, (π, π).

2. On a sheet of graph paper, graph each of the following points: $(3, 5)$, $(4, -3)$, $(-3, 4)$, $(2, 2)$, $\left(\frac{-3}{2}, 2\right)$, $(0, -2)$, $(\sqrt{2}, 0)$.

3. On a sheet of graph paper, graph each of the following points: $(3, 5)$, $(4, 0)$, $(-4, 1)$, $(-2, -2)$, $\left(2, \frac{9}{2}\right)$, $(0, -3)$.

4. What are the coordinates of the origin?

5. What are the coordinates of each of the points *A* through *G* below?

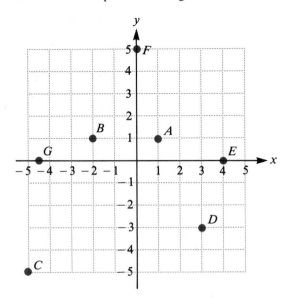

6. What are the coordinates of each of the points *A* through *G* below?

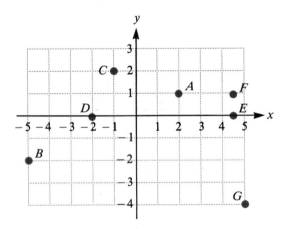

7. What are the coordinates of each of the points *A* through *G* below?

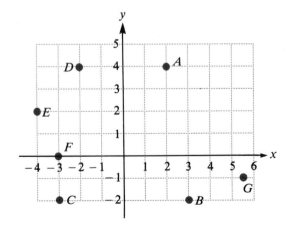

8. In Exercise 7, what is the abscissa of point *A* ?

9. In Exercise 7, what is the abscissa of point *B* ?

10. In Exercise 7, what is the abscissa of point C?

11. In Exercise 7, what is the ordinate of point D?

12. In Exercise 7, what is the ordinate of point E?

13. In Exercise 7, what is the ordinate of point F?

14. In Exercise 7, name the quadrant in which each of the points A through G lies.

15. In what quadrant will we find the points whose abscissas are negative and whose ordinates are positive?

16. (a) On a sheet of graph paper, graph the points (2, 1), (2, −3), (2, 0), (2, 5), and (2, 2).

    (b) What can be said about all the points whose abscissa is 2?

17. (a) On a sheet of graph paper, graph the points (3, −1), (4, −1), (−2, −1), (1, −1), and (0, −1).

    (b) What can be said about all the points whose ordinate is −1?

18. (a) On a sheet of graph paper, graph the points (1, 1), (2, 2), (−4, −4), (0, 0), (3, 3), (−1, −1), and (−2, −2).

    (b) What can be said about all the points whose abscissa and ordinate are equal?

19. The points whose abscissas are positive but whose ordinates are negative lie in which quadrant?

20. Draw a triangle by connecting the points (−2, 4), (−2, −3), and (−7, −3). What kind of triangle appears to be represented?

Supplementary Exercises

1. On a sheet of graph paper, graph each of the following points: (−2, 6), (3, −5), (0, −4), (−5, 0), (4, 2), (1, −3), and (−2, −3).

2. On a sheet of graph paper, graph each of the following points: $\left(\frac{1}{2}, 3\right)$, $\left(-\frac{2}{3}, -1\right)$, (3.2, 3), $\left(-3\frac{1}{4}, -2\right)$, (0, −1.5), and $\left(2\frac{1}{2}, -2\frac{1}{2}\right)$.

3. What are the coordinates of each of the points A through G below?

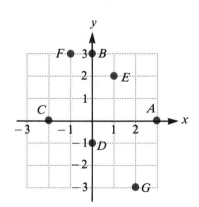

4. What can be said about all the points whose abscissas are equal to 4?

5. What can be said about all the points whose abscissas are equal to −7?

6. What can be said about all the points whose ordinates are equal to 6?

7. What can be said about all the points whose ordinates are equal to −8?

8. What can be said about all the points whose first coordinates are equal to 0?

9. What can be said about all the points whose second coordinates are equal to 0?

10. Draw a four-sided figure by connecting the points (2, −1), (2, 7), (9, 7), and (9, −1). What kind of quadrangle appears to be represented?

8.2   Graphing linear equations in two variables

We can make two observations from Section 8.1.

1. An equation in two variables, like $x - y = 4$, has *pairs* of numbers as solutions.
2. Points in the plane are represented by *pairs* of numbers.

By combining these two observations, we see that we can represent the solutions of equations as points in the plane and, thus, *graph* the solutions. We develop this idea in the next three examples.

Example 1

Graph the solutions of $y - x = 3$.

Solution

We will find several pairs of numbers that satisfy the equation. To make our calculations a bit easier, we first solve $y - x = 3$ for y.

$$y - x = 3$$

$$y = x + 3 \qquad \text{We isolate the } y \text{ by adding}$$
$$x \text{ to both sides of the equation.}$$

There will be many pairs of numbers, x and y, that are solutions to the equation $y = x + 3$. To obtain such solutions, we select *any* x value and then calculate the y value from $y = x + 3$. When $x = 2$, we see that $y = 2 + 3$, so $y = 5$. Hence one solution is the pair $x = 2$, $y = 5$. We write the solution as the *ordered pair* of numbers $(2, 5)$. When $x = 0$, $y = 0 + 3 = 3$, so $(0, 3)$ is another solution. When $x = -4$, $y = -4 + 3 = -1$, so $(-4, -1)$ is another possible solution.

How many solutions do we need? We later find that a minimum of two solutions will be necessary in order to graph the solution in this case. We often use a third solution as a check. In this example, we will choose five solutions. When $x = -3$, $y = -3 + 3 = 0$, and when $x = 3$, $y = 3 + 3 = 6$, so $(-3, 0)$ and $(3, 6)$ are our fourth and fifth solutions. Let's list these solutions in table form and graph the five solutions as five points in the rectangular coordinate system.

x	y
2	5
0	3
-4	-1
-3	0
3	6

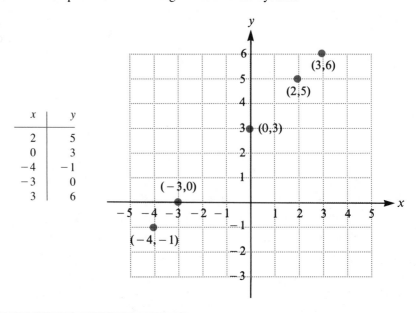

The graph just shown gives us a good idea where other solutions of $y = x + 3$ will be—on the straight line that connects the five points above. If that is the case, we can graph $y = x + 3$ as follows.

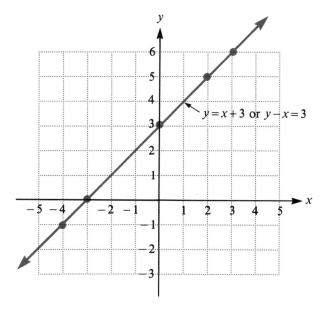

Notice three things:

1. We label the graph with its equation.
2. The line pictured here should be thought of as extending infinitely in both directions, as indicated by the arrows on both ends of the line.
3. The line represents all the solutions to $y - x = 3$.

All equations that will be graphed in this section are called **linear equations,** or **first-degree equations.** Their graphs will be straight lines. More formally, we can define a linear equation in two variables, x and y, as an equation in which x and y appear to the first power only. No variables appear in denominators, and the product of x and y cannot occur.

Since first-degree equations graph as straight lines, we need only as many points as will determine a straight line. A straight line can be drawn through any two points, so we need a minimum of two points to graph the solution to a first-degree equation in two variables. However, we usually like to include a third point as a check, and the more points we graph, the more obvious the line through the points becomes.

Example 2

Graph the linear equation $2x + y = 6$.

Solution

We say ''graph the equation'' instead of the more accurate ''graph the solutions of'' for brevity's sake. In the previous example, we found five solutions and then graphed those solutions as ordered number pairs. Then we drew a straight line through those five points to obtain the graph of the linear equation. In this example, we will find just three solutions: two points to determine a unique line and a third point as a check. First, we solve the given linear equation for y.

$$2x + y = 6$$

$$y = -2x + 6 \qquad \text{Add } -2x \text{ to both sides of the equation.}$$

When $x = 0$, $y = -2 \cdot 0 + 6$ or 6. So one solution is (0, 6). When $x = 3$, $y = -2 \cdot 3 + 6$ or 0. Another solution is (3, 0). When $x = -1$, $y = -2 \cdot -1 + 6$ or 8. A third solution is $(-1, 8)$. These solutions are listed as follows:

x	y
0	6
3	0
-1	8

So the graph of $2x + y = 6$ looks like this:

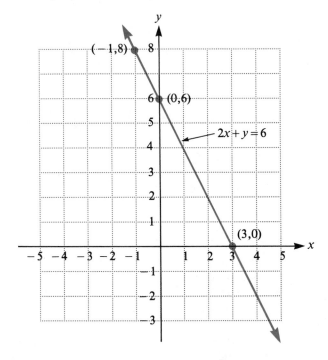

Example 3

Graph the linear equation $2x + 3y = 6$.

Solution

First we solve the given equation for y.

$$2x + 3y = 6$$

$$2x + 3y = -2x + 6 \qquad \text{Add } -2x \text{ to both sides of the equation.}$$

$$y = \frac{-2x + 6}{3} \qquad \text{Divide both sides of the equation by 3.}$$

$$y = -\frac{2}{3}x + \frac{6}{3}$$

$$y = -\frac{2}{3}x + 2$$

Therefore:

When $x = 0$, $y = -\dfrac{2}{3} \cdot 0 + 2$ or 2.

When $x = 3$, $y = -\dfrac{2}{3} \cdot 3 + 2$ or 0.

When $x = 6$, $y = -\dfrac{2}{3} \cdot 6 + 2$ or -2.

These solutions can be tabulated as follows:

x	y
0	2
3	0
6	−2

Notice that it is convenient to use an x value that is a multiple of 3, because of the $\frac{2}{3}$ in $y = -\frac{2}{3}x + 2$.

The graph is:

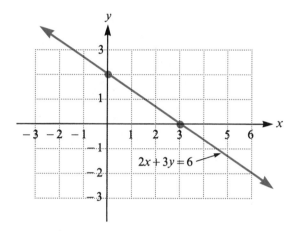

Let's summarize what we have been saying. Given a linear equation in two variables, its graph is obtained as follows:

1. First, solve the linear equation for y.
2. Second, use the result from Step 1 to obtain at least three solutions. This is done by selecting any three x values and then calculating their corresponding y values.
3. Plot (that is, graph) the three solutions obtained in Step 2 as ordered number pairs in a rectangular coordinate system. (Remember, x is listed first and y is listed second in the ordered number pairs.)
4. Draw a straight line through the three points graphed in Step 3 and put arrows on both ends. This is the graph of the given linear equation in two variables.
5. Label the graph.

We can write a linear equation in two variables in this **general form:**

$$Ax + By = C$$

where A, B, and C represent real numbers (A and B cannot both be zero). In $2x + y = 6$, for example, $A = 2$, $B = 1$, and $C = 6$.

When graphing the solutions of $Ax + By = C$ or, more briefly, graphing $Ax + By = C$, two special cases can occur. They are the special cases when $A = 0$ and when $B = 0$. These special cases also have straight-line solutions as the next two examples illustrate.

Example 4

Graph the linear equation $y = 4$.

Solution

The equation $y = 4$ is a special case of $Ax + By = C$ where $A = 0$. This equation says y *must* equal 4. As long as y *does* equal 4, we will have a solution. In other words, the

x value can be any real number. In particular, for x values of -2, 0, or 3, the solutions can be listed as follows:

x	y
-2	4
0	4
3	4

The graph is a horizontal line.

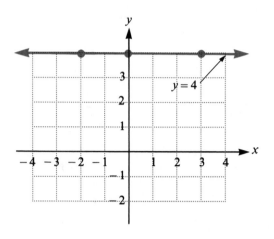

Example 5

Graph $x + 3 = 0$.

Solution

The equation $x + 3 = 0$ is a special case of $Ax + By = C$ where $B = 0$. For the equation $x + 3 = 0$, the variable y can take on any value as long as $x = -3$. We choose three specific values of y, -2, 0, and 4. The list of solutions and graph of the vertical line $x = -3$ follows.

x	y
-3	-2
-3	0
-3	4

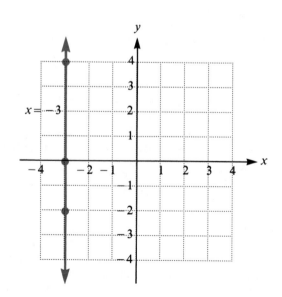

Section 8.2 Exercises

In Exercises 1 through 30, graph each equation.

1. $y = x + 1$

2. $x + y = 3$

3. $x + y = 9$

4. $x - y = 9$

5. $x - y = 3$

6. $y = x$

7. $2x + y = 5$

8. $2x + y = 1$

9. $2x + 4y = 8$

10. $3x + 3y = 9$

11. $y = 3x - 1$

12. $y = 3x$

13. $y = 3x + 1$

14. $y = x - 1$

15. $3y = x - 1$

16. $y = -x$

17. $2x + 2y = 5$

18. $3x - y = -2$

19. $-3x + y = 2$

20. $2x + 5y = 10$

21. $3x + 4y - 12 = 0$

22. $x = 3$

23. $y + 6 = 3$

24. $2x - 3 = 5$

25. $x + 4 = -1$

26. $5y = 2$

27. $y + 3 = 0$

28. $y = 0$

29. $x = 0$

30. $x - 4 = 0$

31. Which of the following are linear equations?

   (a) $x + y = 0$

   (b) $3x + 4y = 5$

   (c) $3x + 1 = 4y + 5$

   (d) $x^2 + 4y = 2$

   (e) $xy = 3$

In Exercises 32 through 35, answer true or false.

32. The point $(0, -2)$ is on the line $x - 2y = 4$.

33. The point $(0, 2)$ is on the line $x - 2y = 4$.

34. The point $(-4, 0)$ is on the line $x - 2y = 4$.

35. The point $\left(-3, \dfrac{1}{2}\right)$ is on the line $x - 2y = 4$.

36. What equation represents the graph:

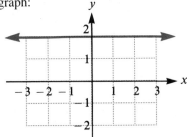

37. What equation represents the graph:

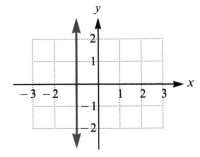

38. Using a calculator, graph $1.5x - 2.3y = 7.5$.

Supplementary Exercises

In Exercises 1 through 19, graph each equation.

1. $y = x + 4$

2. $y = 3x - 1$

3. $y = 2x - 4$

4. $y = 5 - x$

5. $x + y = 4$

6. $x - y = 7$

7. $y = \dfrac{x}{2} + 1$

8. $2x - y = 5$

9. $5x - y = 10$

10. $x + 2y = 4$

11. $x - 3y = 9$

12. $3x - 3y = 0$

13. $\dfrac{x + y}{2} = 4$

14. $\dfrac{x - y}{3} = 2$

15. $y + 4 = 5$

16. $x = -2$

17. $x - 3 = 2x + 4$

18. $4x - 1 = 7$

19. $2y - 1 = 5$

8.3   Finding intercepts and slopes

From Section 8.2 we can make the following observations about the graph of $Ax + By = C$.

If $A = 0$ (and $B \neq 0$), the graph will be a horizontal line.

If $B = 0$ (and $A \neq 0$), the graph will be a vertical line.

If $A \neq 0$ and $B \neq 0$, the graph will be an oblique (or slanted) line that crosses both axes.

The third case, where a graph crosses or *intercepts* both the x-axis and the y-axis, gives rise to two special points called the **x-intercept** and **y-intercept.** The x-intercept is the point where the graph crosses the x-axis; the y-coordinate of the x-intercept must be 0. The y-intercept is the point where the graph crosses the y-axis; the x-coordinate of the y-intercept must be 0.

In the next two examples, we make use of the intercepts in graphing some straight lines.

Example 1

(a) What is the x-intercept for $3x - y = 6$?

(b) What is the y-intercept for $3x - y = 6$?

(c) Use the two intercepts to graph $3x - y = 6$.

Solution

(a) The x-intercept of $3x - y = 6$ is the point where the graph will cross the x-axis. Hence the y value at the x-intercept is 0. To find the x value at that point, we substitute 0 for y in $3x - y = 6$.

$$3x - y = 6$$
$$3x - 0 = 6$$
$$3x = 6$$
$$x = 2$$

So the x-intercept is (2, 0).

(b) The y-intercept is found by substituting 0 for x in $3x - y = 6$, since all points on the y-axis have 0 abscissas.

$$3x - y = 6$$
$$3 \cdot 0 - y = 6$$
$$-y = 6$$
$$y = -6$$

So the y-intercept is $(0, -6)$.

(c) We use the intercepts, $(2, 0)$ and $(0, -6)$, to plot the graph. As a check, we also plot the point $(3, 3)$ because it is also a solution to $3x - y = 6$ and, as we expect, we see that it also lies on the same line.

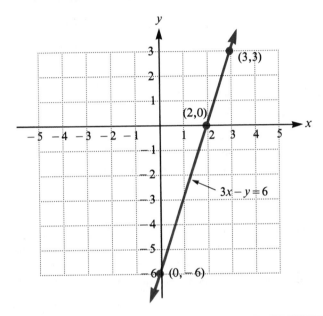

Notice in the previous example that, if we solve $3x - y = 6$ for y, we obtain $y = 3x - 6$ and when $x = 0$, $y = -6$. That is, the y-intercept is obvious from the equation when it is solved for y. So in the graphs of the equations, $y = 2x - 1$, $y = \frac{1}{3}x - 1$, $y = 10x - 1$, and $y = -3x - 1$, the y intercept is -1 in each case. Notice that when we say the y-intercept is -1, we really mean that the y-intercept is the point $(0, -1)$.

Example 2

Graph $y = -3x - 1$ using its intercepts.

Solution

The y-intercept is $(0, -1)$. The x-intercept is found by substituting 0 for y in the given equation.

$$0 = -3x - 1$$
$$1 = -3x$$
$$-\frac{1}{3} = x$$

Thus, the x-intercept is $\left(-\frac{1}{3}, 0\right)$. As a check, we let $x = 1$ and find that

$$y = -3(1) - 1 = -4$$

Thus $(1, -4)$ is a solution for $y = -3x - 1$ since the coordinates of that point satisfy the equation $y = -3x - 1$.

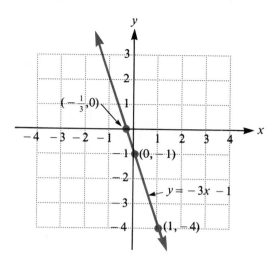

As we said at the end of Example 1, it is sometimes more convenient to write a linear equation in a general form that makes the y-intercept obvious. This general form is written as

$$y = mx + b$$

We have already seen that b is called the y-intercept. That is, the point $(0, b)$ is the y-intercept. Now suppose we have two points on the line $y = mx + b$; call them (x_1, y_1) and (x_2, y_2). Since both (x_1, y_1) and (x_2, y_2) lie on the line, both (x_1, y_1) and (x_2, y_2) satisfy the equation $y = mx + b$. Substituting these values into $y = mx + b$, we have

$$y_1 = mx_1 + b \qquad (1)$$
$$y_2 = mx_2 + b \qquad (2)$$

Notice what happens when we subtract equation (2) from equation (1).

$$y_1 - y_2 = (mx_1 + b) - (mx_2 + b)$$

$$y_1 - y_2 = mx_1 - mx_2$$

$$y_1 - y_2 = m(x_1 - x_2) \qquad \text{Factor the righthand side.}$$

$$\frac{y_1 - y_2}{x_1 - x_2} = m \qquad \text{Assuming } x_1 \neq x_2,$$
$$\text{divide both sides by}$$
$$(x_1 - x_2).$$

What do we have when we solve for m? We can see that m is the coefficient of x in the general form of the equation $y = mx + b$, and it represents the difference of the ordinates divided by the corresponding difference of the abscissas for *any two points on the line*. We call this ratio,

$$m = \frac{y_1 - y_2}{x_1 - x_2}$$

the **slope** of the line; it is a measure of a line's steepness.

For example, let's consider the graph of the straight line connecting the points $(3, 7)$ and $(0, 1)$. Think of $(3, 7)$ as (x_1, y_1) and $(0, 1)$ as (x_2, y_2). Then the slope of the line is

$$m = \frac{y_1 - y_2}{x_1 - x_2} = \frac{7 - 1}{3 - 0} = \frac{6}{3} = 2$$

The y-intercept is $b = 1$ since the given point $(0, 1)$ indicates where the line crosses the y-axis. So, knowing that $m = 2$ and $b = 1$, we can write the equation of the straight line passing through $(3, 7)$ and $(0, 1)$ as

$$y = mx + b$$
$$\downarrow \quad \downarrow$$

or

$$y = 2x + 1$$

The graph looks like this:

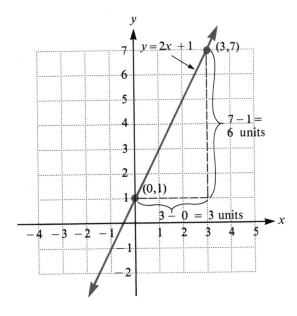

If a line is vertical, we say its slope is *undefined*. For example, consider the line $x = 4$. Two points on the line are $(4, 5)$ and $(4, 2)$; the slope is

$$m = \frac{y_1 - y_2}{x_1 - x_2} = \frac{5 - 2}{4 - 4} = \frac{3}{0} \quad \text{(which is undefined)}$$

If a line is horizontal, its slope is 0. For example, consider the line $y = -3$. Two points on the line are $(2, -3)$ and $(-1, -3)$; its slope is

$$m = \frac{y_1 - y_2}{x_1 - x_2} = \frac{-3 - (-3)}{2 - (-1)} = \frac{0}{3} = 0$$

The next two examples further examine the notion of slope as a measure of the steepness of a line.

Example 3

What is the slope of the straight line connecting each of the following pairs of points?

    (a) $(4, 2)$ and $(-1, -3)$

    (b) $(-3, 6)$ and $(4, -1)$

    (c) $(3, 2)$ and $(5, 2)$

Solution

    (a) Think of $(4, 2)$ as (x_1, y_1) and $(-1, -3)$ as (x_2, y_2). The slope is

$$m = \frac{y_1 - y_2}{x_1 - x_2} = \frac{2 - (-3)}{4 - (-1)} = \frac{5}{5} = 1$$

The slope is 1; we graph the line as follows. Notice that a line with a slope of 1 is less steep than the previous graph of a line with a slope of 2.

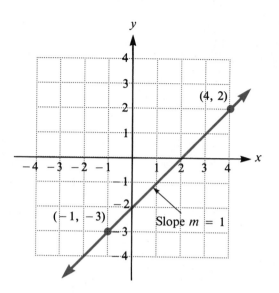

(b) Using $(-3, 6)$ as (x_1, y_1) and $(4, -1)$ as (x_2, y_2), we obtain

$$m = \frac{y_1 - y_2}{x_1 - x_2} = \frac{6 - (-1)}{-3 - 4} = \frac{7}{-7} = -1$$

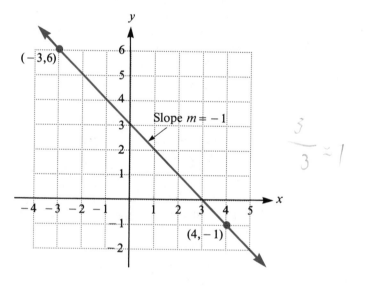

Compare this line with a negative slope to those with positive slopes. Lines with a positive slope slant *upward* going from left to right and lines with a negative slope slant *downward* going from left to right.

(c) The slope of the line connecting $(3, 2)$ and $(5, 2)$ is

$$\frac{2 - 2}{3 - 5} = \frac{0}{-2} = 0$$

This line with zero slope is graphed as follows:

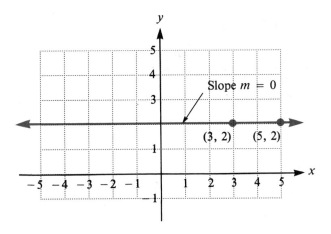

Example 4

Find the slope and y-intercept for the lines that are represented by the equations:

(a) $y = 3x + 1$ (b) $2y - x = 6$ (c) $3x + 4y + 7 = 0$

Solution

(a) Since $y = 3x + 1$ is already solved for y and in the form $y = mx + b$, $m = 3$, $b = 1$. So the slope is 3 and the graph crosses the y-axis at the point $(0, 1)$.

(b) Solving $2y - x = 6$ for y, we obtain

$$2y = x + 6$$

$$y = \frac{1}{2}x + 3$$

So the slope is $m = \frac{1}{2}$ and the y-intercept is $(0, 3)$.

(c) Solving $3x + 4y + 7 = 0$ for y, we obtain:

$$4y = -3x - 7$$

$$y = -\frac{3}{4}x - \frac{7}{4}$$

So the slope is $m = -\frac{3}{4}$ and the y intercept is $b = -\frac{7}{4}$.

Let's look again at the graphs of the three lines from the preceding examples to compare the various slopes.

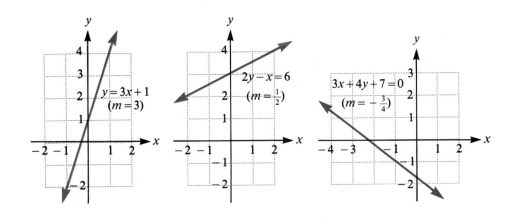

Section 8.3 Exercises

In Exercises 1 through 24, find the x-intercept (if any), the y-intercept (if any), and the slope of the straight line represented by the given equation.

1. $y = 2x + 4$

2. $y = \frac{1}{2}x - 1$

3. $x + y = 7$

4. $x - y = 7$

5. $2x + 4y = 8$

6. $y = 2x + 3$

7. $y = -3x + 5$

8. $2x + y = 5$

9. $y = 2x$

10. $x + 2y = 5$

11. $x + y = 0$

12. $0.1x + 0.2y = 0.5$

13. $2x + 2y = 4$

14. $y + 1 = 2$

15. $y = -3$

16. $x = 4$

17. $3y = x - 1$

18. $x = y - 1$

19. $y = x - 1$

20. $x + 7 = 10$

21. $x = 7$

22. $y = 1$

23. $y = -0.2x - 0.3$

24. $y = mx + b$

In Exercises 25 through 40, find the slope of the line passing through the given pair of points.

25. $(6, 7), (5, 5)$

26. $(5, -5), (7, 6)$

27. $(2, 9), (1, 5)$

28. $(-1, -4), (0, -7)$

29. $(4, 6), (-3, 6)$

30. $(3, 2), (3, -1)$

31. $(1, 1), (3, -2)$

32. $(8, 1), (-8, 0)$

33. $(8, 1), (8, 0)$

34. $(-3, 3), (3, 3)$

35. $(6, 2), (-4, -3)$

36. $(-3, 7), (-5, -5)$

37. $(1, 1), \left(\frac{1}{4}, -\frac{2}{3}\right)$

38. $(2, 5), \left(-\frac{1}{3}, \frac{5}{3}\right)$

39. $\left(-\frac{1}{2}, 3\right), \left(2, \frac{1}{3}\right)$

40. $\left(\frac{1}{2}, \frac{1}{3}\right), \left(\frac{1}{4}, \frac{1}{5}\right)$

Supplementary Exercises

In Exercises 1 through 5, find the x-intercept, the y-intercept, and the slope, if any, for each of the following lines.

1. $y = 4x - 8$

2. $y = \frac{1}{2}x + 1$

3. $2x - 3y = 6$

4. $y = 5$

5. $x = 2$

In Exercises 6 through 10, find the slope, if any, of each line passing through the given pairs of points.

6. $(2, 3), (5, -6)$

7. $(-1, 4), (-5, 12)$

8. $(3, 4), (3, -6)$

9. $(-2, 6), (4, 6)$

10. $\left(\frac{1}{2}, 3\right), \left(\frac{3}{2}, 2\right)$

8.4   Equations of lines

Having learned another general form for a linear equation in the last section, $y = mx + b$, we were able to find the equation for a line once we knew its slope and y-intercept. We call this general form the **slope-intercept form** of a straight line.

Slope–intercept form of a straight line

The form

$$y = mx + b$$

is called the slope–intercept form of a straight line, where m stands for the slope and b represents the y-intercept of the line.

Example 1

Find the equation of the line with slope 4 and y-intercept -3.

Solution

Given $m = 4$ and $b = -3$, we can write the equation as

$$y = 4x - 3$$

Example 2

Find the equation of the line with slope -2 through the point $(0, 5)$.

Solution

We are given $m = -2$. Since the point $(0, 5)$ has $x = 0$, that point must be the y-intercept. Thus $b = 5$ and the equation is

$$y = -2x + 5$$

Given a point on a line and its slope, we can also determine the line's equation as the next example illustrates.

Example 3

Suppose the point $(2, 11)$ falls on the line that has slope 7. Find the equation of the line.

Solution

In the slope–intercept form, we substitute 7 for m.

$$y = 7x + b$$

Next, we can find b by substituting 2 for x and 11 for y.

$$y = 7x + b$$
$$11 = 7(2) + b$$
$$11 = 14 + b$$
$$-3 = b$$

Thus, the desired equation is $y = 7x - 3$.

The *generalized* form for a line, given a point on that line and its slope, is called the **point–slope form** of a straight line. (It is derived in Exercise 31 at the end of this section.)

Point–slope form of a straight line

Suppose (x_1, y_1) is a point on a line with slope m. Then the point–slope form of the line is

$$y - y_1 = m(x - x_1)$$

Example 4

Find the equation of the line whose slope is -2 and which passes through $(-2, 5)$.

Solution

Using the point–slope form with $m = -2$, $x_1 = -2$ and $y_1 = 5$, we have

$$y - y_1 = m(x - x_1)$$
$$y - 5 = -2(x + 2)$$
$$y - 5 = -2x - 4$$
$$y = -2x + 1$$

Example 5

Find the equation of the line whose slope is $-\dfrac{2}{3}$ and which passes through $(-6, 2)$.

Solution

Using the point–slope form with $m = -\dfrac{2}{3}$, $x_1 = -6$, and $y_1 = 2$, we have

$$y - y_1 = m(x - x_1)$$

$$y - 2 = -\frac{2}{3}(x + 6)$$

$$y - 2 = -\frac{2}{3}x - \frac{2}{3}(6)$$

$$y - 2 = -\frac{2}{3}x - 4$$

$$y = -\frac{2}{3}x - 2$$

If two points are given, a variation of the point–slope form of a straight line, called the **two-point form,** can be derived. (See Exercise 32 at the end of this section.) It is stated as follows:

Two-point form of a straight line

If (x_1, y_1) and (x_2, y_2) are two points on a line, then its equation in two-point form is given by

$$y - y_1 = \frac{y_1 - y_2}{x_1 - x_2}(x - x_1) \qquad \text{where } x_1 \neq x_2$$

Example 6

Find the equation of the line passing through $(2, 5)$ and $(4, 9)$.

Solution

The information given in the problem tells us that $x_1 = 2$, $y_1 = 5$, $x_2 = 4$, and $y_2 = 9$. Since $x_1 \neq x_2$, we can substitute these values in the two-point form.

$$y - y_1 = \frac{y_1 - y_2}{x_1 - x_2}(x - x_1)$$

Typing Error

$$y - 5 = \frac{5 - 9}{2 - 4}(x - 2)$$

$$y - 5 = 2(x - 2)$$

$$y - 5 = 2x - 4$$

$$y = 2x + 1$$

Example 7

Find the equation of the line passing through $(-2, 3)$ and $(4, 3)$.

Solution

We have two points whose coordinates we write as $x_1 = -2$, $y_1 = 3$, $x_2 = 4$, and $y_2 = 3$. Note that both y-coordinates have the same value, 3. Substituting in the two-point form, we get

$$y - y_1 = \frac{y_1 - y_2}{x_1 - x_2}(x - x_1)$$

$$y - 3 = \frac{3 - 3}{-2 - 4}(x + 2)$$

$$y - 3 = \frac{0}{-6}(x + 2)$$

$$y - 3 = 0$$

$$y = 3$$

In other words, this is an equation for the horizontal line through all points whose y-coordinate is 3.

Example 8

Find the equation of the line passing through $(-3, 7)$ and $(1, 5)$.

Solution

$$y - y_1 = \frac{y_1 - y_2}{x_1 - x_2}(x - x_1)$$

$$y - 7 = \frac{7 - 5}{-3 - 1}(x + 3)$$

$$y - 7 = -\frac{1}{2}(x + 3)$$

$$y - 7 = -\frac{1}{2}x - \frac{3}{2}$$

$$y = -\frac{1}{2}x + \frac{11}{2}$$

Section 8.4 Exercises

In Exercises 1 through 30, find the equation of the line satisfying the given conditions.

1. slope = 5, y-intercept = 3
2. slope = 4, y-intercept = 2
3. slope = -3, y-intercept = -5
4. slope = -7, y-intercept = -3
5. slope = $\frac{2}{3}$, y-intercept = -2
6. slope = $-\frac{1}{4}$, y-intercept = 4
7. slope = 3, passes through $(0, -6)$
8. slope = -2, passes through $(0, 1)$
9. slope = 3, passes through $(2, 4)$
10. slope = 2, passes through $(1, 5)$
11. slope = -4, passes through $(-1, 8)$
12. slope = -1, passes through $(-2, -7)$
13. slope = 0, passes through $(1, -4)$
14. slope = 0, passes through $\left(2, \frac{1}{2}\right)$
15. slope undefined, passes through $(-3, 2)$
16. slope undefined, passes through $\left(\frac{1}{3}, -2\right)$
17. slope = -3, passes through $(-1, -4)$
18. slope = -1, passes through $(-4, 3)$
19. slope = $-\frac{1}{2}$, passes through $\left(-1, -\frac{3}{2}\right)$
20. slope = $-\frac{3}{2}$, passes through $\left(-2, \frac{1}{3}\right)$
21. x-intercept = 4 and y-intercept = -6
22. x-intercept = -2 and y-intercept = 3
23. passes through the points $(2, 9)$ and $(1, 5)$
24. passes through the points $(6, 2)$ and $(-4, -3)$
25. passes through the points $(6, 7)$ and $(5, 5)$
26. passes through the points $(5, -5)$ and $(7, 6)$
27. passes through the points $(8, 1)$ and $(-8, 0)$
28. passes through the points $(8, 1)$ and $(8, 0)$
29. passes through the points $\left(\frac{1}{2}, \frac{1}{3}\right)$ and $\left(\frac{1}{4}, \frac{1}{5}\right)$
30. passes through the points $\left(-\frac{1}{2}, 3\right)$ and $\left(2, \frac{1}{3}\right)$

31. Supply the missing reasons in the following derivation of the point–slope form.

Statement	*Reason*
1. (x_1, y_1) is a point on the line with slope m.	1. (x_1, y_1) and m are given quantities.
2. $y = mx + b$	2. The slope–intercept form of a straight line.
3. $y_1 = mx_1 + b$	3. The point (x_1, y_1) is on the line and, therefore, satisfies the equation.
4. $b = y_1 - mx_1$	4. (a)
5. $y = mx + y_1 - mx_1$	5. (b)
6. $y - y_1 = mx - mx_1$	6. (c)
7. $y - y_1 = m(x - x_1)$	7. (d)

32. Substitute the value of $\dfrac{y_1 - y_2}{x_1 - x_2}$ for the slope, m, in the point–slope form to derive the two-point form.

33. What is the equation of the horizontal line passing through $(5, 6)$?

34. What is the equation of the vertical line passing through $(5, 6)$?

35. What is the equation of the line that represents the y-axis?

Supplementary Exercises

In Exercises 1 through 13, find the equation of the line satisfying the given conditions.

1. slope = 5, y-intercept = $(0, -4)$
2. slope = -7, y-intercept = $(0, 7)$

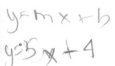

$y = 2x - 4$

3. slope $= 0$, y-intercept $= \left(0, \frac{1}{2}\right)$

$x + y = 4$

4. undefined slope, x-intercept $= (3, 0)$

5. undefined slope, x-intercept $= (-2, 0)$

6. slope $= 6$, passes through $(-2, 1)$

7. slope $= \frac{1}{3}$, passes through $(6, -1)$

8. slope $= 0$, passes through $(-2, 3)$

9. x-intercept $= (-4, 0)$, y-intercept $= (0, 2)$

10. x-intercept $= (6, 0)$, y-intercept $= (0, -3)$

11. passes through the points $(2, 3)$ and $(-1, 6)$

12. passes through the points $(-2, 4)$ and $(-5, -2)$

13. passes through the points $(5, 4)$ and $(-6, 15)$

8.5 Graphing linear inequalities in two variables

Up to this point in the chapter we have looked at graphs of linear equations in two variables. In this section we will examine the graphs of **linear inequalities.** Linear inequalities in two variables may have one of four forms:

$$Ax + By \leq C$$
$$Ax + By \geq C$$
$$Ax + By < C$$
$$Ax + By > C$$

Let us examine the graph of the inequality $2x + 3y \leq 5$. The points that satisfy $2x + 3y = 5$ will certainly be part of the graph because equality is part of the "$\leq$" symbol. The graph of $2x + 3y = 5$ looks like this:

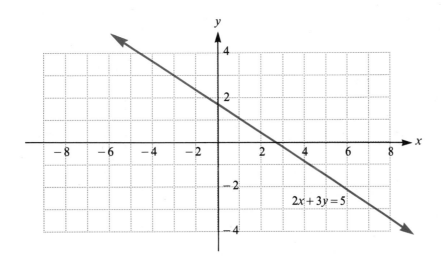

Notice that the straight line $2x + 3y = 5$ divides the plane into two **half-planes,** all points above the line and all points below the line. The graph of $2x + 3y \leq 5$ will include the line $2x + 3y = 5$ and one of the two half-planes. To determine which half-plane is graphed, we use the following method.

To find the correct half-plane

1. Pick *any* point in the plane (provided it isn't *on* the line) as a *test point*.
2. Substitute the x value and y value of the test point in the inequality to see whether it is satisfied.
3. If the point satisfies the inequality, then it and the half-plane it occupies are included in the graph; if the test point does not satisfy the inequality, then the half-plane that does not contain the test point is the correct half-plane.

To illustrate this further, we complete the graph of $2x + 3y \leq 5$ in Example 1.

Example 1

Graph $2x + 3y \leq 5$.

Solution

Since we have already graphed $2x + 3y = 5$, we choose any point not on the line as the test point. We choose $(2, 4)$.

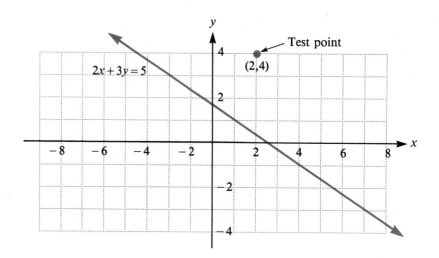

Does $(2, 4)$ satisfy $2x + 3y \leq 5$? We can check it as follows:

$$2x + 3y \leq 5 \quad ?$$

$$2(2) + 3(4) \leq 5 \quad ?$$

$$4 + 12 \leq 5 \quad ?$$

$$16 \leq 5 \quad \text{No!}$$

Therefore, the desired half-plane is the half-plane *not* including $(2, 4)$. The final graph is:

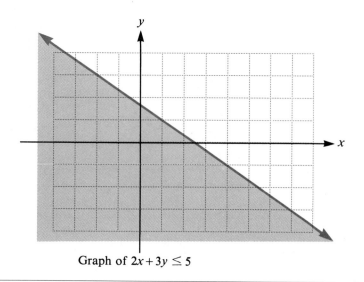

Graph of $2x + 3y \leq 5$

Some examples follow.

Example 2

Graph $y > 3x - 1$.

Solution

This time the line representing $y = 3x - 1$ is not included because equality is not part of the ">" symbol. However, the graph of $y = 3x - 1$ is still critical since it determines the two half-planes. Because it is not included in the final answer, it is represented as a dashed line.

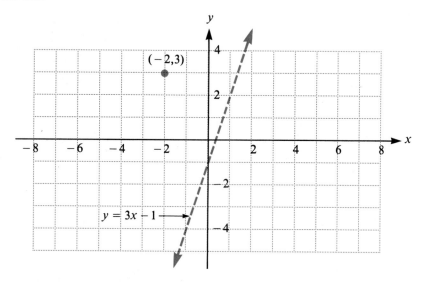

Now, we choose as a test point $(-2, 3)$. Does $(-2, 3)$ satisfy the inequality $y > 3x - 1$?

$$y > 3x - 1$$

$$3 > 3(-2) - 1 \quad ?$$

$$3 > -7 \quad \text{Yes!}$$

Therefore, the half-plane containing $(-2, 3)$ *is* the desired half-plane. The graph of $y > 3x - 1$ is:

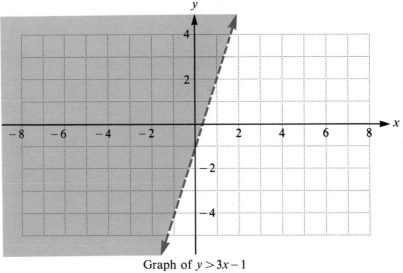

Graph of $y > 3x - 1$

Example 3

Graph $y \geq -x + 2$

Solution

The line is solid (not dashed) because the symbol $\geq$ means equal to or greater than. So the graph is:

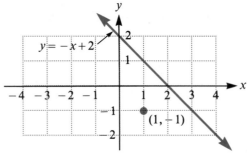

As a test point, we check $(1, -1)$ in $y \geq -x + 2$.

$$y \geq -x + 2$$
$$-1 \geq -1 + 2 \quad ?$$
$$-1 \geq 1 \quad \text{No!}$$

So it is the half-plane *above* the line that is graphed in the answer.

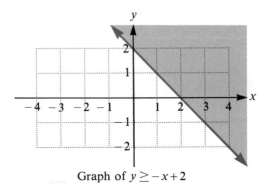

Graph of $y \geq -x + 2$

We conclude this section with a final example of graphing inequalities in the plane.

Example 4

Graph $x - 3 \leq 0$.

Solution

Since $x - 3 \leq 0$ is the same as $x \leq 3$, the vertical line $x = 3$ will be the line separating the two half-planes. Although a test point could still be taken and checked, it should be obvious to the reader that $x \leq 3$ involves abscissas to the left of 3. The graph is:

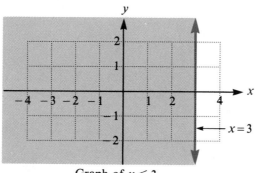

Graph of $x \leq 3$

Section 8.5 Exercises

In Exercises 1 through 20, graph the given inequality.

1. $y \leq 2x + 5$
2. $y > 2x + 5$
3. $y \geq -x + 4$
4. $y < -x + 4$
5. $y < 3x - 1$
6. $y \geq 3x - 1$
7. $y > -2x + 3$
8. $y \leq -2x + 3$
9. $2x - 3y \leq 4$
10. $2x - 3y > 4$
11. $3y - 2x \leq 4$
12. $3y - 2x > 4$
13. $2x - y > 7$
14. $2x - y \leq 7$
15. $2x - y < 7$
16. $x + 3 > 2$
17. $x - 4 \leq 0$
18. $y \geq 6$
19. $2y < 5$
20. $x - 7 \leq 2x + 3$

Supplementary Exercises

In Exercises 1 through 10, graph the given inequality.

1. $x + y \leq 6$
2. $2x - y > 4$
3. $4x - 8 \leq y$
4. $y + 3 < 3x$
5. $x - 2y > -2$
6. $3x + y \geq -6$
7. $x \geq 5$
8. $y + 3 \leq 2y - 1$
9. $-2x > 4$
10. $-3y \leq 6$

8.6   Chapter review

Summary

In this chapter we examined linear equations in two variables and their graphs in the rectangular coordinate system. The equation of a straight line may have several different forms:

$$Ax + By = C \qquad\qquad \text{General form}$$

$$y = mx + b \qquad\qquad \text{Slope–intercept form}$$

$$y - y_1 = m(x - x_1) \qquad\qquad \text{Point–slope form}$$

$$y - y_1 = \frac{y_1 - y_2}{x_1 - x_2}(x - x_1) \qquad\qquad \text{Two-point form}$$

We saw that the slope is the change in y values divided by the change in x values for any two points (x_1, y_1) and (x_2, y_2) on the line. In symbols,

$$m = \frac{y_1 - y_2}{x_1 - x_2} \qquad \text{where } x_1 \neq x_2$$

Linear inequalities in two variables and their half-plane graphs were examined in Section 8.5.

Vocabulary Quiz

Match the expression in Column I with the phrase in Column II that best describes it.

Column I

1. The rectangular (or Cartesian) coordinate system is
2. The origin is
3. The abscissa is
4. The ordinate is
5. The coordinates of a point are
6. The second quadrant contains
7. A linear (or first-degree) equation is
8. The graph of $y = -3$ is
9. The graph of $x = -3$ is
10. The x-intercept of a straight line is
11. The y-intercept of a straight line is
12. $y = mx + b$ is
13. The slope of a straight line is
14. Half-planes are

Column II

a. The point whose x value is zero.
b. The x value of a point.
c. The x and y values of a point.
d. The point $(0, 0)$.
e. What the graph of a straight line divides the plane into.
f. The slope–intercept form of any nonvertical straight line.
g. A graphing system composed of two real number lines intersected at right angles.
h. A horizontal line three units below the x-axis.
i. The ratio of the difference in y values to the difference in x values for any two points on the straight line.
j. The y value of a point.
k. The point whose y value is zero.
l. All points where x is negative and y is positive.
m. A vertical line three units to the left of the y-axis.
n. An equation that can be written as

$$Ax + By = C$$

(where not both A and B are zero).

Chapter 8 Review Exercises

In Exercises 1 through 5, locate each point on the rectangular coordinate system and state in which quadrant it lies.

1. $(7, -1)$
2. $(-8, -8)$
3. $\left(\dfrac{2}{3}, \dfrac{4}{5}\right)$
4. $(-7, 1)$
5. $(1, -7)$

6. On a sheet of graph paper, graph the set of all points that have an abscissa equal to -4.

7. On a sheet of graph paper, graph the set of all points that have an ordinate equal to -4.

8. On a sheet of graph paper, graph the set of all points whose abscissa equals its ordinate.

In Exercises 9 through 15, graph the given equation on a sheet of graph paper.

9. $x + y = 4$
10. $y = 6x - 1$
11. $y + x = 0$
12. $5x + 5y = 0$
13. $x = \dfrac{4}{5}$
14. $y = -3$
15. $x = 0$

In Exercises 16 through 25, find the slope of the line, if possible.

16. The line passing through the points $(3, 4)$ and $(4, 5)$.
17. The line passing through the points $(3, 4)$ and $(5, 4)$.
18. The line passing through the points $(5, 5)$ and $(6, 5)$.
19. The line passing through the points $(5, 5)$ and $(5, 6)$.
20. The line passing through the points $\left(-\dfrac{1}{3}, \dfrac{4}{7}\right)$ and $\left(\dfrac{3}{8}, \dfrac{1}{2}\right)$.
21. $y = 3x + 2$
22. $2x + 4y = 5$
23. $-x - y = \dfrac{2}{3}$
24. $y = \dfrac{3x + 7}{9}$
25. $x + 6 = 5$

26–35. Find the x-intercept and the y-intercept for each of the lines in Exercises 16 through 25.

In Exercises 36 through 50, find the equation of the line satisfying the given conditions.

36. slope $= 5$, y-intercept $= -3$
37. slope $= -4$, y-intercept $= -2$
38. slope $= -\dfrac{2}{3}$, passes through the point $(0, 3)$
39. slope $= \dfrac{1}{2}$, passes through the point $\left(0, -\dfrac{3}{4}\right)$
40. slope $= -3$, passes through the point $(2, -6)$
41. slope $= 2$, passes through the point $(-4, 2)$
42. slope $= -6$, passes through the point $(-1, 0)$
43. slope $= 0$, passes through the point $(-3, -5)$
44. slope $= -\dfrac{1}{4}$, passes through the point $\left(-\dfrac{1}{2}, 2\right)$
45. slope $= -\dfrac{3}{5}$, passes through the point $\left(-4, -\dfrac{3}{4}\right)$

46. passes through the points (4, 5) and (3, 4)

47. passes through the points (3, 3) and (2, 0)

48. passes through the points (3, 3) and (−1, 2)

49. passes through the points $\left(\dfrac{1}{2}, \dfrac{3}{4}\right)$ and $\left(-\dfrac{1}{4}, -\dfrac{1}{2}\right)$

50. passes through the points $(20\dfrac{1}{7}, 6)$ and (3, 6)

In Exercises 51 through 55, graph each inequality.

51. $y \le 3x - 4$

52. $2x + y > 6$

53. $y \ge x - 5$

54. $x \le y$

55. $3x - 2y < -12$

Cumulative review exercises

1. Simplify $5(4 - 3x) - 3(2x + 4)$.   (*Section 2.2*)

2. Simplify $2y^3(y - 5) - y^2(y^3 - 4y^2)$.   (*Section 2.4*)

3. Solve for t: $3xy - 2t = 5$.   (*Section 3.6*)

4. At a certain automobile dealership the ratio of new cars to used cars is 7:5. If there are 252 cars presently at this dealership, how many of these are new cars?   (*Section 4.4*)

5. Simplify $(-2x^2y^3)^3(3x^3y^4)$.   (*Section 5.1*)

6. Multiply $(2x - 3y)(4x^2 + 6xy + 9y^2)$.   (*Section 5.3*)

7. Completely factor $3x^2 - 75y^2$.   (*Section 6.4*)

8. Completely factor $8t^3 + 28t^2 - 16t$.   (*Section 6.5*)

9. Divide

$$\frac{a - b}{4a + 4b} \div \frac{2a^2 - 2b^2}{a^2 + 2ab + b^2} \qquad \text{(\textit{Section 7.2})}$$

10. Add

$$\frac{3}{x^2 - 5x + 6} + \frac{2}{x^2 - 2x - 8} \qquad \text{(\textit{Section 7.4})}$$

Chapter 8 test

Take this test to determine how well you have mastered linear equations with two variables; check your answers using those found at the end of the book.

1. In what quadrant is the point $(-4, -5)$ located?   (*Section 8.1*)

2. Find the coordinates for each point below.   (*Section 8.1*)

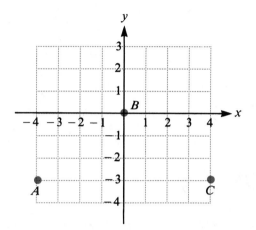

1. _____

2. _____

3. Graph $x + y = 2$.   (*Section 8.2*)

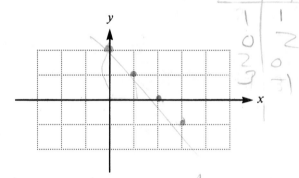

4. Graph $2x + 3y = 6$.   (*Section 8.2*)

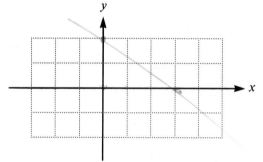

3. _____

4. _____

5. Graph $2x - y = 3$.   (*Section 8.2*)

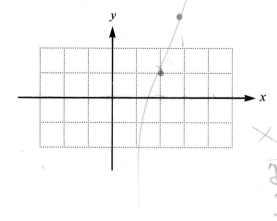

6. Graph $x = -2$.   (*Section 8.2*)

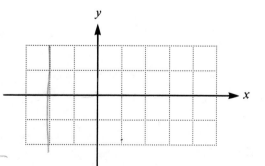

5. _____

6. _____

7. Find the x-intercept and y-intercept of $3x - 2y = 6$.   (*Section 8.3*)   7. _____

$3x = 6 + 2y$

$(2, 0)$
$(0, -3)$

$(0, 2)$

8. Find the slope of the straight line connecting $(4, 6)$ and $(2, -8)$.   (*Section 8.3*)   8. -7

$\frac{8-6}{2-4} = \frac{-14}{2} = -7$

$\frac{y_2 - y_1}{x_2 - x_1}$

$y = mx + b$

9. Find the slope of $2x + 2y = 5$.   (*Section 8.3*)   9. _____

$2x + 2y - 5 = 0$

$-2y = \frac{2x}{2} + \frac{5}{2}$

$y = -\frac{2}{2}x + \frac{5}{2}$

$y = -1x + \frac{5}{2}$

10. Find the slope of the line passing through the points $(3, -4)$ and $(-1, -4)$.   (*Section 8.3*)   10. 0

$\frac{-4 - -4}{3 - -1} = \frac{0}{4}$

x_2, y_2 x y

X-intercept

11. Find the equation of the line with slope 3 passing through the point $(0, -5)$.   (*Section 8.4*)   11. _____

y intercept

$y = 3x - 5$
slope   y

12. Find the equation of the line with slope -2 passing through the point $(1, -6)$.   (*Section 8.4*)   12. _____

$y = -2x - 6$

13. Find the equation of the line passing through the points $(-2, 3)$ and $(2, -1)$.   (*Section 8.4*)   13. _____

$x_2 \, y_2 \, x_1 \, y$

$\frac{-2 - 2}{3 - -1} = \frac{-4}{4} = -4$

$-2 - 2$ -4

$y = -x + 1$

14. Graph the inequality $y \geq -2x + 1$.   (*Section 8.5*)   14. _____

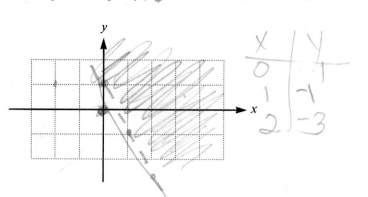

15. Graph the inequality $x + 2y \leq 6$.   (*Section 8.5*)   15. _____

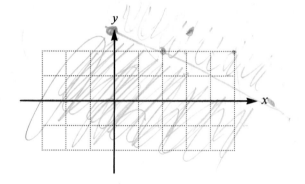

chapter
nine

Systems of equations in two variables

9.1 Graphical solution to a system of linear equations

When we graphed straight lines in Chapter 8, there were infinitely many solutions represented by the points on the line. If we were to examine two linear equations and their graphs in the plane, we would find

1. The straight lines might intersect in one point.
2. The straight lines might be parallel.
3. The straight lines might coincide. (That is, they are the same line.)

In this section we learn to solve a **system of two linear equations** graphically.

Example 1

(a) Graph (on the same set of axes)

$$y = 2x - 4 \quad \text{and} \quad y = -x + 5$$

(b) Do the lines intersect? If so, at what point do they intersect?

Solution

(a) The lines are graphed as shown:

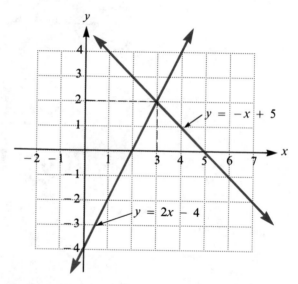

(b) Obviously, the lines intersect; their point of intersection, read from the graph, is (3, 2). That is, although there are infinitely many solutions to $y = 2x - 4$ and infinitely many solutions to $y = -x + 5$, there is only one solution—the point (3, 2)—that satisfies both equations *simultaneously*. This one solution is called the **simultaneous solution.**

So, when presented with a pair of linear equations in x and y, we seek to find the one point (if it exists) that is the simultaneous solution to both equations. In Example 1, $x = 3$ and $y = 2$ is the only pair of values that satisfies $y = 2x - 4$ *and* $y = -x + 5$. The next two examples show that it isn't always possible to find such a solution.

Example 2

(a) Graph $-x + y + 2 = 0$ and $y = x + 2$.

(b) Do the lines intersect? Is there a simultaneous solution to $-x + y + 2 = 0$ and $y = x + 2$?

Solution

A graph of the two lines follows.

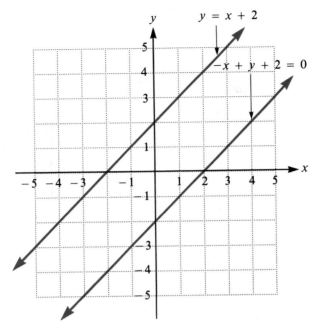

The lines are parallel, so there is no point of intersection. Hence we say there is *no solution* to the system $-x + y + 2 = 0$, $y = x + 2$. The equations are said to be **inconsistent.** We may also say the *system* is inconsistent.

Example 3

Graph $x + y = 3$ and $2x + 2y = 6$.

Solution

These two equations have the same graph.

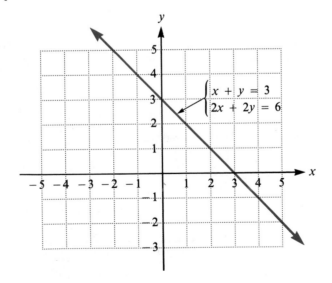

Because their graphs are identical, the equations $x + y = 3$ and $2x + 2y = 6$ are said to be **dependent.** There are infinitely many solutions to the system $x + y = 3$ and $2x + 2y = 6$.

Example 4

Determine if there is a unique simultaneous solution to the equations $x + y + 3 = 0$ and $x - y + 5 = 0$, if they are inconsistent, or if they are dependent.

Solution

Since the graphs of $x + y + 3 = 0$ and $x - y + 5 = 0$ intersect in exactly one point, that point is the unique simultaneous solution to $x + y + 3 = 0$ and $x - y + 5 = 0$. They are graphed as follows.

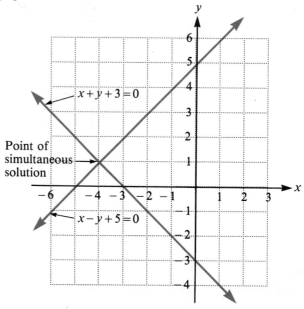

From the graph, we read the simultaneous solution as $(-4, 1)$.

Example 5

Determine the simultaneous solution (if it exists) to the equations $3x + 2y + 2 = 0$ and $3x + 4y + 3 = 0$.

Solution

When we graph the equations, we see that the x value and the y value of the point of intersection fall *between* the integers -1 and 0. Now, the best we can do is approximate those values.

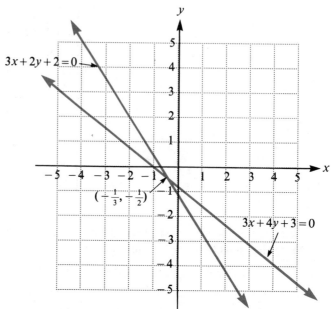

In the next section, we will see how to find these values exactly. We approximate the values as $x = -\frac{1}{3}$, $y = -\frac{1}{2}$ from our graph.

Section 9.1 Exercises

In Exercises 1 through 15, graph the system and determine its unique simultaneous solution, if possible. If no unique solution exists, state whether the equations are inconsistent or dependent.

1. $y = 2x - 3$
   $y = 3x - 6$

2. $x + y = 2$
   $2x + y = 1$

3. $y = x + 1$
   $y = 2x - 2$

4. $x + y + 3 = 0$
   $2x - y = 0$

5. $y = x + 1$
   $y = x + 2$

6. $x + y = 3$
   $x - y = -4$

7. $x + y + 8 = 0$
   $2x - y = 2$

8. $y = x$
   $x + y = 0$

9. $y = x$
   $x - y = 0$

10. $x + y = 5$
    $x - y = 9$

11. $4y = x + 1$
    $2x + 4y = 19$

12. $x + 3y = 7$
    $-2x - 6y = 1$

13. $2x + 3y = 8$
    $4x + 6y = 1$

14. $2x + 3y = 8$
    $4x + 6y = 16$

15. $y = 3x - 6$
    $y = 2x - 4$

16. Each of the following pairs of equations is inconsistent.

    (1) $y = 2x - 3$
        $y = 2x + 1$

    (2) $y = -3x + 1$
        $y = -3x + 2$

    (3) $y = 6x + 7$
        $y = 6x - 7$

    (a) Find the slope for each equation.

    (b) When a system of two equations is inconsistent (that is, when we have parallel lines), what can you say about the slopes of the two lines?

Supplementary Exercises

In Exercises 1 through 8, graph the system and determine its unique simultaneous solution, if possible. If no unique solution exists, state whether the equations are inconsistent or dependent.

1. $y = 2x - 5$
   $y = x - 3$

2. $y = 5 - x$
   $y = x + 7$

3. $2x - y = 4$
   $x - 3y = 7$

4. $3x + 2y = 1$
   $x + y = 0$

5. $x + y = 6$
   $2x = 8 - 2y$

6. $2x - y = 5$
   $4x = 10 + 2y$

7. $x + 2y = 2$
   $4x + y = 8$

8. $y = 2x + 1$
   $2y = x + 5$

9.2 An algebraic solution to a linear system: the elimination method

The graphical method of solving a system of linear equations presented in Section 9.1 has two shortcomings:

1. "Human error" exists in reading—or at best sometimes approximating—unique solutions.

2. In a system of three or more equations involving three or more variables, graphical solutions are awkward or impossible.

In this section we develop, through the first three examples, an algebraic method for solving a system of two linear equations in two variables. The procedure involves the same basic principles presented in Chapter 3. In that chapter we found solutions to equations in one variable by obtaining equivalent equations (equations with the same solution) until we could isolate the variable. We used the addition rule, multiplication rule, division rule, and the concept of simplification to solve equations. The method we present in this chapter extends the use of these rules to equations involving two variables. Although we won't pursue it, the method we develop here is readily adaptable to systems of three or more equations in three or more variables.

Example 1

Find the simultaneous solution (if possible) to the following system of linear equations.

$$2x - y = 4 \qquad \text{and} \qquad x + y = 5$$

Solution

We know that each equation has infinitely many solutions because each involves two variables, x and y. Our goal is to *manipulate* the equations so that we can obtain a new equation with just one variable.

In this case, the second equation tells us that $x + y$ and 5 represent the same number. So, if we were to add $x + y$ to the lefthand side and 5 to the righthand side of the first equation, we would be adding the same quantity to both sides. Doing this would be the same as adding the two equations together. In each case the result is an equation in which only one variable, x, occurs. We have *eliminated* the y variable, since $-y$ and $+y$ add to 0. Symbolically, the addition could be written like this:

$$x + y + 2x - y = 4 + 5 \qquad \text{Add } x + y \text{ to the lefthand side and 5 to the righthand side of the first equation.}$$

$$x + 2x + y - y = 9$$
$$3x = 9 \qquad \text{Simplify.}$$
$$x = 3 \qquad \text{Solve for } x.$$

Or, if we set up the addition vertically:

$$
\begin{array}{rl}
2x - y = 4 & \\
\underline{x + y = 5} & \text{Add } 2x - y = 4 \text{ and } x + y = 5. \\
3x \quad\;\; = 9 & \\
x \quad\;\; = 3 & \text{Solve for } x.
\end{array}
$$

We now know that the simultaneous solution occurs at $x = 3$. To find the y value at that point, return to either equation (say $2x - y = 4$) and substitute 3 for x:

$$2x - y = 4$$
$$2 \cdot 3 - y = 4$$
$$6 - y = 4$$
$$y = 2$$

So the simultaneous solution of $2x - y = 4$ and $x + y = 5$ is $x = 3$, $y = 2$ or simply the point $(3, 2)$. We must check $(3, 2)$ to make sure it satisfies *both* equations.

Check in $2x - y = 4$:

$$2 \cdot 3 - 2 = 4$$
$$6 - 2 = 4$$
$$4 = 4 \checkmark$$

Check in $x + y = 5$:

$$3 + 2 = 5$$
$$5 = 5 \checkmark$$

We usually set up equations in two variables vertically, because it is easier to see what we need to do to find a solution. So we will show this setup only for the rest of the chapter.

Example 2

Find the simultaneous solution (if possible) to the system of linear equations:

$$x + 3y = 5$$
$$2x + 5y = 8$$

Solution

This time, adding the equations won't work—neither variable "falls out." However, our goal of eliminating a variable can be obtained by multiplying the first equation on both sides by -2. If we do, we obtain $-2x - 6y = -10$. So our system now looks like this:

$$-2x - 6y = -10 \qquad \text{This is the result of multiplying}$$
both sides of the first equation above by -2.

and $\qquad 2x + 5y = 8 \qquad$ This is the second equation above, just rewritten.

Next we add and obtain an equation only involving y (the x will be eliminated):

$$
\begin{array}{r}
-2x - 6y = -10 \\
\underline{2x + 5y = 8} \\
-y = -2 \\
y = 2
\end{array}
$$

Add these two equations together.

Solve for y.

So, we have part of our answer, $y = 2$. To find the x value, we return to either *original* equation (say $x + 3y = 5$) to find x:

$$x + 3y = 5$$
$$x + 3 \cdot 2 = 5$$
$$x + 6 = 5$$
$$x = -1$$

So our answer is $(-1, 2)$.

Check in $x + 3y = 5$:

$$-1 + 3 \cdot 2 = 5$$
$$-1 + 6 = 5$$
$$5 = 5 \checkmark$$

Check in $2x + 5y = 8$:

$$2 \cdot -1 + 5 \cdot 2 = 8$$
$$-2 + 10 = 8$$
$$8 = 8 \checkmark$$

Before proceeding with another example, let's summarize the technique for solving a system of equations algebraically.

1. *Eliminate a variable.* We do this by multiplying (if necessary) one or both equations by an appropriate number. Then we add the two equations together.
2. *Solve the newly formed one-variable equation.*

3. *Solve for the second variable.* We do this by returning to one of the original equations and substituting in the solution found in part 2.
4. *Check the answer in BOTH original equations.*

Example 3

Find the simultaneous solution (if possible) to the system of equations:

$$3x - 4y = 6$$
$$2x - 5y = 11$$

Solution

According to Step 1, we must eliminate one of the variables. We concentrate on eliminating x by noticing that, if we multiply the first equation by -2 and the second by 3, x will have coefficients of opposite signs so that, when the equations are added, we get $0 \cdot x$.

Step 1

Eliminate a variable (we choose x).

$$-6x + 8y = -12$$ This is the result of multiplying both sides of the first equation above by -2.

$$6x - 15y = 33$$ This is the result of multiplying both sides of the second equation above by 3.

Step 2

Solve for y.

$$-7y = 21$$ Add the two equations together.
$$y = -3$$ Solve for y.

Step 3

Solve for x. Returning to one of the original equations, $3x - 4y = 6$, substitute -3 for y and solve for x.

$$3x - 4y = 6$$
$$3x - 4(-3) = 6$$
$$3x + 12 = 6$$
$$3x = -6$$
$$x = -2$$

So our answer is $x = -2$, $y = -3$, or simply $(-2, -3)$.

Step 4

Check.

Check in $3x - 4y = 6$:   Check in $2x - 5y = 11$:

$$3(-2) - 4(-3) = 6$$ $$2(-2) - 5(-3) = 11$$
$$-6 + 12 = 6$$ $$-4 + 15 = 11$$
$$6 = 6 \checkmark$$ $$11 = 11 \checkmark$$

So far in this section, we have only examined systems of linear equations that have a unique solution. As we know, two other cases exist, the inconsistent case and the dependent case. The clue as to which type of system we have can be found algebraically as well as graphically. The next two examples investigate the two remaining cases.

Example 4

Demonstrate that there is no solution to the following system of linear equations or, in other words, that the equations are *inconsistent*.

$$x - y = 2$$
$$-x + y = 2$$

Solution

In the previous section, we saw that the graphs of these equations are parallel straight lines. An algebraic investigation follows.

Notice that when we add the equations we obtain:

$$x - y = 2$$
$$\underline{-x + y = 2}$$
$$0 = 4$$

The result, $0 = 4$, is *never* true regardless of the values of x and y. Hence no pair of values can satisfy the two equations simultaneously.

Example 5

Demonstrate that there are infinitely many solutions to the following system of linear equations (or, in other words, that the equations are *dependent*).

$$x + y = 3$$
$$2x + 2y = 6$$

Solution

In Example 3 of Section 9.1, we saw that these equations were dependent, which meant that their graphs were identical. We can also see it algebraically.

$$x + y = 3$$
$$2x + 2y = 6$$

If we proceed as before, we multiply the top equation by -2.

$$-2x - 2y = -6$$
$$\underline{2x + 2y = 6}$$
$$0 = 0 \qquad \text{Add the two equations together.}$$

The result, $0 = 0$ is *always* true regardless of the values of x and y. Thus there are infinitely many solutions to the system of linear equations.

To summarize, when the two linear equations in a system are inconsistent, an attempt to find the simultaneous solution, algebraically, yields a result that is *never* true (an equality of two unequal numbers). On the other hand, when the two linear equations in a system are dependent, an attempt to find the simultaneous solution, algebraically, yields a result that is *always* true (an equality of two numbers that *are* equal).

Section 9.2 Exercises

In Exercises 1 through 21, determine the unique solution to the system of linear equations, if it exists, by using the elimination method of this section. If no unique solution exists, state whether the system is inconsistent or dependent.

1. $x + y = 9$
   $x - y = 5$

2. $3x + 2y = 1$
   $-x + y = 3$

3. $2x - 4y = -4$
   $x + 2y = 1$

4. $3x - 3y = 3$
$2x - 3y = -2$

5. $2x + y = 1$
$-3x - y = 0$

6. $2x + 3y = 11$
$x - 2y = 2$

7. $2x + 3y = 11$
$5x - 10y = 10$

8. $4x + 3y = 9$
$6x + 6y = 17$

9. $3x + 2y = -2$
$3x + 4y = -3$

10. $5x + 7y = 12$
$7x + 5y = 12$

11. $5x + 7y = 12$
$5x + 7y = 12$

12. $5x + 7y = 12$
$5x + 7y = 13$

13. $3x - 2y = 5$
$-6x + 4y = -10$

14. $0.2x + 0.2y = 1$
$x + y = 5$

15. $0.2x + 0.2y = 1$
$x - y = 1$

16. $3x + 5y = 7$
$2x - 9y = 11$

17. $3x + 4y = 9$
$-6x - 8y = 3$

18. $4x + 7y = 10$
$8x + 14y = 20$

19. $x - y = 3$
$3x + 3y = -3$

20. $x + y = 7$
$x + y = 8$

21. $x + y = 7$
$3x + 3y = 21$

The examples in this section were in the general form $Ax + By = C$. In Exercises 22 through 26, first rewrite each equation in the form $Ax + By = C$ and then follow the same directions as for Exercises 1 through 21.

22. $x = y - 7$
$y = x - y$

23. $y = \frac{2}{3}x - 2$
$2x = y$

24. $x + y = 2y - 1$
$x = 7 - y$

25. $y - x = 7$
$x - y = -7$

26. $x = y + 3$
$y = x - 3$

27. (a) For the reader who prefers a formula (that is, a "cure all" for all solvable systems) show that for the system

$$A_1x + B_1y = C_1 \quad \text{and} \quad A_2x + B_2y = C_2$$

that

$$x = \frac{B_2C_1 - B_1C_2}{A_1B_2 - A_2B_1} \quad \text{and} \quad y = \frac{A_1C_2 - A_2C_1}{A_1B_2 - A_2B_1}$$

provided $A_1B_2 - A_2B_1 \neq 0$.

(b) Use the formulas of part a to solve the system

$$3x + 7y = 12$$
$$2x - y = 8$$

28. Use a calculator and the results of Exercise 27 to solve the following system.

$$5.8x - 3.7y = 11.0$$
$$1.9x + 8.1y = -5.7$$

Supplementary Exercises

In Exercises 1 through 16, determine the unique solution to the system of linear equations, if it exists, by using the elimination method of this section. If no unique solution exists, state whether the system is inconsistent or dependent.

1. $x - y = 6$
$x + y = 10$

2. $-x + 3y = 4$
$x - y = -6$

3. $x + y = 10$
$x + y = 7$

4. $x - y = 4$
$2x + y = 2$

5. $2x - y = 4$
$x - y = 5$

6. $x + 2y = 3$
$3x - y = 5$

7. $x + 3y = 1$
$2x - y = -5$

8. $3x + 2y = 1$
$x + y = 0$

9. $\frac{x}{2} + y = 2$
$x - y = 3$

10. $-4x + 3y = 1$
$2x + y = 3$

11. $-3x + 2y = -2$
$2x - 3y = -2$

12. $9x + y = 0$
$-8x - y = 1$

13. $0.3x - 0.2y = 0.1$
    $x + y = 2$

14. $x + y = 5$
    $-3x - 3y = -6$

15. Solve for x and y:

    $x = 2y$
    $y = 2x - 3$

16. Solve for x and y:

    $3x - 2y = 8$
    $y = 1 - x$

9.3   An algebraic solution to a linear system: the substitution method

The **substitution method** is another algebraic method of solving a linear system. It involves solving one equation for one variable and then substituting the expression obtained in the second equation. We illustrate the method, which is particularly useful if one equation is already solved for a variable, in Example 1.

Example 1

Solve the system

$$y = 2x - 3$$
$$x + y = 18$$

Solution

Notice that the first equation is already solved for y. We *substitute* $2x - 3$ for y in the second equation.

$$x + y = 18$$
$$x + (2x - 3) = 18 \qquad 2x - 3 \text{ equals } y \text{ from first equation.}$$
$$x + 2x - 3 = 18$$
$$3x = 21 \qquad \text{Solve for } x.$$
$$x = 7$$

Find y by substituting 7 for x in an original equation.

$$y = 2x - 3$$
$$y = 2(7) - 3$$
$$y = 11$$

Thus, $x = 7$, $y = 11$ is the unique solution. We check in both equations.

Check in $y = 2x - 3$:

$$y = 2x - 3$$
$$11 = 2(7) - 3$$
$$11 = 14 - 3$$
$$11 = 11 \ \checkmark$$

Check in $x + y = 18$:

$$x + y = 18$$
$$7 + 11 = 18$$
$$18 = 18 \ \checkmark$$

If an equation is not already solved for a variable, we must first solve one equation for one variable. To save computation, it is best to choose the variable with the smallest (ignore signs) coefficient. The next two examples illustrate this technique.

Example 2

Solve the system

$$2x + y = -2$$
$$-5x - 2y = 7$$

Solution

We solve the first equation for y because its coefficient is smallest (ignoring signs):

$$2x + y = -2$$
$$y = -2x - 2$$

Now, as in Example 1, we substitute $-2x - 2$ for y in the second equation.

$$-5x - 2y = 7$$
$$-5x - 2(-2x - 2) = 7 \qquad \text{$-2x - 2$ equals y from first equation.}$$
$$-5x + 4x + 4 = 7$$
$$-x = 3 \qquad \text{Solve for x.}$$
$$x = -3$$

Find y by substituting -3 for x in an original equation.

$$-5x - 2y = 7$$
$$-5(-3) - 2y = 7$$
$$15 - 2y = 7$$
$$2y = 8$$
$$y = 4$$

The check of $x = -3$, $y = 4$ in both equations is left to the reader.

Example 3

Solve the system

$$2x + 3y = 29$$
$$14x - 5y = -57$$

Solution

We solve for x in the first equation.

$$2x + 3y = 29$$
$$2x = -3y + 29$$
$$x = \frac{-3y + 29}{2}$$

Next, we substitute $\dfrac{-3y + 29}{2}$ for x in the second equation.

$$14x - 5y = -57$$
$$14\left(\frac{-3y + 29}{2}\right) - 5y = -57$$
$$-21y + 203 - 5y = -57$$
$$-26y = -260$$
$$y = 10$$

Now, we find x by substituting 10 for y in an original equation.

$$2x + 3y = 29$$
$$2x + 3(10) = 29$$
$$2x = -1$$
$$x = -\frac{1}{2}$$

Thus, $x = -\frac{1}{2}$, $y = 10$. We check the solution in each equation:

Check in $2x + 3y = 29$:

$$2x + 3y = 29$$
$$2\left(-\frac{1}{2}\right) + 3(10) = 29$$
$$-1 + 30 = 29$$
$$29 = 29 \ \checkmark$$

Check in $14x - 5y = -57$:

$$14x - 5y = -57$$
$$14\left(-\frac{1}{2}\right) - 5(10) = -57$$
$$-7 - 50 = -57$$
$$-57 = -57 \ \checkmark$$

Inconsistent systems and dependent systems can be identified using the substitution method, as Examples 4 and 5 point out.

Example 4

Solve the system

$$2x - 4y = 6$$
$$3x - 6y = 10$$

Solution

We proceed with the substitution method. Solve the first equation for x.

$$2x - 4y = 6$$
$$2x = 4y + 6$$
$$x = 2y + 3$$

Substitute $2y + 3$ for x in the second equation.

$$3x - 6y = 10$$
$$3(2y + 3) - 6y = 10$$
$$6y + 9 - 6y = 10$$
$$9 = 10$$

$9 = 10$ is *never* true. So there is no solution to the system; the equations are *inconsistent*.

Example 5

Solve the system

$$2x - 4y = 6$$
$$3y - 6y = 9$$

Solution

Solving for x in the first equation, we have:

$$x = 2y + 3$$

Substituting $2y + 3$ for x in the second equation yields:

$$3x - 6y = 9$$
$$3(2y + 3) - 6y = 9$$
$$6y + 9 - 6y = 9$$
$$9 = 9$$

$9 = 9$ is *always* true, regardless of x and y values; the system is *dependent*.

Section 9.3 Exercises

In Exercises 1 through 20, use the substitution method to solve the system, if possible. If no unique solution exists, state whether the system is dependent or inconsistent.

1. $y = 3x - 1$
   $2x + 3y = 8$

2. $y = 4x + 7$
   $2x + 5y = 13$

3. $x = -y + 1$
   $3x - y = 23$

4. $x = -y + 1$
   $x + y = 1$

5. $x = -y + 1$
   $x + y = 2$

6. $y = 7x - 1$
   $5x - 6y = 6$

7. $2x - 3y = -13$
   $y = 2x + 11$

8. $y = x$
   $y = 2x$

9. $x + 2y = 3$
   $3x - 4y = 14$

10. $2x + 4y = 6$
    $-5x - 10y = -30$

11. $2x + 4y = 6$
    $-5x - 10y = -15$

12. $3x - y = 4$
    $9x - 7y = 0$

13. $3x - y = 4$
    $6x + 5y = 29$

14. $6x + 5y = 29$
    $9x - 7y = 0$

15. $8x + 12y = -20$
    $4x - 7y = -10$

16. $3x + 4y = 14$
    $4x - y = 44$

17. $2x - 3y = 12$
    $5x + 4y = -3$

18. $-3x + 2y = 12$
    $4x + 5y = -3$

19. $3x - 3y = 3$
    $2x - 3y = -2$

20. $4x + 3y = 9$
    $6x + 6y = 17$

Supplementary Exercises

In Exercises 1 through 10, use the substitution method to solve the system, if possible. If no unique solution exists, state whether the system is dependent or inconsistent.

1. $y = 2x - 1$
   $x + y = -4$

2. $y = 2 - x$
   $3x + 2y = 3$

3. $x = y + 2$
   $2y - x = 0$

4. $x = y + 1$
   $x - y - 1 = 0$

5. $y = 3x$
   $y = \dfrac{x}{4}$

6. $2x - y = 1$
   $x + y = 2$

7. $x - y = 6$
   $x - 5y = 2$

8. $2x + y = 3$
   $x - y = 0$

9. $2x - 4y = -4$
   $3x - y = 4$

10. $3x - y = 2$
    $y + x = 2$

9.4   Solving word problems using systems of equations

In this last section of the chapter we will show some examples of word problems that can be solved using a system of two equations in two unknowns.

Example 1

When two adults and four children enter a movie theater, the total cost for tickets is $16. When one adult and six children enter the same theater, the cost is $18. Find the price of an adult's ticket and the price of a child's ticket.

Solution

We use our usual problem translation—equation solution—check approach but notice we employ two variables.

Step 1   Problem translation

Let x = the price of one adult's ticket.

Let y = the price of one child's ticket.

From the first sentence of the problem, we get the following equation

$$2x + 4y = 16$$

From the second sentence of the problem, we get the second equation

$$x + 6y = 18$$

So we must solve the following system of linear equations:

$$2x + 4y = 16$$
$$x + 6y = 18$$

Step 2   Solution of equations

Multiplying the second equation by -2 yields

$$2x + 4y = 16$$
$$-2x - 12y = -36$$

Adding eliminates the x's.

$$2x + 4y = 16$$
$$\underline{-2x - 12y = -36}$$
$$-8y = -20$$
$$y = \$2.50$$

Substituting 2.50 for y in the first equation yields:

$$2x + 4y = 16$$
$$2x + 4(2.50) = 16$$
$$2x + 10 = 16$$
$$x = \$3$$

So the price of one adult's ticket is $3 and the price of one child's ticket is $2.50.

Step 3   Check

Two adults' tickets at $3 each, plus four children's tickets at $2.50 each, totals $16. $\checkmark$

One adult's ticket ($3), plus six children's tickets ($15), totals $18. $\checkmark$

Example 2

A man invests money in two stocks, call them stock A and stock B. After one year he sells the stock and makes a total profit of $80. If he gained five times as much on stock A as he *lost* on stock B, how much did he make on stock A? How much did he lose on stock B?

Solution

Step 1   Problem translation

Let x = the amount he gained on stock A.

Let y = the amount he gained on stock B.

So $\qquad\qquad\qquad x + y = 80 \qquad$ (Total profit = $80.)

$\qquad\qquad\qquad\qquad x = -5y \qquad$ (The gain on A is five times the loss on B; note the minus sign to denote loss.)

Step 2   Solution of equations

Since the second equation is already solved for x, we substitute $-5y$ for x in the first equation.

$$x + y = 80$$
$$-5y + y = 80$$
$$-4y = 80$$
$$y = -20$$

Substituting -20 for y in the equation $x + y = 80$ yields

$$x + y = 80$$
$$x - 20 = 80$$
$$x = 100$$

So he gained $100 on stock A and he gained $-$20 (that is, he *lost* $20) on stock B.

Step 3   Check

The total profit is $100 + ($-$20) = $80. $\checkmark$

He gained $100 on stock A, which is five times the amount he lost on stock B, $20. $\checkmark$

Example 3

Judy has inherited $12,000. She invests part of it in stock that pays 8% interest and part of it in bonds that pay 11% interest. After one year, her investments earned a total of $1125 in interest. How much did she invest at each rate?

Solution

Step 1   Problem translation

Let x = the amount invested in stock at 8% interest.

Let y = the amount invested in bonds at 11% interest.

Thus

$$x + y = 12{,}000 \qquad \text{(Total investment = \$12,000)}$$
$$(0.08)x + (0.11)y = 1125 \qquad \text{(Total interest earned on both investments = \$1125.)}$$

Step 2   Solution of equations

Multiply the second equation by 100 to eliminate the decimals.

$$(0.08)x + (0.11)y = 1125$$
$$8x + 11y = 112{,}500$$

If we multiply the first equation by -8 and add it to the rewritten second equation, we will eliminate the x's.

$$-8x - 8y = -96,000$$
$$\underline{8x + 11y = 112,500}$$
$$3y = 16,500$$

$$y = \frac{16,500}{3}$$

$$y = 5500$$

Substituting 5500 for y in the first equation yields

$$x + y = 12,000$$
$$x + 5500 = 12,000$$
$$x = 6500$$

She invested \$6500 in stocks and \$5500 in bonds.

Step 3   Check

The total invested is \$6500 + \$5500 = \$12,000. $\checkmark$

The interest earned on the stock for one year is (\$6500)(0.08) = \$520 and on the bonds is (\$5500)(0.11) = \$605. The total interest earned is \$520 + \$605 = \$1125. $\checkmark$

Example 4

Mrs. Gonzalez purchases \$26 worth of potato salad and cole slaw to serve at the annual family reunion. Potato salad costs \$1.80 per pound while cole slaw costs \$1.40 per pound. If she bought 16 pounds in all, how many pounds of each did she purchase?

Solution

Step 1   Problem translation

Let x = number of pounds of potato salad

y = number of pounds of cole slaw

So

$$x + y = 16 \qquad \text{(16 pounds total)}$$
$$(1.80)x + (1.40)y = 26 \qquad$$

(Total cost of potato salad + total cost of cole slaw = total cost of both)

Step 2   Solution of equations

We will solve by the method of substitution. Solving for x in the first equation, we have

$$x + y = 16$$
$$x = 16 - y$$

Substituting $16 - y$ for x in the second equation yields

$$(1.80)x + (1.40)y = 26$$

$$(1.80)(16 - y) + (1.40)y = 26$$

$$28.8 - 1.8y + 1.4y = 26$$

$$-0.4y = -2.8$$

$$y = \frac{-2.8}{-0.4}$$

$$y = 7$$

Substituting 7 for y in the first equation,

$$x = 16 - 7 = 9$$

She bought 9 pounds of potato salad and 7 pounds of cole slaw.

Step 3 Check

9 pounds + 7 pounds = 16 pounds total $\checkmark$

The cost of 9 pounds of potato salad is 9($1.80) = $16.20.

The cost of 7 pounds of cole slaw is 7($1.40) = $9.80.

The total cost of both is $16.20 + $9.80 = $26. $\checkmark$

Section 9.4 Exercises

In Exercises 1 through 18, solve each of the problems by using a system of two equations in two variables.

1. The sum of two numbers is 124. Their difference is 24. Find the numbers.

2. The sum of two integers is 9. Their difference is 21. Find the integers.

3. Twice one number plus three times a second number is 18. Three times the first number minus twice the second number is 1. Find the numbers.

4. Five times one number plus twice a second number is 99. The second number is three times the first. Find the numbers.

5. Thirty coins in a piggy bank composed only of quarters and half-dollars total $14. How many quarters are there? How many half-dollars are there?

6. A factory has two main costs, labor and material. One week the factory uses 20 tons of material and 30 work-hours of labor for a total weekly cost of $560. The next week it uses 30 tons of material and 20 work-hours of labor for a total weekly cost of $540. What is the cost of 1 ton of material? What is the cost of 1 work-hour of labor?

7. A home construction company has two cost concerns: material (in tons) and labor (in work-hours). January's total cost was $2500, involving 3 tons of material and 100 work-hours. February's total cost was $3500, involving 4 tons of material and 150 work-hours. Find the cost of 1 ton of material and the cost of 1 work-hour.

8. Another home construction company has the same cost concerns as the company in Exercise 7. March's costs were $2700, involving 2 tons of material and 300 work-hours. April's costs were $5400, involving 4 tons of material and 600 work-hours. Can you find the cost of 1 ton of material and the cost of 1 work-hour?

9. Hank has $6000 to invest, part at 12% and part at 9%. After one year, the total interest he receives is $648. How much did he invest at each rate?

10. Lois has twice as much money invested at 14% as she does at 11%. If her yearly return from these investments is $429, how much does she have invested at each rate?

11. A vendor has some $5 bills and some $10 bills. The total value of these bills is $335. If he has 47 bills in all, how many of each type are there?

12. There are 2500 tickets sold for a basketball game. Student tickets were priced at $1.50 each, while nonstudent tickets cost $3 each. If the total gate was $4800, how many students bought tickets?

13. Tony's age is 6 more than three times Lea's age. The sum of their ages is 50. What is the age of each?

14. Twice Marc's age plus three times Courtney's age is 39. If Marc is two years older than Courtney, find each of their ages.

15. Ruby purchased 23 stamps at the post office. She spent $4.56 comprised of 20-cent and 18-cent stamps. How many of each did she purchase?

16. Elvira bought some jeans at $18 each and some blouses at $12 each. If she spent $174 on 12 items, how many of each type did she buy?

17. A car leaves Columbus traveling at 48 mph. A second car leaves Columbus 3 hours later traveling at 64 mph in the same direction. How long will the second car need to overtake the first and what distance will it cover?

18. Two trains, 450 miles apart, head toward each other on parallel tracks. The eastbound train travels twice as fast as the westbound train. If they meet in 3 hours, find the rate of each train.

Supplementary Exercises

In Exercises 1 through 8, solve each of the problems by using a system of two equations in two variables.

1. The sum of two numbers is 4. The difference of the two numbers is -2. Find the numbers.

2. Mary has 9 coins in her bank consisting of only nickels and dimes. The total value of the coins is 70¢. How many of each coin does she have?

3. A factory has two main costs, labor and material. One week the factory uses 2 tons of material and 100 work-hours of labor for a total weekly cost of $2800. The next week it uses 3 tons of material and 250 work-hours of labor for a total weekly cost of $5000. What is the cost of 1 ton of material? What is the cost of 1 work-hour of labor?

4. Leo has only five-dollar and 10-dollar bills in his cash register at closing time. In total, he has 30 bills adding up to $245. How many bills of each denomination are there?

5. Walt has investments in two different accounts, part at 9% and part at 12%. If the amount invested at 12% is $500 more than twice the amount invested at 9% and the total yearly interest he receives is $1380 for both accounts, find the amount invested at each rate.

6. When three adults and one child pay admission to the planetarium, the total cost of the tickets is $5.00. When one adult and three children pay admission to the planetarium, the total cost is $3.00. Determine the admission price of one adult ticket and the admission price of one child's ticket.

7. A father's age is four less than twice his son's age. The difference in their ages is 35 years. Find the age of each.

8. Two cars start from positions 510 miles apart and travel towards each other. One car travels 15 mph faster than the other. If they meet in 6 hours, find the average speed of each car.

9.5   Chapter review

Summary

In this chapter we showed how to solve systems of equations using three methods: a graphical approach, the elimination method, and the substitution method. There are three types of solutions possible for a system of two linear equations in two variables:

1. There may be a unique simultaneous solution, which means the graphs of the equations intersect at one point.
2. There may be no solution, which means the straight line graphs are parallel. We call these equations inconsistent.
3. There may be infinitely many solutions, which means the straight lines coincide. We call these equations dependent.

In the last section of the chapter we showed how word problems can be solved by using a system of equations.

Vocabulary Quiz

Match the expression in Column 1 with the phrase in Column II that best describes it.

Column I

1. Inconsistent equations are
2. Dependent equations are
3. The simultaneous solution to a system of two linear equations in two variables is

Column II

a. The point where the straight-line graphs of the two linear equations of a system intersect.

b. Equations having no common solution.

c. Equations having infinitely many simultaneous solutions.

Chapter 9 Review Exercises

In Exercises 1 through 6, find the solution (if possible) by graphing the system. If there is a unique solution, find it. Otherwise, state whether the equations are inconsistent or dependent.

1. $y = x + 3$
   $y = 3x - 1$

2. $y = 4 - x$
   $y = 3 - x$

3. $y = 3x + 1$
   $3y = 6x + 6$

4. $2x + 2y = 10$
   $x + y = 5$

5. $4x - 2y = 2$
   $2x - y = 4$

6. $2x + y = 7$
   $x - 2y = 3$

In Exercises 7 through 24, solve each system algebraically. If a unique solution exists, find it. Otherwise state whether the equations are inconsistent or dependent.

7. $2x + 3y = 5$
   $2x - 3y = 3$

8. $x + y = 10$
   $x - y = 10$

9. $3x + 2y = 5$
   $y = 5$

10. $4x + y = -8$
    $x = -3$

11. $x - 2y = 5$
    $3x - 5y = 12$

12. $2x + 3y = 1$
    $4x - y = -5$

13. $x = y + 8$
    $x + 3y = 48$

14. $y = x - 7$
    $3x + 5y = 5$

15. $x - 2y = 3$
    $3x - 6y = 9$

16. $x = 3y + 2$
    $2x - 6y = 4$

17. $2x + 5y = 7$
    $3x - 4y = -1$

18. $5x - 4y = -32$
    $4x + 7y = 5$

19. $4x - 5y = 3$
    $3x + y = 7$

20. $2x - y = 12$
    $5x + 4y = 17$

21. $2x + 6y = 9$
    $x + 3y = 4$

22. $4x - 8y = 20$
    $3x - 6y = 15$

23. $x = 3 + 4y$
    $6x + 5y = 11$

24. $y = x - 12$
    $3x + 4y = 1$

25. The sum of two numbers is 9. When twice the smaller is subtracted from the larger, the result is 18. Find the numbers.

26. Five kilograms of sugar plus 10 kilograms of flour cost $10.00. Ten kilograms of sugar plus 5 kilograms of flour cost $12.50. How much is sugar per kilogram? How much is flour per kilogram?

27. Bill and Sandy work in the same factory. When both work 40 hours each, their combined gross income is $340. When Bill works 35 hours and Sandy works 20 hours, their combined gross income is $218.75. Find the hourly wage of each.

28. Elaine invests money at two different rates, 10% and 12%. She has $1000 more invested at 12% than she does at 10%. Her yearly interest from the two investments is $450. How much does she have invested at each rate?

29. The record and tape store offers records at $6 each and cassette tapes at $7 each. Deborah spends $101 on 15 of these items. How many of each type did she buy?

30. Two travelers from Augusta head out in the same direction along the same road. The first travels 18 mph slower than twice the rate of the second. If the second starts out 2 hours later and overtakes the first in 5 hours, find the rate of each.

Cumulative review exercises

1. Simplify $(-2ab^2)(3a^2b)(a^3b^2)$. *(Section 2.3)*

2. Solve for t: $(3t + 2) - 17 = 4 - (9 + 2t)$. *(Section 3.4)*

3. Solve the inequality and graph the solution on a number line.

$$8 - 3(x - 5) > -4(x - 6)$$ *(Section 3.7)*

4. A computer store purchases a computer for \$160. The store marks up the price 35% for sale to its customers. What is the customer price? *(Section 4.5)*

5. Divide $(3x^3 - 10x + 5) \div (x - 2)$. *(Section 5.6)*

6. Completely factor $18x^2 - 69x + 60$. *(Section 6.5)*

7. Multiply

$$\frac{3x^2 - 6x}{6x + 18} \cdot \frac{5 - x}{x^2 - 7x + 10}$$ *(Section 7.2)*

8. Simplify

$$\frac{\dfrac{a}{b} - \dfrac{b}{a}}{\dfrac{1}{a} + \dfrac{1}{b}}$$ *(Section 7.5)*

9. Graph the equation $2x - 5y = 10$. *(Section 8.2)*

10. Find the equation of the line passing through the point $(-2, 3)$ and $(2, -5)$. *(Section 8.4)*

Chapter 9 test

Take this test to determine how well you have mastered systems of linear equations; check your answers with those found at the end of the book.

1. Find the unique solution, if it exists, to the following system of equations by graphing. *(Section 9.1)*

$$y = x + 2$$
$$y = -x + 2$$

2. If there is a unique solution to the following system of linear equations, find it. If not, state whether the equations are inconsistent or dependent. Use the elimination method. *(Section 9.2)*

$$3x - 6y = 8$$
$$-4x + 8y = -8$$

1. _____

2. _____

3. If there is a unique solution to the following system of linear equations, find it. If not, state whether the equations are inconsistent or dependent. Use the substitution method. *(Section 9.3)*

$$x + y = 2$$
$$3x + 2y = 1$$

3. _____

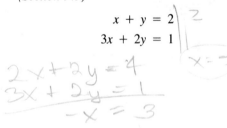

In Problems 4 through 7, use any method to solve the system of equations. If there is a unique solution, find it. If not, state whether the equations are inconsistent or dependent. *(Sections 9.2, 9.3)*

4. $2x - y = 7$
   $x + y = 8$

5. $x = 2y + 4$
   $2x - 4y = -8$

4. _____

5. _____

6. $3x + y = 3$
   $5x - 3y = 19$

7. $2x - 5y = 25$
   $3x + 4y = 3$

6. _____

7. _____

Chapter 9 test   **329**

8. The sum of two numbers is 33. If twice the second is subtracted from 3 times the first, the result is 3. Find the numbers. *(Section 9.4)*

$$x+y = 33 \| 2$$
$$3x-2y = 3 \|$$

$$2x+2y = 66$$
$$\underline{3x-2y = 3}$$
$$5x = 69 \qquad x = \frac{69}{5}$$

10. A sum of money is invested in two types of bonds, one yielding 8% and the other 10%. The amount invested at 10% is twice the amount invested at 8%. If the total interest for both types is $2100, find the amount invested at each rate. *(Section 9.4)*

$$.08a + .10b = 2100$$
$$.20b = .08a$$

$$.20b + .10b = 2100$$
$$.30b = 2100$$
$$\overline{.30} \qquad \overline{.30}$$

$$b = \$7000$$

9. A farmer sells "pick-your-own" strawberries and apples. If 10 quarts of strawberries and 2 bushels of apples cost $17.50 and 4 quarts of strawberries and 1 bushel of apples cost $8, find the cost of 1 quart of strawberries. *(Section 9.4)*

8. _____

9. _____

$$10s + 2a = 17.50 \| \cdot 1$$
$$4s + 1a = 8\$ \| \cdot 2$$

10. _____

$$10s + 2a = 17.50$$
$$8s + 2a = 16.00$$
$$\overline{2s \qquad \qquad 1.50}$$
$$\overline{2} \qquad \qquad \overline{2}$$

$$s = .75\cancel{\phi}$$

chapter
ten

Radical
expressions

10.1 Irrational numbers and the real number system

If an integer is multiplied by itself, the product is called a **perfect square.** For example,

$$3 \cdot 3 = 9 \qquad \text{So 9 is a perfect square.}$$
$$-5 \cdot -5 = 25 \qquad \text{So 25 is a perfect square.}$$
$$6 \cdot 6 = 36 \qquad \text{So 36 is a perfect square.}$$
$$0 \cdot 0 = 0 \qquad \text{So 0 is a perfect square.}$$

Notice that a perfect square can never be negative.

Consider the perfect square 36. It is equal to the product $6 \cdot 6$ as well as the product $-6 \cdot -6$. We call 6 the **positive square root** of 36 and we write it as $6 = \sqrt{36}$. We call -6 the **negative square root** of 36 and we write it as $-6 = -\sqrt{36}$. The symbol $\sqrt{}$ is called a **radical symbol.** Notice the important relationship between a number and its square roots.

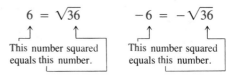

The meaning of a square root then is:

$$b \text{ is called a square root of } a \text{ when } b^2 = a$$

We write $b = \sqrt{a}$ if b is positive and $b = -\sqrt{a}$ if b is negative. Any integer that is a perfect square (except zero) has two square roots, one positive and one negative.

Example 1

(a) What is the positive square root of 4?

(b) What is the negative square root of 9?

(c) What are the square roots of 64?

Solution

(a) The positive square root of 4 is 2, because $2 \cdot 2 = 4$.

(b) The negative square root of 9 is -3, because $-3 \cdot -3 = 9$.

(c) There are two square roots of 64; one is positive and one is negative. The two square roots of 64 are 8 and -8, written collectively as ± 8.

Example 2

Find the value of each of the following.

(a) $\pm\sqrt{16}$ (b) $-\sqrt{49}$ (c) $\sqrt{49}$ (d) $\sqrt{0}$

Solution

(a) $\pm\sqrt{16}$ denotes *two* numbers, the positive square root of 16 and the negative square root of 16. So, $\pm\sqrt{16} = \pm 4$.

(b) $-\sqrt{49} = -7$

(c) $\sqrt{49} = 7$

(d) $\sqrt{0} = 0$

In Chapter 1 we saw that the square root of a positive integer that is *not* a perfect square is, itself, an irrational number. That is, $\sqrt{a}$ and $-\sqrt{a}$ are both irrational numbers whenever the integer a is not a perfect square and $a > 0$. An irrational number, such as $\sqrt{2}$, does not have an exact decimal equivalent. A number such as this must be approximated when used in any kind of applied work. One might ask: "What is a good approximation to $\sqrt{2}$?" For our purposes in this book, "good approximations" will be approximations to 3 decimal places. Most computers can express irrational numbers to 16 or more decimal places, and pocket calculators usually show 6 or 7 decimal places. We shall use the table of squares and square roots in Appendix B for our approximations. The next two examples show how to use that table.

Example 3

Find $\sqrt{2}$ using the table of squares and square roots in Appendix B. Show that the value read in the table is only an approximation.

Solution

In the table, we go down the "n" column until we see a 2. Then go across to the "$\sqrt{n}$" column to read: 1.414. So

$$\sqrt{2} \approx 1.414$$

We can show that 1.414 is only an approximation to $\sqrt{2}$ because $(1.414)^2 = (1.414)(1.414) = 1.999396 \approx 2$.

Example 4

Find the square roots of 11 using the table of squares and square roots.

Solution

We read 3.317 for $\sqrt{11}$. This is our approximate value for the positive square root of 11. The approximation for the negative square root of 11 is -3.317. So our answer is ± 3.317.

One important irrational number is 3.141592653. . . . It is so special, in fact, that it is represented by the symbol π (the Greek letter **pi**) and is equal to the ratio, in any circle, of the circle's circumference to its diameter. Two formulas involving π are

$$C = 2\pi r$$
$$A = \pi r^2$$

where C is the circumference of a circle of radius r and A is its area.

Example 5

Find the circumference and area of a circle of radius 3 inches.

Solution

Using the formulas yields:

$$
\begin{aligned}
C &= 2\pi r & A &= \pi r^2 \\
&= 2\pi(3) & &= \pi(3)^2 \\
&= 6\pi \text{ inches} & &= 9\pi \text{ square inches}
\end{aligned}
$$

Now, 6π inches and 9π square inches are *exact* representations for the circumference and area. If, however, we are interested in an *approximation* for these quantities, we could use an *approximate* value of π; two common such approximations are 3.14 and $\frac{22}{7}$. If we choose 3.14 as our approximation, the circumference and area expressions become

$$C = 6\pi \text{ inches} \qquad\qquad A = 9\pi \text{ square inches}$$
$$C \approx 6(3.14) \text{ or } 18.84 \text{ inches} \qquad A \approx 9(3.14) \text{ or } 28.26 \text{ square inches}$$

Summarizing what we have discussed so far:

1. The special number $\pi = 3.141592653\ldots$ and its multiples $\left(\text{such as } 2\pi, -3\pi, \frac{\pi}{2}, \frac{\pi}{10}\right)$ are all irrational numbers.

2. Numbers of the form $\sqrt{a}$ or $-\sqrt{a}$ (where a is *not* a perfect square and $a > 0$) are called **radical numbers.**

Section 10.1 Exercises

1. What is the positive square root of 25?
2. What is the negative square root of 25?
3. What are the square roots of 49?
4. What is the value of $\pm\sqrt{36}$?
5. What is the value of $-\sqrt{81}$?
6. What is the value of $\pm\sqrt{100}$?
7. What is the value of $\sqrt{121}$?
8. *True or False:* 3 is a square root of 9 because $3 \cdot 3 = 9$.
9. *True or False:* -1 is a square root of 1 because $-1 \cdot -1 = 1$.
10. How many square roots does 81 have?

In Exercises 11 through 22, use the table of squares and square roots in Appendix B to approximate each of the following irrational numbers.

11. $\sqrt{14}$
12. $-\sqrt{5}$
13. $\sqrt{10}$
14. $\sqrt{40}$
15. $-\sqrt{91}$
16. $\sqrt{12}$
17. $\sqrt{7}$
18. $\sqrt{3}$
19. $\pm\sqrt{6}$
20. $\sqrt{99}$
21. $\sqrt{7} - \sqrt{11}$
22. $\sqrt{2} + \sqrt{3}$
23. What is the value of $\sqrt{0}$?
24. Is $\sqrt{2} + \sqrt{3} = \sqrt{5}$?
25. Is $\sqrt{2} \cdot \sqrt{3} = \sqrt{6}$?

In Exercises 26 through 28 use 3.14 as an approximation to π.

26. What is the circumference of a circle of radius 40 centimeters?
27. What is the area of a circle of radius 4 feet?
28. What is the area of a circle whose circumference is 31.4 inches?
29. Find the circumference and area of a circle of radius 4 meters as directed below.
    (a) Express your answer in terms of π.
    (b) Use 3.14 as an approximation to π to find the circumference and area.
    (c) Use $\frac{22}{7}$ as an approximation to π to find the circumference and area.
30. The formula for the volume (V) of a cylinder of radius (r) and height (h) is given by

$$V = \pi r^2 h$$

If $h = 2$ inches and $r = 3$ inches, find V when
    (a) The answer is expressed in terms of π.
    (b) The answer is approximated by using 3.14 for π.

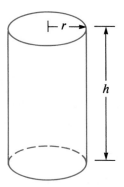

31. (a) Show that π, 3.14, and $\frac{22}{7}$ are all different numbers.

    (b) Which is the largest: π, 3.14, $\frac{22}{7}$, or $\sqrt{10}$?

    (c) Which is the smallest: π, 3.14, $\frac{22}{7}$, or $\sqrt{10}$?

    (d) Which are rational numbers: π, 3.14, $\frac{22}{7}$, or $\sqrt{10}$?

32. The formula for the volume of a sphere of radius r is given by

$$V = \frac{4}{3}\pi r^3$$

    If $r = 3$ feet, find V. $\left(\text{Use } \pi \approx \frac{22}{7}.\right)$

33. The extraordinary property of the circle, that the ratio of the circumference to the diameter is *always* the same number (π), should not be overlooked. To begin to appreciate this, measure (perhaps using a string) the circumference and diameter of several circular objects. For each object, divide the circumference by the diameter. How close are you to the approximation 3.14?

34. Another interpretation of $\sqrt{A}$ is the length of each side of a square of A square units. For example, if the area of a square is 4 square units, the length of each side is $\sqrt{4} = 2$ units.

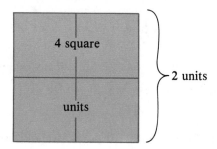

A square with 9 square units of area has $\sqrt{9} = 3$ units for its side. More formally, if the area of a square is A square units, the length of one side can be obtained by using the following pictorial and algebraic models:

$$s^2 = A$$

or

We find s by *taking the square root of both sides of the equation.* For example, if $A = 7$, we have

$$s^2 = 7$$

$$s = \underbrace{\pm\sqrt{7}}$$

└──── the square roots of 7

Since s represents a length, we ignore the negative value. Solve each of the following equations. (In each case there should be two answers.)

(a) $s^2 = 9$ (b) $s^2 = 10$ (c) $x^2 = 2$

(d) $x^2 = 3$ (e) $x^2 = A$ (f) $(x - 1)^2 = 2$

35. (a) Show that, if $x = 3$, the equation $\sqrt{x^2} = x$ is true.

    (b) Show that, if $x = -3$, the equation $\sqrt{x^2} = x$ is false.

Supplementary Exercises

In Exercises 1 through 15, evaluate each of the following. Use the table of squares and square roots in Appendix B as necessary.

1. $\sqrt{9}$ 2. $\sqrt{36}$ 3. $-\sqrt{4}$ 4. $-\sqrt{16}$

5. $\pm\sqrt{144}$ 6. $\pm\sqrt{81}$ 7. $\sqrt{17}$ 8. $\sqrt{29}$

9. $-\sqrt{39}$ 10. $-\sqrt{41}$ 11. $\pm\sqrt{83}$ 12. $\pm\sqrt{69}$

13. $\sqrt{9} - \sqrt{4}$ 14. $-\sqrt{121} + \sqrt{100}$ 15. $\sqrt{21} - \sqrt{10}$

16. What is the length of the side of a square if its area is 64 sq ft?

17. What is the radius of a circle if the area of the circle is 28.26 sq ft? Use $A = \pi r^2$ and π as 3.14.

18. What is the length of the side of a square if its area is 100 sq yd?

10.2   Simplifying radical expressions

For the remainder of this chapter, we will call any expression involving a radical symbol a **radical expression.** Hence $-\sqrt{5}$, $3 + \sqrt{18}$, and $\dfrac{\sqrt{7}}{4}$ are radical expressions, as are $\sqrt{x + y}$, $\dfrac{1}{\sqrt{x} - \sqrt{y}}$, and $\sqrt{15xy} - \sqrt{10a^3}$. In this section we will be concerned with simplifying radical expressions. To do so, we will use two rules of radicals that are developed in the next two examples.

Example 1

Verify that $\sqrt{x} \cdot \sqrt{y} = \sqrt{xy}$ by substituting the following values for x and y:

(a) $x = 4, y = 9$ (b) $x = 25, y = 4$ (c) $x = 2, y = 3$

Solution

(a) $\sqrt{x} \cdot \sqrt{y} = \sqrt{xy}$

$\sqrt{4} \cdot \sqrt{9} = \sqrt{36}$

$\downarrow$ $\downarrow$ $\downarrow$

$2 \cdot 3 = 6$

$6 = 6 \;\checkmark$

(b) $\sqrt{x} \cdot \sqrt{y} = \sqrt{xy}$

$\qquad \sqrt{25} \cdot \sqrt{4} = \sqrt{100}$

$\qquad \qquad 5 \cdot 2 = 10$

$\qquad \qquad \quad 10 = 10 \ \checkmark$

(c) For $\sqrt{2}$, $\sqrt{3}$, and $\sqrt{6}$, we must use approximations from the table in Appendix B.

$$\sqrt{x} \cdot \sqrt{y} = \sqrt{xy}$$

$$\sqrt{2} \cdot \sqrt{3} = \sqrt{6}$$

$$1.414 \cdot 1.732 \approx 2.449$$

$$2.449048 \approx 2.449 \ \checkmark$$

In general, we can say:

Radical Rule 1

$$\sqrt{x} \cdot \sqrt{y} = \sqrt{xy} \qquad \text{(where } x \text{ and } y \text{ are both nonnegative)}$$

Example 2

Verify that $\dfrac{\sqrt{x}}{\sqrt{y}} = \sqrt{\dfrac{x}{y}}$ by substituting the following values for x and y.

(a) $x = 4$, $y = 1$ (b) $x = 16$, $y = 4$ (c) $x = 81$, $y = 9$

Solution

(a) $\dfrac{\sqrt{x}}{\sqrt{y}} = \sqrt{\dfrac{x}{y}}$

$\dfrac{\sqrt{4}}{\sqrt{1}} = \sqrt{\dfrac{4}{1}}$

$\dfrac{2}{1} = \sqrt{4}$

$2 = 2 \ \checkmark$

(b) $\dfrac{\sqrt{x}}{\sqrt{y}} = \sqrt{\dfrac{x}{y}}$

$\dfrac{\sqrt{16}}{\sqrt{4}} = \sqrt{\dfrac{16}{4}}$

$\dfrac{4}{2} = \sqrt{4}$

$2 = 2 \ \checkmark$

(c) $\dfrac{\sqrt{x}}{\sqrt{y}} = \sqrt{\dfrac{x}{y}}$

$\dfrac{\sqrt{81}}{\sqrt{9}} = \sqrt{\dfrac{81}{9}}$

$\dfrac{9}{3} = \sqrt{9}$

$3 = 3 \ \checkmark$

In general, this gives us Radical Rule 2.

Radical Rule 2

$$\frac{\sqrt{x}}{\sqrt{y}} = \sqrt{\frac{x}{y}} \qquad \text{(where } x \geq 0 \text{ and } y > 0\text{)}$$

The next six examples show two different ways of using the radical rules for simplification. Notice that the first two examples involve *numerical radical expressions* whereas the last four examples involve *algebraic radical expressions*.

Example 3

Simplify the following.

(a) $\sqrt{200}$ (b) $\sqrt{75}$ (c) $\sqrt{175}$ (d) $\sqrt{385}$

Solution

(a) To simplify an expression like $\sqrt{200}$ means to write 200 as a product of two numbers, one of which is a perfect square, and then apply Radical Rule 1. First, we express 200 as $10 \cdot 10 \cdot 2$. So

$$\begin{aligned}
\sqrt{200} &= \sqrt{10^2 \cdot 2} \\
&= \sqrt{10^2} \cdot \sqrt{2} \qquad \text{Radical Rule 1} \\
&= 10 \cdot \sqrt{2} \\
&= 10\sqrt{2} \qquad \text{Omit the } \cdot \text{ for multiplication,} \\
&\qquad\qquad\quad \text{as is customary.}
\end{aligned}$$

Hence $\sqrt{200} = 10\sqrt{2}$. One advantage of rewriting $\sqrt{200}$ as $10\sqrt{2}$ is that, if we needed an approximation to $\sqrt{200}$, we could rewrite $\sqrt{200}$ as $10\sqrt{2}$ and then approximate $\sqrt{200}$ by approximating $\sqrt{2} \approx 1.414$ and multiplying by 10. So

$$\sqrt{200} = 10\sqrt{2} \approx 10 \cdot 1.414 \approx 14.14$$

(b) $$\begin{aligned}
\sqrt{75} &= \sqrt{5 \cdot 5 \cdot 3} \\
&= \sqrt{5^2 \cdot 3} \\
&= \sqrt{5^2} \cdot \sqrt{3} \qquad \text{Radical Rule 1} \\
&= 5\sqrt{3}
\end{aligned}$$

Notice if we use $5\sqrt{3}$ for $\sqrt{75}$ and approximate $\sqrt{3}$ as 1.732, we obtain

$$\sqrt{75} = 5\sqrt{3} \approx 5 \cdot 1.732 \approx 8.660$$

which is the value for $\sqrt{75}$ given in Appendix Table B.

(c) $$\begin{aligned}
\sqrt{175} &= \sqrt{5^2 \cdot 7} \\
&= \sqrt{5^2} \cdot \sqrt{7} \qquad \text{Radical Rule 1} \\
&= 5\sqrt{7}
\end{aligned}$$

An approximation to $\sqrt{175}$ is $5\sqrt{7} \approx 5 \cdot 2.646 \approx 13.230$.

(d) We cannot simplify $\sqrt{385}$ because there is no perfect square factor of 385. ($385 = 5 \cdot 7 \cdot 11$, none of which is a perfect square.)

Example 4

Simplify the following.

(a) $\sqrt{\dfrac{4}{9}}$ (b) $\sqrt{0.04}$ (c) $\sqrt{0.12}$

Solution

(a) $\sqrt{\dfrac{4}{9}}$ can be rewritten, according to Radical Rule 2, as

$$\frac{\sqrt{4}}{\sqrt{9}} = \frac{2}{3}$$

(b) In order to use Radical Rule 2 for $\sqrt{0.04}$, we rewrite $\sqrt{0.04}$ as a fraction:

$$\sqrt{0.04} = \sqrt{\frac{4}{100}}$$

$$= \frac{\sqrt{4}}{\sqrt{100}} \qquad \text{Radical Rule 2}$$

$$= \frac{2}{10}$$

$$= 0.2 \text{ or } \frac{1}{5}$$

(c)
$$\sqrt{0.12} = \sqrt{\frac{12}{100}}$$

$$= \frac{\sqrt{12}}{\sqrt{100}} \qquad \text{Radical Rule 2}$$

$$= \frac{\sqrt{4}\cdot\sqrt{3}}{10} \qquad \begin{array}{l}\text{Here we apply Radical Rule 1}\\\text{to the numerator.}\end{array}$$

$$= \frac{2\sqrt{3}}{10}$$

$$= \frac{\sqrt{3}}{5}$$

The two rules of radicals are valid when variables appear under the radical sign but only when those variables represent positive numbers. For example, $\sqrt{10^2} = \sqrt{100} = 10$ but $\sqrt{(-10)^2} = \sqrt{100} = 10 \neq -10$. In general, we have $\sqrt{x^2} = x$ whenever x is a nonnegative number. We will assume that all variables are positive for the remainder of this chapter. The next four examples involve simplifying algebraic radical expressions.

Example 5

Simplify $\sqrt{9x^2}$.

Solution

$$\sqrt{9x^2} = \sqrt{9}\cdot\sqrt{x^2} \qquad \text{Radical Rule 1}$$
$$\phantom{\sqrt{9x^2}} = \;\;3\;\;\cdot\;\; x$$
$$\phantom{\sqrt{9x^2}} = 3x$$

Check: We claim that $3x$ is the (positive) square root of $9x^2$; therefore, if $3x$ is the right answer, we should get $9x^2$ when it is squared.

$$(3x)^2 = 3x \cdot 3x$$
$$(3x)^2 = 9x^2 \;\checkmark$$

Example 6

Simplify $-\sqrt{49y^5}$.

Solution

$$-\sqrt{49y^5} = -\sqrt{49} \cdot \sqrt{y^5} \qquad \text{Radical Rule 1}$$

$$= -\sqrt{49} \cdot \sqrt{y^2 \cdot y^2 \cdot y} \qquad y^5 = y^2 \cdot y^2 \cdot y$$

$$= -\sqrt{49} \cdot \sqrt{y^2} \cdot \sqrt{y^2} \cdot \sqrt{y} \qquad \text{Radical Rule 1}$$

$$= -7 \cdot y \cdot y \cdot \sqrt{y}$$

$$= -7y^2\sqrt{y}$$

Check: $\quad (-7y^2\sqrt{y})^2 = (-7y^2\sqrt{y})(-7y^2\sqrt{y})$

$$= \underbrace{-7 \cdot -7}_{\downarrow} \cdot \underbrace{y^2 \cdot y^2}_{\downarrow} \cdot \underbrace{\sqrt{y} \cdot \sqrt{y}}_{\downarrow}$$

$$= \quad 49 \quad \cdot \quad y^4 \quad \cdot \quad \sqrt{y^2}$$

$$= \quad 49 \quad \cdot \quad y^4 \quad \cdot \quad y$$

$$= 49y^5 \;\checkmark$$

Example 7

Simplify $\sqrt{80x^3y^5z}$.

Solution

We factor the expression so that as many perfect squares appear as possible.

$$\sqrt{80x^3y^5z} = \sqrt{16 \cdot 5 \cdot x^2 \cdot x \cdot y^2 \cdot y^2 \cdot y \cdot z}$$

$$= \sqrt{16} \cdot \sqrt{x^2} \cdot \sqrt{y^2} \cdot \sqrt{y^2} \cdot \sqrt{5xyz}$$

$$= 4 \cdot x \cdot y \cdot y \cdot \sqrt{5xyz}$$

$$= 4xy^2\sqrt{5xyz}$$

Check: $\quad (4xy^2\sqrt{5xyz})^2 = \underbrace{4xy^2 \cdot 4xy^2}_{} \cdot \underbrace{\sqrt{5xyz} \cdot \sqrt{5xyz}}_{}$

$$= \quad 16x^2y^4 \quad \cdot \quad \sqrt{(5xyz)^2}$$

$$= \quad 16x^2y^4 \quad \cdot \quad 5xyz$$

$$= 80x^3y^5z \;\checkmark$$

Example 8

Simplify $\pm\dfrac{x}{2}\sqrt{4x^2}$.

Solution

$$\pm\frac{x}{2}\sqrt{4x^2} = \pm\frac{x}{2} \cdot \sqrt{4 \cdot x^2}$$

$$= \pm\frac{x}{2} \cdot \sqrt{4} \cdot \sqrt{x^2} \qquad \text{Radical Rule 1}$$

$$= \pm\frac{x}{2} \cdot 2 \cdot x \;\leftarrow \text{This part is needed in the check.}$$

$$= \pm\frac{x}{2} \cdot \frac{2}{1} \cdot x$$

$$= \pm x^2$$

Check: We must show that $(2x)^2 = 4x^2$ (*Why?*)

$$(2x)^2 = 2x \cdot 2x$$
$$= 4x^2 \checkmark$$

Section 10.2 Exercises

In Exercises 1 through 18, simplify the expressions and then use Appendix Table B to approximate your answer.

1. $\sqrt{220}$ 2. $\sqrt{2000}$ 3. $\sqrt{500}$

4. $\sqrt{250}$ 5. $\sqrt{450}$ 6. $\sqrt{320}$

7. $\sqrt{1000}$ 8. $\pm\sqrt{288}$ 9. $\sqrt{10,400}$

10. $-\sqrt{243}$ 11. $\sqrt{\dfrac{81}{16}}$ 12. $\sqrt{0.09}$

13. $\sqrt{0.16}$ 14. $\sqrt{0.48}$ 15. $\sqrt{\dfrac{-4}{25}}$

16. $\sqrt{0.81}$ 17. $-\sqrt{\dfrac{16}{9}}$ 18. $\sqrt{0.0016}$

In Exercises 19 through 35, simplify each expression, if possible. Assume all variables are positive.

19. $\sqrt{36x^2}$ 20. $\sqrt{x^4}$ 21. $\sqrt{16x^2}$

22. $\sqrt{32x^2}$ 23. $\sqrt{x^4y^3}$ 24. $\sqrt{25x^6}$

25. $\sqrt{x^{600}}$ 26. $\sqrt{x^{601}}$ 27. $\sqrt{81x^3y^2z^5}$

28. $\sqrt{243x^3}$ 29. $\sqrt{\dfrac{243x^3}{y^2}}$ 30. $-5\sqrt{54x^3}$

31. $\sqrt{120x^8}$ 32. $\sqrt{120x^{25}}$ 33. $\pm\sqrt{x^2y^4z^{17}}$

34. $\dfrac{x^2}{y^3}\sqrt{\dfrac{y^3}{x^2}}$ 35. $\pm\dfrac{5}{x}\sqrt{\dfrac{x^2}{25}}$

Supplementary Exercises

In Exercises 1 through 9, simplify the expressions and then use Appendix Table B to approximate your answer.

1. $\sqrt{200}$ 2. $\sqrt{810}$ 3. $-\sqrt{432}$

4. $-\sqrt{\dfrac{100}{49}}$ 5. $\sqrt{0.04}$ 6. $\sqrt{-64}$

7. $\sqrt{0.0009}$ 8. $-\sqrt{\dfrac{121}{36}}$ 9. $-\sqrt{125}$

In Exercises 10 through 17, simplify each expression, if possible. Assume all variables are positive.

10. $\sqrt{4x^4}$ 11. $\sqrt{16x^6}$ 12. $-\sqrt{9x^3y^3}$

13. $\sqrt{x^{400}}$ 14. $\sqrt{\dfrac{x^5}{y^3}}$ 15. $\sqrt{125x^3y^7}$

16. $\pm\sqrt{18x^4y^5}$ 17. $\sqrt{\dfrac{121x^3}{y^2}}$

10.3   Adding and subtracting radical expressions

Just as there is no way of simplifying expressions like $x + y$, or $x + x^2$, or $x^2 - 2xy + y$, there is no way of combining expressions like $\sqrt{2} + \sqrt{3}$, or $\sqrt{5} \pm \sqrt{7}$, or $7 + \sqrt{7}$. There are some radical expressions, however, that can be rewritten so that like terms can be collected; and this is what the next four examples deal with.

Example 1

Add $\sqrt{8} + \sqrt{32}$.

Solution

We caution the reader against a common error here. It is tempting to say that $\sqrt{8} + \sqrt{32} = \sqrt{40}$, which is *incorrect*. A check with Appendix Table B will verify that such a procedure is incorrect. Now, let us see how to proceed correctly.

$$\sqrt{8} + \sqrt{32} = \sqrt{4 \cdot 2} + \sqrt{16 \cdot 2}$$
$$= \sqrt{4} \cdot \sqrt{2} + \sqrt{16} \cdot \sqrt{2} \qquad \text{Radical Rule 1}$$
$$= 2\sqrt{2} + 4\sqrt{2}$$

We now combine (add) the like terms (just like $2x + 4x = 6x$).

$$2\sqrt{2} + 4\sqrt{2} = 6\sqrt{2}$$

Thus,
$$\sqrt{8} + \sqrt{32} = 6\sqrt{2}$$

Example 2

Simplify $2\sqrt{3} - \sqrt{27}$.

Solution

We can rewrite $\sqrt{27}$ as $\sqrt{9 \cdot 3} = 3\sqrt{3}$. So

$$2\sqrt{3} - \sqrt{27} = 2\sqrt{3} - 3\sqrt{3}$$
$$= -\sqrt{3}$$

Example 3

Simplify: $\sqrt{8} - \sqrt{5} - \sqrt{20}$.

Solution

$$\sqrt{8} - \sqrt{5} - \sqrt{20} = \sqrt{4 \cdot 2} - \sqrt{5} - \sqrt{4 \cdot 5}$$
$$= \sqrt{4} \cdot \sqrt{2} - \sqrt{5} - \sqrt{4} \cdot \sqrt{5} \qquad \text{Radical Rule 1}$$
$$= 2\sqrt{2} - \underbrace{\sqrt{5} - 2\sqrt{5}}$$
$$= 2\sqrt{2} - 3\sqrt{5}$$

Notice that we cannot simplify $2\sqrt{2} - 3\sqrt{5}$ any further.

Example 4

Simplify $\sqrt{49} + 4\sqrt{63} - 10\sqrt{28}$.

Solution

$$\sqrt{49} + 4\sqrt{63} - 10\sqrt{28} = 7 + 4\sqrt{9 \cdot 7} - 10\sqrt{4 \cdot 7}$$

$$= 7 + 4\sqrt{9} \cdot \sqrt{7} - 10\sqrt{4} \cdot \sqrt{7} \qquad \text{Radical Rule 1}$$

$$= 7 + 4 \cdot 3 \cdot \sqrt{7} - 10 \cdot 2 \cdot \sqrt{7}$$

$$= 7 + \underbrace{12\sqrt{7} - 20\sqrt{7}}$$

$$= \qquad 7 - 8\sqrt{7}$$

Once we have combined as many like terms as we can, it is sometimes possible to factor expressions. This will be especially handy in Chapter 11 where we are curious as to whether or not the fraction can be reduced when radical expressions appear in the numerator of a fraction. The next three examples illustrate this.

Example 5

Simplify $\dfrac{12 + 2\sqrt{3}}{2}$.

Solution

As in any fraction, we can reduce it if we can find a factor common to both numerator and denominator. If we view $12 + 2\sqrt{3}$ as a binomial, there is a common factor of 2. We have

$$\frac{12 + 2\sqrt{3}}{2} = \frac{2(6 + \sqrt{3})}{2} \qquad \text{Factor numerator.}$$

$$= 6 + \sqrt{3} \qquad \begin{array}{l}\text{Divide by 2 in numerator and} \\ \text{denominator.}\end{array}$$

Example 6

Simplify $\dfrac{5 - \sqrt{50}}{15}$.

Solution

First notice that $\sqrt{50}$ is $5\sqrt{2}$. So

$$\frac{5 - \sqrt{50}}{15} = \frac{5 - 5\sqrt{2}}{15}$$

$$= \frac{5(1 - \sqrt{2})}{5 \cdot 3} \qquad \begin{array}{l}\text{Factor the numerator and} \\ \text{denominator.}\end{array}$$

$$= \frac{1 - \sqrt{2}}{3} \qquad \text{Divide by 5.}$$

Example 7

Simplify $\dfrac{-32 \pm \sqrt{72}}{6}$.

Solution

$$\frac{-32 \pm \sqrt{72}}{6} = \frac{-32 \pm 6\sqrt{2}}{6} \qquad \sqrt{72} = \sqrt{36 \cdot 2} \text{ or } 6\sqrt{2}$$

$$= \frac{2(-16 \pm 3\sqrt{2})}{2 \cdot 3} \qquad \begin{array}{l}\text{Factor the numerator} \\ \text{and denominator.}\end{array}$$

$$= \frac{-16 \pm 3\sqrt{2}}{3}$$

The reader is urged to note that the above answer really represents two numbers because of the presence of the "$\pm$" sign. These two numbers are $\dfrac{-16 + 3\sqrt{2}}{3}$ and $\dfrac{-16 - 3\sqrt{2}}{3}$.

Section 10.3 Exercises

In Exercises 1 through 30, simplify each expression, if possible.

1. $\sqrt{6} + \sqrt{24}$

2. $\sqrt{8} - \sqrt{18}$

3. $\sqrt{18} + \sqrt{16}$

4. $\sqrt{8a} \pm \sqrt{18a}$

5. $\sqrt{144} - \sqrt{100}$

6. $5\sqrt{5} - \sqrt{20}$

7. $\sqrt{3} + \sqrt{12}$

8. $\sqrt{8} - \sqrt{18} - \sqrt{32}$

9. $\sqrt{8} - \sqrt{18} + \sqrt{2}$

10. $\sqrt{45} + \sqrt{80} - \sqrt{20}$

11. $\sqrt{44} + \sqrt{200} - \sqrt{99} - \sqrt{50}$

12. $\sqrt{200} + \sqrt{300} + \sqrt{400} + \sqrt{500}$

13. $2\sqrt{3} + \sqrt{13}$

14. $\sqrt{2x} + \sqrt{32x}$

15. $\sqrt{9x} + \sqrt{16x}$

16. $\sqrt{200} - \sqrt{32} + \sqrt{12}$

17. $\dfrac{-5 \pm \sqrt{50}}{5}$

18. $\dfrac{1 \pm 2\sqrt{2}}{2}$

19. $\dfrac{2 \pm 2\sqrt{2}}{2}$

20. $\dfrac{4 + 8\sqrt{3}}{2}$

21. $\dfrac{2 + \sqrt{8}}{2}$

22. $\dfrac{6 - \sqrt{12}}{2}$

23. $\dfrac{6 - 3\sqrt{12}}{2}$

24. $\dfrac{4 \pm \sqrt{8}}{4}$

25. $\dfrac{-7 \pm 2\sqrt{98}}{14}$

26. $\dfrac{-9 \pm \sqrt{72}}{6}$

27. $\dfrac{-2 \pm \sqrt{60}}{2}$

28. $\dfrac{-18 \pm \sqrt{180}}{18}$

29. $\dfrac{-20 \pm \sqrt{700}}{35}$

30. $\dfrac{-4 \pm \sqrt{128}}{16}$

Supplementary Exercises

In Exercises 1 through 15, simplify each expression, if possible.

1. $\sqrt{8} + \sqrt{18}$

2. $\sqrt{125} + \sqrt{180}$

3. $\sqrt{32} - \sqrt{8}$

4. $\sqrt{75} - \sqrt{48} + \sqrt{27}$

5. $\sqrt{112} + \sqrt{18} - \sqrt{28} + \sqrt{27}$

6. $\sqrt{300} - \sqrt{200}$

7. $\sqrt{500} + 2\sqrt{48} - \sqrt{300}$

8. $\dfrac{5 + 10\sqrt{6}}{5}$

9. $\dfrac{-6 \pm 12\sqrt{7}}{2}$

10. $\sqrt{12} \pm \sqrt{27}$

11. $\dfrac{-8 \pm 4\sqrt{6}}{4}$

12. $\dfrac{5 \pm \sqrt{125}}{5}$

13. $\dfrac{-2 \pm \sqrt{8}}{2}$

14. $\dfrac{6 \pm \sqrt{27}}{-3}$

15. $-7 \pm \sqrt{49}$

10.4   Rationalizing denominators

In this section, we examine expressions involving radical signs in denominators. The expressions $\dfrac{1}{\sqrt{3}}$, $\dfrac{7}{1 - \sqrt{2}}$, and $\dfrac{5}{\sqrt{a} - \sqrt{b}}$ are examples.

To approximate $\dfrac{1}{\sqrt{3}}$ directly to three decimals would require doing a long division problem: dividing 1 by the three-digit-decimal $1.732(\sqrt{3} \approx 1.732)$.

$$
\begin{array}{r}
0.577 \\
1.732.\overline{)\,1.000.000} \\
\underline{866\ 0} \\
134\ 00 \\
\underline{121\ 24} \\
12\ 760 \\
\underline{12\ 124} \\
636
\end{array}
$$

So $\dfrac{1}{\sqrt{3}} \approx .577$.

There is an easier way. We build up the fraction $\dfrac{1}{\sqrt{3}}$ by multiplying both the numerator and denominator by $\sqrt{3}$.

$$\frac{1}{\sqrt{3}} = \frac{1}{\sqrt{3}} \cdot \frac{\sqrt{3}}{\sqrt{3}}$$

Notice that, by multiplying numerator and denominator by $\sqrt{3}$, we obtain a fraction *without* a radical sign in the denominator.

$$= \frac{\sqrt{3}}{\sqrt{9}}$$

$$= \frac{\sqrt{3}}{3}$$

Now, to approximate $\dfrac{\sqrt{3}}{3}$ is an easy matter.

$$\frac{\sqrt{3}}{3} \approx \frac{1.732}{3} \approx .577$$

The process of eliminating a radical sign in the denominator of a fraction is called **rationalizing the denominator.** The next four examples further illustrate this process.

Example 1

Rationalize the denominator of $\dfrac{4}{\sqrt{2}}$ in order to approximate it to three decimal places.

Solution

We can eliminate the radical sign in the denominator by multiplying by $\sqrt{2}$

$$\frac{4}{\sqrt{2}} = \frac{4}{\sqrt{2}} \cdot \frac{\sqrt{2}}{\sqrt{2}}$$

$$= \frac{4\sqrt{2}}{\sqrt{4}}$$

$$= \frac{4\sqrt{2}}{2}$$

$$= 2\sqrt{2}$$

Now we continue by referring to Appendix Table B.

$$\frac{4}{\sqrt{2}} = 2\sqrt{2} \approx 2 \cdot 1.414$$

So we get 2.828.

Example 2

Rewrite $\dfrac{\sqrt{8}}{\sqrt{6}}$ *without a radical sign in the denominator.*

Solution

$$\frac{\sqrt{8}}{\sqrt{6}} = \frac{\sqrt{8}}{\sqrt{6}} \cdot \frac{\sqrt{6}}{\sqrt{6}}$$

$$= \frac{\sqrt{48}}{\sqrt{36}}$$

$$= \frac{\sqrt{48}}{6}$$

Now, using the technique of Section 10.2, we can further simplify the numerator as follows.

$$\frac{\sqrt{48}}{6} = \frac{\sqrt{16 \cdot 3}}{6}$$

$$= \frac{\sqrt{16} \cdot \sqrt{3}}{6}$$

$$= \frac{4\sqrt{3}}{6}$$

$$= \frac{2\sqrt{3}}{3}$$

Example 3

Rationalize the denominator $\dfrac{\sqrt{2}}{\sqrt{5} + \sqrt{3}}$.

Solution

To rationalize the denominator of $\dfrac{\sqrt{2}}{\sqrt{5} + \sqrt{3}}$, we multiply both the numerator and denominator by $\sqrt{5} - \sqrt{3}$. [The reason is this: We know that $(x + y)(x - y) = x^2 - y^2$ from Chapter 5. So $(\sqrt{a} + \sqrt{b})(\sqrt{a} - \sqrt{b}) = (\sqrt{a})^2 - (\sqrt{b})^2 = a - b$, an expression *without* radical signs.]

$$\frac{\sqrt{2}}{\sqrt{5}+\sqrt{3}} = \frac{\sqrt{2}}{\sqrt{5}+\sqrt{3}} \cdot \frac{\sqrt{5}-\sqrt{3}}{\sqrt{5}-\sqrt{3}}$$

$$= \frac{\sqrt{2}(\sqrt{5}-\sqrt{3})}{(\sqrt{5})^2 - (\sqrt{3})^2}$$

$$= \frac{\sqrt{2}(\sqrt{5}-\sqrt{3})}{5-3}$$

$$= \frac{\sqrt{2}(\sqrt{5}-\sqrt{3})}{2}$$

Example 4

Rationalize the denominator $\dfrac{4}{10-\sqrt{2}}$.

Solution

We must rationalize the denominator of

$$\frac{4}{10-\sqrt{2}}$$

We will multiply the numerator and denominator by $10 + \sqrt{2}$.

$$\frac{4}{10-\sqrt{2}} = \frac{4}{10-\sqrt{2}} \cdot \frac{10+\sqrt{2}}{10+\sqrt{2}}$$

$$= \frac{4(10+\sqrt{2})}{(10)^2 - (\sqrt{2})^2}$$

$$= \frac{4(10+\sqrt{2})}{100-2}$$

$$= \frac{4(10+\sqrt{2})}{98}$$

$$= \frac{2(10+\sqrt{2})}{49}$$

Section 10.4 Exercises

In Exercises 1 through 15, rationalize the denominator and then use Appendix Table B to approximate the number to three decimal places.

1. $\dfrac{2}{\sqrt{2}}$

2. $\dfrac{5}{\sqrt{5}}$

3. $\dfrac{20}{\sqrt{8}}$

4. $\dfrac{30}{\sqrt{45}}$

5. $\dfrac{9}{\sqrt{3}}$

6. $\dfrac{1+\sqrt{2}}{\sqrt{2}}$

7. $\dfrac{2}{\sqrt{3}-\sqrt{2}}$

8. $\dfrac{2}{\sqrt{2}-\sqrt{3}}$

9. $\dfrac{12-\sqrt{3}}{\sqrt{3}}$

10. $\dfrac{\sqrt{5}}{1+\sqrt{5}}$

11. $\dfrac{\sqrt{5}+\sqrt{10}}{\sqrt{20}}$

12. $\dfrac{\sqrt{20}}{\sqrt{10}+\sqrt{5}}$

13. $\dfrac{1}{\sqrt{8}-\sqrt{6}}$

14. $\dfrac{\sqrt{11}}{\sqrt{5}-\sqrt{7}}$

15. $\dfrac{\sqrt{12}}{\sqrt{8}+\sqrt{3}}$

In Exercises 16 through 19, rationalize the denominator.

16. $\dfrac{3}{\sqrt{7} - \sqrt{2}}$

17. $\dfrac{3}{\sqrt{8} - \sqrt{2}}$

18. $\dfrac{\sqrt{3}}{\sqrt{5} + \sqrt{6}}$

19. $\dfrac{\sqrt{3}}{6 + \sqrt{3}}$

20. Using a calculator with a $\boxed{\sqrt{}}$ key, find the value of $\dfrac{\sqrt{12}}{\sqrt{8} + \sqrt{3}}$; compare with your answer to Exercise 15.

Supplementary Exercises

In Exercises 1 through 10, rationalize the denominator and then use Appendix Table B to approximate the number to three decimal places.

1. $\dfrac{5}{\sqrt{5}}$

2. $\dfrac{10}{\sqrt{2}}$

3. $\dfrac{-6}{\sqrt{3}}$

4. $\dfrac{-7}{\sqrt{14}}$

5. $\dfrac{15}{\sqrt{3}}$

6. $\dfrac{3}{\sqrt{2} - \sqrt{3}}$

7. $\dfrac{-2}{\sqrt{5} - \sqrt{2}}$

8. $\dfrac{4}{\sqrt{6} + \sqrt{2}}$

9. $\dfrac{1}{3 - \sqrt{2}}$

10. $\dfrac{6}{\sqrt{3} - \sqrt{2}}$

10.5   Equations involving radicals

In this section we will examine some techniques for solving equations in which a variable occurs within a radical sign. Such equations are called **radical equations.** Examples of some radical equations are $\sqrt{2x} = 10$, $\sqrt{x - 7} + 2 = 4$, and $\sqrt{5x} = \sqrt{6x - 10}$. We solve them in the examples that follow using a technique called *squaring both sides of an equation.* First, however, we will illustrate this technique.

Consider the simple equation $y = 4$ (whose solution is the obvious number 4). When we square both sides of that equation we get $y^2 = 16$, whose solutions are 4 and -4. Notice that, when we squared both sides, our original solution, 4, remained a solution in the new equation. However, another number, -4, also "appeared." This other number *is* a solution to the new equation ($y^2 = 16$) but *is not* a solution to the original equation ($y = 4$). It is important to realize that squaring both sides of an equation *might* produce another equation that, when solved, has one or more solutions that are *not* solutions to the original equation. Such numbers are called **extraneous solutions** to the original equation. The reader should be on guard—an extraneous solution is *not* a solution.

Since the process of squaring both sides of an equation will be used in the examples that follow, the "check" portion of each example is very important. We do not want to include any extraneous solutions as part of our answer.

Example 1

Solve for x: $\sqrt{2x} = 10$.

Solution

First we square both sides of the equation. Notice that, by doing so, we eliminate the radical sign.

$$\sqrt{2x} = 10$$
$$(\sqrt{2x})^2 = (10)^2 \qquad \text{Square both sides.}$$
$$2x = 100$$
$$x = 50$$

$$\text{Check:} \qquad \sqrt{2x} = 10$$
$$\sqrt{2 \cdot 50} = 10$$
$$\sqrt{100} = 10$$
$$10 = 10 \; \checkmark$$

Example 2

Solve for x: $12 - \sqrt{x} = 15$.

Solution

Before we square both sides, we isolate $\sqrt{x}$.

$$12 - \sqrt{x} = 15$$
$$-\sqrt{x} = 3 \qquad \begin{array}{l}\text{Isolate } \sqrt{x} \text{ by subtracting 12} \\ \text{from both sides.}\end{array}$$
$$(-\sqrt{x})^2 = (3)^2 \qquad \text{Square both sides.}$$
$$x = 9$$

$$\text{Check:} \quad 12 - \sqrt{x} = 15$$
$$12 - \sqrt{9} = 15$$
$$12 - 3 = 15$$
$$9 = 15 \; \times \qquad \text{Does not check.}$$

The value 9 is an extraneous solution. Since it was the only solution to the squared equation, *there is no solution to the original equation.*

Example 3

Solve $y = \sqrt{y}$.

Solution

$$y = \sqrt{y}$$
$$y^2 = y \qquad \text{Square both sides.}$$

Since $y^2 = y$ is a factorable equation, we proceed as we did in Section 6.6.

$$y^2 = y$$
$$y^2 - y = 0$$
$$y(y - 1) = 0$$
$$y = 0 \qquad y - 1 = 0$$
$$y = 1 \longleftarrow \text{Answers}$$

Check for $y = 0$:   **Check** for $y = 1$:

$$y = \sqrt{y}$$
$$0 = \sqrt{0}$$
$$0 = 0 \checkmark$$

$$y = \sqrt{y}$$
$$1 = \sqrt{1}$$
$$1 = 1 \checkmark$$

Hence both numbers, 0 and 1, are solutions to the original equation.

Example 4

Solve $\sqrt{5x} = \sqrt{6x - 10}$.

Solution

$$\sqrt{5x} = \sqrt{6x - 10}$$

$$5x = 6x - 10 \qquad \text{Square both sides.}$$

$$-x = -10 \qquad \text{Subtract } 6x \text{ from both sides.}$$

$$x = 10$$

Check:

$$\sqrt{5x} = \sqrt{6x - 10}$$
$$\sqrt{5 \cdot 10} = \sqrt{6 \cdot 10 - 10}$$
$$\sqrt{50} = \sqrt{50} \checkmark$$

Now that we have had practice in solving radical equations, we will apply that knowledge to some practical problems from various life situations in the next three examples.

Example 5

The equation $s = 16t^2$ is an algebraic model for a free-falling body (an object falling under the influence of gravity alone) where s represents the distance fallen (in feet) and t represents the time that elapses (in seconds).

(a) Solve this equation for t.

(b) How long will it take a free-falling body to fall 64 feet?

Solution

(a) We desire to solve $s = 16t^2$ for t. We have seen that, when we are working with equations, we must do the same thing to both sides of the equation. We continue that policy now.

$$s = 16t^2$$

$$\frac{s}{16} = \frac{\cancel{16}t^2}{\cancel{16}} \qquad \text{Divide both sides of the equation by 16.}$$

$$\frac{s}{16} = t^2$$

$$\pm\sqrt{\frac{s}{16}} = t \qquad \text{Take the square root of both sides of the equation.}$$

However, since the time, t, cannot be negative, we *reject* the negative solution. So

$$t = \sqrt{\frac{s}{16}}$$

(b) We have just found $t = \sqrt{\frac{s}{16}}$, which expresses the time, t, a free-falling body takes to fall a certain distance, s. Thus if the distance is 64 feet,

$$t = \sqrt{\frac{s}{16}}$$

$$= \sqrt{\frac{64}{16}}$$

$$= \sqrt{4}$$

$$= 2$$

Thus it takes 2 seconds for a free-falling body to fall 64 feet.

Example 6

The equation $A = \pi r^2$ is an algebraic model for the area A of any circle whose radius is r. (If r is expressed in inches, then A is expressed in square inches; if r is expressed in centimeters, then A is expressed in square centimeters, and so on.)

(a) Solve this equation for r.

(b) If a circle has an area 15.7 square centimeters, what is its radius? (Use $\pi \approx 3.14$.)

Solution

(a) As in the previous example, we will do the same thing to both sides of the equation.

$$A = \pi r^2$$

$$\frac{A}{\pi} = \frac{\cancel{\pi} r^2}{\cancel{\pi}} \qquad \text{Divide both sides of the equation by the number } \pi.$$

$$\frac{A}{\pi} = r^2$$

$$\pm\sqrt{\frac{A}{\pi}} = r \qquad \text{Take the square root of both sides of the equation.}$$

However, since the radius, r, cannot be negative, we *reject* the negative solution. So

$$r = \sqrt{\frac{A}{\pi}}$$

(b) We need to find r, when A has the value 15.7 square centimeters.

$$r = \sqrt{\frac{A}{\pi}}$$

$$\approx \sqrt{\frac{15.7}{3.14}}$$

$$\approx \sqrt{5}$$

$$\approx 2.236$$

So the radius is approximately 2.236 cm.

Example 7

The equation $E = 8LWH^2$ is an algebraic model for the energy, E (in foot-pounds), that can be expected from sea waves, where:

L is the length (in feet) of the wave (distance between successive crests).

H is the height (in feet) of the wave (vertical distance from trough to crest).

W is the width (in feet) of the wave.

(a) Solve this equation for H.

(b) If a certain wave that is 50 feet long and 20 feet wide has 400,000 foot-pounds of energy, how high is that wave?

Solution

(a)
$$E = 8LWH^2$$

$$\frac{E}{8LW} = \frac{\cancel{8LW}H^2}{\cancel{8LW}}$$ Divide both sides of the equation by $8LW$.

$$\frac{E}{8LW} = H^2$$

$$\pm\sqrt{\frac{E}{8LW}} = H$$ Take the square root of both sides of the equation.

Since the height of the wave, H, cannot be negative, we *reject* the negative solution. So

$$H = \sqrt{\frac{E}{8LW}}$$

(b) We need to determine H, when $E = 400,000$, $L = 50$, and $W = 20$.

$$H = \sqrt{\frac{E}{8LW}}$$

$$= \sqrt{\frac{400,000}{8 \cdot 50 \cdot 20}}$$

$$= \sqrt{\frac{400,000}{8,000}}$$

$$= \sqrt{50}$$

$$\approx 7.071$$

So the height of the wave is approximately 7.071 feet.

We conclude this section with an application of radicals to a formula from plane geometry, the **Pythagorean Formula.** It expresses the relationship between the two *legs* (labeled a and b in the triangle) of a *right triangle* and the *hypotenuse* (labeled c). The hypotenuse is always the side opposite the right angle.

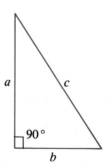

The Pythagorean Formula states that, in any right triangle, *the square of the hypotenuse is equal to the sum of the squares of the other two sides.* In symbols, we have

$$c^2 = a^2 + b^2$$

Example 8

In the triangle below, $a = 3$ inches and $b = 4$ inches. Find c.

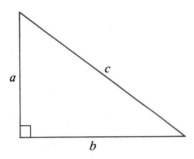

Solution

$$c^2 = a^2 + b^2$$
$$c^2 = 3^2 + 4^2$$
$$c^2 = 9 + 16$$
$$c^2 = 25$$
$$c = \pm\sqrt{25} \qquad \text{Take the square root of both sides}$$
$$\text{of the equation.}$$
$$c = \pm 5$$

The solution $c = -5$ is *rejected* because the hypotenuse must be positive. Hence the hypotenuse is 5 inches.

Example 9

If $a = 1$ meter and $b = 3$ meters, find the hypotenuse c, for the right triangle below.

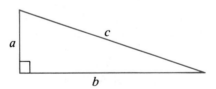

Solution

$$c^2 = a^2 + b^2$$
$$c^2 = 1^2 + 3^2$$
$$c^2 = 1 + 9$$
$$c^2 = 10$$
$$c = \pm\sqrt{10} \qquad \text{Take the square root of both sides}$$
$$\text{of the equation.}$$

But the value $-\sqrt{10}$ is *rejected* because the hypotenuse must be positive. So the hypotenuse is $\sqrt{10} \approx 3.162$ meters.

Example 10

A support wire runs from the top of a 20-foot vertical pole to a point on the ground 20 feet from its base. How long must the wire be?

Solution

First we draw a pictorial model.

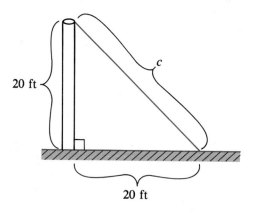

Since we have a right triangle, we can find c by using the Pythagorean Formula.

$$c^2 = a^2 + b^2$$
$$c^2 = 20^2 + 20^2$$
$$c^2 = 800$$
$$c = \pm\sqrt{800}$$
$$c = \sqrt{800} \qquad \text{We } reject \text{ the negative solution.}$$
$$c = \sqrt{400 \cdot 2}$$
$$c = \sqrt{400} \cdot \sqrt{2}$$
$$c = 20\sqrt{2}$$

So the wire is $20\sqrt{2} \approx 28.28$ feet long.

Example 11

The diagonal of a rectangle is $2\sqrt{5}$ inches long. If the length of the rectangle is twice its width, find the dimensions of the rectangle.

Solution

According to our pictorial model, we see that this rectangle's length, width, and diagonal compose a right triangle.

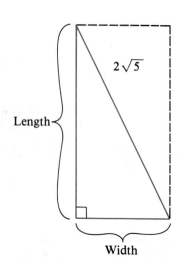

Now, let x = the width (in inches) of the rectangle. So the length is $2x$. We have

$$\text{leg}^2 + \text{leg}^2 = \text{hypotenuse}^2$$
$$x^2 + (2x)^2 = (2\sqrt{5})^2$$
$$x^2 + 4x^2 = 2^2(\sqrt{5})^2$$
$$5x^2 = 20$$
$$x^2 = 4$$
$$x = \pm 2$$
$$x = 2 \qquad \text{Again, the solution } -2 \text{ has}$$
$$\text{no significance in this problem.}$$

So the dimensions of the rectangle are 2 inches and 4 inches.

Section 10.5 Exercises

In Exercises 1 through 20, solve each equation. Be sure to check for extraneous solutions.

1. $\sqrt{x} = 10$
2. $\sqrt{5x} = 10$
3. $\sqrt{3x} = 10$
4. $\sqrt{3x} = 9$
5. $\sqrt{6x} = 0.1$
6. $3 - \sqrt{x} = 1$
7. $1 - \sqrt{x} = 3$
8. $3 - \sqrt{2x} = 1$
9. $1 - \sqrt{2x} = 3$
10. $y + \sqrt{y} = 0$
11. $\sqrt{x - 7} + 2 = 4$
12. $\sqrt{8x + 3} = 3$
13. $\sqrt{5x} = \sqrt{3x + 1}$
14. $\sqrt{4x - 1} = 0$
15. $1 + \sqrt{2x - 3} = 8$
16. $2\sqrt{x} = 4x$
17. $\sqrt{x} = 7x$
18. $1 + \sqrt{2x - 3} = x - 2$
19. $1 + \sqrt{5x - 1} = 1 + \sqrt{19}$
20. $\sqrt{3x + 1} = \sqrt{x} + 3$ (*Hint:* Square both sides, simplify, and isolate the radical; then square both sides again.)

21. The ability of an automobile to turn is limited by its resistance to skidding. The maximum speed, v, that an auto can maintain around a curb of radius, r, without skidding is given (approximately) by the algebraic model

$$r = \frac{2}{5}v^2$$

where r is in feet and v is in miles per hour.

(a) Solve this equation for v.

(b) If a curb has a radius of 10 feet, what is the maximum speed that a car can maintain, without skidding, while driving around the curb?

22. The time, T, that it takes a pendulum of length, ℓ, to swing through one complete oscillation is given in the algebraic model

$$T = 2\pi\sqrt{\frac{\ell}{9.8}}$$

where T is in seconds and ℓ is in meters.

(a) Solve this equation for ℓ.

(b) How long would a pendulum have to be for it to have an oscillation time of 1 second?

23. Undersea earthquakes can give rise to huge ocean waves called **tsunami.** Tsunami can travel thousands of miles without suffering much loss of energy. It is important to be able to predict the arrival times

of these large waves at land sites. The equation $v = \sqrt{32h}$ is an algebraic model for the speed of these waves, v, measured in feet/second, and h is the depth of the water, measured in feet.

(a) With what speed do tsunami travel in the Pacific Ocean, where the average depth is 15,800 feet?

(b) Solve the equation $v = \sqrt{32h}$ for h.

(c) Determine the ocean depth if tsunami travel at 300 miles per hour (440 ft/sec).

24. An equation from chemistry that can be used to find the velocity of a molecule of gas is given by

$$v = \sqrt{\frac{3RT}{M}}$$

(a) Solve this formula for T.

(b) If $v = 3\sqrt{30}$, $R = 9.6$, and $M = 32$, find T.

25. The equation $V = \pi r^2 h$ is an algebraic model for the volume, V, of a cylinder of radius r and height h. (If r and h are measured in centimeters, then V is measured in cubic centimeters, etc.)

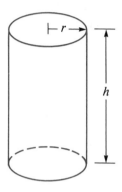

(a) Solve this equation for r.

(b) If the volume is 188.4 cubic centimeters and the height is 10 centimeters, what is the radius?

26. The equation $s = \dfrac{1}{20} v^2$ is an algebraic model for the length of a skid mark, s (measured in feet), for a certain automobile traveling at a speed, v, measured in miles per hour.

(a) Solve this equation for v.

(b) If the length of a skid mark for this automobile is 50 feet, how fast was the automobile traveling?

In Exercises 27 through 35, assume a and b are legs of a right triangle and c is the hypotenuse.

27. $a = 2$, $b = 3$, find c 28. $b = 5$, $c = \sqrt{41}$, find a

29. $a = \dfrac{1}{4}$, $c = \dfrac{1}{2}$, find b 30. $a = 6$, $b = 8$, find c

31. $a = 8$, $b = 6$, find c 32. $a = 1$, $b = 7$, find c

33. $a = 6$, $c = 12$, find b 34. $a = 1$, $c = \sqrt{3}$, find b

35. $a = 5$, $b = 12$, find c

36. A ladder 16 feet long has its base placed 4 feet from the foot of a building. How high will the ladder reach up the building?

37. In Little League baseball, the distance between bases is 60 feet. Find the straight-line distance from second base to home plate. (Think of the bases as corners in a square.)

38. Do Exercise 37 for major league baseball where the distance from base to base is 90 feet.

39. The area, A, of any triangle can be found by the formula

$$A = \frac{1}{2} bh$$

where h represents the perpendicular distance from an angle to a base:

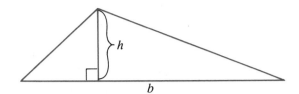

(a) If the base is 6 inches and the height 4 inches, what is the area?

(b) If the base is twice the height and the area is 12 square meters, find the base and height.

(c) If the base is six times the height and the area is 12 square meters, find the base and height.

Supplementary Exercises

In Exercises 1 through 20, solve each equation. Be sure to check for extraneous solutions.

1. $\sqrt{y} = 9$

2. $\sqrt{3x} = 6$

3. $\sqrt{2y} = 8$

4. $\sqrt{\dfrac{x}{3}} = 2$

5. $\sqrt{\dfrac{y}{5}} = 3$

6. $3\sqrt{x} = 9$

7. $2\sqrt{3x} = 6$

8. $2 - \sqrt{x} = 1$

9. $\sqrt{x} - 2 = 3$

10. $\sqrt{5x - 1} = 2$

11. $\sqrt{2x} = \sqrt{x - 1}$

12. $\sqrt{x} = 3x$

13. $2 + \sqrt{3x + 1} = 6$

14. $2x + \sqrt{x} = 0$

15. $\sqrt{3x - 1} + 4 = 5$

16. $2 + \sqrt{x - 1} = 6$

17. $5 - \sqrt{1 - x} = 3$

18. $\sqrt{\dfrac{2x}{3}} = 4$

19. $\sqrt{5x} = \sqrt{2x - 3}$

20. $\sqrt{x} = \sqrt{x + 1} - 2$

10.6   Chapter review

Summary

We began this chapter by examining some irrational numbers. A special group of numbers in the class of irrational numbers are called the radical numbers. These are numbers of the form $\pm\sqrt{a}$ (where a is nonnegative and not a perfect square). We also saw that $\pi = 3.141592653\ldots$ is an irrational number. When we combine the irrational numbers with the rational numbers, we get the set of real numbers, which can be represented on a real number line.

The generalization of arithmetic to algebra was continued in this chapter as our concern shifted from radical numbers to algebraic radical expressions. In particular, two rules for radicals were examined:

$$\text{Radical Rule 1:}\quad \sqrt{x} \cdot \sqrt{y} = \sqrt{xy}$$

$$\text{Radical Rule 2:}\quad \frac{\sqrt{x}}{\sqrt{y}} = \sqrt{\frac{x}{y}}$$

We used these rules and Appendix Table B to approximate square roots of numbers. For example, to approximate $\sqrt{300}$ we write

$$\sqrt{100 \cdot 3} = \sqrt{100} \cdot \sqrt{3} = 10 \cdot \sqrt{3} \approx 10 \cdot 1.732 \approx 17.32$$

The two rules for radicals were also used to simplify algebraic radical expressions. When doing so, it is often necessary to use the fact that $\sqrt{x^2} = x$ (which is true, provided that x is nonnegative).

In Section 10.3, we learned that, in order to add or subtract radical expressions, we must have "like" terms. For example, to add $\sqrt{8} + \sqrt{18}$, we must write

$$\sqrt{8} + \sqrt{18} = \sqrt{4 \cdot 2} + \sqrt{9 \cdot 2} = \sqrt{4} \cdot \sqrt{2} + \sqrt{9} \cdot \sqrt{2} = 2\sqrt{2} + 3\sqrt{2} = 5\sqrt{2}$$

A fraction whose denominator contains a radical expression can be changed so that there are no longer any radical signs in the denominator. This process is called rationalizing the denominator, and it was the topic of Section 10.4. For example, we rationalize the denominator of $\dfrac{3}{\sqrt{7}}$ by proceeding as follows:

$$\frac{3}{\sqrt{7}} = \frac{3}{\sqrt{7}} \cdot \frac{\sqrt{7}}{\sqrt{7}} = \frac{3\sqrt{7}}{\sqrt{49}} = \frac{3\sqrt{7}}{7}$$

In order to rationalize the denominator of $\dfrac{2}{3 - \sqrt{2}}$, we write

$$\frac{2}{3 - \sqrt{2}} \cdot \frac{3 + \sqrt{2}}{3 + \sqrt{2}} = \frac{2(3 + \sqrt{2})}{(3)^2 - (\sqrt{2})^2} = \frac{2(3 + \sqrt{2})}{9 - 2} = \frac{2(3 + \sqrt{2})}{7}$$

When radical signs appear in equations, we saw that they can be eliminated by squaring both sides of the original equation. Sometimes, this idea of squaring both sides of an equation can produce extraneous solutions, that is, numbers that solve the squared equation but not the original equation. We also dealt with equations containing radicals in various life situations, including the applications of the Pythagorean Formula to any right triangle.

Vocabulary Quiz

Match the expression in Column I with the phrase in Column II that *best* describes it.

Column I

1. The positive square root of a is
2. The square roots of a are
3. π is
4. The real number line is
5. A radical expression is
6. By Radical Rule 1, $\sqrt{x} \cdot \sqrt{y} =$
7. By Radical Rule 2, $\dfrac{\sqrt{x}}{\sqrt{y}} =$
8. Rationalizing a denominator means
9. A radical equation is
10. An extraneous solution is
11. $a^2 + b^2 = c^2$ represents
12. Provided $x \geq 0$, $\sqrt{x^2} =$

Column II

a. The Pythagorean Formula for a right triangle with legs denoted by a and b and hypotenuse denoted by c.

b. x.

c. $\sqrt{a}$.

d. $\sqrt{xy}$.

e. The ratio of any circle's circumference to its diameter.

f. Converting a fraction containing a radical sign in its denominator to an equivalent fraction with no radical sign in its denominator.

g. $\pm\sqrt{a}$.

h. A graphical representation of the real numbers.

i. $\sqrt{\dfrac{x}{y}}$.

j. An expression involving a radical symbol.

k. An equation in which a variable appears within a radical sign.

l. A solution to an equation derived from some original equation that does not solve the original equation.

Chapter 10 Review Exercises

In Exercises 1 through 6, answer true or false.

1. $\sqrt{a}$ is only defined when a is not a negative number.

2. $\pi = 3.14$

3. $\pi = \dfrac{22}{7}$

4. Every irrational number is a real number.

5. Every real number is either rational or irrational.

6. $\sqrt{x^2} = x$ if $x \geq 0$.

In Exercises 7 through 15, use Appendix Table B to approximate each of the numbers to three decimal places.

7. $\sqrt{17}$

8. $\sqrt{68}$

9. $\sqrt{200}$

10. $\sqrt{\dfrac{15}{16}}$

11. $\sqrt{0.4}$

12. $\pm\sqrt{180}$

13. $-\sqrt{1200}$

14. $3 \pm \sqrt{80}$

15. $-2 \pm \sqrt{50}$

In Exercises 16 through 30, simplify the expressions, if possible. (Assume all variables are positive.)

16. $\sqrt{y^4}$

17. $\sqrt{4x^3}$

18. $\sqrt{49x^2y^2z^2}$

19. $\sqrt{\dfrac{49x^3}{16y^2}}$

20. $\sqrt{\dfrac{x^2}{25y}}$

21. $5\sqrt{5} + \sqrt{20}$

22. $\sqrt{6} + \sqrt{24}$

23. $\sqrt{7} - 3\sqrt{28} + \sqrt{63}$

24. $\sqrt{4x} + \sqrt{16x}$

25. $\sqrt{12} - \sqrt{27} + \sqrt{8}$

26. $\dfrac{2 - \sqrt{8}}{2}$

27. $\dfrac{-9 \pm \sqrt{18}}{6}$

28. $\dfrac{-3 \pm \sqrt{45}}{3}$

29. $\dfrac{2 \pm \sqrt{2}}{2}$

30. $\dfrac{4 \pm \sqrt{8}}{4}$

In Exercises 31 through 35, rationalize the denominator.

31. $\dfrac{5}{\sqrt{5}}$

32. $\dfrac{10 + \sqrt{5}}{\sqrt{5}}$

33. $\dfrac{2}{1 - \sqrt{2}}$

34. $\dfrac{\sqrt{5}}{\sqrt{10} + \sqrt{5}}$

35. $\dfrac{\sqrt{3}}{\sqrt{2} - \sqrt{8}}$

In Exercises 36 through 45, solve each equation for x.

36. $\sqrt{x} = 5$

37. $-\sqrt{x} = -5$

38. $-\sqrt{x} = 5$

39. $12 - \sqrt{x} = 3$

40. $\sqrt{6x - 1} = 0$

41. $1 + \sqrt{3x - 1} = 8$

42. $x + \sqrt{7x} = 0$

43. $1 + \sqrt{3x - 2} = x - 9$

44. $5 + \sqrt{3x + 1} = x + 2$

45. $10 + \sqrt{2x + 1} = 2x - 1$

46. The equation $v = 8\sqrt{h}$ is an algebraic model for the speed, v, of the exiting water at the fire hydrant in the diagram. The distance h is shown in feet and v is in feet/second.

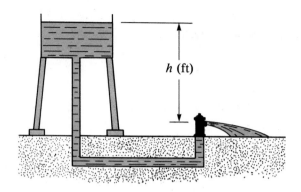

h (ft)

(a) Suppose h is 100 feet. What is the velocity of the water at the moment the hydrant is opened?

(b) Solve the equation for h.

(c) If we wanted the water speed to be 100 ft/sec, what height, h, would be needed?

47. At 12 noon, a boat left a small island and headed north at a rate of 10 km/hr (kilometers/hour). At 1:00 P.M. another boat left the island and headed west at a rate of 8 km/hr. What is the distance between the boats at 5:00 P.M.?

48. The diagonal of a rectangle is 15 cm. If the ratio of the rectangle's length to width is 4:3, find its dimensions.

49. A formula for finding the area of any triangle with sides a, b, and c is given by

$$A = \sqrt{s(s - a)(s - b)(s - c)}$$

where

$$s = \frac{a + b + c}{2}$$

Use the formula to find the area of a triangle with sides 4, 6, and 8 inches.

Cumulative review exercises

1. Simplify $x^2(x^3 - 3x + 2) - x(x^3 + 2x^2 - 4x - 3)$.   (*Section 2.4*)

2. Given the formula $s = 2h^2 + 4hb$, find b when $s = 80$ and $h = 4$.   (*Section 3.6*)

3. Multiply $(2x + 3)(x^2 - 3x + 7)$.   (*Section 5.5*)

4. Completely factor $4x^3 + 36x$.   (*Section 6.4*)

5. A rectangular parking lot has an area of 7500 square feet. The length of the lot is 5 more than twice the width. Find the dimensions of the lot. (*Section 6.6*)

6. Solve for t: $1 - \dfrac{3}{t} = \dfrac{t}{t - 3}$ (*Section 7.6*)

7. Find the x-intercept, y-intercept, and slope of the straight line with the equation $x - 3y = -6$. (*Section 8.3*)

8. Graph the inequality: $3x - 2y \le 8$ (*Section 8.5*)

9. Solve: $3x - 2y = 9$
   $2x + y = 13$ (*Section 9.2*)

10. Solve: $y = 3x - 7$
    $6x - 2y = 4$ (*Section 9.3*)

Chapter 10 test

Take this test to determine how well you have mastered radical expressions; check your answers with those found at the end of the book.

1. What are the square roots of 49?   (*Section 10.1*)

1. _____

2. *True or False:* Every irrational number is a real number.   (*Section 10.1*)

2. _____

In Exercises 3 through 9, simplify the given expression.

3. $\sqrt{180}$ (*Section 10.2*)

4. $\sqrt{0.16}$ (*Section 10.2*)

3. _____

4. _____

5. $\sqrt{18x^2y^3}$ (*Section 10.2*)

6. $\sqrt{5} + \sqrt{20}$ (*Section 10.3*)

5. _____

6. _____

7. $\sqrt{6x} - \sqrt{24x}$ (*Section 10.3*)

8. $\dfrac{2 + 2\sqrt{2}}{2}$ (*Section 10.3*)

7. _____

8. _____

9. $\dfrac{-2 \pm \sqrt{8}}{4}$ (*Section 10.3*)

9. _____

10. _____ 10. Rationalize the denominator $\dfrac{8}{\sqrt{5}}$. (*Section 10.4*)

11. _____ 11. Rationalize the denominator $\dfrac{5}{\sqrt{7} - \sqrt{2}}$. (*Section 10.4*)

12. _____ 12. Solve for t: $\sqrt{5t} = 10$. (*Section 10.5*)

13. _____ 13. Solve for x: $3 + \sqrt{2x - 3} = 8$. (*Section 10.5*)

14. _____ 14. Solve for x: $J = \dfrac{7x^2 y}{8}$. (*Section 10.5*)

15. _____ 15. If one side of a right triangle is 9 inches long and the hypotenuse is 15 inches long, find the length of the third side. (*Section 10.5*)

chapter eleven

Quadratic equations

11.1 Quadratic equations: factoring

Much of the foundation for solving quadratic equations has already been laid in Section 6.6 where we saw the importance of factoring for solving *certain* (factorable) quadratic equations. This section will serve as a brief review of that material.

Recall the definition of a **general quadratic equation.**

A quadratic equation

Any equation that has the form

$$ax^2 + bx + c = 0$$

where a, b, and c are real numbers ($a \neq 0$) is called a **quadratic equation.** We call $ax^2 + bx + c = 0$ the **general form** of a quadratic equation in the single variable, x.

The underlying principle used in the factoring solution to quadratic equations is

Property of zero

If $p \cdot q = 0$ then $p = 0$ or $q = 0$

Thus, given a quadratic equation like

$$x^2 - 10x + 24 = 0$$

for example, we begin by factoring it.

$$x^2 - 10x + 24 = 0$$
$$(x - 6)(x - 4) = 0$$

Now, thinking of $x - 6$ as p and $x - 4$ as q in the Property of Zero just stated, we deduce that x will solve the given equation provided

$$x - 6 = 0 \quad \text{or} \quad x - 4 = 0$$
$$x = 6 \quad \text{or} \quad x = 4$$

So $x = 6$ and $x = 4$ are the two solutions to $x^2 - 10x + 24 = 0$. Each is checked as follows:

Check for $x = 6$:   *Check* for $x = 4$:

$$x^2 - 10x + 24 = 0 \qquad\qquad x^2 - 10x + 24 = 0$$
$$6^2 - 10(6) + 24 = 0 \qquad\qquad 4^2 - 10(4) + 24 = 0$$
$$36 - 60 + 24 = 0 \qquad\qquad 16 - 40 + 24 = 0$$
$$0 = 0 \checkmark \qquad\qquad\qquad\qquad 0 = 0 \checkmark$$

Some examples follow.

Example 1

Solve $2x^2 + 7x - 15 = 0$.

Solution

$$2x^2 + 7x - 15 = 0$$

$$(2x - 3)(x + 5) = 0 \qquad \text{Factor } 2x^2 + 7x - 15.$$

$$2x - 3 = 0 \quad \text{or} \quad x + 5 = 0 \qquad \text{Set each factor equal to 0.}$$

$$2x = 3 \quad \text{or} \quad x = -5$$

$$x = \frac{3}{2}$$

The solutions are $x = \frac{3}{2}$, $x = -5$.

Check for $x = \frac{3}{2}$:

$$2x^2 + 7x - 15 = 0$$

$$2\left(\frac{3}{2}\right)^2 + 7\left(\frac{3}{2}\right) - 15 = 0$$

$$2\left(\frac{9}{4}\right) + \frac{21}{2} - 15 = 0$$

$$\frac{9}{2} + \frac{21}{2} - \frac{30}{2} = 0$$

$$0 = 0 \ \checkmark$$

Check for $x = -5$:

$$2x^2 + 7x - 15 = 0$$

$$2(-5)^2 + 7(-5) - 15 = 0$$

$$50 - 35 - 15 = 0$$

$$0 = 0 \ \checkmark$$

Example 2

Solve $2x^2 + 64 = 24x$.

Solution

First, we rewrite the equation in the general form of a quadratic equation. We must have 0 on the righthand side of the equation, so we subtract $24x$ from both sides.

$$2x^2 + 64 = 24x$$

$$2x^2 - 24x + 64 = 0$$

$$2(x^2 - 12x + 32) = 0$$

$$2(x - 8)(x - 4) = 0 \qquad \text{Factor the lefthand side.}$$

Since 2 cannot be zero, we set each of $x - 8$ and $x - 4$ equal to zero.

$$x - 8 = 0 \qquad x - 4 = 0$$

$$x = 8 \qquad x = 4$$

Check for $x = 8$:

$$2x^2 + 64 = 24x$$

$$2(8)^2 + 64 = 24(8)$$

$$128 + 64 = 192$$

$$192 = 192 \ \checkmark$$

Check for $x = 4$:

$$2x^2 + 64 = 24x$$

$$2(4)^2 + 64 = 24(4)$$

$$32 + 64 = 96$$

$$96 = 96 \ \checkmark$$

We conclude this section with a word problem that involves a quadratic equation.

Example 3

The Pixelux Corporation manufactures video monitors for the computer industry. They have determined that the profit, P (in dollars), they make by selling x monitors is given by the equation

$$P = x^2 - 980x + 1000$$

How many monitors must be sold to realize a $21,000 profit?

Solution

We must find x when $P = 21,000$.

$$x^2 - 980x + 1000 = 21,000$$

$$x^2 - 980x - 20,000 = 0 \qquad \text{Subtract 21,000 from both sides.}$$

$$(x - 1000)(x + 20) = 0 \qquad \text{Factor the lefthand side.}$$

$$x - 1000 = 0 \qquad x + 20 = 0 \qquad \text{Set each factor equal to 0.}$$

$$x = 1000 \qquad\qquad x = -20$$

This solution must be *rejected* since it is meaningless to sell -20 monitors!

Thus, selling 1000 monitors will result in a $21,000 profit for the company.

Section 11.1 Exercises

In Exercises 1 through 30, solve each equation.

1. $x^2 - 5x - 6 = 0$ 2. $x^2 - 5x + 6 = 0$ 3. $x^2 + 5x - 6 = 0$

4. $x^2 + 5x + 6 = 0$ 5. $x^2 - 7x + 6 = 0$ 6. $x^2 + 7x + 6 = 0$

7. $x^2 - 6x - 7 = 0$ 8. $x^2 + 6x - 7 = 0$ 9. $2x^2 - 5x + 2 = 0$

10. $2x^2 + 5x + 2 = 0$ 11. $2x^2 - 3x - 2 = 0$ 12. $2x^2 + 3x - 2 = 0$

13. $4x^2 + 3 = 13x$ 14. $4x^2 + 11x = 3$ 15. $6x^2 + 17x = 3$

16. $6x^2 = 17x + 3$ 17. $6x^2 + 19x = -3$ 18. $6x^2 + 3 = 19x$

19. $6x^2 + 7x = 3$ 20. $6x^2 = 11x - 3$ 21. $x + 2 = 6x^2$

22. $6x^2 = 7x - 2$ 23. $6x^2 + 2 = 8x$ 24. $x^2 - 9 = 0$

25. $2x^2 + 7x = 0$ 26. $5x^2 = 10x$ 27. $4x^2 = 81$

28. $4x^2 - 12x + 9 = 0$ 29. $4x^2 + 12x + 9 = 0$ 30. $(2x + 11)^2 = 0$

31. The cost, C, in dollars, of producing a certain article is given by

$$C = x^2 + 10x + 25$$

How many articles can be produced for a total cost of $100?

32. The product of two consecutive positive integers is 110. What are the integers?

Supplementary Exercises

In Exercises 1 through 16, solve each equation.

1. $x^2 - 5x - 36 = 0$ 2. $x^2 + 5x - 36 = 0$ 3. $x^2 - 6x + 8 = 0$

4. $x^2 - 9x - 10 = 0$

5. $2x^2 - 3x - 2 = 0$

6. $3x^2 - 2x - 1 = 0$

7. $5x^2 + 7x + 2 = 0$

8. $4x^2 + 4x + 1 = 0$

9. $x^2 - 2x = 0$

10. $3x^2 - 6x = 0$

11. $5x^2 - 25x = 0$

12. $x^2 = 4x$

13. $x^2 - 9 = 0$

14. $x^2 - 100 = 0$

15. $x^2 - x = 6$

16. $9x^2 = 100$

11.2 Completing the square

Sections 6.6 and 11.1 point out that when $ax^2 + bx + c$ is factorable, we can use factoring to solve the quadratic equation $ax^2 + bx + c = 0$. However, even if the expression $ax^2 + bx + c$ is not factorable, we may still be able to find solutions. One process we might try is called *completing the square*.

Consider the equation

$$x^2 + 6x + 1 = 0$$

To complete the square, we isolate $x^2 + 6x$ on one side of the equation and then try to rewrite that part of the equation as a *perfect square trinomial*. So we first write $x^2 + 6x + 1 = 0$ as

$$x^2 + 6x = -1$$
$$x^2 + 6x + \square = -1 + \square$$

Now we want to add some number, represented by $\square$ above, that will make $x^2 + 6x + \square$ a perfect square. As we know, if $x^2 + 6x + \square$ is to be a perfect square, it should be of the form $(x + a)^2$. Since $(x + a)^2 = x^2 + 2ax + a^2$ and $x^2 + 6x + \square$ should be of that form, then

$$2ax = 6x$$
$$2a = 6$$
$$a = 3$$

and $a^2 = \square = 9$ Notice that the number we must add is half the coefficient of x, 3, squared, which is 9.

So, replacing $\square$ with 9 in our original equation, we have

$$x^2 + 6x + 9 = -1 + 9 \qquad \square = 9$$
$$(x + 3)^2 = 8 \qquad \text{Factor the lefthand side.}$$
$$x + 3 = \pm\sqrt{8} \qquad \text{Take the square root of both sides.}$$
$$x = -3 \pm \sqrt{8} \qquad \text{Solve for } x \text{ by subtracting 3 from both sides.}$$

The two solutions are $x = -3 + \sqrt{8}$ and $x = -3 - \sqrt{8}$.

Check for $x = -3 + \sqrt{8}$:

$$x^2 + 6x + 1 = 0$$
$$(-3 + \sqrt{8})^2 + 6(-3 + \sqrt{8}) + 1 = 0$$
$$9 - 6\sqrt{8} + 8 - 18 + 6\sqrt{8} + 1 = 0$$
$$0 = 0 \checkmark$$

Check for $x = -3 - \sqrt{8}$:

$$x^2 + 6x + 1 = 0$$
$$(-3 - \sqrt{8})^2 + 6(-3 - \sqrt{8}) + 1 = 0$$
$$9 + 6\sqrt{8} + 8 - 18 - 6\sqrt{8} + 1 = 0$$
$$0 = 0 \checkmark$$

Some examples follow.

Example 1

Use the method of completing the square to solve $x^2 - 8x - 1 = 0$.

Solution

$$x^2 - 8x - 1 = 0$$
$$x^2 - 8x + \square = 1 + \square$$

As we know, $\square$ is half the coefficient of x, squared. So

$$\square = \left[\tfrac{1}{2}(-8)\right]^2 = 16$$

Then, $x^2 - 8x + 16 = 1 + 16$ $\square = 16$

$(x - 4)^2 = 17$ Factor the lefthand side.

$x - 4 = \pm\sqrt{17}$ Take the square root of both sides.

$x = 4 \pm \sqrt{17}$ Add 4 to both sides.

The two solutions of x are $x = 4 + \sqrt{17}$ and $x = 4 - \sqrt{17}$.

Check for $x = 4 + \sqrt{17}$:

$$x^2 - 8x - 1 = 0$$
$$(4 + \sqrt{17})^2 - 8(4 + \sqrt{17}) - 1 = 0$$
$$16 + 8\sqrt{17} + 17 - 32 - 8\sqrt{17} - 1 = 0$$
$$0 = 0 \checkmark$$

Check for $x = 4 - \sqrt{17}$:

$$x^2 - 8x - 1 = 0$$
$$(4 - \sqrt{17})^2 - 8(4 - \sqrt{17}) - 1 = 0$$
$$16 - 8\sqrt{17} + 17 - 32 + 8\sqrt{17} - 1 = 0$$
$$0 = 0 \checkmark$$

Example 2

Solve $x^2 - 9x - 7 = 0$.

Solution

$$x^2 - 9x - 7 = 0$$
$$x^2 - 9x + \square = 7 + \square$$
$$x^2 - 9x + \frac{81}{4} = 7 + \frac{81}{4} \qquad \square = \left[\tfrac{1}{2}(-9)\right]^2 = \frac{81}{4}$$
$$\left(x - \frac{9}{2}\right)^2 = \frac{109}{4}$$
$$x - \frac{9}{2} = \pm\frac{\sqrt{109}}{2}$$
$$x = \frac{9}{2} \pm \frac{\sqrt{109}}{2}$$

The two solutions are $x = \dfrac{9 + \sqrt{109}}{2}$ and $x = \dfrac{9 - \sqrt{109}}{2}$. The checks are left as an exercise.

Example 3

Use the method of completing the square to solve $4x^2 + 12x + 5 = 0$.

Solution

The process of completing the square discussed so far only works when the coefficient of x^2 is 1. So we first divide both sides of the equation by 4 and then proceed with the method.

$$4x^2 + 12x + 5 = 0$$

$$x^2 + 3x + \frac{5}{4} = 0$$

$$x^2 + 3x + \square = -\frac{5}{4} + \square$$

$$x^2 + 3x + \frac{9}{4} = \frac{-5}{4} + \frac{9}{4} \qquad \square = \left[\frac{1}{2}(3)\right]^2 = \frac{9}{4}$$

$$\left(x + \frac{3}{2}\right)^2 = 1$$

$$x + \frac{3}{2} = \pm 1$$

$$x = -\frac{3}{2} \pm 1$$

Hence, $x = -\dfrac{3}{2} + 1 = -\dfrac{1}{2}$ and $x = -\dfrac{3}{2} - 1 = -\dfrac{5}{2}$.

Check for $x = -\dfrac{1}{2}$:

$$4x^2 + 12x + 5 = 0$$

$$4\left(-\frac{1}{2}\right)^2 + 12\left(-\frac{1}{2}\right) + 5 = 0$$

$$1 - 6 + 5 = 0$$

$$0 = 0 \checkmark$$

Check for $x = -\dfrac{5}{2}$:

$$4x^2 + 12x + 5 = 0$$

$$4\left(-\frac{5}{2}\right)^2 + 12\left(-\frac{5}{2}\right) + 5 = 0$$

$$25 - 30 + 5 = 0$$

$$0 = 0 \checkmark$$

Section 11.2 Exercises

In Exercises 1 through 20, use the method of completing the square to solve each equation. Check your answers.

1. $x^2 + 2x - 7 = 0$

2. $x^2 + 2x - 11 = 0$

3. $x^2 - 6x + 2 = 0$

4. $x^2 - 6x + 3 = 0$

5. $x^2 - 6x + 4 = 0$

6. $x^2 - 6x + 5 = 0$

7. $x^2 - 6x + 6 = 0$

8. $x^2 - 6x + 7 = 0$

9. $x^2 - 8x + 1 = 0$

10. $x^2 - 8x + 14 = 0$

11. $x^2 - 7x + 1 = 0$

12. $x^2 + 7x + 1 = 0$

13. $x^2 - 7x + 2 = 0$

14. $x^2 + 7x + 2 = 0$

15. $2x^2 - 8x - 1 = 0$

16. $2x^2 - 8x + 1 = 0$ 17. $2x^2 - 6x - 1 = 0$ 18. $2x^2 - 6x - 5 = 0$

19. $2x^2 - 6x - 8 = 0$ 20. $3x^2 - 7x + 1 = 0$

21. Check the two solutions to $x^2 - 9x - 7 = 0$ that were found in Example 2.

In Exercises 22 through 25, use a calculator and the process of completing the square to solve each equation.

22. $7x^2 - 2x - 3 = 0$ 23. $10x^2 - 19x - 41 = 0$

24. $1.5x^2 - 6.7x + 0.1 = 0$ 25. $-1.5x^2 - 1.89x + 12.5 = 0$

Supplementary Exercises

In Exercises 1 through 13, use the method of completing the square to solve each equation. Check your answers.

1. $x^2 - 4x + 1 = 0$ 2. $x^2 + 6x + 9 = 0$ 3. $x^2 - 4x + 2 = 0$

4. $x^2 + 2x + 1 = 0$ 5. $x^2 - 8x + 6 = 0$ 6. $x^2 + 8x + 7 = 0$

7. $x^2 + 2x - 1 = 0$ 8. $2x^2 - 8x + 2 = 0$ 9. $3x^2 - 9x + 6 = 0$

10. $3x^2 + 6x - 1 = 0$ 11. $\frac{x^2}{2} - x - 1 = 0$ 12. $\frac{x^2}{3} + 2x - 1 = 0$

13. $\frac{x^2}{4} + 2x - 1 = 0$

11.3   The quadratic formula

The process of completing the square is very useful for solving many types of quadratic equations. If we had several such equations to solve, however, the process would become quite tedious. An alternative is to apply the process to the *general* quadratic equation

$$ax^2 + bx + c = 0$$

The result will be a *generalized* solution, which we can use for *any* quadratic equation. We will complete the square on $ax^2 + bx + c = 0$ as follows.

$$ax^2 + bx + c = 0$$

$$x^2 + \frac{b}{a}x + \frac{c}{a} = 0 \qquad \text{Divide by } a, \text{ assuming that } a \neq 0.$$

$$x^2 + \frac{b}{a}x + \square = -\frac{c}{a} + \square$$

$$x^2 + \frac{b}{a}x + \frac{b^2}{4a^2} = -\frac{c}{a} + \frac{b^2}{4a^2} \qquad \square = \left[\frac{1}{2}\left(\frac{b}{a}\right)\right]^2 = \frac{b^2}{4a^2}$$

$$\left(x + \frac{b}{2a}\right)^2 = \frac{b^2 - 4ac}{4a^2}$$

Factor the lefthand side; combine $-\frac{c}{a} + \frac{b^2}{4a^2}$ by getting a common denominator of $4a^2$.

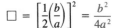

$$x + \frac{b}{2a} = \pm\frac{\sqrt{b^2 - 4ac}}{2a} \qquad \text{Take the square root of both sides.}$$

$$x = \frac{-b}{2a} \pm \frac{\sqrt{b^2 - 4ac}}{2a} \qquad \text{Subtract } \frac{b}{2a} \text{ from both sides.}$$

$$x = \frac{-b \pm \sqrt{b^2 - 4ac}}{2a}$$

That is, the two solutions are $x = \dfrac{-b + \sqrt{b^2 - 4ac}}{2a}$ and $x = \dfrac{-b - \sqrt{b^2 - 4ac}}{2a}$. We call this generalized solution, the **quadratic formula.**

The quadratic formula

If $ax^2 + bx + c = 0$ $(a \neq 0)$, then

$$x = \frac{-b \pm \sqrt{b^2 - 4ac}}{2a}$$

The quadratic formula can now be used to solve *any* quadratic equation. The next four examples illustrate its use.

Example 1

Use the quadratic formula to solve

$$2x^2 - x - 10 = 0$$

Solution

$2x^2 - x - 10 = 0$ is already in the form of $ax^2 + bx + c = 0$ where $a = 2$, $b = -1$, and $c = -10$. These values are now substituted into the quadratic formula

$$x = \frac{-b \pm \sqrt{b^2 - 4ac}}{2a}$$

With $a = 2$, $b = -1$, and $c = -10$ we get

$$x = \frac{-(-1) \pm \sqrt{(-1)^2 - 4(2)(-10)}}{2(2)}$$

$$= \frac{1 \pm \sqrt{1 + 80}}{4}$$

$$= \frac{1 \pm \sqrt{81}}{4}$$

$$= \frac{1 \pm 9}{4}$$

Since $x = \dfrac{1 \pm 9}{4}$ represents *two* values for x, we write

$$x = \frac{1 + 9}{4} \qquad \text{and} \qquad x = \frac{1 - 9}{4}$$

$$= \frac{10}{4} \qquad\qquad\qquad = \frac{-8}{4}$$

$$= \frac{5}{2} \qquad\qquad\qquad = -2$$

Check for $x = \dfrac{5}{2}$:

$$2x^2 - x - 10 = 0$$

$$2\left(\frac{5}{2}\right)^2 - \frac{5}{2} - 10 = 0$$

$$\frac{50}{4} - \frac{5}{2} - 10 = 0$$

$$\frac{25}{2} - \frac{5}{2} - 10 = 0$$

$$\frac{20}{2} - 10 = 0$$

$$10 - 10 = 0$$

$$0 = 0 \checkmark$$

Check for $x = -2$:

$$2x^2 - x - 10 = 0$$

$$2(-2)^2 - (-2) - 10 = 0$$

$$2(4) + 2 - 10 = 0$$

$$8 + 2 - 10 = 0$$

$$10 - 10 = 0$$

$$0 = 0 \checkmark$$

Example 2

Solve $x^2 - 8x = -15$.

Solution

First we rewrite $x^2 - 8x = -15$ as $x^2 - 8x + 15 = 0$ and notice that $a = 1$, $b = -8$, and $c = 15$. Now, using the quadratic formula with these values,

$$x = \frac{-b \pm \sqrt{b^2 - 4ac}}{2a}$$

$$= \frac{-(-8) \pm \sqrt{(-8)^2 - 4(1)(15)}}{2(1)}$$

$$= \frac{8 \pm \sqrt{64 - 60}}{2}$$

$$= \frac{8 \pm \sqrt{4}}{2}$$

$$= \frac{8 \pm 2}{2}$$

The two values of x are

$$x = \frac{8 + 2}{2} \quad \text{and} \quad x = \frac{8 - 2}{2}$$

$$= 5 \qquad\qquad\qquad = 3$$

Check for $x = 5$:

$$x^2 - 8x = -15$$

$$(5)^2 - 8(5) = -15$$

$$25 - 40 = -15$$

$$-15 = -15 \checkmark$$

Check for $x = 3$:

$$x^2 - 8x = -15$$

$$(3)^2 - 8(3) = -15$$

$$9 - 24 = -15$$

$$-15 = -15 \checkmark$$

The next two examples solve quadratic equations with irrational number solutions.

Example 3

Solve $x^2 + 2x - 1 = 0$.

Solution

$x^2 + 2x - 1 = 0$ is already in the general form with $a = 1$, $b = 2$, and $c = -1$. So

$$x = \frac{-b \pm \sqrt{b^2 - 4ac}}{2a}$$

$$= \frac{-2 \pm \sqrt{(2)^2 - 4(1)(-1)}}{2(1)}$$

$$= \frac{-2 \pm \sqrt{4 + 4}}{2}$$

$$= \frac{-2 \pm \sqrt{8}}{2}$$

$$= \frac{-2 \pm 2\sqrt{2}}{2} \qquad \sqrt{8} = \sqrt{4} \cdot \sqrt{2} = 2\sqrt{2}$$

$$= \frac{\cancel{2}(-1 \pm \sqrt{2})}{\cancel{2}} \qquad \text{Factor the numerator.}$$

$$= -1 \pm \sqrt{2}$$

The two values of x are

$$x = -1 + \sqrt{2} \approx .414 \qquad \text{and} \qquad x = -1 - \sqrt{2} \approx -2.414$$

The checks for $-1 + \sqrt{2}$ and $-1 - \sqrt{2}$ are left as an exercise.

Example 4

Solve $3t^2 = 2t + 2$.

Solution

First we rewrite $3t^2 = 2t + 2$ as $3t^2 - 2t - 2 = 0$ and find $a = 3$, $b = -2$, and $c = -2$. Applying the quadratic formula, we get

$$t = \frac{-b \pm \sqrt{b^2 - 4ac}}{2a}$$

$$= \frac{-(-2) \pm \sqrt{(-2)^2 - 4(3)(-2)}}{2(3)}$$

$$= \frac{2 \pm \sqrt{4 + 24}}{6}$$

$$= \frac{2 \pm \sqrt{28}}{6}$$

$$= \frac{2 \pm 2\sqrt{7}}{6} \qquad \sqrt{28} = \sqrt{4} \cdot \sqrt{7} = 2\sqrt{7}$$

$$= \frac{\cancel{2}(1 \pm \sqrt{7})}{\cancel{2} \cdot 3} \qquad \text{Factor the numerator.}$$

$$= \frac{1 \pm \sqrt{7}}{3}$$

The two values of t are

$$t = \frac{1 + \sqrt{7}}{3} \approx 1.215 \qquad \text{and} \qquad t = \frac{1 - \sqrt{7}}{3} \approx -.549$$

Section 11.3 Exercises

In Exercises 1 through 20, solve each quadratic equation by using the quadratic formula.

1. $x^2 + x - 12 = 0$ 2. $2x^2 + 2x - 24 = 0$ 3. $x^2 + 10 = 7x$

4. $x^2 + 9 = 6x$ 5. $x^2 + 5 = 6x$ 6. $6r^2 = 5r - 1$

7. $y^2 - 7y - 12 = 0$

8. $-\frac{1}{3}x^2 - \frac{2}{3}x - \frac{1}{3} = 0$ (*Hint*: Multiply both sides by -3 first.)

9. $x^2 + 3x - 1 = 0$ 10. $t^2 - 2t - 2 = 0$ 11. $3y^2 - 5y - 2 = 0$

12. $t^2 = 2t + 1$ 13. $x^2 - 3x - 2 = 0$ 14. $4m^2 - 3 = 0$

15. $x^2 = x$ 16. $x^2 = 5x$ 17. $3x^2 - 6x - 6 = 0$

18. $x^2 - x - 21 = 0$ 19. $3m^2 + 1 = 5m$ 20. $7q^2 = 2q + 2$

In Exercises 21 through 25, use a calculator and the quadratic formula to solve each quadratic equation.

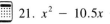 21. $x^2 - 10.5x - 1 = 0$ 22. $3.1x^2 + 7.3x - 8.9 = 0$

23. $0.001x^2 - 86.9x + 20 = 0$ 24. $0.001x^2 - 86.9x - 20 = 0$

25. $x^2 - 0.5x - 0.25 = 0$

Supplementary Exercises

In Exercises 1 through 13, solve each quadratic equation using the quadratic formula.

1. $x^2 - 3x + 1 = 0$ 2. $x^2 + 4x + 1 = 0$ 3. $x^2 + 5x = 2$

4. $x^2 - 1 = 3x$ 5. $2x^2 - 5x + 1 = 0$ 6. $3x^2 - 9x + 2 = 0$

7. $y^2 + 3y + 7 = 0$ 8. $2y^2 - 3y - 3 = 0$ 9. $-x^2 + 3x + 1 = 0$

10. $4m^2 - 2 = 6m$ 11. $\frac{x^2}{2} - 2x + 1 = 0$ 12. $5s^2 + 2s - 3 = 0$

13. $-t^2 + 4t + 2 = 0$

11.4 Graphing $y = ax^2 + bx + c$

In Chapter 8 we graphed linear equations in two variables and saw that their graphs were straight lines. In this section, we will examine the graph of a quadratic equation in two variables that has the general form

$$y = ax^2 + bx + c \qquad (a \neq 0)$$

We proceed with two examples.

Example 1

Graph $y = x^2 + 2x - 8$.

Solution

The graph of this equation will *not* be a straight line, so we will need to graph more than three points before we will know the general shape of the solution through these points. We begin by choosing seven values for x: -5, -4, -3, -1, 0, 1, and 3. We find the corresponding y values by substituting these values for x into the equation $y = x^2 + 2x - 8$ and present them in tabular form, as follows.

x	x^2	$+$	$2x$	-8	$=$	y
-5	$(-5)^2$	$+$	$2(-5)$	-8	$=$	7
-4	$(-4)^2$	$+$	$2(-4)$	-8	$=$	0
-3	$(-3)^2$	$+$	$2(-3)$	-8	$=$	-5
-1	$(-1)^2$	$+$	$2(-1)$	-8	$=$	-9
0	$(0)^2$	$+$	$2(0)$	-8	$=$	-8
1	$(1)^2$	$+$	$2(1)$	-8	$=$	-5
3	$(3)^2$	$+$	$2(3)$	-8	$=$	7

So our seven solutions are $(-5, 7)$, $(-4, 0)$, $(-3, -5)$, $(-1, -9)$, $(0, -8)$, $(1, -5)$ and $(3, 7)$. If we were to continue to find points that solved the equation, we would find that the graph of all these points would be a smooth curve. So when we graph these seven solutions as points in a rectangular coordinate system, we connect them with a smooth curve.

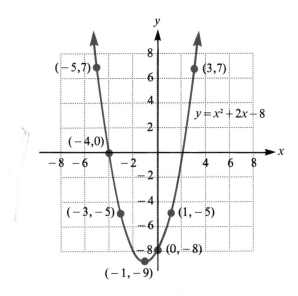

The graph presented in Example 1 is called a **parabola.** Although the shape and position of the parabola with respect to the axes change as a, b, and c change, the quadratic equation $y = ax^2 + bx + c$ will always be represented by a parabolically shaped graph. In the case of the graph in Example 1, we say the parabola *opens upward*. Also, the point $(-1, -9)$ is the "turning point" of this parabola and is called its **vertex.** An example of a parabola *opening downward* follows.

Example 2

Graph $y = 4 - x^2$.

Solution

We again find seven solutions that will be graphed in order to help us determine the location and shape of the graph of $y = 4 - x^2$ (or equivalently, $y = -x^2 + 4$). We choose x values -3, -2, -1, 0, 1, 2, and 3 and find the y values in the tabular form that follows.

x	4	$-$	x^2	$=$	y
-3	4	$-$	$(-3)^2$	$=$	-5
-2	4	$-$	$(-2)^2$	$=$	0
-1	4	$-$	$(-1)^2$	$=$	3
0	4	$-$	$(0)^2$	$=$	4
1	4	$-$	$(1)^2$	$=$	3
2	4	$-$	$(2)^2$	$=$	0
3	4	$-$	$(3)^2$	$=$	-5

So now we graph the solutions $(-3, -5)$, $(-2, 0)$, $(-1, 3)$, $(0, 4)$, $(1, 3)$, $(2, 0)$, and $(3, -5)$ as points in the rectangular coordinate system and draw the parabola.

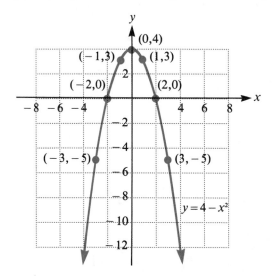

Notice that the parabola opens downward and its vertex is the point $(0, 4)$.

The general form $y = ax^2 + bx + c$ reveals two facts about the parabola that are useful for graphing purposes. We now state (without proof) these two facts:

1. Whenever a is positive, the parabola opens upward; when a is negative, the parabola opens downward.

2. The x value of the vertex is always $\dfrac{-b}{2a}$.

So, when dealing with an equation of the form $y = ax^2 + bx + c$, we can immediately determine which way the parabola opens and where its vertex is. To get a more definite idea of its shape, we choose three x values on each side of the vertex. The next example illustrates this.

Example 3

Graph $y = x^2 - 8x + 10$.

Solution

Notice that $a = 1$, $b = -8$, and $c = 10$. Since a is positive, the parabola opens upward. The x value of the vertex is

$$\frac{-b}{2a} = \frac{-(-8)}{2(1)} = 4$$

When $x = 4$, $y = (4)^2 - 8(4) + 10 = -6$. So the vertex is $(4, -6)$.

For the other x values we take 1, 2, 3, 5, 6 and 7. We leave it to the reader to verify the y values of the following solutions: $(1, 3)$, $(2, -2)$, $(3, -5)$, $(4, -6)$, $(5, -5)$, $(6, -2)$, $(7, 3)$. Next, we graph these solutions and draw the parabola.

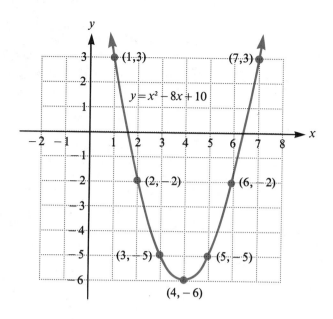

We conclude this section with an application.

Example 4

When an object is dropped from a height of 400 feet, its height (h) in feet after t seconds is given by the equation

$$h = -16t^2 + 400$$

Graph this relationship.

Solution

First, notice that we must label the horizontal axis with a t and the vertical axis with an h. Also, the parabola will open downward, and the t value of the vertex, $\frac{-b}{2a}$, is $\frac{0}{-32} = 0$. So the coordinates of the vertex are $(0, 400)$. Since it would take an awfully large sheet of paper to graph that point with the axes scaled the same, we "compress" the scale on the vertical axis (the h axis). Actually, the scales shown in the graph exhibit a 100-to-1 *scale ratio*. That is, 100 units on the vertical axis are represented by the same length as 1 unit on the horizontal axis. We leave it to the reader to verify that the following pairs of values are solutions for $h = -16t^2 + 400$: $(-5, 0)$, $(-3, 256)$, $(-1, 384)$, $(0, 400)$, $(2, 336)$, $(5, 0)$, and $(6, -176)$.

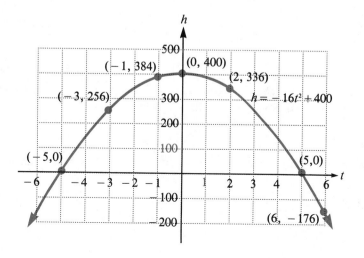

Notice that this problem is physical in nature, and values of t less than zero have no meaning. Also, since we are dropping an object, the whole process comes to an end when $h = 0$ (see $t = 5$ on the graph). This information tells us that, for this problem of dropping an object, all that is necessary is the first quadrant portion of the graph above. We draw it below and interpret three points.

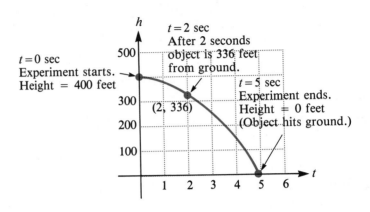

Section 11.4 Exercises

In Exercises 1 through 15, (a) state whether the parabola opens upward or downward, (b) state the coordinates of the vertex, and (c) graph the equation on a sheet of graph paper.

1. $y = x^2 - 6x + 5$
2. $y = -x^2 + 6x - 5$
3. $y = x^2 - 1$
4. $y = 1 - x^2$
5. $y = x^2 + 6x + 5$
6. $y = -x^2 - 6x - 5$
7. $y = x^2 + 6x$
8. $y = x^2 - 6x$
9. $y = 6x - x^2$
10. $y = x^2 + 4$
11. $y = x^2$
12. $y = 9 - x^2$
13. $y = 2x^2 - x - 10$
14. $y = 2x^2 + 8x - 10$
15. $y = \dfrac{x^2 - 4x}{2}$

16. If a ball is thrown upward from the roof of a 96-foot building with an initial velocity of 80 ft/sec, its height (h) in feet above the ground after t seconds is given by

$$h = -16t^2 + 80t + 96$$

   (a) Using a 16-to-1 scale ratio, graph the equation on a sheet of graph paper.

   (b) What part of the graph in part (a) pertains to the physical aspects of the problem?

   (c) After how many seconds will the ball hit the ground? (That is, what is t when $h = 0$?)

17. Substitute $\dfrac{-b}{2a}$ for x in the general equation $y = ax^2 + bx + c$ to show that the coordinates of the vertex are $\left(\dfrac{-b}{2a}, c - \dfrac{b^2}{4a}\right)$.

18. The x-intercepts of $y = ax^2 + bx + c$ occur when $y = 0$, and hence are the solutions to the special case $0 = ax^2 + bx + c$. So one way of approximating the solutions to $x^2 - 2x - 15 = 0$, for example, would be to graph $y = x^2 - 2x - 15$ and read its x-intercepts from the graph.

   (a) Graph $y = x^2 - 2x - 15$ on a sheet of graph paper.

   (b) From your graph, approximate solutions to $x^2 - 2x - 15 = 0$. (That is, approximate the x-intercepts.)

   (c) Use the quadratic formula to solve $x^2 - 2x - 15 = 0$. Compare with your answer to part (b).

19. (a) Graph $y = -x^2 + 8x - 9$ on a sheet of graph paper.

(b) From your graph, approximate solutions to $-x^2 + 8x - 9 = 0$. (That is, approximate the x-intercepts.)

(c) Use the quadratic formula to solve $-x^2 + 8x - 9 = 0$. Compare with your answer to part (b).

20. (a) Graph $y = x^2 - 6x + 1$ on a sheet of graph paper.

(b) From your graph, read the x-intercepts.

(c) Solve, using the quadratic formula, $x^2 - 6x + 1 = 0$.

(d) Use a calculator to approximate your answer to part (c). Compare it with part (b).

Supplementary Exercises

In Exercises 1 through 10, (a) state whether the parabola opens upward or downward, (b) state the coordinates of the vertex, and (c) graph the equation on a sheet of graph paper.

1. $y = x^2 - 6x - 7$

2. $y = x^2 - 6x + 9$

3. $y = -x^2 - x + 2$

4. $y = -x^2 + 2x + 3$

5. $y = x^2 - 2x$

6. $y = x^2 + 6x$

7. $y = -x^2 + 2x$

8. $y = -x^2 + 1$

9. $y = 2x^2 - 3x - 2$

10. $y = 3x^2 - 2x - 1$

11.5   Applications of quadratic equations

We have already examined some applications of quadratic equations in Sections 6.6, 11.1, and the last section. In this section, we examine more applications involving quadratic equations in one variable (Examples 1, 2, and 3) and two variables (Examples 4, 5, 6 and 7).

Example 1

The length of a rectangle is 1 meter more than its width. If the diagonal of the rectangle is 3 meters, find the length and width (to the nearest tenth of a meter).

Solution

Step 1   Problem translation

First we display a pictorial model.

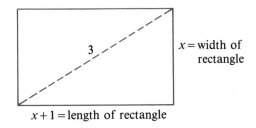

$x+1 =$ length of rectangle

By using the Pythagorean Formula,

$$(x + 1)^2 + (x)^2 = (3)^2 \qquad \leftarrow \text{This is our algebraic model.}$$

Step 2   Equation solution

$$(x + 1)^2 + (x)^2 = (3)^2$$
$$x^2 + 2x + 1 + x^2 = 9$$
$$2x^2 + 2x + 1 = 9$$
$$2x^2 + 2x - 8 = 0 \qquad \leftarrow \text{This is a quadratic equation,}$$

with $a = 2$, $b = 2$, and $c = -8$.

$$x = \frac{-2 \pm \sqrt{2^2 - 4(2)(-8)}}{2(2)}$$

$$= \frac{-2 \pm \sqrt{68}}{4}$$

$$\approx \frac{-2 \pm 8.246}{4}$$

So

$$x \approx \frac{-2 - 8.246}{4} \qquad \text{and} \qquad x \approx \frac{-2 + 8.246}{4}$$

$$\approx \frac{-10.246}{4} \qquad\qquad\qquad \approx \frac{6.246}{4}$$

$$\approx -2.561 \qquad\qquad\qquad\quad \approx 1.561$$

Both of these numbers are solutions to the quadratic equation; however, we must reject the negative solution, -2.561, since the physical meaning of x is the width of the rectangle. Thus the rectangle's dimensions (to the nearest tenth of a meter) are 2.6 m (length) by 1.6 m (width).

Step 3   Check

The original pictorial model looks like this:

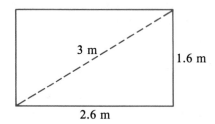

3 m

1.6 m

2.6 m

An approximate check of the Pythagorean Formula follows.

$$(2.6)^2 + (1.6)^2 \approx (3)^2$$
$$6.76 + 1.96 \approx 9$$
$$8.72 \approx 9 \ \checkmark$$

Example 2

A page for a book must be designed so that there is a 1-inch margin on each side and also on the top and bottom. The height of the page is 2 inches more than the width. The area of the printed matter is to be 20 square inches. What is the width of the page (to the nearest tenth of an inch)?

Solution

Step 1   Problem translation

Here we display a pictorial model.

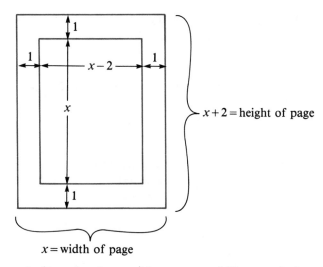

The inner rectangle (the printed matter) has an area of 20 square inches. So

$$x(x - 2) = 20 \qquad \leftarrow \text{This is our algebraic model.}$$

Step 2   Equation solution

$$x(x - 2) = 20$$
$$x^2 - 2x = 20$$
$$x^2 - 2x - 20 = 0 \qquad \leftarrow \text{This is a quadratic equation with}$$
$$a = 1, b = -2, \text{ and } c = -20.$$

$$x = \frac{-(-2) \pm \sqrt{(2)^2 - 4(1)(-20)}}{2}$$

$$= \frac{2 \pm \sqrt{84}}{2} \approx \frac{2 \pm 9.165}{2}$$

So

$$x \approx \frac{2 - 9.165}{2} \qquad \text{and} \qquad x \approx \frac{2 + 9.165}{2}$$

$$\approx \frac{-7.165}{2} \qquad\qquad\qquad \approx \frac{11.165}{2}$$

$$\approx -3.583 \qquad\qquad\qquad \approx 5.583$$

Both of these numbers are solutions to the quadratic equation. However, we must reject the negative solution, -3.583, since the physical meaning of x is the width of the page. Thus the width of the page (to the nearest tenth of an inch) is 5.6 inches.

Step 3   Check

The original pictorial model looks like this:

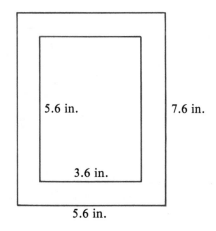

An approximate check of the area of the printed matter follows.

$$3.6 \cdot 5.6 \approx 20$$

$$19.16 \approx 20 \; \checkmark$$

Also note that the height of the page (7.6 in.) *is* 2 inches more than the width of the page (5.6 in.). $\checkmark$

Example 3

A farmer desires to enclose a rectangular region. One side is bounded by a river and needs no fencing. The other three sides need fencing. If she has 400 feet of fencing and she wants the rectangular region to have an area of 15,000 square feet, what should the dimensions of the field be?

Solution

Step 1   Problem translation

We first display a pictorial model.

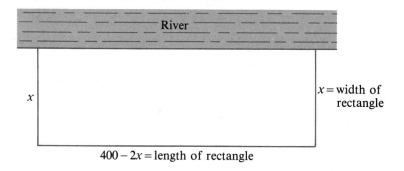

$400 - 2x =$ length of rectangle

The rectangular region must have an area of 15,000 square feet. So

$$x(400 - 2x) = 15,000 \qquad \leftarrow \text{This is our algebraic model.}$$

Step 2   Equation solution

$$x(400 - 2x) = 15,000$$

$$400x - 2x^2 = 15,000$$

$$2x^2 - 400x + 15,000 = 0$$

$$x^2 - 200x + 7500 = 0 \qquad \leftarrow \text{This is a quadratic equation with } a = 1, \\ b = -200, \text{ and} \\ c = 7500.$$

$$x = \frac{-b \pm \sqrt{b^2 - 4ac}}{2a}$$

$$= \frac{-(-200) \pm \sqrt{(-200)^2 - 4(1)(7500)}}{2(1)}$$

$$= \frac{200 \pm \sqrt{10,000}}{2}$$

$$= \frac{200 \pm 100}{2}$$

So
$$x = \frac{200 + 100}{2} \qquad \text{and} \qquad x = \frac{200 - 100}{2}$$

$$= 150 \qquad\qquad\qquad = 50$$

In this problem, there are two allowable solutions for x (width of rectangle). If $x = 150$, the length of the rectangle is $400 - 2x = 400 - 300$, or 100 feet. If $x = 50$, the length of the rectangle is $400 - 2x = 400 - 100$, or 300 feet. The farmer can make a rectangle of 150 feet in width and 100 feet in length, or she can make a rectangle of 50 feet in width and 300 feet in length.

Step 3 Check

Check for $x = 150$:

Our pictorial model is

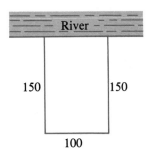

Check for $x = 50$:

Our pictorial model is

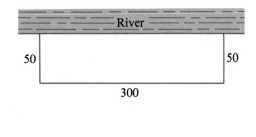

400 feet of fencing is being used, and the area is 15,000. ✓

400 feet of fencing is being used, and the area is 15,000. ✓

Example 4

The number of different possible pairs of people (t) that can be selected from a group of n people to fill the rooms of a dormitory is given by the formula

$$t = \frac{n(n-1)}{2}$$

(a) The director of a dormitory that houses 50 people wants to know how many different possible pairs of people exist. How many pairs are possible?

(b) Suppose a dormitory director knows that 780 different pairs of people are possible. How many people are there to live in the dormitory?

Solution

(a) Here we use the formula $t = \frac{n(n-1)}{2}$ with $n = 50$.

$$t = \frac{n(n-1)}{2}$$

$$= \frac{50(49)}{2}$$

$$= 1225$$

So there are 1225 different possible pairs of people.

(b) Here we use the formula $t = \frac{n(n-1)}{2}$ with $t = 780$.

$$t = \frac{n(n-1)}{2}$$

$$780 = \frac{n(n-1)}{2}$$

$$1560 = n(n-1)$$

$$1560 = n^2 - n$$

$$n^2 - n - 1560 = 0 \qquad \leftarrow \text{This is a quadratic equation with } a = 1, b = -1,$$
$$\text{and } c = -1560.$$

$$n = \frac{-b \pm \sqrt{b^2 - 4ac}}{2a}$$

$$= \frac{-(-1) \pm \sqrt{(-1)^2 - 4(1)(-1560)}}{2(1)}$$

$$= \frac{1 \pm \sqrt{6241}}{2}$$

$$= \frac{1 \pm 79}{2}$$

$$n = \frac{1 - 79}{2} \quad \text{and} \quad n = \frac{1 + 79}{2}$$

$$= \frac{-78}{2} \qquad\qquad = \frac{80}{2}$$

$$= -39 \qquad\qquad = 40$$

↑
This solution is *rejected* because
n represents a number of people.

So there are 40 people.

Example 5

The equation $A = \frac{4}{7}\sqrt{14T - T^2}$ has been used to estimate the number of accidents (A) occurring where T is the number of cars per day on a two-lane road. (T is measured in thousands; that is, $T = 1$ means 1000 cars per day on a two-lane road, and so on.)

(a) If the number of cars per day on a two-lane road is 2000, what number of accidents can be expected?

(b) If the number of accidents is 4, how many cars could we expect to have been traveling?

Solution

(a) We must substitute $T = 2$ into the equation.

$$A = \frac{4}{7}\sqrt{14T - T^2}$$

$$= \frac{4}{7}\sqrt{14 \cdot 2 - 2^2}$$

$$= \frac{4}{7}\sqrt{24}$$

$$\approx \frac{4}{7} \cdot 4.899$$

$$\approx 2.799$$

So 2 or 3 accidents can be expected.

(b) We must substitute $A = 4$ into the equation.

$$A = \frac{4}{7}\sqrt{14T - T^2}$$

$$4 = \frac{4}{7}\sqrt{14T - T^2}$$

$$7 = \sqrt{14T - T^2} \qquad \text{Multiply both sides by } \frac{7}{4}.$$

$$49 = 14T - T^2 \qquad \text{Square both sides of the equation.}$$

$$T^2 - 14T + 49 = 0 \qquad \text{Rearrange terms. This is a quadratic equation with } a = 1,\ b = -14,\ \text{and } c = 49.$$

$$T = \frac{-b \pm \sqrt{b^2 - 4ac}}{2a}$$

$$= \frac{-(-14) \pm \sqrt{(-14)^2 - 4(1)(49)}}{2}$$

$$= \frac{14 \pm \sqrt{196 - 196}}{2}$$

$$= \frac{14 \pm 0}{2}$$

$$= 7$$

So about 7000 cars would have been traveling.

Example 6

The equation

$$d = \frac{2}{9}t^2 - \frac{40}{3}t$$

gives the distance (d) in feet that a computer-controlled diving bell will descend in t seconds. (The process starts at the surface level of the water when $t = 0$.)

(a) Using an appropriate scale on each axis, graph the equation.

(b) What part of the graph in part (a) pertains to the physical aspects of the problem?

(c) What is the deepest point during the dive?

(d) What is the time for the whole diving experiment?

Solution

(a) We label the horizontal axis t and the vertical axis d. The graph will be a parabola because $d = \frac{2}{9}t^2 - \frac{40}{3}t$ is a quadratic equation containing two variables where $a = \frac{2}{9}$, $b = -\frac{40}{3}$, and $c = 0$. The parabola will open upward because a is positive. Furthermore, the t value of the vertex, $\frac{-b}{2a}$, is

$$\frac{-\left(-\frac{40}{3}\right)}{2\left(\frac{2}{9}\right)} = 30$$

So the coordinates of the vertex are $(30, -200)$. We leave it to the reader to verify that the following pairs of values are solutions: $(-30, 600)$, $(0, 0)$, $(15, -150)$, $(30, -200)$, $(45, -150)$, $(60, 0)$, and $(90, 600)$.

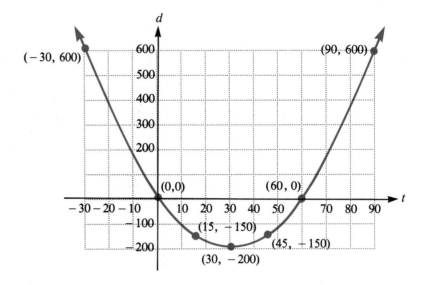

(b) The diving process starts at $t = 0$. Hence negative values of t have no meaning in this problem. Furthermore, when the diving bell reaches the surface again [see the point $(60, 0)$ on the graph], the experiment comes to an end. So values of t greater than 60 have no meaning in this problem either. The portion of the graph that pertains to this problem follows.

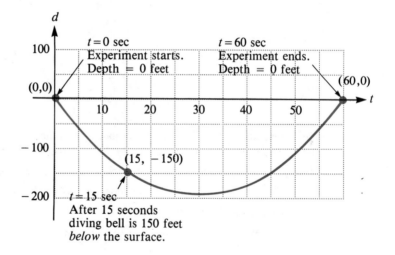

(c) The deepest point during the dive can be found by reading the coordinates of the vertex $(30, -200)$. It is the point that has the d value indicating the greatest depth. Hence, the deepest point during the dive is 200 feet *below* surface level. (Incidentally, the t value of the vertex tells us that it took 30 seconds for the diving bell to reach its greatest depth.)

(d) The time required for the whole diving experiment is read from the t value of the point $(60, 0)$. That time is 60 seconds.

In our next example we consider a modification of Example 3.

Example 7

A farmer desires to enclose a rectangular region. One side is bounded by a river and needs no fencing. The other three sides need fencing. The farmer has 400 feet of fencing available and she is *undecided* on the area of the rectangle to enclose.

(a) Express the area (A) of the rectangle she can enclose in terms of its dimensions.

(b) Graph the equation that is your answer to part (a).

(c) What part of the graph in part (b) pertains to the physical aspects of this problem?

(d) What dimensions should she choose for the field to attain a maximum area?

Solution

(a) A pictorial model follows.

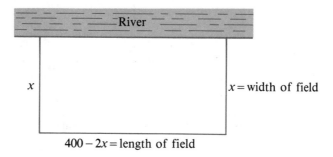

The area (A) is the product of width with length. So

$$A = x(400 - 2x)$$

(b) $A = x(400 - 2x)$ must now be graphed. First, we rewrite the equation as:

$$A = 400x - 2x^2$$
$$A = -2x^2 + 400x \leftarrow \text{This is a quadratic equation}$$

containing two variables where
$a = -2$, $b = 400$, and $c = 0$.

We label the horizontal axis x and the vertical axis A. The graph will be a parabola because $A = -2x^2 + 400x$ is a quadratic equation containing two variables. Since a is negative, the parabola will open downward. Furthermore, the x value of the vertex, $\dfrac{-b}{2a}$, is $\dfrac{-400}{2(-2)} = 100$. So the coordinates of the vertex are (100, 20,000). We leave it to the reader to verify that the following pairs of values are solutions: $(-50, -25,000)$, $(0, 0)$, $(50, 15,000)$, $(100, 20,000)$, $(150, 15,000)$, $(200, 0)$, $(250, -25,000)$.

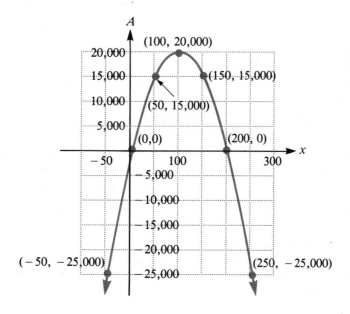

(c) The part of the graph that pertains to the physical aspects of this problem is the portion between (0, 0) and (200, 0). (This is where the area (A) is nonnegative.) We display that portion:

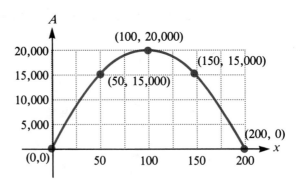

(d) The highest point (largest A value) will correspond to the area being a maximum. This occurs at (100, 20,000).

↑

This is the width of the field needed to produce a maximum area of 20,000 square feet.

So the dimensions are:

$$x = 100 \text{ feet (width)}$$
$$400 - 2x = 200 \text{ feet (length)}$$

Section 11.5 Exercises

$(-11, 13)$

1. The length of a rectangle is twice its width. If the diagonal of the rectangle is 10 yards, find the length and width (to the nearest tenth of a yard).

2. The length of a rectangle is 1 meter longer than its width. The diagonal is 2 meters. Find the length and width of the rectangle (to the nearest hundredth of a meter).

3. A page of a book is to be designed so that there are $1\frac{1}{2}$-inch margins on the left, right, top, and bottom.

   The height of the page is to be 4 inches more than the width. The area of the printed matter is to be 45 square inches. Find the length and width of the page.

4. Solve the problem in Exercise 3 if the area of the printed matter is to be 20 square inches. (Express your answer to the nearest tenth of an inch.)

5. A farmer desires to enclose a rectangular region. He has 400 feet of fencing available for the four sides and wants the rectangular region to have an area of 7500 square feet. Find the dimensions of the rectangle.

6. The length of a rectangle is 1 centimeter more than its width. If the diagonal is 4 centimeters, find the length and width of the rectangle (to the nearest tenth of a centimeter).

7. The length of a rectangle is 2 feet more than twice its width. If the diagonal is 5 feet, find the dimensions of the rectangle.

8. As shown in the figure, a portion of land is to have a rectangular garden planted on it. The garden will be bounded by a walk on each side. The width of the walk is to be 1 meter. The length of the outer rectangle is 3 meters more than its width. The area of the garden is to be 270 square meters.

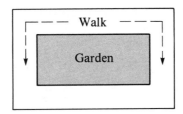

(a) Find the dimensions of the outer rectangle.

(b) Find the dimensions of the garden.

9. Solve the problem in Exercise 8 if the area of the garden is to be 46.75 square meters.

10. A cattle rancher will use 500 feet of fencing to enclose a rectangular region. He wants the rectangular field to have an area of 15,000 square feet. Find the dimensions of the rectangle.

11. A cattle rancher has 500 feet of fencing available to enclose a rectangular region. He wants the length to be 50 feet more than its width and the area to be 12,600 square feet. How much fencing will he have left over?

12. A dog-breeding kennel has n male dogs available for breeding purposes. On a certain day another breeder brings $n - 10$ female dogs to the kennel to be bred. The equation $p = n^2 - 10n$ represents the number of possible matings that can be made.

(a) How many matings are possible if there are 20 male dogs?

(b) If the number of matings is 1200, how many female dogs were brought to be bred?

13. A record store offers a special "two for one sale." The equation

$$n = \frac{x(x - 1)}{2}$$

gives the number of possible pairs of records that could be selected (n) when the total number of records in stock is x.

(a) If there are 5000 records in stock, how many possible pairs of records are there?

(b) If there are 10,000 records in stock, how many possible pairs of records are there?

14. The equation $p = n^2 + 10n$ represents the number of possible pairs of people, p (one male with one female), that can be formed by a computer dating service that has applications on file from n different males and $n + 10$ different females.

(a) How many pairs are possible if there are applications on file from 50 males and 60 females?

(b) If the number of possible pairs is 9000, how many females have applications on file?

15. The equation $D = 0.05v^2 + 1.1v$ gives the minimum safe stopping distance (D) for a motor vehicle traveling with speed v. (D is measured in feet and v is measured in mph.)

(a) What is the minimum safe stopping distance for a motor vehicle traveling at 50 mph?

(b) In order to stop your vehicle in a distance of 100 feet, what speed cannot be exceeded?

16. The equation

$$h = -\frac{5}{8}t^2 + 50t$$

gives the height (h) in feet that a computer-controlled rocket will ascend in t seconds. (The process starts at ground level when $t = 0$.)

(a) Using an appropriate scale on each axis, graph the equation.

(b) What part of the graph in part (a) pertains to the physical aspects of this problem?

(c) What is the maximum height attained?

(d) What is the total time of flight?

(e) At what times will the rocket be 750 feet high?

17. How far would it be to the horizon from the top of a mountain that is one mile high? (*Hint*: Determine the distance, d, shown in the figure.)

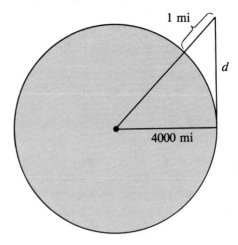

18. A farmer desires to enclose a rectangular region using 900 feet of fencing. He is undecided on the area of the rectangle to enclose.

    (a) Express the area (A) of the rectangle he can enclose in terms of its dimensions.

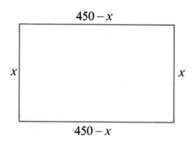

    (b) On a sheet of graph paper, graph the equation that is your answer to part (a). (Depict A on the vertical axis.)

    (c) What part of the graph pertains to the physical aspects of this problem?

    (d) What dimensions should be chosen for the field to attain a maximum area?

19. In a test for the concentration of a certain drug in a person's blood, the amount of the drug present, D (in parts per million), was found to be given by the equation

$$D = -0.2t^2 + 0.4t$$

    (a) On a sheet of graph paper, graph this equation. (Depict D on the vertical axis.)

    (b) What part of the graph pertains to the physical aspects of this problem?

    (c) At what time, t, is the concentration a maximum?

20. The cost, C (in dollars), of producing x FM stereo receivers is given by

$$C = 200 + 10x + \frac{1}{4}x^2$$

    (a) On a sheet of graph paper, graph this equation. (Depict C on the vertical axis.)

    (b) What part of the graph pertains to the monetary aspects of this problem?

21. An object is thrown upward from the roof of a building 100 feet high. The equation

$$d = -16t^2 + 32t + 100$$

gives the object's height, d (in feet), above the ground at any time, t, where t is measured in hours.

(a) On a sheet of graph paper, graph this equation. (Depict d on the vertical axis.)

(b) What part of the graph pertains to the physical aspects of this problem?

(c) At what time, t, does the object hit the ground?

(d) At what time, t, is the object's height above the ground a maximum?

Supplementary Exercises

1. The length of a rectangle is 3 feet more than its width. If the diagonal of the rectangle is $\sqrt{29}$, find its length and its width.

2. The area of a rectangle is 63 square feet. Its length is 18 feet more than its width. Find the dimensions of the rectangle.

3. The area of a square is 49 square feet. Find the length of a side.

4. The width of a rectangle is 7 feet less than its length. Its diagonal is 13 feet. Find the length and the width of the rectangle.

5. A farmer wishes to fence in an area of 30,000 square feet. The length is to be 50 feet more than the width. Find the dimensions.

6. The area of a right triangle is 20 square feet. If the base is 3 feet less than the height, find the dimensions of the triangle. $\left(Hint\text{: Use } A = \dfrac{1}{2}bh.\right)$

11.6   An introduction to imaginary and complex numbers

In Section 11.3 we saw that the general quadratic equation,

$$ax^2 + bx + c = 0$$

has solutions given by

$$x = \frac{-b \pm \sqrt{b^2 - 4ac}}{2a}$$

The quantity $b^2 - 4ac$ (which appears under the radical sign) is called the **discriminant.** When the discriminant is positive, there are two distinct real number solutions for the quadratic equation. They are

$$x = \frac{-b + \sqrt{b^2 - 4ac}}{2a} \qquad \text{and} \qquad x = \frac{-b - \sqrt{b^2 - 4ac}}{2a}$$

When the discriminant is zero, there is just one real number solution to the quadratic equation. It is

$$x = \frac{-b}{2a}$$

It can also happen that the discriminant has a negative value. When this occurs, the quadratic equation has *no real number solution*. We now look further into this situation.

Example 1

The equation $x^2 = -1$ does not have a real number solution because no real number, when squared, will equal a *negative* 1. What happens when we use the quadratic formula to solve $x^2 = -1$?

Solution

Rewriting $x^2 = -1$ as $x^2 + 1 = 0$, where $a = 1$, $b = 0$, and $c = 1$, we get

$$x = \frac{-b \pm \sqrt{b^2 - 4ac}}{2a}$$

$$x = \frac{-0 \pm \sqrt{0 - 4(1)(1)}}{2}$$

$$x = \frac{\pm \sqrt{-4}}{2}$$

Notice that the discriminant is negative.

$$x = \frac{\pm 2\sqrt{-1}}{2}$$

We rewrite the numerator as
$$\sqrt{-4} = \sqrt{4 \cdot -1}$$
$$= \sqrt{4} \cdot \sqrt{-1}$$
$$= 2\sqrt{-1}$$

So $x = \sqrt{-1}$ and $x = -\sqrt{-1}$

But these expressions *do not* represent real numbers. In fact, $\sqrt{-1}$ and $-\sqrt{-1}$ belong to another classification of numbers known as **imaginary numbers.** In particular, we usually reserve the letter i to represent $\sqrt{-1}$. That is, $i = \sqrt{-1}$.

Basically, imaginary numbers involve taking square roots of negative real numbers. Furthermore, when we add (or subtract) a real number to an imaginary number, we obtain still another type of number, a **complex number.** Such numbers look like $3 + i$, $4 - 2i$, $\frac{1}{3} + \frac{i}{2}$, and so on. The reader, in higher level courses, will undoubtedly see the significance of imaginary and complex numbers and also learn about the special arithmetic properties of these numbers. The amazing thing about these strange-looking numbers is that there are numerous applications of them in various life situations. Our goal here is only to acquaint the reader with their existence.

Section 11.6 Exercises

For the quadratic equations in Exercises 1 through 10, find (a) the value of the discriminant ($b^2 - 4ac$), and (b) the number of real number solutions to the quadratic equation. [*Hint*: Use your answer from part (a).]

1. $x^2 - 10x + 25 = 0$

2. $x^2 - 12x = -36$

3. $2x^2 - 5x - 3 = 0$

4. $x^2 + x + 1 = 0$

5. $2x^2 + x + 4 = 0$

6. $2x^2 + x - 4 = 0$

7. $4x^2 + x - 2 = 0$

8. $x^2 + 7 = 0$

9. $x = x^2 - 2$

10. $2 = -x^2$

11. *True or False:* Every quadratic equation has real number solutions.

12. *True or False:* $\sqrt{-1}$, $2\sqrt{-3}$, and $\sqrt{-5}$ are imaginary numbers.

13. *True or False:* $3 + 2i$, $-4 + \frac{2}{3}i$, $-1 - i$, and $\frac{1}{2} + \frac{1}{2}i$ are complex numbers.

Supplementary Exercises

For the quadratic equations in Exercises 1 through 7, find (a) the value of the discriminant ($b^2 - 4ac$), and (b) the number of real number solutions to the quadratic equation. [*Hint*: Use your answer from part (a).]

1. $x^2 - 5x + 7 = 0$ 2. $x^2 + 4x + 4 = 0$ 3. $x^2 - 9 = 0$

4. $3x^2 = x - 2$ 5. $2x^2 - 5 = 3x$ 6. $-4x^2 - 4x - 4 = 0$

7. $-5 - 2x + x^2 = 0$

11.7   Chapter review

Summary

In this chapter we returned to the study of quadratic equations, which we had begun in Section 6.6. In particular, we saw that quadratic equations can be solved not only by factoring but also by completing the square or by using the quadratic formula:

$$x = \frac{-b \pm \sqrt{b^2 - 4ac}}{2a}$$

We also saw that the graph of $y = ax^2 + bx + c$ (a quadratic equation in two variables) is a parabola opening upward (if a is positive) or downward (if a is negative). The x value of the vertex (turning point) of the parabola is $\frac{-b}{2a}$. The y value of the vertex is found by substituting the x value into the quadratic equation and then evaluating for y. We examined various life situations that have, as an algebraic model, quadratic equations of either one or two variables.

Finally, we were introduced to imaginary numbers and complex numbers and saw that real numbers aren't always "enough" to solve a given equation.

Vocabulary Quiz

Match the expression in Column I with the phrase in Column II that *best* describes it.

Column I

1. The general form of the quadratic equation (in one variable) is

2. The quadratic formula states that, if $ax^2 + bx + c = 0$, then $x =$

3. The discriminant of $ax^2 + bx + c = 0$ is

4. The graph of $y = ax^2 + bx + c$ ($a \neq 0$) is a

5. The turning point of $y = ax^2 + bx + c$ is called its

6. The x value of the turning point is

7. If a is positive, then the graph of $y = ax^2 + bx + c$ opens

8. The square root of a negative real number is

9. A real number plus an imaginary number is

Column II

a. $ax^2 + bx + c = 0$.

b. $b^2 - 4ac$.

c. $\frac{-b}{2a}$.

d. upward.

e. parabola.

f. $\frac{-b \pm \sqrt{b^2 - 4ac}}{2a}$

g. vertex.

h. a complex number.

i. an imaginary number.

Chapter 11 Review Exercises

In Exercises 1 through 11, use (a) the process of completing the square and then (b) the quadratic formula to solve each of the quadratic equations.

1. $x^2 - 7x + 12 = 0$ 2. $x^2 = 8x + 9$ 3. $x^2 = 8x$

4. $2x^2 - 8x + 8 = 0$ 5. $2x^2 + x = 1$ 6. $x^2 + 3x - 1 = 0$

7. $x^2 + 2x - 1 = 0$ 8. $3x^2 + 5x + 1 = 0$ 9. $6x^2 - 1 = 0$

10. $5x^2 + 5x - 1 = 0$ 11. $-\dfrac{1}{2}x^2 + 4x + \dfrac{9}{2} = 0$

In Exercises 12 through 22, determine whether the parabola opens upward or downward. Find the coordinates of its vertex and graph the equation on a sheet of graph paper. Also estimate the x-intercepts from the graph. Compare with Exercises 1 through 11.

12. $y = x^2 - 7x + 12$ 13. $y = x^2 - 8x - 9$ 14. $y = x^2 - 8x$

15. $y = 2x^2 - 8x + 8$ 16. $y = 2x^2 + x - 1$ 17. $y = x^2 + 3x - 1$

18. $y = x^2 + 2x - 1$ 19. $y = 3x^2 + 5x + 1$ 20. $y = 6x^2 - 1$

21. $y = 5x^2 + 5x - 1$ 22. $y = -\dfrac{1}{2}x^2 + 4x + \dfrac{9}{2}$

23. Let $x_1 = \dfrac{-b + \sqrt{b^2 - 4ac}}{2a}$ and $x_2 = \dfrac{-b - \sqrt{b^2 - 4ac}}{2a}$.

     (a) Show that $x_1 + x_2 = \dfrac{-b}{a}$.

     (b) Show that $x_1 \cdot x_2 = \dfrac{c}{a}$.

24. One leg of a right triangle is three times the length of the other leg. If the hypotenuse is 10 cm, find (a) the lengths of the two legs and (b) the area of the triangle.

25. If the length of a rectangle is 2 inches more than its width and the diagonal of the rectangle is $\dfrac{1}{2}$ ft, find the dimensions of the rectangle.

Cumulative review exercises

1. Find the volume of the right circular cylinder with radius $= 4$ inches and height $= 2$ inches. Use $\pi \approx 3.14$.   (*Section 2.5*)

2. Solve for y: $3(y + 1) - 2(y - 4) = 8y - 17$.   (*Section 3.4*)

3. Simplify by rewriting the expression using only positive exponents. Assume all variables cannot be zero.

$$\frac{-96x^{-3}y^{-2}z^3}{-64x^{-5}y^{-3}z^{-2}}$$ (*Section 5.2*)

4. Completely factor $2x^3 - 12y^2 + 18xy^2$.   (*Section 6.5*)

5. Add $\dfrac{5x}{x^2 - 3x} + \dfrac{3}{x^2 + 4x - 21}$.   (*Section 7.4*)

6. Find the equation of the line passing through the points $(5, -7)$ and $(2, 2)$.   (*Section 8.4*)

7. Solve

$$2x - 3y = -18$$
$$x + 2y = 5 \qquad\qquad (Section\ 9.3)$$

8. Antonio has three times as much money invested at 12% as he does at 10%. If the total yearly interest from these investments is $1150, how much does he have invested at each rate?   (*Section 9.4*)

9. Simplify $\sqrt{98x^3y^2}$.   (*Section 10.2*)

10. Solve $3 + \sqrt{3x + 7} = 8$.   (*Section 10.5*)

Chapter 11 test

Take this test to see how well you have mastered quadratic equations; check your answers with those at the end of the book.

1. Use factoring to solve $2x^2 - 3x - 14 = 0$.   (*Section 11.1*)

1. _____

2. Use the process of completing the square to solve $x^2 - 8x + 3 = 0$.   (*Section 11.2*)

2. _____

3. Use the quadratic formula to solve $y^2 = 4y + 4$.   (*Section 11.3*)

3. _____

4. Use the quadratic formula to solve $2x^2 - 3x - 6 = 0$.   (*Section 11.3*)

4. _____

5. Graph $y = x^2 - 7x + 10$ and state whether the parabola opens upward or downward and state the vertex.   (*Section 11.4*)

5. _____

6. Determine whether $y = 12 - x - x^2$ opens upward or downward. What are the coordinates of the vertex?   (*Section 11.4*)

6. _____

7. The equation $s = -16t^2 + 32t + 160$ gives an object's height, s (in feet), above the ground at any time, t (in seconds), for an object thrown upward from the top of a building 160 feet high.
   (a) Graph $s = -16t^2 + 32t + 160$.

7. (a) _____

   (b) _____

   (b) What part of the graph pertains to the physical aspects of this problem?   (*Section 11.5*)

8. In Problem 7, at what time t will the object hit the ground?   (*Section 11.5*)   8. _____

9. (a) Evaluate the discriminant of $3x^2 + x + 1 = 0$.   (*Section 11.6*)   9. (a) _____

    (b) Are the solutions to $3x^2 + x + 1 = 0$ real or complex?   (b) _____

10. *True or False:*   The square root of a negative integer is an imaginary number.   (*Section 11.6*)   10. _____

Appendix A

Decimals

Our present-day system of representing numbers is called the *decimal* system. In the decimal system the position, or place, of the digit within a number gives that digit its value. The number 10 is the *base* of the system. Numbers are indicated by using a combination of the *digits* 0, 1, 2, 3, 4, 5, 6, 7, 8, 9 and a *decimal point* that reflects the digit's value as a power of 10. The following place value chart is useful in dealing with decimals.

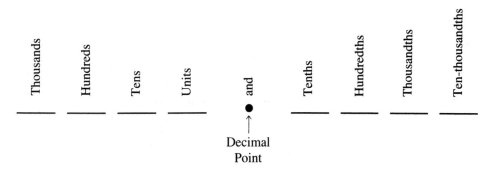

For example, the number

$$712.34$$

is read "seven hundred twelve *and* thirty-four hundredths."

The correct reading of a decimal according to the place value chart above enables us to change the decimal to a fraction easily.

Example 1

Write each of the following decimals in words and as a fraction.

(a) 0.47   (b) 1.17   (c) 18.9

Solution

(a) $0.47 = $ forty-seven hundredths $= \dfrac{47}{100}$

(b) $1.17 = $ one and seventeen hundredths $= 1\dfrac{17}{100}$ or $\dfrac{117}{100}$

(c) $18.9 = $ eighteen and 9 tenths $= 18\dfrac{9}{10}$ or $\dfrac{189}{10}$

It is also important to be able to change a fraction to a decimal. Since the fraction bar means division, we convert a fraction to a decimal by dividing.

Example 2

Write the following fractions as decimals.

(a) $\dfrac{2}{5}$ (b) $\dfrac{7}{20}$ (c) $\dfrac{4}{25}$

Solution

(a) $\dfrac{2}{5}$ means $2 \div 5$, so by dividing we get

$$
\begin{array}{r}
0.4 \\
5\overline{)2.0} \\
\underline{2\ 0}
\end{array}
$$

So $\dfrac{2}{5} = 0.4$.

(b) $\dfrac{7}{20}$ means $7 \div 20$, so we have

$$
\begin{array}{r}
0.35 \\
20\overline{)7.00} \\
\underline{6\ 0} \\
1\ 00 \\
\underline{1\ 00}
\end{array}
$$

Thus $\dfrac{7}{20} = 0.35$.

(c) $\dfrac{4}{25}$ means $4 \div 25$, so division yields

$$
\begin{array}{r}
0.16 \\
25\overline{)4.00} \\
\underline{2\ 5} \\
1\ 50 \\
\underline{1\ 50}
\end{array}
$$

Hence $\dfrac{4}{25} = 0.16$.

Each of the fractions in Example 2 resulted in decimals that terminated after a certain number of digits to the right of the decimal. But for certain fractions division results in a repeating pattern of one digit or a block of digits. For example, the decimal representation of $\dfrac{2}{9}$ is $0.2222\ldots$; the three dots indicate that the digit 2 repeats indefinitely. We shorten this notation by placing a horizontal bar over the repeating digit. Thus we have $\dfrac{2}{9} = 0.2222\ldots, = 0.\overline{2}$.

Example 3

Write the following fractions in decimal form.

(a) $\dfrac{2}{3}$ (b) $\dfrac{7}{11}$ (c) $\dfrac{16}{99}$

Solution

(a) $\dfrac{2}{3} = 0.6666\ldots = 0.\overline{6}$ The digit 6 repeats.

(b) $\dfrac{7}{11} = 0.636363\ldots = 0.\overline{63}$ The digits 63 repeat.

(c) $\dfrac{16}{99} = 0.161616\ldots = 0.\overline{16}$ The digits 16 repeat.

Appendix B

Table of Squares and Square Roots

n	n^2	$\sqrt{n}$	n	n^2	$\sqrt{n}$
1	1	1.000	51	2,601	7.141
2	4	1.414	52	2,704	7.211
3	9	1.732	53	2,809	7.280
4	16	2.000	54	2,916	7.348
5	25	2.236	55	3,025	7.416
6	36	2.449	56	3,136	7.483
7	49	2.646	57	3,249	7.550
8	64	2.828	58	3,364	7.616
9	81	3.000	59	3,481	7.681
10	100	3.162	60	3,600	7.746
11	121	3.317	61	3,721	7.810
12	144	3.464	62	3,844	7.874
13	169	3.606	63	3,969	7.937
14	196	3.742	64	4,096	8.000
15	225	3.873	65	4,225	8.062
16	256	4.000	66	4,356	8.124
17	289	4.123	67	4,489	8.185
18	324	4.243	68	4,624	8.246
19	361	4.359	69	4,761	8.307
20	400	4.472	70	4,900	8.367
21	441	4.583	71	5,041	8.426
22	484	4.690	72	5,184	8.485
23	529	4.796	73	5,329	8.544
24	576	4.899	74	5,476	8.602
25	625	5.000	75	5,625	8.660
26	676	5.099	76	5,776	8.718
27	729	5.196	77	5,929	8.775
28	784	5.292	78	6,084	8.832
29	841	5.385	79	6,241	8.888
30	900	5.477	80	6,400	8.944
31	961	5.568	81	6,561	9.000
32	1,024	5.657	82	6,724	9.055
33	1,089	5.745	83	6,889	9.110
34	1,156	5.831	84	7,056	9.165
35	1,225	5.916	85	7,225	9.220
36	1,296	6.000	86	7,396	9.274
37	1,369	6.083	87	7,569	9.327
38	1,444	6.164	88	7,744	9.381
39	1,521	6.245	89	7,921	9.434
40	1,600	6.325	90	8,100	9.487
41	1,681	6.403	91	8,281	9.539
42	1,764	6.481	92	8,464	9.592
43	1,849	6.557	93	8,649	9.644
44	1,936	6.633	94	8,836	9.695
45	2,025	6.708	95	9,025	9.747
46	2,116	6.782	96	9,216	9.798
47	2,209	6.856	97	9,409	9.849
48	2,304	6.928	98	9,604	9.899
49	2,401	7.000	99	9,801	9.950
50	2,500	7.071	100	10,000	10.000

Appendix C   The metric system

The common units of measurement in the metric system are summarized in the following table:

English Units and Metric Units

Length

1 millimeter (mm) $\approx$ 0.04 inch (in.)*	or 1 in. = 25.4 mm
1 centimeter (cm) $\approx$ 0.4 inch (in.)	or 1 in. = 2.54 cm
1 meter (m) $\approx$ 1.1 yards (yd)	or 1 yd $\approx$ 0.92 m
1 kilometer (km) $\approx$ 0.62 mile (mi)	or 1 mi $\approx$ 1.6 km

Mass

1 gram (g) $\approx$ 0.035 ounce (oz)	or 1 oz $\approx$ 28.3 g
1 kilogram (kg) $\approx$ 2.2 pounds (lb)	or 1 lb $\approx$ 0.45 kg

Volume

1 milliliter (mL) $\approx$ 0.03 fluid ounce (fl oz)	or 1 fl oz $\approx$ 30 mL
1 liter (L) $\approx$ 1.06 quarts (qt)	or 1 qt $\approx$ 0.95 L
1 cubic meter (m^3) $\approx$ 1.3 cubic yards (yd^3)	or 1 yd^3 $\approx$ 0.76 m^3

Area

1 square centimeter (cm^2) $\approx$ 0.16 square inch (in^2)	or 1 $in.^2$ $\approx$ 6.5 cm^2
1 square meter (m^2) $\approx$ 1.2 square yards (yd^2)	or 1 yd^2 $\approx$ 0.8 m^2
1 square kilometer (km) $\approx$ 0.4 square mile (mi^2)	or 1 mi^2 $\approx$ 2.6 km^2

Common Comparisons within the Metric System

Length

1 millimeter (mm) $= \dfrac{1}{1000}$ meter (m)

1 centimeter (cm) $= \dfrac{1}{100}$ meter (m)

1 kilometer (km) = 1000 meters (m)

Mass

1 milligram (mg) $= \dfrac{1}{1000}$ gram (g)

1 kilogram (kg) = 1000 grams (g)

Volume

1000 milliliters (mL) = 1 liter (L)

*The symbol $\approx$ means "is approximately equal to." The information in this table is from the United States Department of Commerce, National Bureau of Standards, Special Publication 330, revised 1974.

Answers to Selected Exercises

Chapter 1

Section 1.1 Exercises (p. 6)

1. 7, 8, 9, 10, 11, 12   **3.** . . . , −4, −2, 0, 2, 4, . . .

5.

7.

9.   **11.** The natural number is 2.   **13.** The rational

numbers are 0, −3, 2, $\frac{3}{4}$, and $-5\frac{1}{8}$.   **15.** The real numbers are 0, −3, 2, $\frac{3}{4}$, $-5\frac{1}{8}$, and π.   **17.** The

intergers are −5, 0, and 3.   **19.** The irrational number is $-\pi$.

21.

	Natural numbers	Integers	Rational numbers	Irrational numbers	Real numbers
−4		✓	✓		✓
0		✓	✓		✓
$\frac{2}{3}$			✓		✓
600	✓	✓	✓		✓

23. False   **25.** False   **27.** True   **29.** True   **31.** True   **33.** False   **35.** False   **37.** $-7 \leq -6$
39. $4 \geq 0$ **41.** $\frac{1}{2} < \frac{3}{4}$ **43.** $\geq$ **45.** $\geq$

Section 1.1 Supplementary Exercises (p. 8)

1.

	Natural numbers	Integers	Rational numbers	Irrational numbers	Real numbers
−10		✓	✓		✓
$\frac{4}{9}$			✓		✓
$-\pi$				✓	✓
12	✓	✓	✓		✓
6 + 9	✓	✓	✓		✓

3. $129 < 149$ **5.** $-3 < 9$ **7.** $0 > -12$ **9.** $-21 > -26$ **11.** $-39 \geq -43$ **13.** $-2 < -1$
15. $0 \geq 9 - 9$ **17.** $-3 < 8$ **19.** $-23 \geq -24$

Section 1.2 Exercises (p. 13)

1. Since the given numbers have different signs we subtract the smaller absolute value from the larger absolute value. Thus $|-3| = 3$, $|4| = 4$, and $4 - 3 = 1$. Since the positive number, 4, has the larger absolute value, the sum must be positive. Thus $-3 + 4 = 1$.

3. Since both numbers have the same sign we add their absolute values $|7| = 7$, $|3| = 3$, and $7 + 3 = 10$. Since both numbers are positive, the sum must also be positive. Thus $7 + 3 = 10$.

5. The numbers to be added have different signs. We subtract the smaller absolute value from the larger: $|9| = 9$, $|-5| = 5$, and $9 - 5 = 4$. Since the positive number, 9, has the largest absolute value, the sum must be positive. Thus $9 + (-5) = 4$.

7. Both numbers have the same sign so we add their absolute values: $|-11| = 11$, $|-7| = 7$, and $11 + 7 = 18$. Since both numbers are negative, the sum must be negative. So $(-11) + (-7) = -18$.

9. The numbers have different signs so we subtract the smaller absolute value from the larger: $|-7| = 7$, $|3| = 3$, and $7 - 3 = 4$. Since the negative number, -7, has the larger absolute value, the sum must be negative. Thus $(-7) + 3 = -4$.

11. The numbers have different signs. We subtract the smaller absolute value from the larger: $|2| = 2$, $|-9| = 9$, and $9 - 2 = 7$. Since the negative number, -9, has the larger absolute value, the sum must be negative. Thus $2 + (-9) = -7$.

13. Both numbers are negative, so we add their absolute values and keep the negative sign for our sum: $|-9| = 9$, $|-2| = 2$, and $9 + 2 = 11$. Thus $(-9) + (-2) = -11$.

15. The numbers have different signs; subtract the smaller absolute value from the larger absolute value: $|9| = 9$, $|-2| = 2$, and $9 - 2 = 7$. Since the positive number, 9, has the larger absolute value the sum is positive. Thus $9 + (-2) = 7$.

17. Both numbers are negative; add their absolute values and keep the negative sign for the sum: $|-15| = 15$, $|-12| = 12$, and $15 + 12 = 27$. Thus $(-15) + (-12) = -27$.

19. The numbers have different signs; subtract the smaller absolute value from the larger: $|23| = 23$, $|-28| = 28$, and $28 - 23 = 5$. Since the negative number, -28, has the larger absolute value, the sum is negative. Thus $23 + (-28) = -5$.

21. 8   23. -7 25. -20 27. 14   29. -4 31. -3 33. -23 35. -21 37. -3.737

39. 0.14493

Section 1.2 Supplementary Exercises (p. 13)

1. Both numbers positive; add their absolute values and make the sum positive: $|24| = 24$, $|46| = 46$, and $24 + 46 = 70$. Thus $24 + 46 = 70$.

3. Both numbers negative; add their absolute values and make the sum negative: $|-23| = 23$, $|-42| = 42$, and $23 + 42 = 65$. So $(-23) + (-42) = -65$.

5. Different signs; subtract the smaller absolute value from the larger and use the sign of the number with the larger absolute value—in this case positive. $|-3| = 3$, $|28| = 28$, and $28 - 3 = 25$. So $(-3) + 28 = 25$.

7. Different signs; subtract the smaller absolute value from the larger and use the sign of the number with the larger absolute value—in this case positive. $|37| = 37$, $|-32| = 32$, and $37 - 32 = 5$. Thus $37 + (-32) = 5$.

9. All three numbers are negative. Add their absolute values and keep the negative sign. $|-32| = 32$, $|-45| = 45$, $|-21| = 21$ and $32 + 45 + 21 = 98$. Thus $(-32) + (-45) + (-21) = -98$.

11. $(-32) + (-24) + (-4) = (-56) + (-4)$
$\qquad\qquad\qquad\qquad = -60$

13. $-7 + (-3) + 12 = (-10) + 12$
$\qquad\qquad\qquad\qquad = 2$

15. $94 + (-94) + 12 = 0 + 12$
$\qquad\qquad\qquad\qquad = 12$

17.
$$\begin{array}{r} -32 \\ -22 \\ -73 \\ \hline -127 \end{array}$$

19. $(-21) + 43 + (-32) = 22 + (-32)$
$\qquad\qquad\qquad\qquad = -10$

Section 1.3 Exercises (p. 16)

1. -8 is the opposite of 8   3. 6 is the opposite of -6 5. $-\dfrac{3}{8}$ is the opposite of $\dfrac{3}{8}$

7. 0 is the opposite of 0   9. $7 - 12 = 7 + (-12)$ (Subtracting 12 is the same as adding -12.)
$\qquad\qquad\qquad = -5$

11. $(-7) - 6 = -7 + (-6) = -13$ (Subtracting 6 is the same as adding -6.)

13. $(-4) - (-5) = (-4) + (+5)$ (Subtracting -5 is the same as adding $+5$.)
$\qquad\qquad = (-4) + 5 = 1$

15. $16 - 7 = 16 + (-7) = 9$ (Subtracting 7 is the same as adding -7.)

17. $(-20) - 6 = -20 + (-6)$ (Subtracting 6 is the same as adding -6.)
$$= -26$$

19. $(-14) - (-10) = (-14) + (+10)$ (Subtracting -10 is the same as adding $+10$.)
$$= (-14) + 10$$
$$= -4$$

21. $6 - 20 = 6 + (-20)$ (Subtracting 20 is the same as adding -20.)
$$= -14$$

23. 17   **25.** 17   **27.** -21 **29.** -7 **31.** 185   **33.** 42   **35.** 183   **37.** $-\$4$ **39.** 766.3406

Section 1.3 Supplementary Exercises (p. 17)

1. $15 - 8 = 15 + (-8)$ (Subtracting 8 is the same as adding -8.)
$$= 7$$

3. $86 - 94 = 86 + (-94)$ (Subtracting 94 is the same as adding -94.)
$$= -8$$

5. $-25 - 21 = -25 + (-21)$ (Subtracting 21 is the same as adding -21.)
$$= -46$$

7. $-23 - (-24) = -23 + (+24)$ (Subtracting -24 is the same as adding $+24$.)
$$= -23 + 24$$
$$= 1$$

9. $-45 - (-24) = -45 + (+24)$ (Subtracting -24 is the same as adding $+24$.)
$$= -21$$

11. $-9 - 0 = -9 + 0$ (Subtracting 0 is the same as adding 0.)
$$= -9$$

13. $-8 - 2 - 15 = -8 + (-2) - 15$ (Subtracting 2 is the same as adding -2.)
$$= -10 - 15$$
$$= -10 + (-15)$$ (Subtracting 15 is the same as adding -15.)
$$= -25$$

15. $12 - 23 - (-2) = 12 + (-23) - (-2)$ (Subtracting 23 is the same as adding -23.)
$$= -11 - (-2)$$
$$= -11 + (+2)$$ (Subtracting -2 is the same as adding $+2$.)
$$= -11 + 2$$
$$= -9$$

17. -245
$\underline{-134}$ (Subtracting 134 is the same as adding -134.)
-379

19.   24
$\underline{-9}$ (Subtracting 9 is the same as adding -9.)
15

Section 1.4 Exercises (p. 21)

1. $5 \cdot 7 = 35$ $|5| = 5, |7| = 7$, and $5 \cdot 7 = 35$
Since both numbers are positive, the product must be positive.

3. $(-6) \cdot 5 = -30$ $|-6| = 6, |5| = 5$, and $6 \cdot 5 = 30$
Since the numbers have different signs, the product must be negative.

5. $(-8)(-6) = 48$ $|-8| = 8, |-6| = 6$, and $8 \cdot 6 = 48$
Since both numbers are negative, the product must be positive.

7. $5(-9) = -45$ $|5| = 5, |-9| = 9$, and $5 \cdot 9 = 45$
The product is negative, since one number is positive and the other is negative.

9. $(-3)(-7) = 21$ $|-3| = 3, |-7| = 7$, and $3 \cdot 7 = 21$
Since both numbers are negative, the product must be positive.

11. -81 **13.** -120 **15.** -375 **17.** 320   **19.** 252   **21.** -168 **23.** -46.3 **25.** -13.16
27. -21.36288

29. $\dfrac{28}{2} = 14$ $|28| = 28, |2| = 2$, and $\dfrac{28}{2} = 14$

The quotient is positive, since both numbers are positive.

31. $(-15) \div 3 = -5$ $|-15| = 15, |3| = 3$, and $\dfrac{15}{3} = 5$

The quotient is negative, since the dividend is negative and the divisor is positive.

33. $\dfrac{-42}{-6} = 7$ $|-42| = 42, |-6| = 6$, and $\dfrac{42}{6} = 7$

The quotient is positive, since both the numerator and denominator are negative.

35. $21 \div (-7) = -3$ $|21| = 21, |-7| = 7$, and $\dfrac{21}{7} = 3$

Since one of the numbers is positive and the other negative, the quotient must be negative.

37. $\dfrac{-56}{8} = -7$ $|-56| = 56, |8| = 8$, and $\dfrac{56}{8} = 7$

The quotient is negative, since the numerator is negative and the denominator is positive.

39. 9   **41.** -5 **43.** -10 **45.** -18 **47.** -8 **49.** -57.6 **51.** -12.869847

Section 1.4 Supplementary Exercises (p. 21)

1. $12(-2) = -24$ $|12| = 12, |-2| = 2$, and $12 \cdot 2 = 24$

The product is negative, since the factors have different signs.

3. $-3(5) = -15$ $|-3| = 3, |5| = 5$, and $3 \cdot 5 = 15$

Since one factor is positive and the other negative, the product is negative.

5. $-2(-3)(3) = 6(3)$ Product of the first two negative factors is positive.
$ = 18$

7. $2(3)(5)(0) = 0$ The product of 0 and any number is 0.

9. $-3(-2)(-4) = 6(-4)$ Product of the first two negative factors is positive. The product of a positive
$ = -24$ and negative number is negative.

11. $(-45) \div (-9) = 5$ $|-45| = 45, |-9| = 9$, and $\dfrac{45}{9} = 5$

The quotient of two negative numbers must be positive.

13. $-121 \div (-11) = 11$ $|-121| = 121, |-11| = 11$, and $\dfrac{121}{11} = 11$

The quotient of two negative numbers is positive.

15. $-85 \div 5 = -17$ $|-85| = 85, |5| = 5$, and $\dfrac{85}{5} = 17$

Since the numbers have different signs, the quotient is negative.

17. $\dfrac{200}{10} = 20$ $|200| = 200, |10| = 10$, and $\dfrac{200}{10} = 20$

The quotient of two positive numbers is always positive.

19. $7\overline{\smash{\big)}{-56}}$ $|-56| = 56, |7| = 7$, and $\dfrac{56}{7} = 8$

The dividend is negative and the divisor is positive, so the quotient is negative.

21. 120   **23.** -70 **25.** -110

Section 1.5 Exercises (p. 27)

1. True. Associative Property for Addition   (grouping changed)   **3.** True. Distributive Property
5. True. Commutative Property for Addition   (order changed)   **7.** False. Subtraction is *not* commutative.
9. True. Associative Property for Multiplication   (grouping changed)   **11.** True. Distributive Property
13. True. Associative Property for Multiplication   (grouping changed)   **15.** True. Commutative Property
for Addition   (order changed)   **17.** True. Distributive Property   **19.** True. Associative Property for
Multiplication   (grouping changed)   **21.** True. Distributive Property   **23.** $8 + (-5)$ **25.** $6[x \cdot (-3)]$
27. $(-6)4 + (-6)12$ **29.** $(6 + 17) + y$ **31.** $x + (-3)$ **33.** $(-4)(x + 6)$

Section 1.5 Supplementary Exercises (p. 28)

1. $[-3 + 5] = [5 + (-3)]$
$\,2 = 2$
True. Commutative Property for Addition

3. $8 - 9 = 9 - 8$
$\,-1 = 1$
False. Subtraction is *not* commutative.

5. $4 + (-5 + 23) = [4 + (-5)] + 23$
$4 + 18 = [-1] + 23$
$22 = 22$
True. Associative Property for Addition

7. $[3(-2)](-5) = 3[(-2)(-5)]$
$[-6](-5) = 3[10]$
$30 = 30$
True. Associative Property for Multiplication

9. $0 - 9 = 9 - 0$
$\qquad -9 = 9$
False. Subtraction is *not* commutative.

11. $-6[5 + (-2)] = -6(5) + (-6)(-2)$
$\qquad -6[3] = -30 + 12$
$\qquad -18 = -18$
True. Distributive Property

13. $2(8 \div 4) = 2(8) \div [2(4)]$
$\qquad 2(2) = 16 \div [8]$
$\qquad 4 = 2$
False. Multiplication does not distribute over division.

15. $-7(5 - 9) = -7(5) - (-7)(9)$
$\qquad -7(-4) = -35 - (-63)$
$\qquad 28 = -35 + 63$
$\qquad 28 = 28$
True. Distributive Property

17. $0 + 21 = 21 + 0$
$\qquad 21 = 21$
True. Commutative Property for Addition

Section 1.6 Exercises (p. 31)

1. $(-13) + 0 = -13$ Identity Property for Addition   **3.** $\frac{0}{7} = 0$ Dividing 0 by a nonzero number is always 0.   **5.** $8 \cdot 1 = 8$ Identity Property for Multiplication   **7.** $16 - 0 = 16$
9. $0 + (-10) = -10$ Identity Property for Addition   **11.** $-\frac{16}{0}$ is undefined   **13.** $\frac{0}{0}$ cannot be determined   **15.** $1 \cdot (-15) = -15$ Identity Property for Multiplication   **17.** $\pi + 0 = \pi$ Identity Property for Addition   **19.** $\left(\frac{3}{4}\right) \cdot 1 = \frac{3}{4}$ Identity Property for Multiplication   **21.** undefined
23. 0   **25.** -126 Identity Property for Addition   **27.** undefined   **29.** error

Section 1.6 Supplementary Exercises (p. 32)

1. $0 + 181 = 181$ Identity Property for Addition   **3.** $0 - 9 = 0 + (-9)$ **5.** $0(5) = 0$
$\qquad\qquad\qquad = -9$
7. $\frac{5}{0}$ is undefined   **9.** $\frac{0}{0}$ cannot be determined   **11.** $1(-9) = -9$ Identity Property for Multiplication   **13.** $-9 \div 1 = -9$ **15.** $1573(-4)(23)(7)0 = 0$ (Any number multiplied by 0 is 0).

Section 1.7 Exercises (p. 37)

1. $3^3 = 3 \cdot 3 \cdot 3 = 27$ **3.** $9^2 = 9 \cdot 9 = 81$ **5.** $(-6)^3 = (-6)(-6)(-6) = -216$
7. $(-1)^5 = (-1)(-1)(-1)(-1)(-1) = -1$ **9.** $0^5 = 0 \cdot 0 \cdot 0 \cdot 0 \cdot 0 = 0$ **11.** $3^5 = 3 \cdot 3 \cdot 3 \cdot 3 \cdot 3 = 243$
13. $(-3)^4 = (-3)(-3)(-3)(-3) = 81$ **15.** $(-10)^3 = (-10)(-10)(-10) = -1000$
17. $-6^3 = -(6 \cdot 6 \cdot 6) = -216$ **19.** $-2^4 = -(2 \cdot 2 \cdot 2 \cdot 2) = -16$ **21.** 16   **23.** -729
25. 2143.69   **27.** 289   **29.** $\sqrt{64} = 8$ since $8^2 = 64$ **31.** $\sqrt[3]{64} = 4$ since $4^3 = 64$
33. $\sqrt[3]{-8} = -2$ since $(-2)^3 = -8$ **35.** $\sqrt{81} = 9$ since $9^2 = 81$ **37.** $\sqrt[4]{81} = 3$ since $3^4 = 81$
39. $\sqrt[3]{-64} = -4$ since $(-4)^3 = -64$ **41.** $\sqrt[5]{-1} = -1$ since $(-1)^5 = -1$ **43.** $\sqrt[4]{-16}$ is not real since any number raised to the 4th power must be positive.   **45.** $-\sqrt{81} = -9$ since $-\sqrt{81}$ indicates the negative square root of 81.   **47.** -6 **49.** -4 **51.** not real   **53.** 10   **55.** -12 **57.** 26   **59.** 19.9
61. 36.43   **63.** 9.31

Section 1.7 Supplementary Exercises (p. 37)

1. $5^3 = 5 \cdot 5 \cdot 5 = 125$ **3.** $-5^3 = -(5 \cdot 5 \cdot 5) = -125$ **5.** $(-2)^4 = (-2)(-2)(-2)(-2) = 16$
7. $0^2 = 0 \cdot 0 = 0$ **9.** $25^1 = 25$ **11.** $(-7)^3 = (-7)(-7)(-7) = -343$ **13.** $(-6)^2 = (-6)(-6) = 36$
15. $1^5 = 1 \cdot 1 \cdot 1 \cdot 1 \cdot 1 = 1$ **17.** $\sqrt{121} = 11$ since $11^2 = 121$ **19.** $\sqrt{100} = 10$ since $10^2 = 100$
21. not real   **23.** -2 **25.** 1   **27.** -5 **29.** 2   **31.** -1

Section 1.8 Exercises (p. 41)

1. $36 \div 4 \cdot 3 = 9 \cdot 3 = 27$ **3.** $2\sqrt{49} - 6 = 2 \cdot 7 - 6$ **5.** $12 - 6 + 3 - 8 = 12 + (-6) + 3 + (-8)$
$\qquad\qquad\qquad\qquad\quad = 14 - 6 \qquad\qquad\qquad\qquad\qquad\qquad = 6 + 3 + (-8)$
$\qquad\qquad\qquad\qquad\quad = 8 \qquad\qquad\qquad\qquad\qquad\qquad\quad = 9 + (-8)$
$\qquad\qquad\qquad\qquad\qquad\qquad\qquad\qquad\qquad\qquad\qquad\quad = 1$

7. $8 - 3 \cdot 6 = 8 - 18$ **9.** $5 \cdot 4 + 21 \div 3 = 20 + 7$ **11.** $-16 + 4(-2) \div 2^3 = -16 + 4(-2) \div 8$
$\qquad\qquad = 8 + (-18)$ $\qquad\qquad\qquad\quad = 27$ $\qquad\qquad\qquad\qquad\qquad\quad = -16 + (-8) \div 8$
$\qquad\qquad = -10$ $\qquad\qquad\qquad\qquad\qquad\qquad\qquad\qquad\qquad = -16 + (-1)$
$\qquad\qquad\qquad\qquad\qquad\qquad\qquad\qquad\qquad\qquad\qquad\qquad\quad = -17$

13. $(-9 + 6)^2 \div 3 + 5\sqrt{4} = (-3)^2 \div 3 + 5\sqrt{4}$ **15.** $(-12) \cdot 4 \div (-4)^2 - 5 \cdot 3^2 = (-12) \cdot 4 \div 16 - 5 \cdot 9$
$\qquad\qquad\qquad\qquad = 9 \div 3 + 5(2)$ $\qquad\qquad\qquad\qquad\qquad\qquad = -48 \div 16 - 5 \cdot 9$
$\qquad\qquad\qquad\qquad = 3 + 10$ $\qquad\qquad\qquad\qquad\qquad\qquad = -3 - 5 \cdot 9$
$\qquad\qquad\qquad\qquad = 13$ $\qquad\qquad\qquad\qquad\qquad\qquad = -3 - 45$
$\qquad\qquad\qquad\qquad\qquad\qquad\qquad\qquad\qquad\qquad\qquad = -3 + (-45)$
$\qquad\qquad\qquad\qquad\qquad\qquad\qquad\qquad\qquad\qquad\qquad = -48$

17. $16 - [8 - (5 - 7)] = 16 - [8 - (-2)]$ **19.** $10 - 3[(18 \div 3) - 12] = 10 - 3[6 - 12]$
$\qquad\qquad\qquad\quad = 16 - [8 + 2]$ $\qquad\qquad\qquad\qquad\qquad\quad = 10 - 3[-6]$
$\qquad\qquad\qquad\quad = 16 - 10$ $\qquad\qquad\qquad\qquad\qquad\quad = 10 + 18$
$\qquad\qquad\qquad\quad = 6$ $\qquad\qquad\qquad\qquad\qquad\quad = 28$

21. -2 **23.** 17 **25.** 8 **27.** 5 **29.** 5 **31.** 5 **33.** 26.112258 **35.** 54.5398 **37.** 5
39. 0.40501563

Section 1.8 Supplementary Exercises (p. 42)

1. $16 \div 2(8) = 8(8)$ **3.** $5 - (2 + 3^2) = 5 - (2 + 9)$ **5.** $-(5 - 9) - 7 \div (-1) = -(-4) - 7 \div (-1)$
$\qquad\qquad\quad = 64$ $\qquad\qquad\qquad\quad = 5 - (11)$ $\qquad\qquad\qquad\qquad\qquad\quad = -(-4) - (-7)$
$\qquad\qquad\qquad\qquad\qquad\quad = 5 + (-11)$ $\qquad\qquad\qquad\qquad\qquad\quad = 4 + 7$
$\qquad\qquad\qquad\qquad\qquad\quad = -6$ $\qquad\qquad\qquad\qquad\qquad\quad = 11$

7. $8 + 5(3 - 7) = 8 + 5(-4)$ **9.** $-3^2 + 7 = -9 + 7$ **11.** $1 - 2^2 + 5 = 1 - 4 + 5$
$\qquad\qquad\quad = 8 - 20$ $\qquad\qquad\qquad\quad = -2$ $\qquad\qquad\qquad\quad = 1 + (-4) + 5$
$\qquad\qquad\quad = 8 + (-20)$ $\qquad\qquad\qquad\qquad\qquad\qquad = -3 + 5$
$\qquad\qquad\quad = -12$ $\qquad\qquad\qquad\qquad\qquad\qquad = 2$

13. $6 + 5(9 - 10 \div 5) = 6 + 5(9 - 2)$ **15.** $2^2 - 3^2 = 4 - 9$ **17.** $4\sqrt{9} - 12 = 4(3) - 12$
$\qquad\qquad\qquad\qquad = 6 + 5(7)$ $\qquad\qquad\qquad = 4 + (-9)$ $\qquad\qquad\qquad\quad = 12 - 12$
$\qquad\qquad\qquad\qquad = 6 + 35$ $\qquad\qquad\qquad = -5$ $\qquad\qquad\qquad\quad = 0$
$\qquad\qquad\qquad\qquad = 41$

19. $15 - 3^2 \div (-1) + 4 = 15 - 9 \div (-1) + 4$
$\qquad\qquad\qquad\qquad = 15 - (-9) + 4$
$\qquad\qquad\qquad\qquad = 15 + 9 + 4$
$\qquad\qquad\qquad\qquad = 28$

Vocabulary Quiz (p. 42)

1. d   **2.** j   **3.** e   **4.** i   **5.** a   **6.** k   **7.** c   **8.** b   **9.** f   **10.** g   **11.** l   **12.** h

---

Chapter 1 Review Exercises (p. 43)

1. $1, 2, 3, 4, 5, 6$ **2.** $\ldots, -9, -6, -3, 0, 3, 6, 9, \ldots$ **3.** (a) $0, -5, 7$ (b) $-\frac{3}{4}, 0, -5, 7, \frac{1}{2}$

(c) $\sqrt{3}, -\sqrt{2}$ (d) $-\frac{3}{4}, 0, -5, 7, \frac{1}{2}, \sqrt{3}, -\sqrt{2}$ **4.** False   **5.** True   **6.** True   **7.** False

8. True   **9.** True   **10.** -9 **11.** 21 **12.** -20 **13.** -10 **14.** 4 **15.** -12 **16.** -7
17. -11 **18.** 33 **19.** 7 **20.** -72 **21.** 2 **22.** -9 **23.** 48 **24.** 70 **25.** 6 **26.** 12
27. -23 **28.** 51 **29.** 25 **30.** 49 **31.** 31 **32.** 52 **33.** -65 **34.** -204 **35.** -17
36. -31 **37.** 17 **38.** 29 **39.** -5 **40.** -6 **41.** not real   **42.** not real   **43.** 8 **44.** 41
45. -75 **46.** -31 **47.** -4 **48.** 15 **49.** -11 **50.** -10 **51.** -96 **52.** -36 **53.** 10
54. 8 **55.** -2.5814696 **56.** 2.0178253 **57.** 22.926067 **58.** 17.935725 **59.** True. Distributive Property   **60.** True. Commutative Property for Addition   **61.** True. Associative Property for Multiplication   **62.** False.   **63.** True. Commutative Property for Multiplication   **64.** True. Identity Property for Addition   **65.** True. Identity Property for Multiplication   **66.** True. Distributive Property   **67.** False.   **68.** True. Associative Property for Addition

Chapter 1 Test (p. 45)

1. 5   **2.** $-7, 0, 5$ **3.** $-7, \frac{4}{5}, 0, 5, -\frac{2}{3}$ **4.** $-7, \frac{4}{5}, -\pi, 0, 5, -\frac{2}{3}$ **5.** -7 **6.** -13 **7.** -2

8. 6   **9.** 5   **10.** -8 **11.** -12 **12.** -3 **13.** 11   **14.** -4 **15.** undefined   **16.** 0

17. 12   **18.** -40 **19.** undetermined   **20.** 2   **21.** True. Associative Property for Addition

22. True. Identity Property for Multiplication   **23.** False.   **24.** True. Distributive Property   **25.** 27

26. -4 **27.** 25   **28.** -1 **29.** 7   **30.** -6 **31.** not real   **32.** -2 **33.** -2 **34.** 18

35. 24   **36.** -16 **37.** -4 **38.** 6   **39.** 10   **40.** -51

Chapter 2

Section 2.1 Exercises (p. 53)

1. $3j - 7$ **3.** $x + n = 3 + 5$ **5.** $x - n = 3 - 5$ **7.** $-3(x + 2) = -3(3 + 2)$
$$\qquad\qquad = 8 \qquad\qquad\quad = 3 + (-5) \qquad\qquad\qquad = -3(5)$$
$$\qquad\qquad\qquad\qquad\qquad\qquad = -2 \qquad\qquad\qquad\qquad = -15$$

9. $x[d - 5(n + d)] = 3[(-3) - 5(5 + -3)]$ **11.** $2x + dx = 2(3) + (-3)(3)$
$$\qquad\qquad\qquad = 3[(-3) - 5(2)] \qquad\qquad\qquad = 6 + -9$$
$$\qquad\qquad\qquad = 3[(-3) - 10] \qquad\qquad\qquad\; = -3$$
$$\qquad\qquad\qquad = 3[(-3) + (-10)]$$
$$\qquad\qquad\qquad = 3[-13]$$
$$\qquad\qquad\qquad = -39$$

13. $nx(x - d) = 5 \cdot 3[3 - (-3)]$ **15.** $x \cdot x \cdot x = 3 \cdot 3 \cdot 3$ **17.** $y \div n = 0 \div 5$
$$\qquad\qquad = 5 \cdot 3[3 + 3] \qquad\qquad\quad = 9 \cdot 3 \qquad\qquad\qquad = 0$$
$$\qquad\qquad = 5 \cdot 3[6] \qquad\qquad\qquad = 27$$
$$\qquad\qquad = 15[6]$$
$$\qquad\qquad = 90$$

19. $p + 8 = 2 + 8$ **21.** 6   **23.** 3   **25.** -24 **27.** 19   **29.** 5   **31.** 13   **33.** -48 **35.** -11
$$\qquad\quad = 10$$

37. 33   **39.** $25,000 + 2000n$ **41.** $2j + 5$ **43.** 47   **45.** $300c$ **47.** $11k$ **49.** Add 2 to x

51. Subtract y from the product of 3 and x.   **53.** Divide x by y.   **55.** Add 4 to the product of 2 and x and then subtract this result from 6.   **57.** Take the absolute value of the sum of 4 and two times s.

59. Subtract three times y from the product of 100 and x.

Section 2.1 Supplementary Exercises (p. 54)

1. $-x = -(-6)$ **3.** $5x - 2t + r = 5(-6) - 2(-8) + 2$ **5.** $r(x + t) = 2[(-6) + (-8)]$
$$\qquad = 6 \qquad\qquad\qquad = -30 + 16 + 2 \qquad\qquad\qquad = 2[-14]$$
$$\qquad\qquad\qquad\qquad\qquad\qquad = -14 + 2 \qquad\qquad\qquad\qquad = -28$$
$$\qquad\qquad\qquad\qquad\qquad\qquad = -12$$

7. $x \div q = -6 \div 0$ (undefined)   **9.** $-2rst = -2(2)(3)(-8)$ **11.** $4(r + s) = 4(2 + 3)$
$$\qquad\qquad\qquad\qquad\qquad\qquad = -4(3)(-8) \qquad\qquad\qquad = 4(5)$$
$$\qquad\qquad\qquad\qquad\qquad\qquad = (-12)(-8) \qquad\qquad\quad = 20$$
$$\qquad\qquad\qquad\qquad\qquad\qquad = 96$$

13. $sss = 3 \cdot 3 \cdot 3$ **15.** $(t + s) - (t + x) = (-8 + 3) - (-8 + -6)$
$$\qquad = 9 \cdot 3 \qquad\qquad\qquad\qquad\qquad = (-5) - (-14)$$
$$\qquad = 27 \qquad\qquad\qquad\qquad\qquad\; = -5 + 14 = 9$$

17. $|x - s| + 3 = |-6 - 3| + 3$ **19.** $s[2r - (t + 4s)] = 3[2 \cdot 2 - (-8 + 4 \cdot 3)]$
$$\qquad\qquad\quad = |-6 + (-3)| + 3 \qquad\qquad\qquad = 3[2 \cdot 2 - (-8 + 12)]$$
$$\qquad\qquad\quad = |-9| + 3 \qquad\qquad\qquad\qquad = 3[2 \cdot 2 - (4)]$$
$$\qquad\qquad\quad = 9 + 3 \qquad\qquad\qquad\qquad\; = 3[4 - 4]$$
$$\qquad\qquad\quad = 12 \qquad\qquad\qquad\qquad\qquad = 3[0] = 0$$

21. $3s + 1000$ **23.** \$13,000   **25.** $200 + 5d$ **27.** q subtracted from the product of 5 times t.

29. The sum of twice r and 3 subtracted from q.

Section 2.2 Exercises (p. 58)

1. $(t + 4) + 9 = t + 4 + 9$ **3.** $6 + (2x + 5) = 6 + 2x + 5$ **5.** $10 + (3 - 6x) + 5 = 10 + 3 - 6x + 5$
$\qquad\qquad\qquad = t + 13$ $\qquad\qquad\qquad\quad = 2x + 6 + 5$ $\qquad\qquad\qquad\qquad\quad = -6x + 10 + 3 + 5$
$\qquad\qquad\qquad\qquad\qquad\qquad\qquad\quad = 2x + 11$ $\qquad\qquad\qquad\qquad\qquad\qquad\quad = -6x + 18$

7. $7 + (4p - 2) = 7 + 4p - 2$ **9.** $(v - 9) + 27 = v - 9 + 27$
$\qquad\qquad\qquad = 4p + 7 - 2$ $\qquad\qquad\qquad\quad = v + 18$
$\qquad\qquad\qquad = 4p + 5$

11. $-2y - 4(y + 3) = -2y - 4y - 12$ **13.** $3(a + 20) - 2(a + 6) = 3a + 60 - 2a - 12$
$\qquad\qquad\qquad\quad\; = -6y - 12$ $\qquad\qquad\qquad\qquad\qquad\;\; = 3a - 2a + 60 - 12$
$\qquad\qquad\qquad\qquad\qquad\qquad\qquad\qquad\qquad\qquad\; = a + 48$

15. $5a + 4j - 2a - j = 5a - 2a + 4j - j$ **17.** $6(x - 4) + 24 = 6x - 24 + 24$
$\qquad\qquad\qquad\qquad = 3a + 3j$ $\qquad\qquad\qquad\qquad\quad = 6x$

19. $8(12x + 1) + 2(x + 5) = 96x + 8 + 2x + 10$ **21.** $3(2x + y) + 3[x - (y + 2z)] = 6x + 3y + 3[x - y - 2z]$
$\qquad\qquad\qquad\qquad\qquad = 96x + 2x + 8 + 10$ $\qquad\qquad\qquad\qquad\qquad\qquad\qquad\quad = 6x + 3y + 3x - 3y - 6z$
$\qquad\qquad\qquad\qquad\qquad = 98x + 18$ $\qquad\qquad\qquad\qquad\qquad\qquad\qquad\quad = 6x + 3x + 3y - 3y - 6z$
$\qquad\qquad\qquad\qquad\qquad\qquad\qquad\qquad\qquad\qquad\qquad\qquad\quad = 9x - 6z$

23. 30   **25.** $5x - 46$ **27.** $-5x + 16y + 2z$ **29.** (a) $29; 29$ (b) $-7; -7$ **31.** (a) Smaller rectangle has area ab; larger rectangle has area ac.   (b) $a(b + c)$ (c) $ab + ac = a(b + c)$, since the sum of the area of the separate rectangles must be equal to the area of the rectangle formed by adjoining the two rectangles.   **33.** $2x + 14$ **35.** Thirteen 15-cent stamps, eight 20-cent stamps for a total of twenty-one stamps.   **37.** $5x + 4y$ **39.** $15x + 12y$

Section 2.2 Supplementary Exercises (p. 60)

1. $(3t - 2) + (4 - 5t) = 3t - 2 + 4 - 5t$ **3.** $6 - 2(3 + a) = 6 - 6 - 2a$
$\qquad\qquad\qquad\qquad = 3t - 5t - 2 + 4$ $\qquad\qquad\qquad\qquad\;\; = -2a$
$\qquad\qquad\qquad\qquad = -2t + 2$

5. $(2x + y - 5) - (x - y - 3) + (3y - 2x + 1) = 2x + y - 5 - x + y + 3 + 3y - 2x + 1$
$\qquad\qquad\qquad\qquad\qquad\qquad\qquad\qquad\qquad\qquad = 2x - x - 2x + y + y + 3y - 5 + 3 + 1$
$\qquad\qquad\qquad\qquad\qquad\qquad\qquad\qquad\qquad\qquad = -x + 5y - 1$

7. $3[x - 2(x + 1)] = 3[x - 2x - 2]$ **9.** $2(3x + y - 1) - (x - 2y - 1) = 6x + 2y - 2 - x + 2y + 1$
$\qquad\qquad\qquad\;\; = 3[-x - 2]$ $\qquad\qquad\qquad\qquad\qquad\qquad\qquad = 6x - x + 2y + 2y - 2 + 1$
$\qquad\qquad\qquad\;\; = -3x - 6$ $\qquad\qquad\qquad\qquad\qquad\qquad\qquad = 5x + 4y - 1$

11. $3x - [2(x - 1) - 4x] = 3x - [2x - 2 - 4x]$ **13.** $3 - [2 - (3x - 1)] = 3 - [2 - 3x + 1]$
$\qquad\qquad\qquad\qquad\qquad = 3x - [2x - 4x - 2]$ $\qquad\qquad\qquad\qquad\qquad\;\; = 3 - [2 + 1 - 3x]$
$\qquad\qquad\qquad\qquad\qquad = 3x - [-2x - 2]$ $\qquad\qquad\qquad\qquad\qquad\;\; = 3 - [3 - 3x]$
$\qquad\qquad\qquad\qquad\qquad = 3x + 2x + 2$ $\qquad\qquad\qquad\qquad\qquad\;\; = 3 - 3 + 3x$
$\qquad\qquad\qquad\qquad\qquad = 5x + 2$ $\qquad\qquad\qquad\qquad\qquad\;\; = 3x$

15. $-2 - 5(a + 1) + 4a = -2 - 5a - 5 + 4a$ **17.** $(3x + y) + (x - 3y) = 3x + y + x - 3y$
$\qquad\qquad\qquad\qquad\;\; = -5a + 4a - 2 - 5$ $\qquad\qquad\qquad\qquad\qquad\;\; = 3x + x + y - 3y$
$\qquad\qquad\qquad\qquad\;\; = -a - 7$ $\qquad\qquad\qquad\qquad\qquad\;\; = 4x - 2y$

19. $39(3x + 1) + 75(x - 1) = 117x + 39 + 75x - 75$
$\qquad\qquad\qquad\qquad\qquad\quad = 117x + 75x + 39 - 75$
$\qquad\qquad\qquad\qquad\qquad\quad = 192x - 36$

Section 2.3 Exercises (p. 64)

1. $y^4 = 2^4 = 2 \cdot 2 \cdot 2 \cdot 2 = 16$ **3.** $x^4 + y^4 = 1^4 + 2^4 = 1 \cdot 1 \cdot 1 \cdot 1 + 2 \cdot 2 \cdot 2 \cdot 2 = 1 + 16 = 17$
5. $(-3y)^2 = (-3 \cdot 2)^2 = (-6)^2 = (-6)(-6) = 36$ **7.** $(xyz)^2 = (1 \cdot 2 \cdot 3)^2 = 6^2 = 36$
9. $2(3y - 4z)^3 = 2(3 \cdot 2 - 4 \cdot 3)^3 = 2(6 - 12)^3$
$\qquad\qquad\qquad\qquad\qquad\qquad = 2(-6)^3$
$\qquad\qquad\qquad\qquad\qquad\qquad = 2 \cdot (-216)$
$\qquad\qquad\qquad\qquad\qquad\qquad = -432$

11. $a^2 + b^2 = (-2)^2 + (-4)^2 = (-2)(-2) + (-4)(-4) = 4 + 16 = 20$

13. $(a+b)^2 = (-2+-4)^2 = (-6)^2 = (-6)(-6) = 36$ **15.** $5a^2 - 3b^2 = 5(-2)^2 - 3(-4)^2 = 5 \cdot 4 - 3 \cdot 16$
$$= 20 - 48$$
$$= -28$$

17. $|a^3 + c^2| = |(-2)^3 + 5^2| = |-8 + 25| = |17| = 17$

19. $|a^3| - c^2 = |(-2)^3| - 5^2 = |-8| - 25 = 8 - 25 = -17$ **21.** $6x^5$ **23.** $40x^2y^3$ **25.** $x^3y^4z^5$

27. $11a^3b^3c$ **29.** $10x^{16}$ **31.** $(3x-2y)^6$ **33.** $x^a + x^b$ **35.** $6a^5b^8c$ **37.** 3^{12}

39. $-447.54884x^7$

Section 2.3 Supplementary Exercises (p. 65)

1. $t^2 = (-1)^2 = (-1) \cdot (-1) = 1$ **3.** $-r^3 = -(-2)^3 = -(-2)(-2)(-2) = -(-8) = 8$

5. $s^3 \div r^2 = 2^3 \div (-2)^2 = 8 \div 4 = 2$

7. $(2bc)^2 = (2bc)(2bc) = 4b^2c^2 = 4(2)^2(-2)^2 = 4 \cdot 4 \cdot 4 = 64$

9. $a^2 - 2b^2 = (-1)^2 - 2 \cdot 2^2 = 1 - 2 \cdot 4 = 1 - 8 = -7$

11. $(-3a^2)(5a^7) = -3 \cdot 5a^2a^7 = -15a^{2+7} = -15a^9$

13. $(2x^3)(x^5)^2 = 2x^3 x^{5 \cdot 2} = 2x^3 x^{10} = 2x^{3+10} = 2x^{13}$

15. $(-2x^7y)(5x^3y^2)(3y^2x) = -2 \cdot 5 \cdot 3x^7 \cdot x^3 \cdot x \cdot y \cdot y^2 \cdot y^2$
$$= -30x^{7+3+1}y^{1+2+2}$$
$$= -30x^{11}y^5$$

17. $x^a x^b x^c = x^{a+b+c}$ **19.** $x^5 x^3 x^y = x^{5+3+y} = x^{8+y}$

Section 2.4 Exercises (p. 67)

1. $3x(x^2 - 2) = 3x \cdot x^2 - 3x \cdot 2 = 3x^3 - 6x$ **3.** $5z^2(3z - 6) = 5z^2 \cdot 3z - 5z^2 \cdot 6 = 15z^3 - 30z^2$

5. $-3t^2(t^2 - 4t + 1) = -3t^2 \cdot t^2 - 3t^2(-4t) - 3t^2 \cdot 1$
$$= -3t^4 + 12t^3 - 3t^2$$

7. $30b^3(b^2 - 3b + 1) = 30b^3 \cdot b^2 - 30b^3 \cdot 3b + 30b^3 \cdot 1$
$$= 30b^5 - 90b^4 + 30b^3$$

9. $6a(a^2 - 2) + 7(a - 2) = 6a \cdot a^2 - 6a \cdot 2 + 7 \cdot a - 7 \cdot 2$
$$= 6a^3 - 12a + 7a - 14$$
$$= 6a^3 - 5a - 14$$

11. $3x(x^2 - x) + 5(x - 4) = 3x \cdot x^2 - 3x \cdot x + 5 \cdot x - 5 \cdot 4$
$$= 3x^3 - 3x^2 + 5x - 20$$

13. $7x(x^2 - 5x + 1) + 3x(x^2 - 2) = 7x^3 - 35x^2 + 7x + 3x^3 - 6x$
$$= 7x^3 + 3x^3 - 35x^2 + 7x - 6x$$
$$= 10x^3 - 35x^2 + x$$

15. $2a(a^2b - b) + 4b(a^3 + 9a) = 2a^3b - 2ab + 4a^3b + 36ab$
$$= 2a^3b + 4a^3b - 2ab + 36ab$$
$$= 6a^3b + 34ab$$

17. $3c^2(2c - 5) - 4c(c^2 + 7c) = 6c^3 - 15c^2 - 4c^3 - 28c^2$
$$= 6c^3 - 4c^3 - 15c^2 - 28c^2$$
$$= 2c^3 - 43c^2$$

19. $4x - x(2x - 1) + 10 = 4x - 2x^2 + x + 10$ **21.** $-5x^2 + 29x$ **23.** $107x^4 - 93x^3 - 30x^2$
$$= -2x^2 + 4x + x + 10$$
$$= -2x^2 + 5x + 10$$

25. $10x_1x_2 - 15x_1 - 12x_2$ **27.** $15T_F - 27T_FT_C - 63T_C$ **29.** $6x_1^2 + 2x_1$ **31.** $x^2 + 2xy + y^2$

33. $5ab + 5b^2$ **35.** $12x^2 + 11xy - 6x - 5y^2 - 5y$

Section 2.4 Supplementary Exercises (p. 67)

1. $15y(2 - 3y^2) = 15y \cdot 2 - 15y \cdot 3y^2$
$$= 30y - 45y^3$$
$$= -45y^3 + 30y$$

3. $-4x^3(5 - x + 2x^2 - x^3) = -4x^3 \cdot 5 - 4x^3 \cdot (-x) - 4x^3(2x^2) - 4x^3(-x^3)$
$$= -20x^3 + 4x^4 - 8x^5 + 4x^6$$
$$= 4x^6 - 8x^5 + 4x^4 - 20x^3$$

5. $5 - 2x - (3x - 2) = 5 - 2x - 3x - (-2)$
$$= 5 - 2x - 3x + 2$$
$$= -2x - 3x + 5 + 2$$
$$= -5x + 7$$

7. $15a - (b - a) + (4b - 7a) = 15a - b + a + 4b - 7a$
$$= 15a + a - 7a - b + 4b$$
$$= 9a + 3b$$

9. $-5(x^2)^4 + 2x^4(3 - x^4) = -5x^8 + 6x^4 - 2x^8$
$$= -5x^8 - 2x^8 + 6x^4$$
$$= -7x^8 + 6x^4$$

11. $2t \left\{ 4 - [(4t - 3) - (t - 8)] \right\} - 6t = 2t \left\{ 4 - [4t - 3 - t + 8] \right\} - 6t$
$$= 2t \left\{ 4 - [3t + 5] \right\} - 6t$$
$$= 2t \left\{ 4 - 3t - 5 \right\} - 6t$$
$$= 2t \left\{ -3t - 1 \right\} - 6t$$
$$= -6t^2 - 2t - 6t$$
$$= -6t^2 - 8t$$

13. $6 - 4(2x - 8) = 6 - 8x + 32$
$$= -8x + 6 + 32$$
$$= -8x + 38$$

15. $(9x - 3y + 7)2xy^2 = 9x \cdot 2xy^2 - 3y \cdot 2xy^2 + 7 \cdot 2xy^2$
$$= 18x^2y^2 - 6xy^3 + 14xy^2$$

17. $5(a - 2b) - 2(a + b) = 5a - 10b - 2a - 2b$
$$= 5a - 2a - 10b - 2b$$
$$= 3a - 12b$$
When $a = 2$ and $b = -1$, $3a - 12b = 3 \cdot 2 - 12(-1) = 6 + 12 = 18$.

19. $(a^2 - b^2) - 3(a^2 + b^2) = a^2 - b^2 - 3a^2 - 3b^2 = a^2 - 3a^2 - b^2 - 3b^2$
$$= -2a^2 - 4b^2$$
When $a = 2$ and $b = -1$, $-2a^2 - 4b^2 = -2 \cdot 2^2 - 4 \cdot (-1)^2$
$$= -2 \cdot 4 - 4 \cdot 1$$
$$= -8 - 4$$
$$= -12$$

Section 2.5 Exercises (p. 71)

1. (a) $P = 2l + 2w$
$$= 2 \cdot 2 \text{ in.} + 2 \cdot 4 \text{ in.}$$
$$= 4 \text{ in.} + 8 \text{ in.}$$
$$= 12 \text{ in.}$$
   (b) $A = l \cdot w$
$$= (2 \text{ in.})(4 \text{ in.})$$
$$= 8 \text{ in.}^2$$

3. (a) $P = 2l + 2w$
$$= 2 \cdot 5 \text{ ft.} + 2 \cdot 4 \text{ ft.}$$
$$= 10 \text{ ft.} + 8 \text{ ft.}$$
$$= 18 \text{ ft.}$$
   (b) $A = l \cdot w$
$$= (5 \text{ ft.})(4 \text{ ft.})$$
$$= 20 \text{ ft.}^2$$

5. (a) $P = 2l + 2w$
$$= 2 \cdot (120 \text{ cm}) + 2(200 \text{ cm}) \quad \text{Note: } 2m = 200 \text{ cm}$$
$$= 240 \text{ cm} + 400 \text{ cm}$$
$$= 640 \text{ cm}$$
   (b) $A = l \cdot w$
$$= (120 \text{ cm})(200 \text{ cm})$$
$$= 24,000 \text{ cm}^2$$

7. (a) $C = 2\pi r$
$$\approx 2(3.14)(4 \text{ in.})$$
$$\approx 25.12 \text{ in.}$$
   (b) $A = \pi r^2$
$$\approx (3.14)(4 \text{ in.})^2$$
$$\approx (3.14)(16 \text{ in.}^2)$$
$$\approx 50.24 \text{ in.}^2$$

9. (a) $C = 2\pi r$
$\approx 2(3.14)(7 \text{ ft.})$
$\approx 43.96 \text{ ft.}$

(b) $A = \pi r^2$
$\approx (3.14)(7 \text{ ft.})^2$
$\approx (3.14)(49 \text{ ft}^2)$
$\approx 153.86 \text{ ft}^2$

11. (a) $C = 2\pi r$
$\approx 2(3.14)(10 \text{ km})$
$\approx 62.8 \text{ km}$

(b) $A = \pi r^2$
$\approx (3.14)(10 \text{ km})^2$
$\approx (3.14)(100 \text{ km}^2)$
$\approx 314 \text{ km}^2$

13. $A = \frac{1}{2} bh$
$= \frac{1}{2} (4 \text{ in.})(5 \text{ in.})$
$= 10 \text{ in.}^2$

15. $A = \frac{1}{2} bh$
$= \frac{1}{2} (3m)(5m)$
$= 7.5 \ m^2$

17. (a) 1 foot = 12 inches $\quad A = \frac{1}{2} bh$
$= \frac{1}{2} (18 \text{ in.})(12 \text{ in.})$
$= 108 \text{ in.}^2$

(b) 18 inches = 1.5 ft. $\quad A = \frac{1}{2} bh$
$= \frac{1}{2} (1.5 \text{ ft.})(1 \text{ ft.})$
$= 0.75 \text{ ft.}^2$

19. $V = l \cdot w \cdot h$
$= (10 \text{ cm})(8 \text{ cm})(4 \text{ cm})$
$= 320 \text{ cm}^3$

21. 15 ft.^3 or $25,920 \text{ in.}^3$

23. $V = \pi r^2 h$
$\approx (3.14)(4 \text{ ft.})^2 (3 \text{ ft.})$
$\approx (3.14)(16 \text{ ft.}^2)(3 \text{ ft.})$
$\approx 150.72 \text{ ft.}^3$

25. 15.7 cm^3

27. $V = \frac{1}{3} \pi r^2 h$
$\approx \frac{1}{3} (3.14)(5 \text{ in.})^2 (4 \text{ in.})$
$\approx \frac{1}{3} (3.14)(25 \text{ in.}^2)(4 \text{ in.})$
$\approx 104.67 \text{ in.}^3$

29. 1205.76 in.^3 or 0.6977 ft.^3

31. $A = bh$
$= (4 \text{ in.})(5 \text{ in.})$
$= 20 \text{ in.}^2$

$P = 2a + 2b$
$= 2(2 \text{ in.}) + 2(4 \text{ in.})$
$= 4 \text{ in.} + 8 \text{ in.}$
$= 12 \text{ in.}$

33. $A = 0.5 \text{ ft.}^2$ or 72 in.^2
$P = 2\frac{2}{3} \text{ ft.}$ or 32 in.

35. 86.590149 in.^2

37. $10,582.877 \text{ in.}^3$ 39. Depends on your choice of object   41. Depends on your choice of object

43. 108 ft.^2 45. 50.24 ft. 47. 60 ft.^3 49. 24 ft.^2

Section 2.5 Supplementary Exercises (p. 73)

1. (a) $P = 2l + 2w$
$= 2(4 \text{ yd.}) + 2(5 \text{ yd.})$
$= 8 \text{ yd.} + 10 \text{ yd.}$
$= 18 \text{ yd.}$

(b) $A = lw$
$= (4 \text{ yd.})(5 \text{ yd.})$
$= 20 \text{ yd.}^2$

3. (a) $P = 2l + 2w$
$= 2(4 \text{ ft.}) + 2(1.5 \text{ ft.})$
$= 8 \text{ ft.} + 3 \text{ ft.}$
$= 11 \text{ ft.}$

(b) $A = l \cdot w$
$= (4 \text{ ft.})(1.5 \text{ ft.})$
$= 6 \text{ ft.}^2$

5. (a) $C = 2\pi r$
$\approx 2(3.14)(15 \text{ yd.})$
$\approx 94.2 \text{ yd.}$

(b) $A = \pi r^2$
$\approx (3.14)(15 \text{ yd.})^2$
$\approx (3.14)(225 \text{ yd.}^2)$
$\approx 706.5 \text{ yd.}^2$

7. (a) $C = 2\pi r$
$\approx 2(3.14)(9 \text{ km})$
$\approx 56.52 \text{ km}$

(b) $A = \pi r^2$
$\approx (3.14)(9 \text{ km})^2$
$\approx (3.14)(81 \text{ km}^2)$
$\approx 254.34 \text{ km}^2$

9. $A = \frac{1}{2} bh$

$= \frac{1}{2} (9 \text{ ft.})(8 \text{ ft.})$

$= 36 \text{ ft.}^2$

11. $A = \frac{1}{2} bh$

$= \frac{1}{2} (9 \text{ ft.})(6 \text{ ft.})$

$= 27 \text{ ft.}^2$

13. $V = l \cdot w \cdot h$

$= (4 \text{ ft.})(3 \text{ ft.})(2 \text{ ft.})$

$= 24 \text{ ft.}^3$

15. $V = l \cdot w \cdot h$

$= (2 \text{ ft.})(3 \text{ ft.})(3 \text{ ft.})$

$= 18 \text{ ft.}^3$

17. $V = \pi r^2 h$

$\approx (3.14)(4 \text{ ft.})^2 (7 \text{ ft.})$

$\approx (3.14)(16 \text{ ft.}^2)(7 \text{ ft.})$

$\approx 351.68 \text{ ft.}^3$

19. $V = \pi r^2 h$

$\approx (3.14)(5 \text{ ft.})^2 (40 \text{ ft.})$

$\approx (3.14)(25 \text{ ft.}^2)(40 \text{ ft.})$

$\approx 3140 \text{ ft.}^3$

21. 150.72 ft.^3 23. 904.32 ft.^3 25. 50.24 ft.^2

Vocabulary Quiz (p. 74)

1. b   2. c   3. a   4. g   5. f   6. e   7. d

Chapter 2 Review Exercises (p. 74)

1. 10   2. 7   3. 16   4. 1   5. 72   6. -87 7. -92 8. -4 9. -144 10. 0

11. $-\frac{21}{5}$ 12. $2k - 23$ 13. $2x + 3$ 14. $7x + 9$ 15. $-5y - 12$ 16. $-5y + 12$ 17. $5t - 13$

18. 15   19. $-2a + 27b$ 20. $33x - 30$ 21. 13   22. 5   23. 1   24. 47   25. 9   26. 29

27. 9   28. 27   29. $20x^2 y^3$ 30. $35a^3 b^3$ 31. $-6x^3 y^5 z$ 32. $25x^6$ 33. $3x^3 + 12x^2$

34. $-2a^4 + 10a^3 - 14a^2$ 35. $84c^5 - 14c^3$ 36. $4y^3 - 3y^2 + 23y$ 37. $z^3 - 2z^2 + 35z$

38. $-12z^3 - 3z^2 + 66z - 25$ 39. $P = 70 \text{ cm}, A = 250 \text{ cm}^2$ 40. $C = 62.8 \text{ in.}, A = 314 \text{ in.}^2$

Cumulative Review Exercises (p. 75)

1. $\dots, -3, -2, -1, 0, 1, 2, 3, \dots$ 2. -7 3. -16 4. 60   5. Distributive Property

6. undefined   7. -81 8. -2 9. 0   10. -1

Chapter 2 Test (p. 76)

1. 10   2. -5 3. -11 4. 576   5. 36   6. 5   7. $5x - 6$ 8. $11x + 16$ 9. $-27y + 17$

10. $12x^3 y^4 z$ 11. $-2x^3 - 5x^2$ 12. $10a^3 - 5a^2 - 2a$ 13. 18 m

14. $C = 18.84 \text{ ft.}, A = 28.26 \text{ ft.}^2$ 15. 4000 cm^3

Chapter 3

Section 3.1 Exercises (p. 83)

1. $\begin{aligned} x - 1 &= 2 \\ +1 &= +1 \\ \hline x &= 3 \end{aligned}$ 3. $\begin{aligned} x - 7 &= 9 \\ +7 &= +7 \\ \hline x &= 16 \end{aligned}$ 5. $\begin{aligned} x - 9 &= 7 \\ +9 &= +9 \\ \hline x &= 16 \end{aligned}$ 7. $\begin{aligned} y - 1 &= 0 \\ +1 &= +1 \\ \hline y &= 1 \end{aligned}$ 9. $\begin{aligned} m - 9 &= 99 \\ +9 &= +9 \\ \hline m &= 108 \end{aligned}$

11. $\begin{aligned} x - 15 &= 10 \\ +15 &= +15 \\ \hline x &= 25 \end{aligned}$ 13. $\begin{aligned} y - 7 &= 7 \\ +7 &= +7 \\ \hline y &= 14 \end{aligned}$ 15. $\begin{aligned} m - 4 &= -5 \\ +4 &= +4 \\ \hline m &= -1 \end{aligned}$ 17. $\begin{aligned} x + 3 &= 8 \\ -3 &= -3 \\ \hline x &= -5 \end{aligned}$ 19. $\begin{aligned} a + 9 &= 1 \\ -9 &= -9 \\ \hline a &= -8 \end{aligned}$

21. 8   23. 50   25. -49 27. -12 29. 12   31. 3   33. 10   35. 8   37. -8 39. 8

41. 137   43. 1   45. -4 47. 3.9999   49. -16.1694 51. 8.3691

Section 3.1 Supplementary Exercises (p. 84)

1. $\begin{aligned} m - 9 &= 7 \\ +9 &= +9 \\ \hline m &= 16 \end{aligned}$ 3. $\begin{aligned} w - 6 &= -9 \\ +6 &= +6 \\ \hline w &= -3 \end{aligned}$ 5. $\begin{aligned} p + 3 &= -9 \\ -3 &= -3 \\ \hline p &= -12 \end{aligned}$ 7. $\begin{aligned} m - 15 &= 15 \\ +15 &= +15 \\ \hline m &= 30 \end{aligned}$ 9. $\begin{aligned} t - 17 &= -63 \\ +17 &= +17 \\ \hline t &= -46 \end{aligned}$

11. $\begin{aligned} 5 + x &= 8 \\ -5 &= -5 \\ \hline x &= 3 \end{aligned}$ **13.** $\begin{aligned} -8 + y &= -8 \\ +8 &= +8 \\ \hline y &= 0 \end{aligned}$ **15.** $\begin{aligned} 12 + x &= 49 \\ -12 &= -12 \\ \hline x &= 37 \end{aligned}$ **17.** $\begin{aligned} 10 &= t - 14 \\ +14 &= +14 \\ \hline 24 &= t \end{aligned}$ **19.** $\begin{aligned} -19 &= x - 7 \\ +7 &= +7 \\ \hline -12 &= x \end{aligned}$

Section 3.2 Exercises (p. 87)

1. $\begin{aligned} 5x &= 30 \\ \frac{5x}{5} &= \frac{30}{5} \\ x &= 6 \end{aligned}$ **3.** $\begin{aligned} -6x &= 48 \\ \frac{-6x}{-6} &= \frac{48}{-6} \\ x &= -8 \end{aligned}$ **5.** $\begin{aligned} \frac{x}{3} &= 7 \\ 3\frac{x}{3} &= 3 \cdot 7 \\ x &= 21 \end{aligned}$ **7.** $\begin{aligned} \frac{x}{-4} &= 7 \\ (-4)\frac{x}{-4} &= (-4) \cdot 7 \\ x &= -28 \end{aligned}$ **9.** $\begin{aligned} 2t &= 28 \\ \frac{2t}{2} &= \frac{28}{2} \\ t &= 14 \end{aligned}$

11. $\begin{aligned} -2t &= 28 \\ \frac{-2t}{-2} &= \frac{28}{-2} \\ t &= -14 \end{aligned}$ **13.** $\begin{aligned} \frac{x}{5} &= -4 \\ 5 \cdot \frac{x}{5} &= 5(-4) \\ x &= -20 \end{aligned}$ **15.** $\begin{aligned} \frac{x}{-3} &= -8 \\ -3 \cdot \frac{x}{-3} &= -3(-8) \\ x &= 24 \end{aligned}$ **17.** $\begin{aligned} 8y &= 0 \\ \frac{8y}{8} &= \frac{0}{8} \\ y &= 0 \end{aligned}$ **19.** $\begin{aligned} -5x &= -35 \\ \frac{-5x}{-5} &= \frac{-35}{-5} \\ x &= 7 \end{aligned}$

21. -300 **23.** 250 **25.** -6.3691845 **27.** 27.980368

Section 3.2 Supplementary Exercises (p. 88)

1. $\begin{aligned} 4x &= 12 \\ \frac{4x}{4} &= \frac{12}{4} \\ x &= 3 \end{aligned}$ **3.** $\begin{aligned} -t &= 9 \\ \frac{-t}{-1} &= \frac{9}{-1} \\ t &= -9 \end{aligned}$ **5.** $\begin{aligned} 100 &= -25t \\ \frac{100}{-25} &= \frac{-25t}{-25} \\ -4 &= t \end{aligned}$ **7.** $\begin{aligned} -121 &= -11t \\ \frac{-121}{-11} &= \frac{-11t}{-11} \\ 11 &= t \end{aligned}$ **9.** $\begin{aligned} \frac{x}{8} &= 11 \\ 8\frac{x}{8} &= 8 \cdot 11 \\ x &= 88 \end{aligned}$

11. $\begin{aligned} 21 &= \frac{x}{5} \\ 5 \cdot 21 &= 5 \cdot \frac{x}{5} \\ 105 &= x \end{aligned}$ **13.** $\begin{aligned} 6 &= \frac{-t}{5} \\ 5 \cdot 6 &= 5 \cdot \frac{-t}{5} \\ 30 &= -t \\ \frac{30}{-1} &= \frac{-t}{-1} \\ -30 &= t \end{aligned}$ **15.** $\begin{aligned} \frac{5m}{2} &= 10 \\ 2 \cdot \frac{5m}{2} &= 2 \cdot 10 \\ 5m &= 20 \\ \frac{5m}{5} &= \frac{20}{5} \\ m &= 4 \end{aligned}$ **17.** $\begin{aligned} \frac{4x}{3} &= -12 \\ 3 \cdot \frac{4x}{3} &= 3(-12) \\ 4x &= -36 \\ \frac{4x}{4} &= \frac{-36}{4} \\ x &= -9 \end{aligned}$

19. $\begin{aligned} \frac{-x}{-7} &= 8 \\ -7 \cdot \frac{-x}{-7} &= -7 \cdot 8 \\ -x &= -56 \\ \frac{-x}{-1} &= \frac{-56}{-1} \\ x &= 56 \end{aligned}$

Section 3.3 Exercises (p. 93)

1. $\begin{aligned} 12 - x &= 19 \\ -12 &= -12 \\ \hline -x &= 7 \\ \frac{-x}{-1} &= \frac{7}{-1} \\ x &= -7 \end{aligned}$ Check: $\begin{aligned} 12 - x &= 19 \\ 12 - (-7) &= 19 \\ 12 + 7 &= 19 \\ 19 &= 19 \checkmark \end{aligned}$ **3.** $\begin{aligned} 5 + 2x &= 11 \\ -5 &= -5 \\ \hline 2x &= 6 \\ \frac{2x}{2} &= \frac{6}{2} \\ x &= 3 \end{aligned}$ Check: $\begin{aligned} 5 + 2x &= 11 \\ 5 + 2 \cdot 3 &= 11 \\ 5 + 6 &= 11 \\ 11 &= 11 \checkmark \end{aligned}$

5. $\dfrac{3z}{8} = -9$

$8 \cdot \dfrac{3z}{8} = 8 \cdot (-9)$

$3z = -72$

$\dfrac{3z}{3} = \dfrac{-72}{3}$

$z = -24$

Check: $\dfrac{3z}{8} = -9$

$\dfrac{3(-24)}{8} = -9$

$\dfrac{-72}{8} = -9$

$-9 = -9\checkmark$

7. $\dfrac{3x}{2} = 15$

$2 \cdot \dfrac{3x}{2} = 2 \cdot 15$

$3x = 30$

$\dfrac{3x}{3} = \dfrac{30}{3}$

$x = 10$

Check: $\dfrac{3x}{2} = 15$

$\dfrac{3 \cdot 10}{2} = 15$

$\dfrac{30}{2} = 15$

$15 = 15\checkmark$

9. $\begin{aligned} 3z + 1 &= 22 \\ -1 &= -1 \\ \hline 3z &= 21 \end{aligned}$

$\dfrac{3z}{3} = \dfrac{21}{3}$

$z = 7$

Check: $3z + 1 = 22$

$3 \cdot 7 + 1 = 22$

$21 + 1 = 22$

$22 = 22\checkmark$

11. $\begin{aligned} 7 - 3y &= -11 \\ -7 &= -7 \\ \hline -3y &= -18 \end{aligned}$

$\dfrac{-3y}{-3} = \dfrac{-18}{-3}$

$y = 6$

Check: $7 - 3y = -11$

$7 - 3 \cdot 6 = -11$

$7 - 18 = -11$

$-11 = -11\checkmark$

13. $\begin{aligned} 7 &= \dfrac{3x}{4} - 2 \\ +2 &= \quad +2 \\ \hline 9 &= \dfrac{3x}{4} \end{aligned}$

$4 \cdot 9 = 4 \cdot \dfrac{3x}{4}$

$36 = 3x$

$\dfrac{36}{3} = \dfrac{3x}{3}$

$12 = x$

Check: $7 = \dfrac{3x}{4} - 2$

$7 = \dfrac{3(12)}{4} - 2$

$7 = \dfrac{36}{4} - 2$

$7 = 9 - 2$

$7 = 7\checkmark$

15. $\begin{aligned} \dfrac{5x}{3} + 2 &= -8 \\ -2 &= -2 \\ \hline \dfrac{5x}{3} &= -10 \end{aligned}$

$3 \cdot \dfrac{5x}{3} = 3(-10)$

$5x = -30$

$\dfrac{5x}{5} = \dfrac{-30}{5}$

$x = -6$

Check: $\dfrac{5x}{3} + 2 = -8$

$\dfrac{5(-6)}{3} + 2 = -8$

$\dfrac{-30}{3} + 2 = -8$

$-10 + 2 = -8$

$-8 = -8\checkmark$

17. $\begin{aligned} 17 &= 3x - 19 \\ +19 &= \quad +19 \\ \hline 36 &= 3x \end{aligned}$

$\dfrac{36}{3} = \dfrac{3x}{3}$

$12 = x$

Check: $17 = 3x - 19$

$17 = 3 \cdot 12 - 19$

$17 = 36 - 19$

$17 = 17\checkmark$

19. $\begin{aligned} 5t + 18 &= -32 \\ -18 &= -18 \\ \hline 5t &= -50 \end{aligned}$

$\dfrac{5t}{5} = \dfrac{-50}{5}$

$t = -10$

Check: $5t + 18 = -32$

$5(-10) + 18 = -32$

$-50 + 18 = -32$

$-32 = -32\checkmark$

21. -6　　23. -4　　25. 5　　27. -1　　29. -19　　31. 2　　33. 1　　35. -3　　37. $-\dfrac{7}{3}$

39. $\dfrac{1}{2}$

Section 3.3 Supplementary Exercises (p. 93)

1. $2x + 5 = 19$

$ \underline{- 5 = -5}$

$2x = 14$

$x = 7$

Check: $2x + 5 = 19$

$2 \cdot 7 + 5 = 19$

$14 + 5 = 19$

$19 = 19 \checkmark$

3. $\dfrac{2x}{3} = 12$

$3 \cdot \dfrac{2x}{3} = 3 \cdot 12$

$2x = 36$

$\dfrac{2x}{2} = \dfrac{36}{2}$

$x = 18$

Check: $\dfrac{2x}{3} = 12$

$\dfrac{2 \cdot 18}{3} = 12$

$\dfrac{36}{3} = 12$

$12 = 12 \checkmark$

5. $\dfrac{4x}{5} = 16$

$5 \cdot \dfrac{4x}{5} = 5 \cdot 16$

$4x = 80$

$\dfrac{4x}{4} = \dfrac{80}{4}$

$x = 20$

Check: $\dfrac{4x}{5} = 16$

$\dfrac{4 \cdot 20}{5} = 16$

$\dfrac{80}{5} = 16$

$16 = 16 \checkmark$

7. $2 - 4x = 18$

$\underline{-2 = -2}$

$-4x = 16$

$\dfrac{-4x}{-4} = \dfrac{16}{-4}$

$x = -4$

Check: $2 - 4x = 18$

$2 - 4(-4) = 18$

$2 + 16 = 18$

$18 = 18 \checkmark$

9. $\dfrac{2x}{3} - 5 = 1$

$\underline{\phantom{\dfrac{2x}{3}} 5 = 5}$

$\dfrac{2x}{3} = 6$

$3 \cdot \dfrac{2x}{3} = 3 \cdot 6$

$2x = 18$

$\dfrac{2x}{2} = \dfrac{18}{2}$

$x = 9$

Check: $\dfrac{2x}{3} - 5 = 1$

$\dfrac{2 \cdot 9}{3} - 5 = 1$

$\dfrac{18}{3} - 5 = 1$

$6 - 5 = 1$

$1 = 1 \checkmark$

11. $2x + 1 = x - 7$

$\underline{-x = -x}$

$x + 1 = -7$

$\underline{-1 = -1}$

$x = -8$

Check: $2x + 1 = x - 7$

$2(-8) + 1 = -8 - 7$

$-16 + 1 = -8 + (-7)$

$-15 = -15 \checkmark$

13. $3x - 9 = x + 7$

$\underline{-x = -x}$

$2x - 9 = 7$

$\underline{+9 = +9}$

$2x = 16$

$\dfrac{2x}{2} = \dfrac{16}{2}$

$x = 8$

Check: $3x - 9 = x + 7$

$3 \cdot 8 - 9 = 8 + 7$

$24 - 9 = 15$

$15 = 15 \checkmark$

15. $5y - 2 = 4y + 11$

$\underline{-4y = -4y}$

$y - 2 = 11$

$\underline{+2 = +2}$

$y = 13$

Check: $5y - 2 = 4y + 11$

$5 \cdot 13 - 2 = 4 \cdot 13 + 11$

$65 - 2 = 52 + 11$

$63 = 63 \checkmark$

17.
$$3x - 10 = x + 7$$
$$\underline{-x \qquad\quad = -x}$$
$$2x - 10 = \qquad 7$$
$$\underline{\quad + 10 = \quad + 10}$$
$$2x \qquad = \qquad 17$$
$$\frac{2x}{2} = \frac{17}{2}$$
$$x = \frac{17}{2} \text{ or } 8.5$$

Check: $3x - 10 = x + 7$
$$3(8.5) - 10 = 8.5 + 7$$
$$25.5 - 10 = 15.5$$
$$15.5 = 15.5 \checkmark$$

19.
$$6x + 9 = 3 - 3x$$
$$\underline{+3x \qquad = \quad + 3x}$$
$$9x + 9 = \quad 3$$
$$\underline{\quad - 9 = -9}$$
$$9x \qquad = -6$$
$$\frac{9x}{9} = \frac{-6}{9}$$
$$x = -\frac{2}{3}$$

Check: $6x + 9 = 3 - 3x$
$$6\left(-\frac{2}{3}\right) + 9 = 3 - 3\left(-\frac{2}{3}\right)$$
$$-4 + 9 = 3 + 2$$
$$5 = 5 \checkmark$$

Section 3.4 Exercises (p. 97)

1.
$$3(4x + 1) = 6x + 9$$
$$12x + 3 = 6x + 9$$
$$\underline{-6x \qquad = -6x}$$
$$6x + 3 = \quad 9$$
$$\underline{\quad - 3 = -3}$$
$$6x \qquad = \quad 6$$
$$\frac{6x}{6} = \frac{6}{6}$$
$$x = 1$$

Check: $3(4x + 1) = 6x + 9$
$$3(4 \cdot 1 + 1) = 6 \cdot 1 + 9$$
$$3(4 + 1) = 6 + 9$$
$$3(5) = 15$$
$$15 = 15 \checkmark$$

3.
$$-2(3 - t) = t - 7$$
$$-6 + 2t = t - 7$$
$$\underline{\quad -t = -t}$$
$$-6 + t = -7$$
$$6 \qquad = \quad 6$$
$$t = -1$$

Check: $-2(3 - t) = t - 7$
$$-2[3 - (-1)] = -1 - 7$$
$$-2[3 + 1] = -1 + (-7)$$
$$-2[4] = -8$$
$$-8 = -8 \checkmark$$

5.
$$3(2x + 8) = 30$$
$$6x + 24 = 30$$
$$\underline{\quad -24 = -24}$$
$$6x \qquad = \quad 6$$
$$\frac{6x}{6} = \frac{6}{6}$$
$$x = 1$$

Check: $3(2x + 8) = 30$
$$3(2 \cdot 1 + 8) = 30$$
$$3(2 + 8) = 30$$
$$3(10) = 30$$
$$30 = 30 \checkmark$$

7.
$$-(x - 3) = 10$$
$$-x + 3 = 10$$
$$\underline{\quad - 3 = -3}$$
$$-x \qquad = \quad 7$$
$$\frac{-x}{-1} = \frac{7}{-1}$$
$$x = -7$$

Check: $-(x - 3) = 10$
$$-(-7 - 3) = 10$$
$$-(-10) = 10$$
$$10 = 10 \checkmark$$

9.
$$
\begin{aligned}
3(x + 2) - x &= 4 \\
3x + 6 - x &= 4 \\
3x - x + 6 &= 4 \\
2x + 6 &= 4 \\
\underline{-6 = -6} & \\
2x &= -2 \\
\frac{2x}{2} &= \frac{-2}{2} \\
x &= -1
\end{aligned}
$$

Check:
$$
\begin{aligned}
3(x + 2) - x &= 4 \\
3(-1 + 2) - (-1) &= 4 \\
3(1) - (-1) &= 4 \\
3 + 1 &= 4 \\
4 &= 4 \checkmark
\end{aligned}
$$

11.
$$
\begin{aligned}
5 - 2(x + 4) &= 7 \\
5 - 2x - 8 &= 7 \\
5 - 8 - 2x &= 7 \\
-3 - 2x &= 7 \\
\underline{+3 \quad\; = +3} & \\
-2x &= 10 \\
\frac{-2x}{-2} &= \frac{10}{-2} \\
x &= -5
\end{aligned}
$$

Check:
$$
\begin{aligned}
5 - 2(x + 4) &= 7 \\
5 - 2(-5 + 4) &= 7 \\
5 - 2(-1) &= 7 \\
7 &= 7 \checkmark
\end{aligned}
$$

13.
$$
\begin{aligned}
3(x + 1) - 4 &= 3 - x \\
3x + 3 - 4 &= 3 - x \\
3x - 1 &= 3 - x \\
\underline{+x \qquad\; = \quad +x} & \\
4x - 1 &= 3 \\
\underline{+ 1 = + 1} & \\
4x &= 4 \\
\frac{4x}{4} &= \frac{4}{4} \\
x &= 1
\end{aligned}
$$

Check:
$$
\begin{aligned}
3(x + 1) - 4 &= 3 - x \\
3(1 + 1) - 4 &= 3 - 1 \\
6 - 4 &= 2 \\
2 &= 2 \checkmark
\end{aligned}
$$

15.
$$
\begin{aligned}
4 - (x + 2) &= 2(x + 1) \\
4 - x - 2 &= 2x + 2 \\
4 - 2 - x &= 2x + 2 \\
2 - x &= 2x + 2 \\
\underline{+x \qquad = +x} & \\
2 &= 3x + 2 \\
\underline{-2 \quad = \qquad - 2} & \\
0 &= 3x \\
\frac{0}{3} &= \frac{3x}{3} \\
0 &= x
\end{aligned}
$$

Check:
$$
\begin{aligned}
4 - (x + 2) &= 2(x + 1) \\
4 - (0 + 2) &= 2(0 + 1) \\
4 - 2 &= 2(1) \\
2 &= 2 \checkmark
\end{aligned}
$$

17.
$$
\begin{aligned}
a + 2(1 - a) &= 3 \\
a + 2 - 2a &= 3 \\
a - 2a + 2 &= 3 \\
-a + 2 &= 3 \\
\underline{-2 = -2} & \\
-a &= 1 \\
\frac{-a}{-1} &= \frac{1}{-1} \\
a &= -1
\end{aligned}
$$

Check:
$$
\begin{aligned}
a + 2(1 - a) &= 3 \\
-1 + 2[1 - (-1)] &= 3 \\
-1 + 2[2] &= 3 \\
-1 + 4 &= 3 \\
3 &= 3 \checkmark
\end{aligned}
$$

19.
$$
\begin{aligned}
9(3y + 8) &= 3(y + 8) \\
27y + 72 &= 3y + 24 \\
\underline{-3y \qquad\; = -3y} & \\
24y + 72 &= 24 \\
\underline{- 72 = -72} & \\
24y &= -48 \\
\frac{24y}{24} &= \frac{-48}{24} \\
y &= -2
\end{aligned}
$$

Check:
$$
\begin{aligned}
9(3y + 8) &= 3(y + 8) \\
9[3(-2) + 8] &= 3(-2 + 8) \\
9[-6 + 8] &= 3(6) \\
9[2] &= 18 \\
18 &= 18 \checkmark
\end{aligned}
$$

21. 6   **23.** -12 **25.** 1   **27.** 1   **29.** -2

Section 3.4 Supplementary Exercises (p. 97)

1.
$$
\begin{aligned}
3 - 2(a + 1) &= 5 \\
3 - 2a - 2 &= 5 \\
3 - 2 - 2a &= 5 \\
1 - 2a &= 5 \\
-1 &= -1 \\
\hline
-2a &= 4 \\
\frac{-2a}{-2} &= \frac{4}{-2} \\
a &= -2
\end{aligned}
$$

Check:
$$
\begin{aligned}
3 - 2(a + 1) &= 5 \\
3 - 2(-2 + 1) &= 5 \\
3 - 2(-1) &= 5 \\
3 + 2 &= 5 \\
5 &= 5 \checkmark
\end{aligned}
$$

3.
$$
\begin{aligned}
4(x - 2) - 7 &= 3x + 1 \\
4x - 8 - 7 &= 3x + 1 \\
4x - 15 &= 3x + 1 \\
-3x &= -3x \\
\hline
x - 15 &= 1 \\
+ 15 &= 15 \\
\hline
x &= 16
\end{aligned}
$$

Check:
$$
\begin{aligned}
4(x - 2) - 7 &= 3x + 1 \\
4(16 - 2) - 7 &= 3 \cdot 16 + 1 \\
4 \cdot 14 - 7 &= 48 + 1 \\
56 - 7 &= 49 \\
49 &= 49 \checkmark
\end{aligned}
$$

5.
$$
\begin{aligned}
4(a + 5) - (13 + 2a) &= a - 9 \\
4a + 20 - 13 - 2a &= a - 9 \\
4a - 2a + 20 - 13 &= a - 9 \\
2a + 7 &= a - 9 \\
-a &= -a \\
\hline
a + 7 &= -9 \\
-7 &= -7 \\
\hline
a &= -16
\end{aligned}
$$

Check:
$$
\begin{aligned}
4(a + 5) - (13 + 2a) &= a - 9 \\
4(-16 + 5) - [13 + 2(-16)] &= -16 - 9 \\
4(-11) - [13 - 32] &= -25 \\
-44 - [-19] &= -25 \\
-44 + 19 &= -25 \\
-25 &= -25 \checkmark.
\end{aligned}
$$

7.
$$
\begin{aligned}
\frac{x + 2}{3} &= 5 \\
3 \cdot \frac{x + 2}{3} &= 3 \cdot 5 \\
x + 2 &= 15 \\
-2 &= -2 \\
\hline
x &= 13
\end{aligned}
$$

Check:
$$
\begin{aligned}
\frac{x + 2}{3} &= 5 \\
\frac{13 + 2}{3} &= 5 \\
\frac{15}{3} &= 5 \\
5 &= 5 \checkmark
\end{aligned}
$$

9.
$$
\begin{aligned}
2x - [1 - (3 - x)] &= 2 + 2x \\
2x - [1 - 3 + x] &= 2 + 2x \\
2x - [-2 + x] &= 2 + 2x \\
2x + 2 - x &= 2 + 2x \\
x + 2 &= 2 + 2x \\
-2x &= -2x \\
\hline
-x + 2 &= 2 \\
-2 &= -2 \\
\hline
-x &= 0 \\
\frac{-x}{-1} &= \frac{0}{-1} \\
x &= 0
\end{aligned}
$$

Check:
$$
\begin{aligned}
2x - [1 - (3 - x)] &= 2 + 2x \\
2 \cdot 0 - [1 - (3 - 0)] &= 2 + 2 \cdot 0 \\
0 - [1 - 3] &= 2 \\
0 - [-2] &= 2 \\
0 + 2 &= 2 \\
2 &= 2 \checkmark
\end{aligned}
$$

11.
$$
\begin{aligned}
\frac{6 - x}{2} &= 7 \\
2 \cdot \frac{6 - x}{2} &= 2 \cdot 7 \\
6 - x &= 14 \\
-6 &= -6 \\
\hline
-x &= 8 \\
\frac{-x}{-1} &= \frac{8}{-1} \\
x &= -8
\end{aligned}
$$

Check:
$$
\begin{aligned}
\frac{6 - x}{2} &= 7 \\
\frac{6 - (-8)}{2} &= 7 \\
\frac{6 + 8}{2} &= 7 \\
\frac{14}{2} &= 7 \\
7 &= 7 \checkmark
\end{aligned}
$$

13.
$$5 - [6 - (x - 2)] = 4x + 3$$
$$5 - [6 - x + 2] = 4x + 3$$
$$5 - [8 - x] = 4x + 3$$
$$5 - 8 + x = 4x + 3$$
$$-3 + x = 4x + 3$$
$$\underline{-x = -x}$$
$$\underline{-3 = 3x + 3}$$
$$\underline{-3 = -3}$$
$$-6 = 3x$$

$$\frac{-6}{3} = \frac{3x}{3}$$
$$-2 = x$$

Check:
$$5 - [6 - (x - 2)] = 4x + 3$$
$$5 - [6 - (-2 - 2)] = 4(-2) + 3$$
$$5 - [6 - (-4)] = -8 + 3$$
$$5 - [6 + 4] = -5$$
$$5 - 10 = -5$$
$$-5 = -5 \checkmark$$

15.
$$\frac{a + 1}{4} = \frac{2a + 3}{4}$$
$$4 \cdot \frac{a + 1}{4} = 4 \cdot \frac{2a + 3}{4}$$
$$a + 1 = 2a + 3$$
$$\underline{-a = -a}$$
$$\underline{1 = a + 3}$$
$$\underline{-3 = -3}$$
$$-2 = a$$

Check:
$$\frac{a + 1}{4} = \frac{2a + 3}{4}$$
$$\frac{-2 + 1}{4} = \frac{2(-2) + 3}{4}$$
$$\frac{-1}{4} = \frac{-4 + 3}{4}$$
$$\frac{-1}{4} = \frac{-1}{4} \checkmark$$

Section 3.5 Exercises (p. 102)

1. Let x = number of people employed by the Carlin Hauling Corporation

$$x + 6 = 15$$
$$\underline{-6 = -6}$$
$$x = 9$$

Carlin employs 9 people.

3. Let x = number
$$x = -9 + 23$$
$$x = 14$$

The number is 14.

5. Let x = number of cards José can punch per minute

$$10 = x - 5$$
$$\underline{+5 = +5}$$
$$15 = x$$

José can punch 15 cards per minute.

7. Let x = Paul's income
$$5x + 1000 = \text{Linda's income}$$
$$x + (5x + 1000) = 13,000$$
$$x + 5x + 1000 = 13,000$$
$$6x + 1000 = 13,000$$
$$\underline{- 1000 = -1,000}$$
$$6x = 12,000$$
$$\frac{6x}{6} = \frac{12,000}{6}$$
$$x = 2,000$$

Paul makes $2,000.

9. Let x = number of identifications made by Philip
$$2x = \text{number made by Erin}$$
$$3x = \text{number made by Sean}$$
$$x + 2x + 3x = 12$$
$$6x = 12$$
$$\frac{6x}{6} = \frac{12}{6}$$
$$x = 2$$

Philip identified 2 correctly, Erin identified 4 correctly and Sean identified 6 correctly.

11. Let x = months Family 1 was on welfare
x = months Family 2 was on welfare
$x - 3$ = months Family 3 was on welfare
$3x$ = months Family 4 was on welfare

$$x + x + (x - 3) + 3x = 21$$
$$x + x + x - 3 + 3x = 21$$
$$6x - 3 = 21$$
$$\underline{+3 = +3}$$
$$6x = 24$$
$$\frac{6x}{6} = \frac{24}{6}$$
$$x = 4$$

Families 1 and 2 were on welfare for 4 months, Family 3 for 1 month and Family 4 for 12 months.

13.
h = height above water level
$8h - 3$ = height below water level
$$h + (8h - 3) = 888$$
$$h + 8h - 3 = 888$$
$$9h - 3 = 888$$
$$\underline{+3 = +3}$$
$$9h = 891$$
$$\frac{9h}{9} = \frac{891}{9}$$
$$h = 99$$
$$8h - 3 = 8 \cdot 99 - 3 = 792 - 3 = 789$$

Thus 789 feet of the iceberg is below water level.

15. Let x = smaller integer
$x + 2$ = next consecutive even integer
$$x + (x + 2) = 30$$
$$x + x + 2 = 30$$
$$2x + 2 = 30$$
$$\underline{-2 = -2}$$
$$2x = 28$$
$$\frac{2x}{2} = \frac{28}{2}$$
$$x = 14$$
$$x + 2 = 16$$

The integers are 14 and 16.

17.
Let x = number of pounds recycled by Ken
Then $2x + 10$ = number of pounds recycled by Louise
$$x + (2x + 10) = 100$$
$$x + 2x + 10 = 100$$
$$3x + 10 = 100$$
$$\underline{-10 = -10}$$
$$3x = 90$$
$$\frac{3x}{3} = \frac{90}{3}$$
$$x = 30$$

Ken recycled 30 pounds and Louise 70 pounds.

19. Let x = amount of money Paul and Eileen invested

$12\frac{1}{4}\% = 12.25\% = 0.1225$. Thus the interest earned on their investment is $0.1225x$.

$$x + 0.1225x = 33{,}675$$
$$1.1225x = 33{,}675$$
$$\frac{1.1225x}{1.1225} = \frac{33{,}675}{1.1225}$$
$$x = 30{,}000$$

The original investment was $30,000.

21. $17,500   **23.** 18   **25.** 7, 49   **27.** John works 12 hours per week, Mark 17 hours, and Joe 9 hours.   **29.** 16, 17, 18   **31.** 111, 113, 115   **33.** 15, 20   **35.** $20

Section 3.5 Supplementary Exercises (p. 104)

1. Let x = number
$$4x - 6 = x + 12$$
$$\underline{-x = -x}$$
$$3x - 6 = 12$$
$$\underline{+6 = +6}$$
$$3x = 18$$
$$\frac{3x}{3} = \frac{18}{3}$$
$$x = 6$$

3. Let x = smallest integer. Then $x + 1$, $x + 2$, and $x + 3$ represent the next three larger consecutive integers.
$$x + (x + 1) + (x + 2) + (x + 3) = 90$$
$$x + x + 1 + x + 2 + x + 3 = 90$$
$$4x + 6 = 90$$
$$\underline{-6 = -6}$$
$$4x = 84$$
$$\frac{4x}{4} = \frac{84}{4}$$
$$x = 21$$

The integers are 21, 22, 23, and 24.

5. Let x = smallest integer. Then $x + 2$ and $x + 4$ will represent the two larger consecutive even integers.

$$x + (x + 2) + (x + 4) = 114$$
$$x + x + 2 + x + 4 = 114$$
$$3x + 6 = 114$$
$$\underline{\quad -6 = -6\quad}$$
$$3x = 108$$
$$\frac{3x}{3} = \frac{108}{3}$$
$$x = 36$$

The integers are 36, 38, and 40.

7. Let x = number

$$17 - x = -10$$
$$\underline{-17\quad = -17}$$
$$-x = -27$$
$$\frac{-x}{-1} = \frac{-27}{-1}$$
$$x = 27$$

9. Let x = interest earned

$$8\frac{1}{2}\% = 8.5\% = 0.085$$
$$x = 2000(0.085)$$
$$x = \$170$$

11. Let x = number of cars held on the second floor
$x + 50$ = number of cars held on the first floor
$\frac{1}{2}x$ = number of cars held on the top floor

$$x + (x + 50) + \frac{1}{2}x = 400$$
$$x + x + 50 + \frac{1}{2}x = 400$$
$$2\frac{1}{2}x + 50 = 400$$
$$2.5x + 50 = 400$$
$$\underline{\quad -50 = -50}$$
$$2.5x = 350$$
$$\frac{2.5x}{2.5} = \frac{350}{2.5}$$
$$x = 140$$

The garage holds 190 cars on the first floor, 140 cars on the second floor, and 70 cars on the top floor.

13. Let x = number of Brand A computers sold
Then $2x + 2$ = number of Brand B computers sold

$$x + (2x + 2) = 44$$
$$x + 2x + 2 = 44$$
$$3x + 2 = 44$$
$$\underline{\quad -2 = -2}$$
$$3x = 42$$
$$\frac{3x}{3} = \frac{42}{3}$$
$$x = 14$$

Harry sold 14 Brand A computers and 30 Brand B computers.

15. Let x = number

$$27 + \frac{2}{3}x = 37$$
$$\underline{-27\qquad = -27}$$
$$\frac{2x}{3} = 10$$
$$3 \cdot \frac{2x}{3} = 3 \cdot 10$$
$$2x = 30$$
$$\frac{2x}{2} = \frac{30}{2}$$
$$x = 15$$

Section 3.6 Exercises (p. 112)

1.
$$x + y = 1$$
$$\underline{-x\quad = \quad -x}$$
$$y = 1 - x$$

3.
$$t - 2k = 0$$
$$\underline{-t\quad = -t}$$
$$-2k = -t$$
$$\frac{-2k}{-2} = \frac{-t}{-2}$$
$$k = \frac{t}{2}$$

5.
$$-2x - 3y = 4$$
$$\underline{+2x\quad = \quad +2x}$$
$$-3y = 4 + 2x$$
$$\frac{-3y}{-3} = \frac{4 + 2x}{-3}$$
$$y = \frac{4 + 2x}{-3}$$

7.
$$5x + 3a = 4$$
$$\underline{-5x\quad = \quad -5x}$$
$$3a = 4 - 5x$$
$$\frac{3a}{3} = \frac{4 - 5x}{3}$$
$$a = \frac{4 - 5x}{3}$$

9. $3ps - ax = y$
$$\underline{\quad + ax = \quad + ax}$$
$$3ps \quad = y + ax$$
$$\frac{3ps}{3s} = \frac{y + ax}{3s}$$
$$p = \frac{y + ax}{3s}$$

11. $P = 2l + 2w$
$$\underline{-2w \quad = \quad -2w}$$
$$P - 2w = 2l$$
$$\frac{P - 2w}{2} = \frac{2l}{2}$$
$$\frac{P - 2w}{2} = l$$

13. $Y + 3x = l$
$$\underline{-3x = \quad -3x}$$
$$Y \quad = l - 3x$$

15. $C = 2\pi r$
$$\frac{C}{2\pi} = \frac{2\pi r}{2\pi}$$
$$\frac{C}{2\pi} = r$$

17. $D = rt$
$$\frac{D}{t} = \frac{rt}{t}$$
$$\frac{D}{t} = r$$

19. $Q = \frac{P}{R}$
$$Q \cdot R = \frac{P}{R} \cdot R$$
$$QR = P$$
$$\frac{QR}{Q} = \frac{P}{Q}$$
$$R = \frac{P}{Q}$$

21. $P = QR$ **23.** $b = \dfrac{3}{ac}$ **25.** -6 **27.** 3 **29.** -2

31. (a) $y = \dfrac{21 - 7x}{4}$ (b) $x = \dfrac{21 - 4y}{7}$ (c) -7 (d) 3 **33.** 96 km/hr   **35.** (a) Smartz

(b) \$40   (c) \$3.75   **37.** 4 hours   **39.** $R = \dfrac{I}{PT}$

Section 3.6 Supplementary Exercises (p. 113)

1. $m - n = 11$
$$\underline{+ n = \quad + n}$$
$$m \quad = 11 + n$$

3. $2t + p = 7$
$$\underline{-p = \quad -p}$$
$$2t \quad = 7 - p$$
$$\frac{2t}{2} = \frac{7 - p}{2}$$
$$t = \frac{7 - p}{2}$$

5. $6a + 4b = 12$
$$\underline{-6a \quad = \quad -6a}$$
$$4b = 12 - 6a$$
$$\frac{4b}{4} = \frac{12 - 6a}{4}$$
$$b = \frac{12 - 6a}{4}$$

7. $A = \dfrac{1}{2}bh$
$$2 \cdot A = 2 \cdot \frac{1}{2}bh$$
$$2A = bh$$
$$\frac{2A}{h} = \frac{bh}{h}$$
$$\frac{2A}{h} = b$$

9. $\qquad P = 2r + b_1 + b_2$
$$\underline{-2r \quad = -2r \quad}$$
$$P - 2r = b_1 + b_2$$
$$\underline{\quad - b_2 = \quad - b_2}$$
$$P - 2r - b_2 = b_1$$

11. $m = 11 + n$ when $n = 9$
$$m = 11 + 9$$
$$m = 20$$

13. $t = \dfrac{7 - p}{2}$ when $p = 15$
$$t = \frac{7 - 15}{2} = \frac{-8}{2} = -4$$

15. $b = \dfrac{12 - 6a}{4}$ when $a = -1$
$$b = \frac{12 - 6(-1)}{4} = \frac{12 + 6}{4} = \frac{18}{4}$$
$$b = \frac{18}{4} = \frac{9}{2} \quad \text{or } 4.5$$

17. $b = \dfrac{2A}{h}$ when $A = 12$ and $h = 3$
$$b = \frac{2(12)}{3} = \frac{24}{3} = 8$$

19. $b_1 = P - 2r - b_2$ when $P = 50, r = 5,$ and $b_2 = 15$
$$b_1 = 50 - 2(5) - 15 = 50 - 10 - 15 = 25$$

21. (a) $b = \dfrac{P - 2h}{2}$ (b) 10

23. (a) $w = \dfrac{V}{lh}$ (b) 30   **25.** (a) $P = \dfrac{I}{RT}$ (b) \$400

Section 3.7 Exercises (p. 118)

1. $x - 2 < 4$
$\underline{+2 \qquad +2}$
$x \qquad < 6$

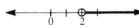

3. $2 + t \geq 9$
$\underline{-2 \qquad -2}$
$t \geq 7$

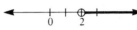

5. $2x + 3 > 7$
$\underline{- 3 \qquad -3}$
$2x \qquad > 4$
$\dfrac{2x}{2} > \dfrac{4}{2}$
$x > 2$

7. $5 - y < 3$
$\underline{-5 \qquad -5}$
$-y < -2$
$\dfrac{-y}{-1} > \dfrac{-2}{-1}$
$y > 2$

9. $3a - 4 \geq 8$
$\underline{+ 4 \qquad +4}$
$3a \qquad \geq 12$
$\dfrac{3a}{3} \geq \dfrac{12}{3}$
$a \geq 4$

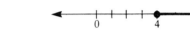

11. $8 - 7y > 29$
$\underline{-8 \qquad\qquad -8}$
$-7y > 21$
$\dfrac{-7y}{-7} < \dfrac{21}{-7}$
$y < -3$

13. $3(2 - x) < 12 - x$
$6 - 3x < 12 - x$
$\underline{+ x \qquad\qquad + x}$
$6 - 2x < 12$
$\underline{-6 \qquad\qquad -6}$
$-2x < 6$
$\dfrac{-2x}{-2} > \dfrac{6}{-2}$
$x > -3$

15. $8 - 7y > y$
$\underline{+ 7y \qquad +7y}$
$8 \qquad > 8y$
$\dfrac{8}{8} > \dfrac{8y}{8}$
$1 > y \quad$ or $\quad y < 1$

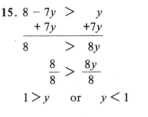

17. $4(z - 1) \leq 5(2z - 8)$
$4z - 4 \leq 10z - 40$
$\underline{-10z \qquad\quad -10z}$
$-6z - 4 \leq \qquad -40$
$\underline{+ 4 \qquad\quad + 4}$
$-6z \qquad \leq \qquad -36$
$\dfrac{-6z}{-6} \geq \dfrac{-36}{-6}$
$z \geq 6$

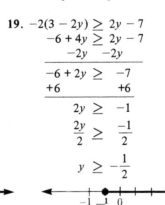

19. $-2(3 - 2y) \geq 2y - 7$
$-6 + 4y \geq 2y - 7$
$\underline{-2y \qquad -2y}$
$-6 + 2y \geq \qquad -7$
$\underline{+6 \qquad\qquad +6}$
$2y \geq -1$
$\dfrac{2y}{2} \geq \dfrac{-1}{2}$
$y \geq -\dfrac{1}{2}$

21. $x \geq -\dfrac{2}{3}$

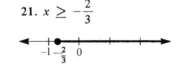

23. at least 3200 pieces   **25.** at least 72   **27.** $y \leq 2.7205$ **29.** $x < 0.4993642$

Section 3.7 Supplementary Exercises (p. 119)

1. $a + 6 \leq 4$
$\underline{- 6 \qquad -6}$
$a \qquad \leq -2$

3. $x - 4 \geq -12$
$\underline{+ 4 \qquad + 4}$
$x \qquad \geq -8$

5. $-3y > 9$

$\dfrac{-3y}{-3} < \dfrac{9}{-3}$

$y < -3$

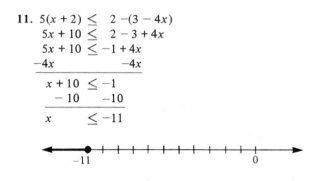

7. $3a + 1 < 10$

$\underline{ -1 \quad -1}$

$3a \quad < \quad 9$

$\dfrac{3a}{3} < \dfrac{9}{3}$

$a < 3$

9. $3(x - 1) \leq 2x - 4$

$3x - 1 \leq 2x - 4$

$\underline{-2x \qquad\quad -2x}$

$x - 3 \leq -4$

$\underline{\quad + 3 \qquad + 3}$

$x \qquad \leq -1$

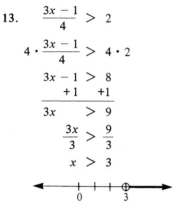

11. $5(x + 2) \leq 2 - (3 - 4x)$

$5x + 10 \leq 2 - 3 + 4x$

$5x + 10 \leq -1 + 4x$

$\underline{-4x \qquad\qquad -4x}$

$x + 10 \leq -1$

$\underline{\quad - 10 \qquad -10}$

$x \qquad\quad \leq -11$

13. $\dfrac{3x - 1}{4} > 2$

$4 \cdot \dfrac{3x - 1}{4} > 4 \cdot 2$

$3x - 1 > 8$

$\underline{\quad +1 \qquad +1}$

$3x \qquad > 9$

$\dfrac{3x}{3} > \dfrac{9}{3}$

$x > 3$

15. Let x = number of miles she must run on the fifth day

$\dfrac{7 + 11 + 8 + 13 + x}{5} \geq 10$

$\dfrac{39 + x}{5} \geq 10$

$5 \cdot \dfrac{39 + x}{5} \geq 5 \cdot 10$

$39 + x \geq 50$

$\underline{-39 \qquad\quad -39}$

$x \geq 11$

Carol must run at least 11 miles.

Vocabulary Quiz (p. 119)

1. a   **2.** d   **3.** b   **4.** c

Chapter 3 Review Exercises (p. 120)

1. 11   **2.** 3   **3.** 7   **4.** 42   **5.** 3   **6.** 4   **7.** 3   **8.** 1   **9.** 2   **10.** 6   **11.** −20
12. −22   **13.** −4   **14.** −6   **15.** 7   **16.** 4   **17.** 648   **18.** −54   **19.** 2   **20.** 3
21. 17 ounces, 21 ounces   **22.** 15 feet by 11 feet   **23.** 74   **24.** 6   **25.** $x \leq 4$ **26.** 14 feet
by 10 feet   **27.** 15 and 17   **28.** (a) $y = \dfrac{1 - 3x}{2}$ (b) −4   **29.** $y = \dfrac{5 - 2x}{-3}$, 5

30. $x \leq 8$ **31.** $y < 5$

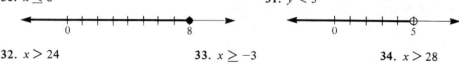

32. $x > 24$ **33.** $x \geq -3$ **34.** $x > 28$

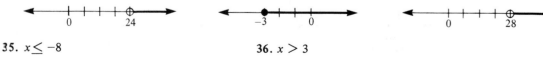

35. $x \leq -8$ **36.** $x > 3$

37. $y \geq -7$ **38.** $x \geq 6$

39. $x < 11$ **40.** at least 162

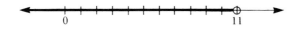

Cumulative Review Exercises (p. 120)

1. -25 **2.** 8   **3.** (a) 0   (b) 0   **4.** d   **5.** 1   **6.** -3 **7.** $10x - 14$
8. $-x^3 - 31x^2 + 39x$ **9.** 50.24 in.²   **10.** 12

Chapter 3 Test (p. 122)

1. 15   **2.** 13   **3.** -7 **4.** -36 **5.** -44 **6.** 6   **7.** 3   **8.** -18 **9.** -5 **10.** -2
11. 250 pounds and 550 pounds   **12.** Barbara is 12 years old, Alice is 24, and Kevin is 10.   **13.** 30 sec.
14. $y = \dfrac{11 - 4x}{7}$ **15.** $C = \dfrac{160 - 5F}{-9}$ **16.** $x \geq -2$ **17.** $x < 3$

18. $x \geq 6$ **19.** $x > \dfrac{19}{8}$

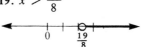

20. $x \geq 14$

Chapter 4

Section 4.1 Exercises (p. 130)

1. Let x = number
$x + 4x = 60$
$5x = 60$
$x = 12$
The number is 12.

3. Let x = number
$2x - 6 = 19$
$2x = 25$
$x = 12\frac{1}{2}$
The number is $12\frac{1}{2}$.

5. Let x = smaller integer
$x + 2$ = next consecutive even integer
$x + (x + 2) = 302$
$x + x + 2 = 302$
$2x + 2 = 302$
$2x = 300$
$x = 150$
The integers are 150 and 152.

7. Let x = number
$3x - 7 = 11$
$3x = 18$
$x = 6$
The number is 6.

9. Let x = smallest integer
$x + 2$ = next consecutive even integer
$x + 4$ = largest even integer
$\frac{1}{2}[x + (x + 2) + (x + 4)] = 183$
$\frac{1}{2}[x + x + 2 + x + 4] = 183$
$\dfrac{3x + 6}{2} = 183$
$3x + 6 = 366$
$3x = 360$
$x = 120$
The integers are 120, 122, and 124.

11. Let x = number
$2(x + 4) = -12$
$2x + 8 = -12$
$2x = -20$
$x = -10$
The number is -10.

13. Let x = number

$$\frac{9 - 3x}{2} = 18$$

$$9 - 3x = 36$$

$$-3x = 27$$

$$x = -9$$

The number is -9.

15. Let x = number of people 12 and over

then $(6 - x)$ = number of people under 12

$$2.25x + 1.00(6 - x) = 9.75$$

$$2.25x + 6 - 1.00x = 9.75$$

$$1.25x + 6 = 9.75$$

$$1.25x = 3.75$$

$$x = \frac{3.75}{1.25}$$

$$x = 3$$

thus $6 - x = 6 - 3 = 3$

There were 3 people under 12 years of age.

17. Let x = amount invested at the 6% rate

then $(3000 - x)$ = amount invested at the $7\frac{1}{4}$ % rate

$$0.06x = 0.0725(3000 - x) + 157$$

$$0.06x = 217.50 - 0.0725x + 157$$

$$0.06x + 0.0725x = 374.50$$

$$0.1325x = 374.50$$

$$x = \frac{374.50}{0.1325}$$

$$x = 2826.42$$

thus $(3000 - x) = 173.58$

He invested \$2826.42 at 6% and \$173.58 at $7\frac{1}{4}$%.

19. Let x = number of 10¢ stamps used

then $2x$ = number of 13¢ stamps used

$$10x + 13(2x) = 180$$

$$10x + 26x = 180$$

$$36x = 180$$

$$x = \frac{180}{36}$$

$$x = 5$$

He used five 10¢ stamps and ten 13¢ stamps.

21. \$600   **23.** There are 300 orchestra seats.   **25.** He invests \$10,000 at 6% and \$15,000 at $7\frac{1}{2}$%.

Section 4.1 Supplementary Exercises (p. 131)

1. Let x = number

$$7x + 9 = 72$$

$$7x = 63$$

$$x = 9$$

The number is 9.

3. Let x = number

$$21 - \frac{1}{2}x = 17$$

$$-\frac{1}{2}x = -4$$

$$-\frac{x}{2} = -4$$

$$-x = -8$$

$$x = 8$$

The number is 8.

5. Let x = number

$$(x + 7) = \frac{1}{2}x - 5$$

$$x + 7 = \frac{1}{2}x - 5$$

$$\frac{1}{2}x + 7 = -5$$

$$\frac{1}{2}x = -12$$

$$x = -24$$

The number is -24.

7. Let x = number of 15¢ stamps

then $(x + 2)$ = number of 20¢ stamps

$$15x + 20(x + 2) = 285$$

$$15x + 20x + 40 = 285$$

$$35x + 40 = 285$$

$$35x = 245$$

$$x = 7$$

She used seven 15¢ stamps and nine 20¢ stamps.

9. Let x = amount spent on food

then $3x$ = amount spent on rent

$$\frac{1}{4}x = \text{amount spent on clothing}$$

$$x + 3x + \frac{1}{4}x = 5100$$

$$4\frac{1}{4}x = 5100$$

$$4.25x = 5100$$

$$x = \frac{5100}{4.25}$$

$$x = 1200$$

She spends \$1200 on food, \$3600 on rent, and \$300 on clothing.

11. Let x = amount invested at 8%
then $(15,000 - x)$ = amount invested at 7%
$$0.08x + 0.07(15,000 - x) = 1140$$
$$0.08x + 1050 - 0.07x = 1140$$
$$0.01x = 90$$
$$x = \frac{90}{0.01}$$
$$x = 9000$$
$$15,000 - x = 6000$$
She has $9000 invested at 8% and $6000 at 7%.

13. Let x = amount loaned to his brother at 5%
$(800 - x)$ = amount invested at 9%
$$0.05x + 0.09(800 - x) = 64$$
$$0.05x + 72 - 0.09x = 64$$
$$-0.04x = -8$$
$$x = \frac{-8}{-0.04}$$
$$x = 200$$
$$800 - x = 600$$
He loaned his brother $200 and deposited $600 in the account.

Section 4.2 Exercises (p. 134)

1. Let t = time needed by John to travel 330 miles
Since $d = 330$ and $r = 55$ we have
$$d = r \cdot t$$
$$330 = 55t$$
$$\frac{330}{55} = t$$
$$6 = t$$
It takes John 6 hours to travel 330 miles.

3. Let r = rate at which Rick travels
Since $t = 4$ and $d = 240$ we get
$$d = rt$$
$$240 = r \cdot 4$$
$$60 = r$$
Rick's rate is 60 mph.

5. Let d = distance Mike jogs
Since $r = 8$ and $t = 2\frac{1}{2}$ we have
$$d = r \cdot t$$
$$d = 8 \cdot 2\frac{1}{2}$$
$$d = 20$$
Mike jogs 20 miles.

7. Let d = distance Danielle walks
We have that $t = \frac{1}{2}$ and $r = 4$ so
$$d = r \cdot t$$
$$d = 4 \cdot \frac{1}{2}$$
$$d = 2$$
Danielle walks 2 miles.

9. Let t = time Katie needs to complete her trip
Since $r = 55$ and $d = 275$ we get
$$d = r \cdot t$$
$$275 = 55t$$
$$5 = t$$
Thus it takes Katie 5 hours to make the trip. She will arrive at her destination at 2:00 P.M.

11.
Train 1	Train 2
$t_1 = 6$ hours	$t_2 = 6$ hours
r_1 = rate	$r_2 = r_1 + 20$
$d_1 = 6r_1$	$d_2 = 6(r_1 + 20)$

Since the trains are 360 miles apart we have that
$$d_1 + d_2 = 360$$
$$6r_1 + 6(r_1 + 20) = 360$$
$$6r_1 + 6r_1 + 120 = 360$$
$$12r_1 = 240$$
$$r_1 = 20$$
The slower train travels at 20 mph and the faster train at 40 mph.

13.
Sam	Nelson	
$r_s = 40$	$r_n = 20$	
t_s = time	$t_n = t_s$	(Travel times are equal)
$d_s = 40t_s$	$d_n = 20t_s$	

Since they are 325 miles apart we have
$$d_s + d_n = 325$$
$$40t_s + 20t_s = 325$$
$$60t_s = 325$$
$$t_s = \frac{325}{60} = 5\frac{25}{60} = 5\frac{5}{12}$$

They each travel for $5\frac{5}{12}$ hours, which is $5 \cdot 60 + \frac{5}{12} \cdot 60$ minutes or $300 + 25 = 325$ minutes.

15.

Sandy	Trooper	
$r_s = 50$	$r_t = 75$	
$t_s = $ time	$t_t = t_s$	(Travel times are equal)
$d_s = 50t_s$	$d_t = 75t_s$	

In order for them to be 50 miles apart we have

$$d_t - d_s = 50$$
$$75t_s - 50t_s = 50$$
$$25t_s = 50$$
$$t_s = 2$$

It will take 2 hours for them to be 50 miles apart.
The time of day will be 3:00 P.M.

17.

Train 1	Train 2	
$r_1 = 45$	$r_2 = 65$	
$t_1 = $ time	$t_2 = t_1$	(Times are equal)
$d_1 = 45t_1$	$d_2 = 65t_1$	

$$d_1 + d_2 = 385$$
$$45t_1 + 65t_1 = 385$$
$$110t_1 = 385$$
$$t_1 = \frac{385}{110}$$
$$t_1 = 3\frac{1}{2}$$

It will take $3\frac{1}{2}$ hours until they crash at 3:30 P.M.

19.

Train 1	Train 2
$t_1 = 6$	$t_2 = 6$
$r_1 = $ rate	$r_2 = r_1 + 20$
$d_1 = 6r_1$	$d_2 = 6(r_1 + 20)$

After 6 hours they are 420 miles apart so

$$d_1 + d_2 = 420$$
$$6r_1 + 6(r_1 + 20) = 420$$
$$6r_1 + 6r_1 + 120 = 420$$
$$12r_1 = 300$$
$$r_1 = 25$$

The slower train travels 25 mph and the faster 45 mph.

21. 55 mph and 60 mph   **23.** 2 hours   **25.** $\frac{1}{3}$ hour or 20 minutes

Section 4.2 Supplementary Exercises (p. 135)

1. Let $r = $ rate Renee rode her bike
Since $t = 2$ and $d = 6$ we have
$$d = rt$$
$$6 = r \cdot 2$$
$$3 = r$$
Her rate is 3 mph.

3. Let $d = $ distance Joe walks
Since $r = 6$ and $t = 2$ we have
$$d = rt$$
$$d = 6 \cdot 2$$
$$d = 12$$
Joe walks 12 miles.

5. Let $t = $ time Joe needs to walk to work
Since $d = 5$ and $r = 3$ we have
$$d = rt$$
$$5 = 3t$$
$$\frac{5}{3} = t$$
Joe takes $1\frac{2}{3}$ hours to walk to work

7.

Jennifer	Brother	
$r_j = 45$	$r_b = 55$	(He travels one
$t_j = $ time	$t_b = t_j - 1$	hour less, since he
$d_j = 45t_j$	$d_b = 55(t_j - 1)$	left one hour later.)

Since the distances traveled by each must be equal,

$$d_j = d_b$$
$$45t_j = 55(t_j - 1)$$
$$45t_j = 55t_j - 55$$
$$-10t_j = -55$$
$$t_j = 5\frac{1}{2}$$

Jennifer travels for $5\frac{1}{2}$ hours so it will take her brother $4\frac{1}{2}$ hours to catch up to her.

9.

Jim	Joe
$r_m = 55$	$r_e = 65$
$t_m = 3\dfrac{1}{2}$	$t_e = 3\dfrac{1}{2}$

$d_m = r_m \cdot t_m = 55(3\tfrac{1}{2}) = 192.5 \qquad d_e = r_e \cdot t_e = 65(3\tfrac{1}{2}) = 227.5$

The distance apart, d, is given by

$$d = d_m + d_e = 192.5 + 227.5 = 420$$

They are 420 miles apart.

11. Spike

$$t_s = 2$$
$$r_s = 55$$
$$d_s = r_s \cdot t_s = 55 \cdot 2 = 110$$

The distance from home to college is 110 miles. To find his sister's rate solve $d = rt$ for r when $d = 110$ and $t = 2\dfrac{1}{2}$.

$$d = rt$$
$$110 = r \cdot (2\tfrac{1}{2})$$
$$\frac{110}{2.5} = r$$
$$44 = r$$

His sister's rate was 44 mph.

Section 4.3 Exercises (p. 138)

1. Let x = pounds of cashews
then $2x$ = pounds of peanuts
$$x + 2x = 15$$
$$3x = 15$$
$$x = 5$$
She mixes 5 pounds of cashews and 10 pounds of peanuts.

3. Let x = gallons of fruit juice
then $3x$ = gallons of soda
$$x + 3x = 12$$
$$4x = 12$$
$$x = 3$$
She needs 9 gallons of soda.

5. Let x = pounds of peanuts
$(x + 1)$ = pounds of pecans
$\dfrac{1}{2}x$ = pounds of walnuts
$$x + (x + 1) + \frac{1}{2}x = 16$$
$$x + x + 1 + \frac{1}{2}x = 16$$
$$2\frac{1}{2}x + 1 = 16$$
$$2.5x = 15$$
$$x = \frac{15}{2.5}$$
$$x = 6$$
He mixes 6 pounds of peanuts, 7 pounds of pecans, and 3 pounds of walnuts.

7. Let x = cups of wheat flour
then $\dfrac{3}{2}x$ = cups of white flour
$$x + \frac{3}{2}x = 20$$
$$x + 1.5x = 20$$
$$2.5x = 20$$
$$x = \frac{20}{2.5}$$
$$x = 8$$
She uses 8 cups of wheat flour and 12 cups of white flour.

9. Total value = number of cans $\cdot$ cost per can
$$= 17 \cdot (0.39)$$
$$= 6.63$$
Her total cost is $6.63.

11.

	Value of one pound	Number of pounds	Total value
Blue cheese	2.19	x	$2.19x$
Cheddar	3.09	$10 - x$	$3.09(10 - x)$
Mixture	2.82	10	$2.82(10)$

$$2.19x + 3.09(10 - x) = 2.82(10)$$
$$2.19x + 30.9 - 3.09x = 28.2$$
$$-0.90x + 30.9 = 28.2$$
$$-0.90x = -2.7$$
$$x = \frac{-2.7}{-0.90}$$
$$x = 3$$

He should blend 3 pounds of blue cheese and 7 pounds of cheddar.

13.

	% solution	Number of gallons	Amount of solution
Original	0.25	5	$(0.25)(5)$
Water	0	x	$0 \cdot x$
Mixture	0.15	$5 + x$	$0.15(5 + x)$

$$(0.25)(5) + 0 \cdot x = 0.15(5 + x)$$
$$1.25 = 0.75 + 0.15x$$
$$0.50 = 0.15x$$
$$\frac{0.50}{0.15} = x$$
$$3\frac{1}{3} = x$$

$3\frac{1}{3}$ gallons of water must be added.

15.

	% solution	Number of gallons	Amount of alcohol
Original	0.60	10	$0.60(10)$.
Alcohol	1.00	x	$1.00x$
New mixture	0.90	$10 + x$	$0.90(10 + x)$

$$0.60(10) + x = 0.90(10 + x)$$
$$6 + x = 9 + 0.90x$$
$$6 + 0.10x = 9$$
$$0.10x = 3$$
$$x = \frac{3}{0.10}$$
$$x = 30$$

30 gallons of pure alcohol must be added.

17.

	% salt	Number of gallons	Amount of salt
Original	0.15	6	$0.15(6)$
Salt	1.00	x	$1.00x$
New mixture	0.20	$6 + x$	$0.20(6 + x)$

$$0.15(6) + x = 0.20(6 + x)$$
$$0.9 + x = 1.2 + 0.20x$$
$$0.9 + 0.80x = 1.2$$
$$0.80x = 0.3$$
$$x = \frac{0.3}{0.8}$$
$$x = \frac{3}{8}$$

$\frac{3}{8}$ of a gallon of salt must be added.

19.

	Value of one pound	Number of pounds	Total value
Cheddar	2.10	x	$2.10x$
Swiss	1.40	$7 - x$	$1.40(7 - x)$
Mixture	1.60	7	$1.60(7)$

$$2.10x + 1.40(7 - x) = 1.60(7)$$
$$2.10x + 9.8 - 1.40x = 11.2$$
$$0.7x + 9.8 = 11.2$$
$$0.7x = 1.4$$
$$x = \frac{1.4}{0.7}$$
$$x = 2$$

He should mix 2 pounds of cheddar cheese and 5 pounds of swiss cheese.

21. $1.80   **23.** 7 pounds of peanuts and 3 pounds of cashews   **25.** $33\frac{1}{3}$ ml

Section 4.3 Supplementary Exercises (p. 139)

1.

	% alcohol	Number of quarts	Amount of alcohol
Original	0.80	15	0.80(15)
Pure alcohol	1.00	x	1.00x
Mixture	0.90	$15 + x$	0.90(15 + x)

$$0.80(15) + x = 0.90(15 + x)$$
$$12 + x = 13.5 + 0.90x$$
$$12 + 0.10x = 13.5$$
$$0.10x = 1.5$$
$$x = 15$$

Fifteen quarts of pure alcohol should be added.

3.

	% gasoline	Number of gallons	Amount of gasoline
Original	0.35	4	0.35(4)
Gasoline	1.00	x	1.00x
Mixture	0.45	$4 + x$	0.45(4 + x)

$$0.35(4) + x = 0.45(4 + x)$$
$$1.4 + x = 1.8 + 0.45x$$
$$1.4 + 0.55x = 1.8$$
$$0.55x = 0.4$$
$$x = \frac{0.4}{0.55}$$
$$x = 0.727$$

He should add 0.727 gallons of gasoline.

5.

	% solution	Number of ml	Amount of solution
Original	0.40	5	0.40(5)
Water	0	x	$0 \cdot x$
Mixture	0.35	$5 + x$	0.35(5 + x)

$$0.40(5) + 0 \cdot x = 0.35(5 + x)$$
$$2 = 1.75 + 0.35x$$
$$0.25 = 0.35x$$
$$\frac{0.25}{0.35} = x$$
$$0.714 = x$$

Add 0.714 ml of water.

7.

	Value of one pound	Number of pounds	Total value
Raisins	1.20	4	1.20(4)
Nuts	1.80	2	1.80(2)
Mixture	x	6	6x

$$1.20(4) + 1.80(2) = 6x$$
$$4.8 + 3.6 = 6x$$
$$8.4 = 6x$$
$$\frac{8.4}{6} = 6x$$
$$1.4 = x$$

The mixture should cost $1.40 per pound.

9.

	Value of one pound	Number of pounds	Total Value
Tea 1	1.08	x	$1.08x$
Tea 2	1.02	$6 - x$	$1.02(6 - x)$
Mixture	1.06	6	$1.06(6)$

$$1.08x + 1.02(6 - x) = 1.06(6)$$
$$1.08x + 6.12 - 1.02x = 6.36$$
$$0.06x + 6.12 = 6.36$$
$$0.06x = 0.24$$
$$x = \frac{0.24}{0.06}$$
$$x = 4$$

Four pounds of the more expensive tea and 2 pounds of the cheaper tea.

11.

	Value of one pound	Number of pounds	Total value
Tea 1	2.20	x	$2.20x$
Tea 2	0.99	$11 - x$	$0.99(11 - x)$
Mixture	1.32	11	$1.32(11)$

$$2.20x + 0.99(11 - x) = 1.32(11)$$
$$2.20x + 10.89 - 0.99x = 14.52$$
$$1.21x + 10.89 = 14.52$$
$$1.21x = 3.63$$
$$x = \frac{3.63}{1.21}$$
$$x = 3$$

Three pounds of the expensive tea and 8 pounds of the cheaper tea.

Section 4.4 Exercises (p. 147)

1. (a) $\dfrac{5000 \text{ adults}}{3000 \text{ children}}$ or   5000:3000   or   5:3

   (b) $\dfrac{3000 \text{ children}}{5000 \text{ adults}}$ or   3000:5000   or   3:5

   (c) $\dfrac{5000 \text{ adults}}{8000 \text{ residents}}$ or   5000:8000   or   5:8

   (d) $\dfrac{1400 \text{ boys}}{1600 \text{ girls}}$ or   1400:1600   or   7:8

   (e) $\dfrac{3000 \text{ men}}{2000 \text{ women}}$ or   3000:2000   or   3:2

3. (a) $\dfrac{40 \text{ new cars}}{60 \text{ used cars}}$ or   40:60   or   2:3

   (b) $\dfrac{60 \text{ used cars}}{40 \text{ new cars}}$ or   60:40   or   3:2

   (c) $\dfrac{40 \text{ new cars}}{100 \text{ cars}}$ or   40:100   or   2:5

5. Let x = weight of the object on earth
Ratio form is moon weight : earth weight
$$\frac{50}{x} = \frac{1}{6}$$
$$50 \cdot 6 = x \cdot 1$$
$$300 = x$$
The object weighs 300 pounds on earth.

7. Let x = number of gallons needed
Ratio form is gallons : acres
$$\frac{100}{1.5} = \frac{x}{55.2}$$
$$1.5x = 100(55.2)$$
$$1.5x = 5520$$
$$x = \frac{5520}{1.5}$$
$$x = 3680$$
3,680 gallons are needed.

9. Let x = number of female voters
Then $(11,106 - x)$ = number of male voters
Ratio form is female voters : male voters
$$\frac{4}{5} = \frac{x}{11,106 - x}$$
$$5x = 4(11,106 - x)$$
$$5x = 44,424 - 4x$$
$$9x = 44,424$$
$$x = 4936$$
There are 4,936 female voters.

11. Let x = number of square miles of solar collectors
Ratio form is people : square miles
$$\frac{50,000}{40} = \frac{350,000}{x}$$
$$50,000x = 40(350,000)$$
$$50,000x = 14,000,000$$
$$x = \frac{14,000,000}{50,000}$$
$$x = 280$$
The city will need 280 square miles of solar collectors.

13. Let x = number of copies
Ratio form is copies : minutes
$$\frac{50}{1} = \frac{x}{2.5}$$
$$1 \cdot x = 50(2.5)$$
$$x = 125$$
125 copies will be made.

15. Let d = distance on the map between Dallas and El Paso
Ratio form is inches : miles
$$\frac{3}{250} = \frac{x}{654}$$
$$250x = 3(654)$$
$$250x = 1962$$
$$x = \frac{1962}{250}$$
$$x = 7.85$$
The map distance between Dallas and El Paso is 7.85 inches.

17. Let d = actual distance from the chin to the top of the head
Ratio form is model height : actual height
$$\frac{1}{12} = \frac{5}{x}$$
$$1 \cdot x = 5 \cdot 12$$
$$x = 60$$
The distance is 60 feet.

19. Let x = cost of 445 kwh
Ratio form is kwh : cost
$$\frac{100}{5} = \frac{455}{x}$$
$$100x = 5(455)$$
$$100x = 2275$$
$$x = \frac{2275}{100}$$
$$x = 22.75$$
The cost is $22.75

21. $4.8 \cdot 10^{12}$ gallons   23. 6000 attorneys   25. 635 mm   27. 78.74 inches   29. 452.8 grams
31. 330 pounds   33. 3.8 liters   35. 2830 ml   37. 88.7 km   39. (a) $212°F$ (b) $32°F$
(c) $68°F$ (d) $86°F$ (e) $-4°F$

Section 4.4 Supplementary Exercises (p. 148)

1. (a) Let x = number of children
Then $(12,000 - x)$ = number of adults
Ratio form is adults : children
$$\frac{3}{1} = \frac{12,000 - x}{x}$$
$$3x = 12,000 - x$$
$$4x = 12,000$$
$$x = \frac{12,000}{4}$$
$$x = 3,000$$
There are 3,000 children.

(b) $12,000 - x = 12,000 - 3,000 = 9,000$
There are 9,000 adults.

3. Let x = number of adults
Ratio form is adults : children

$$\frac{1}{7} = \frac{x}{84}$$
$$7x = 84$$
$$x = \frac{84}{7}$$
$$x = 12$$

There are 12 adults present.

5. Let c = cost of 7 dozen rolls
Ratio form is dozens : cost

$$\frac{15}{16.5} = \frac{7}{c}$$
$$15c = (16.5)(7)$$
$$15c = 115.5$$
$$c = \frac{115.5}{15}$$
$$c = 7.7$$

The rolls cost $7.70.

7. (a) $\dfrac{15 \text{ wins}}{7 \text{ losses}}$ or $15:7$ (b) $\dfrac{15 \text{ wins}}{25 \text{ games}}$ or $15:25$ or $3:5$

9. Let x = number of female teachers
Then $(27 - x)$ = number of male teachers
Ratio form is female : male

$$\frac{7}{2} = \frac{x}{27 - x}$$
$$2x = 7(27 - x)$$
$$2x = 189 - 7x$$
$$9x = 189$$
$$x = \frac{189}{9}$$
$$x = 21$$

There are 21 female teachers.

11. The team lost $\frac{1}{3}$ of 27 or 9 games. They won 18 games.

The ratio is $\dfrac{18 \text{ wins}}{9 \text{ losses}}$ or $18:9$ or $2:1$.

13. 2 yards = 6 feet = 72 inches; 1 inch = 2.54 cm
Let x = number of centimeters in 72 inches and
use the ratio form in. : cm

$$\frac{1}{2.54} = \frac{72}{x}$$
$$x = 72(2.54)$$
$$x = 182.88$$

There are 182.9 cm in 2 yards.

15. 82 g = 0.082 kg; 1 pound = 0.45 kg
Let x = number of pounds in 82 g and use the
ratio form lb : kg

$$\frac{1}{0.45} = \frac{x}{0.082}$$
$$(0.45)x = 0.082$$
$$x = \frac{0.082}{0.45}$$
$$x = 0.182$$

There are 0.182 pounds in 82 grams.

17. 1 in. = 2.54 cm; let x = number of inches in 5 cm
and use the ratio form in. : cm

$$\frac{1}{2.54} = \frac{x}{5}$$
$$2.54x = 5$$
$$x = \frac{5}{2.54}$$
$$x = 1.97$$

There are 1.97 inches in 5 cm.

19. 1 in. = 2.54 cm; let x = number of cm in 5
inches and use the ratio form in. : cm

$$\frac{1}{2.54} = \frac{5}{x}$$
$$x = 5(2.54)$$
$$x = 12.7$$

There are 12.7 cm in 5 inches.

Section 4.5 Exercises (p. 154)

1. $R = 15\% = 0.15, B = 20$
$A = R \cdot B$
$A = (0.15) \cdot 20$
$A = 3$

3. $R = 4\frac{1}{2}\% = 0.045, B = 90$
$A = R \cdot B$
$A = (0.045) \cdot 90$
$A = 4.05$

5. $R = 36\% = 0.36, B = 1200$
$A = R \cdot B$
$A = (0.36) \cdot 1200$
$A = 432$

7. $B = 120, A = 18$

$$R = \frac{A}{B}$$

$$R = \frac{18}{120} = 0.15$$

$$R = 15\%$$

9. $B = 180, A = 36$

$$R = \frac{A}{B}$$

$$R = \frac{36}{180} = 0.20$$

$$R = 20\%$$

11. $B = 300, A = 60$

$$R = \frac{A}{B}$$

$$R = \frac{60}{300} = 0.20$$

$$R = 20\%$$

13. $R = 18\% = 0.18, A = 30.6$

$$B = \frac{A}{R}$$

$$B = \frac{30.6}{0.18}$$

$$B = 170$$

15. $R = 7\% = 0.07, A = 147$

$$B = \frac{A}{R}$$

$$B = \frac{147}{0.07}$$

$$B = 2100$$

17. $A = 288, R = 12\% = 0.12$

$$B = \frac{A}{R}$$

$$B = \frac{288}{0.12}$$

$$B = 2400$$

19. $R = 15\% = 0.15, B = 320$

A = discounted amount

$A = R \cdot B$

$A = (0.15) \cdot 320$

$A = 48$

The discount is $48.

New price = Ticketed price − Discount

= 320 − 48

= 272

The new price is $272.

21. $10,078.72   **23.** $6540   **25.** $2220

Section 4.5 Supplementary Exercises (p. 154)

1. $R = 12\frac{1}{2}\% = 0.125, B = 120$

$A = R \cdot B$

$A = (0.125) \cdot 120$

$A = 15$

3. $R = 0.2\% = 0.002, B = 75$

$A = R \cdot B$

$A = (0.002) \cdot 75$

$A = 0.15$

5. $B = 200, A = 30$

$$R = \frac{A}{B}$$

$$R = \frac{30}{200} = 0.15$$

$$R = 15\%$$

7. $R = 19\% = 0.19, A = 133$

$$B = \frac{A}{R}$$

$$B = \frac{133}{0.19}$$

$$B = 700$$

9. $R = 7\% = 0.07, A = 84$

$$B = \frac{A}{R}$$

$$B = \frac{84}{0.07}$$

$$B = 1200$$

11. Discount = Original price − Sale price

= 60 − 48

= 12

The amount of the discount was $12.

$A = 12, B = 60$

$$R = \frac{A}{B}$$

$$R = \frac{12}{60} = 0.20$$

$$R = 20\%$$

Vocabulary Quiz (p. 155)

1. e   **2.** a   **3.** b   **4.** d   **5.** c   **6.** g   **7.** f

---

Chapter 4 Review Exercises (p. 155)

1. 38 and 39   **2.** 310, 312, and 314   **3.** 5 quarters and 10 dimes   **4.** $200 at 6% and $800 at 8%
5. (a) 4 and 6 are consecutive *even* integers whose sum is 10.   (b) 5 and 7 are consecutive *odd* integers
whose sum is 12.   **6.** Adding water will weaken a 12% acid solution so that the amount of acid is the
same but the percentage is smaller.   **7.** 10   **8.** Joyce walks at 5 km/hr and Cal at 6 km/hr.
9. (a) 60 mph   (b) 50 mph   (c) 300 miles   **10.** 8 pounds of peanuts and 2 pounds of cashews
11. 13.3 ml   **12.** 1400   **13.** 60   **14.** 82.7   **15.** 3.8   **16.** 0.23   **17.** 900   **18.** 1100

19. 165 miles   **20.** −3   **21.** 4   **22.** $1\frac{1}{4}$ hours   **23.** 260 miles apart   **24.** 1.18 ml

25. 6 dimes and 8 nickels   **26.** 70,000   **27.** 18   **28.** 5 nickels, 10 dimes, and 8 quarters
29. $7\frac{1}{2}$ mph   **30.** 1200   **31.** $450   **32.** $1075   **33.** 2.75 *l*   **34.** 0.4 *l*   **35.** $1.60

Cumulative Review Exercises (p. 157)

1. 5   **2.** 9   **3.** -1 **4.** $10x - 14$ **5.** -28 **6.** $3x^3 + 13x^2$ **7.** 24   **8.** -19 **9.** -9

10. $\dfrac{A - P}{PR}$

Chapter 4 Test (p. 158)

1. 20 and 22   **2.** 1000   **3.** \$13,000 at 7% and \$7000 at 9%   **4.** George walks at 3 mph and Leo at $1\frac{1}{2}$ mph.   **5.** 3.6 miles   **6.** 1.875 kg   **7.** 12 pounds of raisins and 18 pounds of nuts   **8.** 50

9. 92 m   **10.** (a) \$129   (b) \$301

Chapter 5

Section 5.1 Exercises (p. 163)

1. $x^2 \cdot x^5 = x^{2+5} = x^7$ **3.** $q^{12} \cdot q^{20} = q^{12+20} = q^{32}$ **5.** $3y^2 \cdot y^4 = 3y^{2+4} = 3y^6$

7. $(6a^2)(5a^4) = 6 \cdot 5 \cdot a^{2+4} = 30a^6$ **9.** $(xyz^2)(5x^2z) = 5x^{1+2}yz^{2+1} = 5x^3yz^3$

11. $(x^7)^8 = x^{7 \cdot 8} = x^{56}$ **13.** $(3x^4)^4 = 3^4 \cdot x^{4 \cdot 4} = 81x^{16}$ **15.** $(2y^2)^3 = 2^3 \cdot y^{2 \cdot 3} = 8y^6$

17. $(3z^2)^4 = 3^4 \cdot z^{2 \cdot 4} = 81z^8$ **19.** $(10abc^2)^3 = 10^3a^3b^3c^{2 \cdot 3} = 1000a^3b^3c^6$ **21.** y **23.** z

25. $6a^2$ **27.** $16y^4$ **29.** $\dfrac{a^3b^6}{c^{12}}$

Section 5.1 Supplementary Exercises (p. 163)

1. $a^3 \cdot a \cdot a^5 = a^{3+1+5} = a^9$ **3.** $4x^3 \cdot x^7 = 4x^{3+7} = 4x^{10}$

5. $(-3x^2y)(4x^3y) = -3 \cdot 4x^{2+3}y^{1+1} = -12x^5y^2$ **7.** $(-2x^3y)^3 = (-2)^3x^{3 \cdot 3}y^3 = -8x^9y^3$

9. $\dfrac{14x^{13}}{-7x^7} = \dfrac{14}{-7} \cdot \dfrac{x^{13}}{x^7} = -2x^{13-7} = -2x^6$ **11.** $\dfrac{-9x^3y^2z^7}{3xy^2z^3} = \dfrac{-9}{3} \cdot \dfrac{x^3}{x} \cdot \dfrac{y^2}{y^2} \cdot \dfrac{z^7}{z^3} = -3x^{3-1}z^{7-3} = -3x^2z^4$

13. $\left(\dfrac{4x^2y^3}{12xy}\right)^2 = \left(\dfrac{4}{12} \cdot \dfrac{x^2}{x} \cdot \dfrac{y^3}{y}\right)^2 = \left(\dfrac{xy^2}{3}\right)^2 = \dfrac{x^2y^{2 \cdot 2}}{3^2} = \dfrac{x^2y^4}{9}$

15. $\dfrac{(5x^3)^9}{(5x^3)^6} = (5x^3)^{9-6} = (5x^3)^3 = 5^3x^{3 \cdot 3} = 125x^9$

Section 5.2 Exercises (p. 166)

1. $2^{-1} = \dfrac{1}{2}$ **3.** $2^{-3} = \dfrac{1}{2^3} = \dfrac{1}{8}$ **5.** $3^{-2} = \dfrac{1}{3^2} = \dfrac{1}{9}$ **7.** $(-4)^2 = (-4)(-4) = 16$

9. $(-4)^{-2} = \dfrac{1}{(-4)^2} = \dfrac{1}{(-4)(-4)} = \dfrac{1}{16}$ **11.** $5^{-3} = \dfrac{1}{5^3} = \dfrac{1}{5 \cdot 5 \cdot 5} = \dfrac{1}{125}$

13. $(-5)^{-3} = \dfrac{1}{(-5)^3} = \dfrac{1}{(-5)(-5)(-5)} = \dfrac{1}{-125} = -\dfrac{1}{125}$ **15.** $7^0 = 1$ **17.** $-7^0 = -1$

19. $[(5^{-1})(2^{-2})16]^0 = 1$ **21.** $\dfrac{1}{x^7}$ **23.** $\dfrac{x^3}{y^5}$ **25.** $\dfrac{q^6}{p^5}$ **27.** $\dfrac{x^8z^4}{y^3}$ **29.** $\dfrac{x^8y^3}{z^4}$ **31.** y^4

33. $\dfrac{y^4}{x^2}$ **35.** $\dfrac{1}{x^2y^4}$ **37.** $\dfrac{5x^3}{2y^7z^2}$ **39.** $\dfrac{3}{4a^3}$ **41.** $\dfrac{2y^2}{3x^3z}$

Section 5.2 Supplementary Exercises (p. 167)

1. $5^{-2} = \dfrac{1}{5^2} = \dfrac{1}{25}$ **3.** $(-6)^{-2} = \dfrac{1}{(-6)^2} = \dfrac{1}{(-6)(-6)} = \dfrac{1}{36}$ **5.** $-8^0 = -1$ **7.** $(-6)^0 = 1$

9. $(2)^0 \cdot (3)^{-2} = 1 \cdot \dfrac{1}{3^2} = \dfrac{1}{9}$ **11.** $z^{-8} = \dfrac{1}{z^8}$ **13.** $x^5 y^{-2} = x^5 \cdot \dfrac{1}{y^2} = \dfrac{x^5}{y^2}$

15. $x^{-2} y^{-7} = \dfrac{1}{x^2} \cdot \dfrac{1}{y^7} = \dfrac{1}{x^2 y^7}$ **17.** $a^{-2} b^3 c^{-5} = \dfrac{1}{a^2} \cdot b^3 \cdot \dfrac{1}{c^5} = \dfrac{b^3}{a^2 c^5}$ **19.** $\dfrac{1}{x^{-7}} = x^7$

21. $\dfrac{5z^3}{4x^4 y^2}$ **23.** $\dfrac{-x^2 y^2 z}{3}$ **25.** $\dfrac{y^{12} z^2}{9x^{10}}$

Section 5.3 Exercises (p. 170)

1. $75.8 = 7.58 \cdot 10$ **3.** $8750 = 8.57 \cdot 10^3$ **5.** $5280 = 5.28 \cdot 10^3$ **7.** $0.0056 = 5.6 \cdot 10^{-3}$

9. $0.000550 = 5.50 \cdot 10^{-4}$ **11.** $205{,}000{,}000 = 2.05 \cdot 10^8$ **13.** $0.000205 = 2.05 \cdot 10^{-4}$

15. $0.0000000000000000000000016734 = 1.6734 \cdot 10^{-24}$

17. $104{,}300{,}000 \cdot 206{,}000{,}000 = 1.043 \cdot 10^8 \cdot 2.06 \cdot 10^8$
$$= 1.043 \cdot 2.06 \cdot 10^8 \cdot 10^8$$
$$= 2.14858 \cdot 10^{16}$$

19. $0.000026 \cdot 0.00000012 = 2.6 \cdot 10^{-5} \cdot 1.2 \cdot 10^{-7}$ **21.** $4.95616 \cdot 10^{15}$
$$= 2.6 \cdot 1.2 \cdot 10^{-5} \cdot 10^{-7}$$
$$= 3.12 \cdot 10^{-12}$$

23. $3.57911 \cdot 10^{14}$ **25.** $4.096 \cdot 10^{-8}$ **27.** $2.496 \cdot 10^{-17}$ **29.** $8.9696257 \cdot 10^{14}$

Section 5.3 Supplementary Exercises (p. 170)

1. $527 = 5.27 \cdot 10^2$ **3.** $4{,}090{,}103.2 = 4.0901032 \cdot 10^6$ **5.** $-12{,}325{,}795 = -1.2325795 \cdot 10^7$

7. $0.0023 = 2.3 \cdot 10^{-3}$ **9.** $-0.002963 = -2.963 \cdot 10^{-3}$ **11.** $0.52196 = 5.2196 \cdot 10^{-1}$

13. $(205{,}000)(1{,}200{,}000) = (2.05 \cdot 10^5)(1.2 \cdot 10^6)$
$$= (2.05)(1.2) \cdot 10^5 \cdot 10^6$$
$$= 2.46 \cdot 10^{11}$$

15. $(0.000052)(210{,}000) = (5.2 \cdot 10^{-5})(2.1 \cdot 10^5)$
$$= (5.2)(2.1) \cdot 10^{-5} \cdot 10^5$$
$$= 10.92 \cdot 10^0$$
$$= 10.92$$
$$= 1.092 \cdot 10$$

Section 5.4 Exercises (p. 174)

1. (a) $-4x^2 + 3x + 7$ (b) degree 2   **3.** (a) $-3x + 8$ (b) degree 1   **5.** (a) 4   (b) degree 0

7. (a) $12y^{12} + 10y - 5$ (b) degree 12   **9.** (a) $2x^3 + 16x^2 - x + 12$ (b) degree 3

11. $(2x^2 - 7x + 12) + (3x^2 + 10x + 20) = 2x^2 + 3x^2 - 7x + 10x + 12 + 20$
$$= 5x^2 + 3x + 32$$

13. $(8x^3 - 2x^2 + x - 6) + (8x^3 - x^2 - 5x - 11) = 8x^3 + 8x^3 - 2x^2 - x^2 + x - 5x - 6 - 11$
$$= 16x^3 - 3x^2 - 4x - 17$$

15. $(x^2 - 7x + 12) + (x^3 - 5x^2 + 10) = x^3 + x^2 - 5x^2 - 7x + 12 + 10$
$$= x^3 - 4x^2 - 7x + 22$$

17. $(5z^2 - 10z + 4) + (3z^3 - 5z^2 + 2z - 8) + (10z^3 - 5z + 11)$
$$= 3z^3 + 10z^3 + 5z^2 - 5z^2 - 10z + 2z - 5z + 4 - 8 + 11$$
$$= 13z^3 - 13z + 7$$

19. $(21x^4 - 5x + 10) + (-20x^4 + x^2 - 10) + (5x - x^2 - x^4)$ **21.** $y^2 - 10y + 15$
$$= 21x^4 - 20x^4 - x^4 + x^2 - x^2 - 5x + 5x + 10 - 10$$
$$= 0$$

23. $2t^2 - 8t + 26$ **25.** $-3x^3 + 3x^2 - 8x - 12$ **27.** $-2x^8 - 9x^4 - 3x - 2$

29. $3y^3 + 2y^2 - 3y - 12$ **31.** $x^2 y + 13xy^2 - 11x + 6$ **33.** $-5x^2 y - xy^2 - 20xy + 5x - 12$

35. $6.372x^2 - 0.001x - 1.814$

Section 5.4 Supplementary Exercises (p. 175)

1. Degree 7 since 7 is the highest exponent of x

3. $(x^2 + 3x - 4) + (2x^2 - 5x + 1) = x^2 + 2x^2 + 3x - 5x - 4 + 1$
$$= 3x^2 - 2x - 3$$

5. $(-a^2 + 2a + 5) + (2a - 3a^2 + 7) = -a^2 - 3a^2 + 2a + 2a + 5 + 7$
$$= -4a^2 + 4a + 12$$

7. $(2a^2 - 9) + (5a^2 - 3a - 4) + (-4a^2 - a + 7) = 2a^2 + 5a^2 - 4a^2 - 3a - a - 9 - 4 + 7$
$$= 3a^2 - 4a - 6$$

9. $(2x^2 - 7x + 3) - (x^2 - 5x - 9) = 2x^2 - 7x + 3 - x^2 + 5x + 9$
$$= 2x^2 - x^2 - 7x + 5x + 3 + 9$$
$$= x^2 - 2x + 12$$

11. $(x^3 - 3x^2 + 2x - 1) - (x^3 + x^2 - 5) + (2x^3 - 4x^2 - 3x + 5)$
$$= x^3 - 3x^2 + 2x - 1 - x^3 - x^2 + 5 + 2x^3 - 4x^2 - 3x + 5$$
$$= x^3 - x^3 + 2x^3 - 3x^2 - x^2 - 4x^2 + 2x - 3x - 1 + 5 + 5$$
$$= 2x^3 - 8x^2 - x + 9$$

13. $-(a^3 - 2a^2 + 3a - 1) - (5a^3 + 4a^2 - a + 4)$
$$= -a^3 + 2a^2 - 3a + 1 - 5a^3 - 4a^2 + a - 4$$
$$= -a^3 - 5a^3 + 2a^2 - 4a^2 - 3a + a + 1 - 4$$
$$= -6a^3 - 2a^2 - 2a - 3$$

15. $(x^2 + 10x - 5) - (3x^2 - 9x + 4) = x^2 + 10x - 5 - 3x^2 + 9x - 4$
$$= x^2 - 3x^2 + 10x + 9x - 5 - 4$$
$$= -2x^2 + 19x - 9$$

17. $(x^2y + 3xy^2 - 4x^2y^2) + (5xy^2 - 4x^2y + 6x^2y^2)$
$$= -4x^2y^2 + 6x^2y^2 + x^2y - 4x^2y + 3xy^2 + 5xy^2$$
$$= 2x^2y^2 - 3x^2y + 8xy^2$$

19. $(a - 2b) - (2b + a) + (3a - 4b) = a - 2b - 2b - a + 3a - 4b$
$$= a - a + 3a - 2b - 2b - 4b$$
$$= 3a - 8b$$

Section 5.5 Exercises (p. 180)

1. $2(4x + 3) = 2 \cdot 4x + 2 \cdot 3 = 8x + 6$ **3.** $x(4x + 3) = x \cdot 4x + x \cdot 3 = 4x^2 + 3x$

5. $5y^2(y - 1) = 5y^2 \cdot y + 5y^2 \cdot (-1) = 5y^3 - 5y^2$

7. $2 + x(x + 1) = 2 + x \cdot x + x \cdot 1 = 2 + x^2 + x = x^2 + x + 2$

9. $4x - x(2x - 1) + 10 = 4x - x(2x) - x(-1) + 10$
$$= 4x - 2x^2 + x + 10$$
$$= -2x^2 + 5x + 10$$

11. $5x(x^2 - 7x + 1) = 5x \cdot x^2 + 5x(-7x) + 5x \cdot 1$
$$= 5x^3 - 35x^2 + 5x$$

13. $4x(x^2 + x - 1) = 4x \cdot x^2 + 4x \cdot x + 4x(-1)$
$$= 4x^3 + 4x^2 - 4x$$

15. $-11y(y^5 - 2y^4 + 6) = -11y \cdot y^5 - 11y(-2y^4) - 11y(6)$
$$= -11y^6 + 22y^5 - 66y$$

17. $20xy^2(3x^2y - x^2 + y - 1) = 20xy^2 \cdot 3x^2y + 20xy^2(-x^2) + 20xy^2 \cdot y + 20xy^2(-1)$
$$= 60x^3y^3 - 20x^3y^2 + 20xy^3 - 20xy^2$$

19. $(x + 5)(x + 2) = x(x + 2) + 5(x + 2)$ **21.** $(x - 6)(x + 3) = x(x + 3) - 6(x + 3)$
$$= x^2 + 2x + 5x + 10 \qquad\qquad = x^2 + 3x - 6x - 18$$
$$= x^2 + 7x + 10 \qquad\qquad\quad = x^2 - 3x - 18$$

23. $(2x + 3)(x + 1) = 2x(x + 1) + 3(x + 1)$ **25.** $3a^2 - a - 4$ **27.** $14t^2 + 17t + 5$
$$= 2x^2 + 2x + 3x + 3$$
$$= 2x^2 + 5x + 3$$

29. $14t^2 - 3t - 5$ **31.** $x^2 - 4$ **33.** $4y^2 - 9$ **35.** $x^4 + 5x^2 - 14$

37. $5x^4 - 2x^3 - 5x^2 - 48x + 20$ **39.** $x^4 - 3x^3 - 3x^2 + 7x + 6$ **41.** (a) $x^2 + 16x + 64$

(b) No, $(x + 8)^2 \neq x^2 + 64$ **43.** (a) $x^3 + 3x^2 + 3x + 1$ (b) No, $(x + 1)^3 \neq x^3 + 1$

45. (a) $2l + 3$ (b) $l + 4$ (c) Volume = (length)(width)(depth) = $l(2l + 3)(l + 4)$ (d) 3,220

Section 5.5 Supplementary Exercises (p. 182)

1. $5(3x - 2) = 5 \cdot 3x + 5 \cdot (-2) = 15x - 10$ **3.** $y(3y - 7) = y \cdot 3y + y \cdot (-7) = 3y^2 - 7y$

5. $2x^3(3x^3 - 2x + 1) = 2x^3 \cdot 3x^3 + 2x^3(-2x) + 2x^3 \cdot 1$
$$= 6x^6 - 4x^4 + 2x^3$$

7. $(2x + 1)(x - 5) = 2x(x - 5) + 1(x - 5)$ **9.** $(4x - 1)(x^2 + 1) = 4x(x^2 + 1) - 1(x^2 + 1)$
$$= 2x^2 - 10x + x - 5 \qquad\qquad = 4x^3 + 4x - x^2 - 1$$
$$= 2x^2 - 9x - 5 \qquad\qquad\quad = 4x^3 - x^2 + 4x - 1$$

11. $5x - x(3x + 4) + 2x^2 = 5x - 3x^2 - 4x + 2x^2$
$$= -x^2 + x$$

13. $(2x + y)(3x + 2y) - (x^2 + 5xy - y^2) = 2x(3x + 2y) + y(3x + 2y) - x^2 - 5xy + y^2$
$$= 6x^2 + 4xy + 3xy + 2y^2 - x^2 - 5xy + y^2$$
$$= 5x^2 + 2xy + 3y^2$$

15. $(a - 2)(a^2 - 3a + 4) = a(a^2 - 3a + 4) - 2(a^2 - 3a + 4)$ **17.** $2 - x(3 - 4x) = 2 - 3x + 4x^2$
$$= a^3 - 3a^2 + 4a - 2a^2 + 6a - 8 \qquad\qquad = 4x^2 - 3x + 2$$
$$= a^3 - 5a^2 + 10a - 8$$

19. $(1 - x)^2 = 1 - x - x + x^2$ **21.** $a^2 - b^2 - 2bc - c^2$ **23.** $x^4 - x^3 - 2x^2 + 3x - 1$
$$= 1 - 2x + x^2$$

25. Volume = $x^3 + 2x^2 - 8x$

Section 5.6 Exercises (p. 187)

1. $\dfrac{4x + 12}{4} = \dfrac{4x}{4} + \dfrac{12}{4} = x + 3$ **3.** $\dfrac{15x - 5}{5} = \dfrac{15x}{5} - \dfrac{5}{5} = 3x - 1$

5. $\dfrac{14y - 21}{7} = \dfrac{14y}{7} - \dfrac{21}{7} = 2y - 3$ **7.** $\dfrac{24x - 32y}{4} = \dfrac{24x}{4} - \dfrac{32y}{4} = 6x - 8y$

9. $\dfrac{24x - 32y}{4x} = \dfrac{24\cancel{x}}{4\cancel{x}} - \dfrac{32y}{4x} = 6 - \dfrac{8y}{x}$

11. $\dfrac{8x^3 + 12x^2 + 4x}{4x} = \dfrac{8x^3}{4x} + \dfrac{12x^2}{4x} + \dfrac{4x}{4x} = 2x^2 + 3x + 1$

13. $\dfrac{16x^2y^2 - 8xy^2 + 32x^2y}{8xy} = \dfrac{16x^2y^2}{8xy} - \dfrac{8xy^2}{8xy} + \dfrac{32x^2y}{8xy}$

$$= \dfrac{16}{8} \cdot \dfrac{x^2}{x} \cdot \dfrac{y^2}{y} - \dfrac{8}{8} \cdot \dfrac{x}{x} \cdot \dfrac{y^2}{y} + \dfrac{32}{8} \cdot \dfrac{x^2}{x} \cdot \dfrac{y}{y}$$

$$= 2xy - y + 4x$$

15.
$$\begin{array}{r}
x + 5 \\
x + 2 \overline{\smash{\big)}\ x^2 + 7x + 10} \\
\underline{x^2 + 2x} \\
5x + 10 \\
\underline{5x + 10} \\
0
\end{array}$$

$(x^2 + 7x + 10) \div (x + 2) = x + 5$

17.
$$\begin{array}{r}
x + 2 \\
x + 3 \overline{\smash{\big)}\ x^2 + 5x + 6} \\
\underline{x^2 + 3x} \\
2x + 6 \\
\underline{2x + 6} \\
0
\end{array}$$

$(x^2 + 5x + 6) \div (x + 3) = x + 2$

19.
$$\begin{array}{r} x - 10 \\ x + 2 \overline{\smash{\big)}\, x^2 - 8x - 20} \\ \underline{x^2 + 2x} \\ -10x - 20 \\ \underline{-10x - 20} \\ 0 \end{array}$$

$(x^2 - 8x - 20) \div (x + 2) = x - 10$

21.
$$\begin{array}{r} 3x + 7 \\ x^2 - 2 \overline{\smash{\big)}\, 3x^3 + 7x^2 + 0x - 5} \\ \underline{3x^3 \qquad - 6x} \\ 7x^2 + 6x - 5 \\ \underline{7x^2 \qquad - 14} \\ 6x + 9 \end{array}$$

$(3x^3 + 7x^2 - 5) \div (x^2 - 2) = 3x + 7 + \dfrac{6x + 9}{x^2 - 2}$

23. $z + 3$ **25.** $x^2 - x + 2$ **27.** $2y^2 + 3y + 1 + \dfrac{5y + 1}{y^2 - y - 3}$ **29.** $x^3 + x^2 + x + 1$

31. (a) $-\dfrac{-12}{-5} \cdot \dfrac{(-1)}{(-1)} = -\dfrac{12}{5}$

(b) The opposite of the negative number $\dfrac{-19}{8}$ is the positive number $\dfrac{19}{8}$.

    Algebraically, this means

$$-\dfrac{-19}{8} = \dfrac{19}{8}$$

(c) The opposite of the negative number $\dfrac{37}{-6}$ is the positive number $\dfrac{37}{6}$.

    Algebraically, this means

$$-\dfrac{37}{-6} = \dfrac{37}{6}$$

33. $1352.6667y^2 - 5358y - 30623.333$

Section 5.6 Supplementary Exercises (p. 188)

1. $\dfrac{8y - 12x}{4} = \dfrac{8y}{4} - \dfrac{12x}{4} = 2y - 3x$ **3.** $\dfrac{-50x + 30y - 60z}{-10} = \dfrac{-50x}{-10} + \dfrac{30y}{-10} - \dfrac{60z}{-10} = 5x - 3y + 6z$

5. $\dfrac{24x^3 - 12x^2 + 3x}{3x} = \dfrac{24x^3}{3x} - \dfrac{12x^2}{3x} + \dfrac{3x}{3x} = 8x^2 - 4x + 1$

7. $\dfrac{-14x^2y^3 - 21xy^2 + 56x^2y^2}{-7xy^2} = \dfrac{-14x^2y^3}{-7xy^2} - \dfrac{21xy^2}{-7xy^2} + \dfrac{56x^2y^2}{-7xy^2}$

$$= \dfrac{-14}{-7} \cdot \dfrac{x^2}{x} \cdot \dfrac{y^3}{y^2} - \dfrac{21}{-7} \cdot \dfrac{x}{x} \cdot \dfrac{y^2}{y^2} + \dfrac{56}{-7} \cdot \dfrac{x^2}{x} \cdot \dfrac{y^2}{y^2}$$

$$= 2xy + 3 - 8x$$

9. $\dfrac{3x^2y^2z - 6xy^3z^2 + 9xyz^2}{3x^2y^2z} = \dfrac{3x^2y^2z}{3x^2y^2z} - \dfrac{6xy^3z^2}{3x^2y^2z} + \dfrac{9xyz^2}{3x^2y^2z}$

$$= 1 - \dfrac{2yz}{x} + \dfrac{3z}{xy}$$

11. $\dfrac{-120x^4 + 60x^3 - 30x^2 + 90x}{30x} = \dfrac{-120x^4}{30x} + \dfrac{60x^3}{30x} - \dfrac{30x^2}{30x} + \dfrac{90x}{30x}$

$$= -4x^3 + 2x^2 - x + 3$$

13. $x + 1 \overline{) x^2 - 2x + 3}$ with quotient $x - 3$

$$\begin{array}{r} x - 3 \\ x + 1 \overline{) x^2 - 2x + 3} \\ \underline{x^2 + x} \\ -3x + 3 \\ \underline{-3x - 3} \\ 6 \end{array}$$

15.

$$\begin{array}{r} y - 9 \\ y + 3 \overline{) y^2 - 6y + 4} \\ \underline{y^2 + 3y} \\ -9y + 4 \\ \underline{-9y - 27} \\ 31 \end{array}$$

$$(x^2 - 2x + 3) \div (x + 1) = x - 3 + \frac{6}{x + 1}$$

$$(y^2 - 6y + 4) \div (y + 3) = y - 9 + \frac{31}{y + 3}$$

17.

$$\begin{array}{r} 5x^2 - 2x + 1 \\ x + 3 \overline{) 5x^3 + 13x^2 - 5x + 3} \\ \underline{5x^3 + 15x^2} \\ -2x^2 - 5x \\ \underline{-2x^2 - 6x} \\ x + 3 \\ \underline{x + 3} \\ 0 \end{array}$$

$$(5x^3 + 13x^2 - 5x + 3) \div (x + 3) = 5x^2 - 2x + 1$$

19.

$$\begin{array}{r} x^3 - x^2 - x \\ x - 1 \overline{) x^4 - 2x^3 + 0 \cdot x^2 + x - 1} \\ \underline{x^4 - x^3} \\ -x^3 + 0 \cdot x^2 \\ \underline{-x^3 + x^2} \\ -x^2 + x \\ \underline{-x^2 + x} \\ -1 \end{array}$$

21. $4 + \dfrac{-3x^2 + 2x + 15}{x^3 - 5}$

$$(x^4 - 2x^3 + x - 1) \div (x - 1) = x^3 - x^2 - x + \frac{-1}{x - 1}$$

23. $-x + 1 + \dfrac{4}{x - 1}$ **25.** $x - 1$

Vocabulary Quiz (p. 189)

1. a   **2.** h   **3.** j   **4.** b   **5.** f   **6.** i   **7.** c   **8.** d   **9.** g   **10.** e

---

Chapter 5 Review Exercises (p. 189)

1. $8x^3$ **2.** $3y^7$ **3.** $8x^6$ **4.** $100a^2 b^4 c^6$ **5.** $27z^6$ **6.** $\dfrac{8x^6 y^3}{z^{12}}$ **7.** $\dfrac{1}{25}$ **8.** $-\dfrac{1}{125}$

9. $\dfrac{64}{9}$ **10.** 1   **11.** $\dfrac{1}{32}$ **12.** undefined   **13.** $\dfrac{1}{z^4}$ **14.** $\dfrac{a^6}{b^3}$ **15.** $\dfrac{y^3}{x^2 z^4}$ **16.** $\dfrac{z^4}{x^2 y^3}$

17. $\dfrac{4ac^9}{5b^2}$ **18.** $\dfrac{15y^5}{x^4}$ **19.** $8.96 \cdot 10$ **20.** $5.78 \cdot 10^2$ **21.** $5.71 \cdot 10^{-2}$ **22.** $5.09 \cdot 10^{-4}$

23. $2.08 \cdot 10^8$ **24.** $5.5 \cdot 10^{11}$ **25.** $8.224 \cdot 10^9$ **26.** $2 \cdot 10^2$ **27.** degree 2
28. degree 3   **29.** degree 1   **30.** degree 4   **31.** $5x^2 - 9x + 17$ **32.** $3x^2 + 17x - 9$
33. $14y^3 + 3y^2 - 3$ **34.** $4x^3 + x^2 - 23x + 37$ **35.** $-2x^2 - 8x + 5$ **36.** $y^4 - 11y^3 + 14y - 6$
37. $3t^3 - 7t^2 + 4t + 7$ **38.** $-3x^3 + 4x^2 - 7x + 2$ **39.** $5x^2 - 12x + 12$ **40.** $6x^3 - 10x^2$
41. $6y^3 - 42y^2 + 18y$ **42.** $-2z^2 + 5z + 10$ **43.** $6x^2 - x - 35$ **44.** $8t^2 - 6t + 1$
45. $9x^2 - 4y^2$ **46.** $6x^3 - 29x^2 + 37x - 5$ **47.** $x - 2$

48. $\dfrac{3}{z^2} - \dfrac{4}{z}$ **49.** $2x - 20 + \dfrac{-1}{x + 2}$ **50.** $3t - 5 + \dfrac{14t - 4}{t^2 - 1}$

Cumulative Review Exercises (p. 191)

1. 7   **2.** -4 **3.** $2P + 1000$ **4.** $-18x - 15$ **5.** $-14y^3$ **6.** 188.49556   **7.** 1

8. 102, 104   **9.** $x \geq -2$

10. 10 quarters, 20 dimes

Chapter 5 Test (p. 192)

1. $6t^7$ **2.** $64x^3 y^6$ **3.** $36s^4$ **4.** $\dfrac{x^3}{y^7}$ **5.** $\dfrac{5a}{8b^3}$ **6.** $\dfrac{4}{3y}$ **7.** (a) $8.09 \cdot 10^5$

(b) $3.05 \cdot 10^{-3}$ **8.** degree 6   **9.** $5x^3 + 3x + 18$ **10.** $5y^3 - 2y^2 - 17y$

11. $t^3 - 8t^2 + 5t + 28$ **12.** $6x^2 - 10x$ **13.** $15v^5 - 33v^4 + 18v^3$ **14.** $21x^2 - 62x + 16$

15. $10y^2 + 29y - 21$ **16.** $4x^2 + 12xy + 9y^2$ **17.** $6x - 5$ **18.** $y - 2 + \dfrac{3}{xy}$

19. $x + 3 + \dfrac{4}{2x + 3}$ **20.** $z + 6 + \dfrac{17z - 52}{z^2 - z + 8}$

Chapter 6

Section 6.1 Exercises (p. 198)

1. $12 = 2 \cdot 2 \cdot 3$ or $2^2 \cdot 3$ **3.** $36 = 2 \cdot 2 \cdot 3 \cdot 3$ or $2^2 \cdot 3^2$

5. $360 = 10 \cdot 36 = 2 \cdot 5 \cdot 2 \cdot 2 \cdot 3 \cdot 3$ or $2^3 \cdot 3^2 \cdot 5$ **7.** $98 = 2 \cdot 7 \cdot 7 = 2 \cdot 7^2$

9. $980 = 10 \cdot 98 = 2 \cdot 5 \cdot 2 \cdot 7 \cdot 7$ or $2^2 \cdot 5 \cdot 7^2$

11. $150 = 10 \cdot 15 = 2 \cdot 5 \cdot 3 \cdot 5$ or $2 \cdot 3 \cdot 5^2$

13. $10 = 2 \cdot 5$ **15.** $42 = 2 \cdot 3 \cdot 7$ **17.** $56 = 2 \cdot 2 \cdot 2 \cdot 7$
 $15 = 3 \cdot 5$ $56 = 2 \cdot 2 \cdot 2 \cdot 7$ $140 = 10 \cdot 14 = 2 \cdot 5 \cdot 2 \cdot 7$
 $\text{GCF} = 5$ $\text{GCF} = 2 \cdot 7 = 14$ $\text{GCF} = 2 \cdot 2 \cdot 7 = 28$

19. $72 = 2 \cdot 2 \cdot 2 \cdot 3 \cdot 3$ **21.** 2   **23.** 50
 $88 = 2 \cdot 2 \cdot 2 \cdot 11$
 $\text{GCF} = 2 \cdot 2 \cdot 2 = 8$

25. (a) $15 = 3 \cdot 5$ (b) $12 = 2 \cdot 2 \cdot 3$ (c) $6 = 2 \cdot 3$ **27.** 15   **29.** 12
 $28 = 2 \cdot 2 \cdot 7$ $24 = 2 \cdot 2 \cdot 2 \cdot 3$ $35 = 5 \cdot 7$
 $\text{GCF} = 1$ $35 = 5 \cdot 7$ $121 = 11 \cdot 11$
  $\text{GCF} = 1$ $\text{GCF} = 1$

Section 6.1 Supplementary Exercises (p. 199)

1. $39 = 3 \cdot 13$ **3.** $125 = 5 \cdot 25 = 5 \cdot 5 \cdot 5$ or 5^3 **5.** $210 = 10 \cdot 21 = 2 \cdot 5 \cdot 3 \cdot 7$

7. $100 = 2 \cdot 2 \cdot 5 \cdot 5$ or $2^2 \cdot 5^2$

9. $8 = 2 \cdot 2 \cdot 2$ **11.** $20 = 2 \cdot 2 \cdot 5$ **13.** $75 = 3 \cdot 5 \cdot 5$ **15.** $85 = 5 \cdot 17$
 $12 = 2 \cdot 2 \cdot 3$ $15 = 3 \cdot 5$ $125 = 5 \cdot 5 \cdot 5$ $119 = 7 \cdot 17$
 $\text{GCF} = 4$ $30 = 2 \cdot 3 \cdot 5$ $50 = 2 \cdot 5 \cdot 5$ $34 = 2 \cdot 17$
  $\text{GCF} = 5$ $\text{GCF} = 5 \cdot 5 = 25$ $\text{GCF} = 17$

Section 6.2 Exercises (p. 202)

1. $6x + 12 = 2 \cdot 3 \cdot x + 2 \cdot 2 \cdot 3$ **3.** $6x + 2 = 2 \cdot 3x + 2$ **5.** $6x^2 + 2x = 2 \cdot 3 \cdot x \cdot x + 2x$
$= 2 \cdot 3(x + 2)$ $= 2(3x + 1)$ $= 2x(3x + 1)$
$= 6(x + 2)$

7. $7z - 14 = 7z - 2 \cdot 7$ **9.** $x^4 + 5x^3 = x \cdot x \cdot x \cdot x + 5 \cdot x \cdot x \cdot x$
$= 7(z - 2)$ $= x \cdot x \cdot x(x + 5)$
$= x^3(x + 5)$

11. $6t^2 - 3t + 12 = 2 \cdot 3 \cdot t \cdot t - 3t + 2 \cdot 2 \cdot 3$
$$= 3(2t^2 - t + 4)$$

13. $-30x^2 + 10x - 15 = -2 \cdot 3 \cdot 5 \cdot x \cdot x + 2 \cdot 5x - 3 \cdot 5$
$$= -5(2 \cdot 3 \cdot x \cdot x - 2x + 3)$$
$$= -5(6x^2 - 2x + 3)$$

15. $-9x^4 - 18x^3 + 27x^2 = -3 \cdot 3 \cdot x \cdot x \cdot x \cdot x - 2 \cdot 3 \cdot 3 \cdot x \cdot x \cdot x + 3 \cdot 3 \cdot 3 \cdot x \cdot x$
$$= -3 \cdot 3 \cdot x \cdot x(x \cdot x + 2x - 3)$$
$$= -9x^2(x^2 + 2x - 3)$$

17. $-x^3 + x^2 - x = -x \cdot x \cdot x + x \cdot x - x$
$$= -x(x \cdot x - x + 1)$$
$$= -x(x^2 - x + 1)$$

19. $-30a^{12}b^2c^5 + 10a^3b^3c^3 - 15ab^2c^2 = -5ab^2c^2(6a^{11}c^3 - 2a^2bc + 3)$　　**21.** $-5xy^2(xy - 3)$
23. $4abc(ac + 3b - 1)$　　**25.** $3tv(3tv - 4 + 5t)$　　**27.** $x(x + 4y)$　　**29.** (a) $C = 20x + 10,000$
(b) $20x + 10,000 = 20(x + 500)$　　(c) $C = 20(x + 500)$　　**31.** $117, 117$

Section 6.2 Supplementary Exercises (p. 204)

1. $5x + 15y = 5(x + 3y)$　　**3.** $5x^2 - 7x = x(5x - 7)$　　**5.** $4x^2y - 8xy = 4xy(x - 2)$
7. $5x^3y - 20x^2y + 5xy = 5xy(x^2 - 4x + 1)$　　**9.** $ab^2c - 4abc^2 + 8a^2bc = abc(b - 4c + 8a)$
11. $30m^4n^2 - 12mn^2 + 18m^2n^2 = 6mn^2(5m^3 - 2 + 3m)$
13. $5x^7y^6 + 15x^3y^5 - 20x^3y^4 = 5x^3y^4(x^4y^2 + 3y - 4)$
15. $-2x^9y^5 + 4x^7y^7 - 8x^7y^5 = -2x^7y^5(x^2 - 2y^2 + 4)$

Section 6.3 Exercises (p. 207)

1. $x^2 + 7x + 10 = (x + 5)(x + 2)$　　**3.** $x^2 - 8x + 12 = (x - 6)(x - 2)$
5. $x^2 - 8x + 16 = (x - 4)(x - 4)$ or $(x - 4)^2$　　**7.** $y^2 - 12y + 20 = (y - 10)(y - 2)$
9. $x^2 + 7x + 6 = (x + 6)(x + 1)$　　**11.** $x^2 + 3x - 10 = (x + 5)(x - 2)$
13. $x^2 - 3x + 10$ is unfactorable　　**15.** $x^2 + 2x - 24 = (x + 6)(x - 4)$
17. $x^2 + 10x - 24 = (x + 12)(x - 2)$　　**19.** $t^2 - 7t - 18 = (t - 9)(t + 2)$　　**21.** $x(x + 4)(x + 3)$
23. $5(x - 1)(x - 1)$ or $5(x - 1)^2$　　**25.** $3(y + 8)(y - 2)$　　**27.** $b^2c^2(a - 6)(a - 3)$
29. $10(a - 4)(a + 3)$　　**31.** $-8y^2(y + 8)(y - 5)$　　**33.** $5x(x - 1)(x - 1)$ or $5x(x - 1)^2$
35. $2m(2m^2 - 14m + 3)$　　**37.** $a^2(a + 4)(a + 4)$ or $a^2(a + 4)^2$　　**39.** $(x + a)(x + b)$

Section 6.3 Supplementary Exercises (p. 208)

1. $a^2 + 6a + 8 = (a + 4)(a + 2)$　　**3.** $x^2 - 7x - 8 = (x - 8)(x + 1)$　　**5.** $m^2 - 7m + 12 = (m - 4)(m - 3)$
7. $a^2 + 3a + 4$ is unfactorable　　**9.** $w^2 - 10w + 16 = (w - 8)(w - 2)$
11. $a^2 - 8a + 15 = (a - 5)(a - 3)$　　**13.** $y^2 - 10y + 24 = (y - 6)(y - 4)$
15. $y^2 + 4y - 24$ is unfactorable　　**17.** $t^2 - 19t + 90 = (t - 9)(t - 10)$
19. $q^2 - 2q + 1 = (q - 1)(q - 1)$ or $(q - 1)^2$　　**21.** $5(a + 2)(a + 1)$　　**23.** $4(p + 10)(p - 1)$
25. $7m^2(x - 4)(x - 8)$

Section 6.4 Exercises (p. 211)

1. $x^2 - 16 = (x)^2 - (4)^2$　　**3.** $z^2 - 49 = (z)^2 - (7)^2$　　**5.** $x^2 - 100 = (x)^2 - (10)^2$
$$= (x - 4)(x + 4) \qquad\qquad = (z - 7)(z + 7) \qquad\qquad = (x - 10)(x + 10)$$

7. $t^2 - 121 = (t)^2 - (11)^2$　　**9.** $4x^2 - 4 = 4(x^2 - 1)$
$$= (t - 11)(t + 11) \qquad\qquad = 4(x - 1)(x + 1)$$

11. $25 - x^2 = (5)^2 - (x)^2$　　**13.** $x^2 - 4y^2 = (x)^2 - (2y)^2$
$$= (5 - x)(5 + x) \qquad\qquad = (x - 2y)(x + 2y)$$

15. $4x^2 - 49 = (2x)^2 - (7)^2$
    $\qquad = (2x - 7)(2x + 7)$

17. $2y^2 - 50 = 2(y^2 - 25)$
    $\qquad = 2(y - 5)(y + 5)$

19. $z^4 - 16 = (z^2)^2 - (4)^2$
    $\qquad = (z^2 - 4)(z^2 + 4)$
    $\qquad = (z - 2)(z + 2)(z^2 + 4)$

21. $(x - y)(x + y)$ 23. $3(x^3 - 3)(x^3 + 3)$

25. unfactorable   27. $(x^2 - y)(x^2 + y)$ 29. $(12a - 1)(12a + 1)$ 31. $(xy - z)(xy + z)$

Section 6.4 Supplementary Exercises (p. 211)

1. $a^2 - 9 = (a)^2 - (3)^2$
   $\qquad = (a - 3)(a + 3)$

3. $m^2 + 121$ is a sum of perfect squares and is not factorable

5. $4y^2 - 9x^2 = (2y)^2 - (3x)^2$
   $\qquad = (2y - 3x)(2y + 3x)$

7. $9x^2 - 1 = (3x)^2 - (1)^2$
   $\qquad = (3x - 1)(3x + 1)$

9. $px^2 - py^2 = p(x^2 - y^2)$
   $\qquad = p(x - y)(x + y)$

11. $x^4 - x^4 = (x^2)^2 - (y^2)^2$
    $\qquad = (x^2 - y^2)(x^2 + y^2)$
    $\qquad = (x - y)(x + y)(x^2 + y^2)$

13. $5x^2 - 125 = 5(x^2 - 25)$
    $\qquad = 5(x - 5)(x + 5)$

15. $7a^2 b^4 - 28 = 7(a^2 b^4 - 4)$
    $\qquad = 7(ab^2 - 2)(ab^2 + 2)$

Section 6.5 Exercises (p. 214)

1. $2x^2 + 7x + 3 = (2x + 1)(x + 3)$ 3. $4x^2 + 4x - 1$ is not factorable

5. $3x^2 + 4x + 1 = (3x + 1)(x + 1)$ 7. $3x^2 + 11x + 8 = (3x + 8)(x + 1)$

9. $4x^2 - x - 4$ is not factorable   11. $7x^2 - 13x - 2 = (7x + 1)(x - 2)$

13. $9x^2 - 1 = (3x - 1)(3x + 1)$ 15. $15x^2 + 16x + 1 = (15x + 1)(x + 1)$

17. $6x^2 - 12x + 4 = 2(3x^2 - 6x + 2)$ 19. $6x^2 - 10x + 1$ is not factorable   21. $x(x + 2)$

23. unfactorable   25. $(x - 1)(x + 1)$ 27. $(5x + 1)(x - 1)$ 29. $(3x - 2)(2x + 1)$

Section 6.5 Supplementary Exercises (p. 215)

1. $2p^2 + 5p + 3 = (2p + 3)(p + 1)$ 3. $5q^2 + 4q - 1 = (5q - 1)(q + 1)$

5. $6a^2 + a - 1 = (3a - 1)(2a + 1)$ 7. $8m^2 - 15m - 2 = (8m + 1)(m - 2)$

9. $7w^2 - 20w - 3 = (7w + 1)(w - 3)$ 11. $4q^2 - 25q - 21 = (4q + 3)(q - 7)$

13. $10t^2 + 25t + 10 = 5(2t^2 + 5t + 2)$ 15. $3c^2 + 10c + 7 = (3c + 7)(c + 1)$
    $\qquad = 5(2t + 1)(t + 2)$

Section 6.6 Exercises (p. 220)

1. $x^2 - 5x + 4 = 0$
   $(x - 4)(x - 1) = 0$
   $x - 4 = 0$ or $x - 1 = 0$
   $x = 4$ or $x = 1$

3. $2x^2 + 10x + 8 = 0$
   $2(x^2 + 5x + 4) = 0$
   $2(x + 4)(x + 1) = 0$
   $x + 4 = 0$ or $x + 1 = 0$
   $x = -4$ or $x = -1$

5. $x^2 - 3x + 2 = 0$
   $(x - 2)(x - 1) = 0$
   $x - 2 = 0$ or $x - 1 = 0$
   $x = 2$ or $x = 1$

7. $x^2 + 8x + 12 = 0$
   $(x + 6)(x + 2) = 0$
   $x + 6 = 0$ or $x + 2 = 0$
   $x = -6$ or $x = -2$

9. $x^2 - 3x = 10$
   $x^2 - 3x - 10 = 0$
   $(x - 5)(x + 2) = 0$
   $x - 5 = 0$ or $x + 2 = 0$
   $x = 5$ or $x = -2$

11. $y^2 = 64$
    $y^2 - 64 = 0$
    $(y - 8)(y + 8) = 0$
    $y - 8 = 0$ or $y + 8 = 0$
    $y = 8$ or $y = -8$

13. $z^2 + 4z = 77$
    $z^2 + 4z - 77 = 0$
    $(z + 11)(z - 7) = 0$
    $z + 11 = 0$ or $z - 7 = 0$
    $z = -11$ or $z = 7$

15. $t^2 = 49$
    $t^2 - 49 = 0$
    $(t - 7)(t + 7) = 0$
    $t - 7 = 0$ or $t + 7 = 0$
    $t = 7$ or $t = -7$

17. $n^2 = 7n + 8$
$n^2 - 7n - 8 = 0$
$(n - 8)(n + 1) = 0$
$n - 8 = 0$ or $n + 1 = 0$
$n = 8$ or $n = -1$

19. $3x^2 + 12x - 63 = 0$
$3(x^2 + 4x - 21) = 0$
$3(x + 7)(x - 3) = 0$
$x + 7 = 0$ or $x - 3 = 0$
$x = -7$ or $x = 3$

21. $x = -4$ or $x = -3$ **23.** 7   **25.** -10 **27.** 14 and 15   **29.** (a) \$644   (b) 9 pins
31. width $= 20$ cm, length $= 25$ cm   **33.** base $= 10$ m, height $= 20$ m   **35.** (a) 40 feet   (b) 3 seconds
37. $x = 0$ or $x = 5$ or $x = 1$

Section 6.6 Supplementary Exercises (p. 222)

1. $x^2 + x - 2 = 0$
$(x + 2)(x - 1) = 0$
$x + 2 = 0$ or $x - 1 = 0$
$x = -2$ or $x = 1$

3. $x^2 + 5x = 14$
$x^2 + 5x - 14 = 0$
$(x + 7)(x - 2) = 0$
$x + 7 = 0$ or $x - 2 = 0$
$x = -7$ or $x = 2$

5. $4 = x^2$
$0 = x^2 - 4$
$0 = (x - 2)(x + 2)$
$x - 2 = 0$ or $x + 2 = 0$
$x = 2$ or $x = -2$

7. $p^2 + p + 1 = 0$
$p^2 + p + 1$ does not factor

9. $p^2 + p - 18 = 2$
$p^2 + p - 20 = 0$
$(p + 5)(p - 4) = 0$
$p + 5 = 0$ or $p - 4 = 0$
$p = -5$ or $p = 4$

11. $(3x + 5)^2 = 16$
$9x^2 + 30x + 25 = 16$
$9x^2 + 30x + 9 = 0$
$3(3x^2 + 10x + 3) = 0$
$3(3x + 1)(x + 3) = 0$
$3x + 1 = 0$ or $x + 3 = 0$
$3x = -1$ or $x = -3$
$x = -\dfrac{1}{3}$ or $x = -3$

13. $5a^2 + 3a - 5 = 4a^2 + a - 2$
$a^2 + 2a - 3 = 0$
$(a + 3)(a - 1) = 0$
$a + 3 = 0$ or $a - 1 = 0$
$a = -3$ or $a = 1$

15. $4x^2 = 2x$
$4x^2 - 2x = 0$
$2x(2x - 1) = 0$
$2x = 0$ or $2x - 1 = 0$
$x = 0$ or $2x = 1$
$x = 0$ or $x = \dfrac{1}{2}$

17. Let x = smaller even integer
then $(x + 2)$ = next consecutive even integer
$x(x + 2) = 440$
$x^2 + 2x = 440$
$x^2 + 2x - 440 = 0$
$(x + 22)(x - 20) = 0$
$x + 22 = 0$ or $x - 20 = 0$
$x = -22$ or $x = 20$
There are two sets of solutions: -22 and -20
or 20 and 22.

19.   Let w = width
then $w - 6$ = length
Since area $(A) = l \cdot w$, we have
$w(w - 6) = 135$
$w^2 - 6w = 135$
$w^2 - 6w - 135 = 0$
$(w - 15)(w + 9) = 0$
$w - 15 = 0$ or $w + 9 = 0$
$w = 15$ or $w = -9$
Since width must be positive, we reject the
solution -9. Thus the width is 15 and the
length must be 9.

Vocabulary Quiz (p. 223)

1. f   **2.** e   **3.** d   **4.** c   **5.** b   **6.** a

Chapter 6 Review Exercises (p. 223)

1. $2 \cdot 43$ **2.** $2 \cdot 2 \cdot 3 \cdot 7$ or $2^2 \cdot 3 \cdot 7$ **3.** $2 \cdot 3 \cdot 3 \cdot 3 \cdot 3$ or $2 \cdot 3^4$
4. $2 \cdot 3 \cdot 3 \cdot 3 \cdot 3 \cdot 5$ or $2 \cdot 3^4 \cdot 5$ **5.** $2 \cdot 3 \cdot 5 \cdot 5$ or $2 \cdot 3 \cdot 5^2$ **6.** 2   **7.** 2
8. 12   **9.** 12   **10.** 24   **11.** $5x(x - 4)$ **12.** $5(x - 2)(x + 2)$ **13.** $5ab(ab^2 - 3c)$
14. $(y + 9)(y - 3)$ **15.** $(t - 6)(t - 5)$ **16.** $t^2(t - 6)(t - 5)$ **17.** $(x - 8)(x + 6)$

18. $(x + 8)(x - 6)$ **19.** not factorable   **20.** $(x - 12)(x + 4)$ **21.** $(x - 12)(x - 4)$
22. $2y(y - 6)(y - 5)$ **23.** $y(2y - 5)(y - 12)$ **24.** $(8x - 1)(8x + 1)$ **25.** $(t - 1)(t + 1)(t^2 + 1)$
26. $(2x - 5)(x + 1)$ **27.** $(2x - 5)(x - 1)$ **28.** $2(3x - 1)(x - 3)$ **29.** $x = 6$ or $x = -1$

30. $x = 3$ or $x = 2$ **31.** $x = 3$ or $x = -9$ **32.** $x = 5$ or $x = 2$
33. $x = \dfrac{1}{2}$ or $x = -3$ **34.** $x = 5$ or $x = -5$ **35.** $x = 4$ or $x = 3$

Cumulative Review Exercises (p. 224)

1. 0 **2.** -16 **3.** $x = 0$ **4.** $x = \dfrac{5 - 3y}{2}$ **5.** $x < 10$ ⟵————⊕——— **6.** 2:00 P.M.
 10

7. $108a^8 b^6 c^6$ **8.** $2.19 \cdot 10^{-3}$ **9.** $6y^2 - 5y - 21$ **10.** $x + 10$

Chapter 6 Test (p. 225)

1. $2 \cdot 2 \cdot 3 \cdot 3 \cdot 5$ or $2^2 \cdot 3^2 \cdot 5$ **2.** 15 **3.** $2(5x + 1)$ **4.** $x^3(x - 3)$
5. $-5t^2(6s^2 - 2s + 3)$ **6.** $(x + 5)(x + 4)$ **7.** $(x - 8)(x + 2)$ **8.** $-2x^2(x - 10)(x + 2)$
9. $3y(y^2 + 9y + 10)$ **10.** $(z - 3)(z + 3)$ **11.** $8(t - 1)(t + 1)(t^2 + 1)$ **12.** $(2x - 5y)(2x + 5y)$
13. $10(2x - 5y)(2x + 5y)$ **14.** $(3y + 2)(y - 4)$ **15.** $2(3t^2 + 11t + 4)$ **16.** $y^2(6y - 5)(2y - 1)$
17. $x = 6$ or $x = -1$ **18.** $y = 8$ or $y = -1$ **19.** $y = 4$ or $y = 2$
20. $x = \dfrac{1}{2}$ or $x = -7$

Chapter 7

Section 7.1 Exercises (p. 234)

1. $\dfrac{3}{4}$ **3.** $\dfrac{-9}{16}$ **5.** $\dfrac{12}{25}$ **7.** $\dfrac{8}{16} = \dfrac{\cancel{2} \cdot \cancel{2} \cdot \cancel{2} \cdot 1}{\cancel{2} \cdot \cancel{2} \cdot \cancel{2} \cdot 2}$ **9.** $\dfrac{12}{42} = \dfrac{\cancel{2} \cdot 2 \cdot \cancel{3}}{\cancel{2} \cdot \cancel{3} \cdot 7}$ **11.** $\dfrac{51}{66} = \dfrac{\cancel{3} \cdot 17}{\cancel{3} \cdot 22}$
$ = \dfrac{1}{2} = \dfrac{2}{7} = \dfrac{17}{22}$

13. $\dfrac{66}{99} = \dfrac{2 \cdot \cancel{3} \cdot \cancel{11}}{3 \cdot \cancel{3} \cdot \cancel{11}}$ **15.** $\dfrac{75x^2}{105x^3} = \dfrac{\cancel{3} \cdot \cancel{5} \cdot 5 \cdot \cancel{x} \cdot \cancel{x}}{\cancel{3} \cdot \cancel{5} \cdot 7 \cdot \cancel{x} \cdot \cancel{x} \cdot x}$ **17.** $\dfrac{t^6}{t^4} = \dfrac{\cancel{t^4} \cdot t^2}{\cancel{t^4}}$ **19.** $\dfrac{4xy}{-2x} = \dfrac{2 \cdot \cancel{2} \cdot \cancel{x} \cdot y}{-1 \cdot \cancel{2} \cdot \cancel{x}}$
$ = \dfrac{2}{3} = \dfrac{5}{7x} = t^2 = \dfrac{2y}{-1}$
$ = -2y$

21. $\dfrac{bc^2}{a}$ **23.** $\dfrac{1}{x + 3}$ **25.** $\dfrac{2}{y + 2}$ **27.** $\dfrac{x - 3}{x + 3}$ **29.** $a + 1$ **31.** $\dfrac{x + 5}{x + 3}$ **33.** $\dfrac{-1}{y + 3}$

35. $\dfrac{-1}{5 + x} = \dfrac{-1}{x + 5}$ **37.** $-1(x + 6) = -x - 6$ **39.** $\dfrac{y - 4}{y + 3}$ **41.** $\dfrac{3x - 1}{x - 1}$ **43.** $\dfrac{x - 5}{5}$

45. Since $a = b$, then $a - b = 0$ and it is not permissible to divide both sides by $(a - b)$. That is, division by zero is not allowed.

Section 7.1 Supplementary Exercises (p. 235)

1. $\dfrac{-75}{125} = \dfrac{-1 \cdot 3 \cdot \cancel{5} \cdot \cancel{5}}{\cancel{5} \cdot \cancel{5} \cdot 5}$ **3.** $\dfrac{7}{9}$ **5.** $\dfrac{-x^7}{x^4} = \dfrac{-1 \cdot x^3 \cdot \cancel{x^4}}{\cancel{x^4}}$ **7.** $\dfrac{-x^3 y^2}{x^2 y} = \dfrac{-xy \cdot \cancel{x^2 y}}{\cancel{x^2 y}}$
$ = \dfrac{-3}{5} = -x^3 = -xy$

9. $\dfrac{-12a^3b^4c^2}{-6a^2b^3c} = \dfrac{-\cancel{X}\cdot 2 \cdot \cancel{2} \cdot \cancel{3} \cdot a \cdot \cancel{a^2} \cdot b \cdot \cancel{b^3} \cdot \cancel{c} \cdot c}{-\cancel{X}\cdot \cancel{2} \cdot \cancel{3} \cdot \cancel{a^2} \cdot \cancel{b^3} \cdot \cancel{c}}$

$= 2abc$

11. $\dfrac{4x^3}{12x^5 + 4x^3} = \dfrac{\cancel{4x^3}}{\cancel{4x^3}(3x^2 + 1)}$

$= \dfrac{1}{3x^2 + 1}$

13. $\dfrac{2x^2 - x - 1}{x^2 - 1} = \dfrac{(2x + 1)\cancel{(x - 1)}}{(x + 1)\cancel{(x - 1)}}$

$= \dfrac{2x + 1}{x + 1}$

15. $\dfrac{u^2 - v^2}{u^2 - uv - 2v^2} = \dfrac{\cancel{(u + v)}(u - v)}{\cancel{(u + v)}(u - 2v)}$

$= \dfrac{u - v}{u - 2v}$

17. $\dfrac{4 - 4x}{x^2 + 6x - 7} = \dfrac{-4\cancel{(x - 1)}}{(x + 7)\cancel{(x - 1)}}$

$= \dfrac{-4}{x + 7}$

19. $\dfrac{5 + x}{5}$ cannot be reduced; there are no common factors.   **21.** $\dfrac{x}{x - 1}$ **23.** $\dfrac{a - 12}{a + 3}$ **25.** $\dfrac{m - 2}{m + 1}$

Section 7.2 Exercises (p. 240)

1. $\dfrac{7}{15} \cdot \dfrac{3}{8} = \dfrac{7}{\cancel{3} \cdot 5} \cdot \dfrac{\cancel{3}}{8}$

$= \dfrac{7}{40}$

3. $\dfrac{3}{8} \cdot \dfrac{7}{9} = \dfrac{\cancel{3}}{8} \cdot \dfrac{7}{\cancel{3} \cdot 3}$

$= \dfrac{7}{24}$

5. $\dfrac{1}{4} \div \dfrac{1}{2} = \dfrac{1}{4} \cdot \dfrac{2}{1}$

$= \dfrac{1}{\cancel{2} \cdot 2} \cdot \dfrac{\cancel{2}}{1}$

$= \dfrac{1}{2}$

7. $\dfrac{1}{9} \div \dfrac{5}{18} = \dfrac{1}{9} \cdot \dfrac{18}{5}$

$= \dfrac{1}{\cancel{9}} \cdot \dfrac{\cancel{9} \cdot 2}{5}$

$= \dfrac{2}{5}$

9. $\dfrac{x^2}{y} \cdot \dfrac{y^2}{2xyz} = \dfrac{x \cdot x}{y} \cdot \dfrac{y \cdot y}{2 \cdot x \cdot y \cdot z}$

$= \dfrac{x}{2z}$

11. $\dfrac{-6ab}{5} \cdot \dfrac{15b^2}{7a^3} = \dfrac{-6 \cdot \cancel{a} \cdot b}{\cancel{5}} \cdot \dfrac{3 \cdot \cancel{5} \cdot b^2}{7 \cdot \cancel{a} \cdot a^2}$

$= \dfrac{-18b^3}{7a^2}$

13. $\dfrac{3x - 12}{2x + 6} \cdot \dfrac{3x + 4}{x - 4} = \dfrac{3\cancel{(x - 4)}}{2(x + 3)} \cdot \dfrac{3x + 4}{\cancel{x - 4}}$

$= \dfrac{3(3x + 4)}{2(x + 3)}$

15. $\dfrac{x^2 - 7x + 12}{16 - x^2} \cdot \dfrac{x^2 + 5x + 4}{x^2 - 2x - 3} = \dfrac{(x - 3)\cancel{(x - 4)}}{-1\cancel{(x + 4)}\cancel{(x - 4)}} \cdot \dfrac{\cancel{(x + 4)}(x + 1)}{\cancel{(x - 3)}\cancel{(x + 1)}}$

$= -1$

17. $\dfrac{3x^2 - 7x + 2}{4 - x^2} \cdot \dfrac{3x^2 + 10x + 8}{1 - 9x^2} = \dfrac{\cancel{(3x - 1)}\cancel{(x - 2)}}{-1(x + 2)\cancel{(x - 2)}} \cdot \dfrac{(3x + 4)\cancel{(x + 2)}}{-1(3x + 1)\cancel{(3x - 1)}}$

$= \dfrac{3x + 4}{3x + 1}$

19. $\dfrac{2x - 10}{3x - 9} \cdot \dfrac{x - 3}{x - 5} \cdot \dfrac{5}{x^2} = \dfrac{2\cancel{(x - 5)}}{3\cancel{(x - 3)}} \cdot \dfrac{\cancel{x - 3}}{\cancel{x - 5}} \cdot \dfrac{5}{x^2}$

$= \dfrac{10}{3x^2}$

21. $\dfrac{14a^2}{27b^2}$ **23.** $\dfrac{x}{x - 1}$ **25.** $\dfrac{x - 1}{x(y + 1)}$ **27.** $\dfrac{1}{2}$ **29.** $\dfrac{2}{x + 1}$ **31.** $\dfrac{2t - 3}{t(t - 4)}$

33. $\dfrac{y(x+1)(x-1)}{x^3(y+1)(y-1)}$ **35.** $15(x+2)$ **37.** $\dfrac{-x(x-3)}{2y^2(x+3)}$ **39.** $\dfrac{2(x+2)^2}{3xy^3(x-2)}$

Section 7.2 Supplementary Exercises (p. 241)

1. $\dfrac{4}{9} \cdot \dfrac{27}{16} = \dfrac{\cancel{4}}{\cancel{9}} \cdot \dfrac{\cancel{9} \cdot 3}{\cancel{4} \cdot 4}$ **3.** $\dfrac{9}{16} \div \dfrac{1}{4} = \dfrac{9}{16} \cdot \dfrac{4}{1}$ **5.** $\dfrac{x^3}{y} \cdot \dfrac{5y^2}{x} = \dfrac{\cancel{x} \cdot x^2}{y} \cdot \dfrac{5 \cdot y \cdot y}{\cancel{x}}$

$\qquad = \dfrac{3}{4}$ $\qquad = \dfrac{9}{\cancel{4} \cdot 4} \cdot \dfrac{\cancel{4}}{1}$ $\qquad = 5x^2 y$

$\qquad\qquad\qquad\qquad = \dfrac{9}{4}$

7. $\dfrac{2x^2}{y} \div \dfrac{4x^3}{1} = \dfrac{2x^2}{y} \cdot \dfrac{1}{4x^3}$ **9.** $\dfrac{-8}{x^2-9} \cdot \dfrac{x^2-6x+9}{4} = \dfrac{-2 \cdot \cancel{4}}{(x+3)\cancel{(x-3)}} \cdot \dfrac{\cancel{(x-3)}(x-3)}{\cancel{4}}$

$\qquad = \dfrac{2\cancel{x^2}}{y} \cdot \dfrac{1}{2\cancel{x^2} \cdot 2x}$ $\qquad\qquad\qquad = \dfrac{-2(x-3)}{x+3}$

$\qquad = \dfrac{1}{2xy}$

11. $\dfrac{x^2+x-2}{x+1} \cdot \dfrac{3x+3}{1-x} = \dfrac{(x+2)\cancel{(x-1)}}{\cancel{x+1}} \cdot \dfrac{3\cancel{(x+1)}}{-1\cancel{(x-1)}}$

$\qquad\qquad = -3(x+2)$

13. $\dfrac{(x+1)(x+2)}{y} \cdot \dfrac{xy^2}{x^2(x+2)} = \dfrac{(x+1)\cancel{(x+2)}}{\cancel{y}} \cdot \dfrac{x \cdot \cancel{y} \cdot y}{x \cdot x\cancel{(x+2)}}$

$\qquad\qquad = \dfrac{y(x+1)}{x}$

15. $\dfrac{-2a^2 b}{5xy^3} \div \dfrac{4ab^2}{15x^2 y} = \dfrac{-2a^2 b}{5xy^3} \cdot \dfrac{15x^2 y}{4ab^2}$

$\qquad\qquad = \dfrac{-2 \cdot \cancel{a} \cdot a \cdot \cancel{b}}{5 \cdot x \cdot \cancel{y} \cdot y^2} \cdot \dfrac{3 \cdot \cancel{5} \cdot \cancel{x} \cdot x \cdot \cancel{y}}{2 \cdot 2 \cdot \cancel{a} \cdot b \cdot \cancel{b}}$

$\qquad\qquad = \dfrac{-3ax}{2by^2}$

Section 7.3 Exercises (p. 244)

1. $6 = 2 \cdot 3$
$15 = 3 \cdot 5$
$\text{LCM} = 2 \cdot 3 \cdot 5 = 30$

3. $6 = 2 \cdot 3$
$12 = 2 \cdot 2 \cdot 3$
$\text{LCM} = 2 \cdot 2 \cdot 3 = 12$

5. $15 = 3 \cdot 5$
$15 = 3 \cdot 5$
$\text{LCM} = 3 \cdot 5 = 15$

7. $64 = 2 \cdot 2 \cdot 2 \cdot 2 \cdot 2 \cdot 2$
$96 = 2 \cdot 2 \cdot 2 \cdot 2 \cdot 2 \cdot 3$
$\text{LCM} = 2 \cdot 2 \cdot 2 \cdot 2 \cdot 2 \cdot 2 \cdot 3 = 192$

9. $30 = 2 \cdot 3 \cdot 5$
$40 = 2 \cdot 2 \cdot 2 \cdot 5$
$\text{LCM} = 2 \cdot 2 \cdot 2 \cdot 3 \cdot 5 = 120$

11. $20 = 2 \cdot 2 \cdot 5$
$30 = 2 \cdot 3 \cdot 5$
$60 = 2 \cdot 2 \cdot 3 \cdot 5$
$\text{LCM} = 2 \cdot 2 \cdot 3 \cdot 5 = 60$

13. $2 = 2$
$4 = 2 \cdot 2$
$6 = 2 \cdot 3$
$\text{LCM} = 2 \cdot 2 \cdot 3 = 12$

15. $45 = 3 \cdot 3 \cdot 5$
$63 = 3 \cdot 3 \cdot 7$
$315 = 3 \cdot 3 \cdot 5 \cdot 7$
$\text{LCM} = 3 \cdot 3 \cdot 5 \cdot 7 = 315$

17. $10 = 2 \cdot 5$
$11 = 11$
$12 = 2 \cdot 2 \cdot 3$
$\text{LCM} = 2 \cdot 2 \cdot 3 \cdot 5 \cdot 11 = 660$

19. $\dfrac{3}{10} + \dfrac{1}{6} = \dfrac{9}{30} + \dfrac{5}{30}$ **21.** 1   **23.** $\dfrac{47}{90}$ **25.** $\dfrac{13}{36}$

$\qquad\qquad = \dfrac{14}{30}$

$\qquad\qquad = \dfrac{7}{15}$

27. $\dfrac{31}{150}$ **29.** $\dfrac{3}{4}$ **31.** $\dfrac{11}{12}$ **33.** $\dfrac{11}{120}$ **35.** $\dfrac{29}{2}$ **37.** $\dfrac{83}{24}$ **39.** $\dfrac{-4}{15}$

Section 7.3 Supplementary Exercises (p. 245)

1.
$$\begin{aligned} 3 &= 3 \\ 4 &= 2 \cdot 2 \\ 5 &= 5 \\ \text{LCM} &= 2 \cdot 2 \cdot 3 \cdot 5 = 60 \end{aligned}$$

3.
$$\begin{aligned} 8 &= 2 \cdot 2 \cdot 2 \\ 12 &= 2 \cdot 2 \cdot 3 \\ 15 &= 3 \cdot 5 \\ \text{LCM} &= 2 \cdot 2 \cdot 2 \cdot 3 \cdot 5 = 120 \end{aligned}$$

5.
$$\begin{aligned} 2 &= 2 \\ 8 &= 2 \cdot 2 \cdot 2 \\ 9 &= 3 \cdot 3 \\ \text{LCM} &= 2 \cdot 2 \cdot 2 \cdot 3 \cdot 3 = 72 \end{aligned}$$

7.
$$\begin{aligned} \frac{1}{4} - \frac{2}{3} &= \frac{3}{12} - \frac{8}{12} \\ &= \frac{-5}{12} \end{aligned}$$

9.
$$\begin{aligned} \frac{2}{5} - \frac{1}{3} &= \frac{6}{15} - \frac{5}{15} \\ &= \frac{1}{15} \end{aligned}$$

11.
$$\begin{aligned} \frac{1}{7} - \frac{2}{3} - \frac{1}{14} &= \frac{6}{42} - \frac{28}{42} - \frac{3}{42} \\ &= \frac{-25}{42} \end{aligned}$$

13.
$$\begin{aligned} \frac{1}{7} + \frac{3}{8} - \frac{1}{4} &= \frac{8}{56} + \frac{21}{56} - \frac{14}{56} \\ &= \frac{15}{56} \end{aligned}$$

15.
$$\begin{aligned} \frac{1}{8} - \frac{11}{12} + \frac{2}{9} &= \frac{9}{72} - \frac{66}{72} + \frac{16}{72} \\ &= \frac{-41}{72} \end{aligned}$$

17.
$$\begin{aligned} \frac{6}{7} - \frac{1}{8} + \frac{3}{4} &= \frac{48}{56} - \frac{7}{56} + \frac{42}{56} \\ &= \frac{83}{56} \end{aligned}$$

19.
$$\begin{aligned} \frac{4}{5} + \frac{3}{8} \div \frac{9}{16} &= \frac{4}{5} + \left(\frac{3}{8} \cdot \frac{16}{9} \right) \\ &= \frac{4}{5} + \frac{2}{3} \\ &= \frac{12}{15} + \frac{10}{15} \\ &= \frac{22}{15} \end{aligned}$$

Section 7.4 Exercises (p. 250)

1.
$$\begin{aligned} \frac{7}{a} + \frac{8}{a} &= \frac{7+8}{a} \\ &= \frac{15}{a} \end{aligned}$$

3.
$$\begin{aligned} \frac{3x}{x+7} + \frac{21}{x+7} &= \frac{3x+21}{x+7} \\ &= \frac{3(x+7)}{x+7} \\ &= 3 \end{aligned}$$

5.
$$\begin{aligned} \frac{5x}{x^2+x-20} + \frac{25}{x^2+x-20} &= \frac{5x+25}{x^2+x-20} \\ &= \frac{5(x+5)}{(x+5)(x-4)} \\ &= \frac{5}{x-4} \end{aligned}$$

7.
$$\begin{aligned} x^4 y &= x \cdot x \cdot x \cdot x \cdot y \\ xy^4 &= x \cdot y \cdot y \cdot y \cdot y \\ \text{LCM} &= x \cdot x \cdot x \cdot x \cdot y \cdot y \cdot y \cdot y = x^4 y^4 \end{aligned}$$

9.
$$\begin{aligned} 5x &= 5 \cdot x \\ 3y &= 3 \cdot y \\ \text{LCM} &= 5 \cdot x \cdot 3 \cdot y = 15xy \end{aligned}$$

11.
$$\begin{aligned} x^2 - 9 &= (x+3)(x-3) \\ x^2 + 4x - 21 &= (x+7)(x-3) \\ \text{LCM} &= (x+3)(x-3)(x+7) \end{aligned}$$

13.
$$\begin{aligned} x^4 - x^2 &= x^2(x+1)(x-1) \\ x^3 + 2x^2 + x &= x(x+1)(x+1) \\ \text{LCM} &= x^2(x+1)(x+1)(x-1) \end{aligned}$$

15.
$$\begin{aligned} 6x - 12 &= 2 \cdot 3(x-2) \\ 9x^2 - 36 &= 3 \cdot 3(x+2)(x-2) \\ x^2 + 4x + 4 &= (x+2)(x+2) \\ \text{LCM} &= 2 \cdot 3 \cdot 3(x+2)(x+2)(x-2) = 18(x+2)^2(x-2) \end{aligned}$$

17.
$$\begin{aligned} \frac{1}{x^2} + \frac{8}{xy^2} &= \frac{y^2}{x^2 y^2} + \frac{8x}{x^2 y^2} \\ &= \frac{8x + y^2}{x^2 y^2} \end{aligned}$$

19.
$$\begin{aligned} \frac{9}{4x+8} + \frac{7}{x+2} &= \frac{9}{4(x+2)} + \frac{28}{4(x+2)} \\ &= \frac{37}{4(x+2)} \end{aligned}$$

21. $\dfrac{x^2 + 10x + 6}{(x + 2)(x - 2)(x + 7)}$ **23.** $\dfrac{4x^2 - x - 2}{(x + 2)(x - 1)(x - 2)(x - 2)}$ **25.** $\dfrac{3x^2 + 2x + 3}{(3x - 1)(x - 1)(x + 1)}$

27. $\dfrac{2x - 1}{(x + 3)(x - 3)(x - 4)}$ **29.** $\dfrac{21x^2 - 2x - 16}{18(x + 2)^2(x - 2)}$ **31.** $\dfrac{x^2 + 4x - 6}{(x + 2)(x - 2)(x + 7)}$

33. $\dfrac{x^2 + 6x + 2}{(x + 3)(x + 2)(x + 3)}$ **35.** $\dfrac{-6(5x + 3)}{(x + 3)^2(x - 3)^2}$ **37.** $\dfrac{a^2 + 6a - 9}{(a + 3)^2(a - 3)}$ **39.** $\dfrac{x + 13}{(x + 2)(x - 2)}$

Section 7.4 Supplementary Exercises (p. 251)

1. $\dfrac{3x}{x + 5} + \dfrac{15}{x + 5} = \dfrac{3x + 15}{x + 5}$

$\qquad\qquad\quad = \dfrac{3(x + 5)}{x + 5}$

$\qquad\qquad\quad = 3$

3. $\dfrac{x}{x^2 + 2x + 1} + \dfrac{1}{x^2 + 2x + 1} = \dfrac{x + 1}{x^2 + 2x + 1}$

$\qquad\qquad\qquad\qquad\qquad\qquad = \dfrac{1}{x + 1}$

5. $a^2 b \; = a \cdot a \cdot b$

$\quad ab^3 \; = a \cdot b \cdot b \cdot b$

$\quad \text{LCM} = a \cdot a \cdot b \cdot b \cdot b = a^2 b^3$

7. $x - 3 = x - 3$

$\quad 9 - x^2 = -1(x - 3)(x + 3)$

$\quad \text{LCM} = -1(x - 3)(x + 3)$

9. $\dfrac{5}{a^2 b} + \dfrac{4}{ab^2} = \dfrac{5b}{a^2 b^2} + \dfrac{4a}{a^2 b^2}$

$\qquad\qquad\quad = \dfrac{4a + 5b}{a^2 b^2}$

11. $\dfrac{3x}{x^2 - 9} + \dfrac{2}{x + 3} = \dfrac{3x}{x^2 - 9} + \dfrac{2(x - 3)}{x^2 - 9}$

$\qquad\qquad\qquad\qquad = \dfrac{5x - 6}{(x + 3)(x - 3)}$

13. $\dfrac{-2}{9 - x^2} + \dfrac{3}{x^2 - 6x + 9} = \dfrac{2(x - 3)}{(x + 3)(x - 3)(x - 3)} + \dfrac{3(x + 3)}{(x - 3)(x - 3)(x + 3)}$

$\qquad\qquad\qquad\qquad\qquad = \dfrac{2x - 6 + 3x + 9}{(x + 3)(x - 3)^2}$

$\qquad\qquad\qquad\qquad\qquad = \dfrac{5x + 3}{(x + 3)(x - 3)^2}$

15. $\dfrac{3x}{x^2 - 1} + \dfrac{2x}{x^2 + 2x + 1} = \dfrac{3x(x + 1)}{(x + 1)(x - 1)(x + 1)} + \dfrac{2x(x - 1)}{(x + 1)(x + 1)(x - 1)}$

$\qquad\qquad\qquad\qquad\qquad = \dfrac{3x^2 + 3x + 2x^2 - 2x}{(x + 1)^2(x - 1)}$

$\qquad\qquad\qquad\qquad\qquad = \dfrac{5x^2 + x}{(x + 1)^2(x - 1)}$

$\qquad\qquad\qquad\qquad\qquad = \dfrac{x(5x + 1)}{(x + 1)^2(x - 1)}$

17. $\dfrac{2}{x + 1} - \dfrac{3}{x^2 - 1} + \dfrac{5}{x - 1} = \dfrac{2(x - 1)}{(x + 1)(x - 1)} - \dfrac{3}{(x + 1)(x - 1)} + \dfrac{5(x + 1)}{(x - 1)(x + 1)}$

$\qquad\qquad\qquad\qquad\qquad\qquad = \dfrac{2x - 2 - 3 + 5x + 5}{(x + 1)(x - 1)}$

$\qquad\qquad\qquad\qquad\qquad\qquad = \dfrac{7x}{(x + 1)(x - 1)}$

19. $\dfrac{3}{x + 1} - \dfrac{2}{x + 2} + \dfrac{-2}{(x + 2)(x + 1)} = \dfrac{3(x + 2) - 2(x + 1) - 2}{(x + 2)(x + 1)}$

$\qquad\qquad\qquad\qquad\qquad\qquad\qquad = \dfrac{3x + 6 - 2x - 2 - 2}{(x + 2)(x + 1)}$

$\qquad\qquad\qquad\qquad\qquad\qquad\qquad = \dfrac{x + 2}{(x + 2)(x + 1)}$

$\qquad\qquad\qquad\qquad\qquad\qquad\qquad = \dfrac{1}{x + 1}$

Section 7.5 Exercises (p. 256)

1. $\dfrac{\dfrac{5}{6}-\dfrac{1}{3}}{\dfrac{5}{9}+\dfrac{1}{6}}\cdot\dfrac{18}{18}=\dfrac{15-6}{10+3}$

$=\dfrac{9}{13}$

3. $\dfrac{\dfrac{7}{12}-\dfrac{1}{2}}{\dfrac{2}{3}+\dfrac{3}{4}}\cdot\dfrac{12}{12}=\dfrac{7-6}{8+9}$

$=\dfrac{1}{17}$

5. $\dfrac{\dfrac{7}{12x}+\dfrac{1}{2x}}{\dfrac{2}{3x}+\dfrac{3}{4x}}\cdot\dfrac{12x}{12x}=\dfrac{7+6}{8+9}$

$=\dfrac{13}{17}$

7. $\dfrac{\dfrac{5}{6x}-\dfrac{1}{3y}}{\dfrac{5}{9x}+\dfrac{1}{6y}}\cdot\dfrac{18xy}{18xy}=\dfrac{15y-6x}{10y+3x}$

$=\dfrac{3(5y-2x)}{3x+10y}$

9. $\dfrac{\dfrac{6xy^2}{5z}}{\dfrac{12xy}{25z^2}}\cdot\dfrac{25z^2}{25z^2}=\dfrac{30xy^2z}{12xy}$

$=\dfrac{5yz}{2}$

11. $\dfrac{\dfrac{1}{x}+\dfrac{1}{y}}{\dfrac{1}{x}}\cdot\dfrac{xy}{xy}=\dfrac{y+x}{y}$

$=\dfrac{x+y}{y}$

13. $\dfrac{\dfrac{1}{x}+\dfrac{1}{y}}{x}\cdot\dfrac{xy}{xy}=\dfrac{y+x}{x^2y}$

$=\dfrac{x+y}{x^2y}$

15. $\dfrac{\dfrac{y^2-9}{4}}{\dfrac{y^2+6y+9}{6}}\cdot\dfrac{12}{12}=\dfrac{3(y+3)(y-3)}{2(y+3)(y+3)}$

$=\dfrac{3(y-3)}{2(y+3)}$

17. $\dfrac{\dfrac{6}{x}-2}{7-\dfrac{21}{x}}\cdot\dfrac{x}{x}=\dfrac{6-2x}{7x-21}$

$=\dfrac{2(3-x)}{7(x-3)}$

$=\dfrac{-2}{7}$

19. $\dfrac{a-\dfrac{1}{a}}{a+\dfrac{1}{a}}\cdot\dfrac{a}{a}=\dfrac{a^2-1}{a^2+1}$

21. $\dfrac{x(x-9)}{(3x^2+1)(x-2)}$ or $\dfrac{x(x-9)}{3x^3-6x^2+x-2}$

Section 7.5 Supplementary Exercises (p. 257)

1. $\dfrac{\dfrac{5}{8}}{\dfrac{-5}{4}}\cdot\dfrac{8}{8}=\dfrac{5}{-10}$

$=-\dfrac{1}{2}$

3. $\dfrac{\dfrac{3}{4}\cdot\dfrac{8}{6}}{\dfrac{7}{8}}=\dfrac{1}{\dfrac{7}{8}}$

$=\dfrac{1}{\dfrac{7}{8}}\cdot\dfrac{8}{8}$

$=\dfrac{8}{7}$

5. $\dfrac{\dfrac{ab^3}{c}}{\dfrac{2a}{c}\cdot\dfrac{b^2}{c^2}}\cdot\dfrac{c^3}{c^3}=\dfrac{ab^3c^2}{2ab^2}$

$=\dfrac{bc^2}{2}$

7. $\dfrac{5-\dfrac{1}{a}}{5+\dfrac{1}{a}}\cdot\dfrac{a}{a}=\dfrac{5a-1}{5a+1}$

9. $\dfrac{x-\dfrac{1}{x}}{x^3}\cdot\dfrac{x}{x}=\dfrac{x^2-1}{x^4}$

11. $\dfrac{\dfrac{x^2-9}{y}}{x-\dfrac{9}{x}}\cdot\dfrac{xy}{xy}=\dfrac{x(x^2-9)}{x^2y-9y}$

$=\dfrac{x(x+3)(x-3)}{y(x+3)(x-3)}$

$=\dfrac{x}{y}$

Section 7.6 Exercises (p. 263)

1. $\dfrac{12}{x} = \dfrac{6}{5}$

$6x = 60$
$x = 10$

3. $\dfrac{4}{x+1} = \dfrac{9}{x^2}$

$4x^2 = 9x + 9$
$4x^2 - 9x - 9 = 0$
$(4x + 3)(x - 3) = 0$
$4x + 3 = 0 \qquad x - 3 = 0$
$$x = -\dfrac{3}{4} \qquad x = 3$$

5. $\dfrac{1}{x} = \dfrac{1}{4}$

$x = 4$

7. $\dfrac{2x}{x-1} = \dfrac{3}{1}$

$3x - 3 = 2x$
$x = 3$

9. $\dfrac{2}{7} + \dfrac{5}{x} = 1$

$\dfrac{5}{x} = \dfrac{5}{7}$
$5x = 35$
$x = 7$

11. $\dfrac{3}{4} + \dfrac{x}{x+1} = \dfrac{3}{2}$

$\dfrac{x}{x+1} = \dfrac{3}{4}$
$4x = 3x + 3$
$x = 3$

13. $\dfrac{1}{x+1} - \dfrac{1}{x-1} - \dfrac{2x}{x^2-1} = 0$

$\dfrac{x-1}{(x+1)(x-1)} - \dfrac{x+1}{(x-1)(x+1)} - \dfrac{2x}{(x+1)(x-1)} = 0$

$\dfrac{-2x-2}{(x+1)(x-1)} = 0$

$2x = -2$
$x = -1$

But $x = -1$ *does not check.* Therefore, there is NO SOLUTION!

15. $\dfrac{x^2}{x-4} = \dfrac{16}{x-4}$

$x^2 = 16$
$x = \pm 4$

But $x = 4$ *does not check.* Therefore, the only solution is $x = -4$.

17. $\dfrac{x}{x+1} - \dfrac{7}{2x+1} = \dfrac{1}{2x^2+3x+1}$

$\dfrac{x(2x+1)}{(x+1)(2x+1)} - \dfrac{7(x+1)}{(2x+1)(x+1)} - \dfrac{1}{(2x+1)(x+1)} = 0$

$\dfrac{2x^2 + x - 7x - 7 - 1}{(2x+1)(x+1)} = 0$

$\dfrac{2x^2 - 6x - 8}{(2x+1)(x+1)} = 0$

$2(x^2 - 3x - 4) = 0$
$2(x - 4)(x + 1) = 0$
$x - 4 = 0 \qquad x + 1 = 0$
$x = 4 \qquad x = -1$

But $x = -1$ *does not check.* Therefore, the only solution is $x = 4$.

19. $\dfrac{x}{3x+5} + \dfrac{1}{x} = \dfrac{9}{3x^2+5x}$ **21.** $x=-2, x=1$ **23.** $x = \dfrac{by}{aby+y-b}$

$\dfrac{x^2+3x+5-9}{x(3x+5)} = 0$

$x^2+3x-4 = 0$
$(x+4)(x-1) = 0$
$x+4 = 0 \qquad x-1 = 0$
$\qquad x = -4 \qquad\quad x = 1$

Both check.

25. $x = \dfrac{1-c^2}{11a-c^2}$ **27.** $x=3$ **29.** $x = -\dfrac{1}{2}$ **31.** $x = \dfrac{a+b}{c}$ **33.** $M = \dfrac{Fr^2}{Gm}$

35. $d = \dfrac{l-a}{n-1}$

Section 7.6 Supplementary Exercises (p. 264)

1. $\dfrac{8}{x} = \dfrac{4}{3}$ **3.** $\dfrac{x}{5} = \dfrac{x}{-6}$ **5.** $\dfrac{3}{x+3} = \dfrac{2}{x}$ **7.** $\dfrac{5-x}{3} = \dfrac{1-x}{2}$ **9.** $-\dfrac{1}{x+1} = \dfrac{1}{4}$

$\quad 4x = 24 \qquad\quad 5x = -6x \qquad\quad 3x = 2x+6 \qquad 10-2x = 3-3x \qquad x+1 = -4$
$\quad\;\; x = 6 \qquad\quad 11x = 0 \qquad\qquad x = 6 \qquad\qquad\quad x = -7 \qquad\qquad\quad x = -5$
$\qquad\qquad\qquad\qquad\;\; x = 0$

11. $x^2 = 25$ **13.** $\dfrac{3(2x+1)}{(x+2)(2x+1)} - \dfrac{1(x+2)}{(2x+1)(x+2)} = \dfrac{2}{(2x+1)(x+2)}$

$\quad\;\; x = \pm 5$

But $x = 5$ *does not check.* $\qquad\qquad\qquad \dfrac{6x+3-x-2-2}{(2x+1)(x+2)} = 0$
Therefore, the only solu-
tion is $x = -5$. $\qquad\qquad\qquad\qquad\qquad\qquad 5x-1 = 0$

$\qquad\qquad\qquad\qquad\qquad\qquad\qquad\qquad\qquad x = \dfrac{1}{5}$

15. $-\dfrac{x+3}{x-9} = 0$ **17.** $\dfrac{x^2-12}{x} = \dfrac{-x}{3}$

$\qquad x+3 = 0 \qquad\qquad\quad 3x^2-36 = -x^2$
$\qquad\quad\; x = -3 \qquad\qquad\quad 4x^2-36 = 0$
$\qquad\qquad\qquad\qquad\qquad\quad 4(x^2-9) = 0$
$\qquad\qquad\qquad\qquad\quad 4(x+3)(x-3) = 0$
$\qquad\qquad\qquad\qquad\quad x+3 = 0 \qquad x-3 = 0$
$\qquad\qquad\qquad\qquad\qquad\; x = -3 \qquad\;\; x = 3$

Vocabulary Quiz (p. 266)

1. d   **2.** f   **3.** a   **4.** c   **5.** b   **6.** e

Chapter 7 Review Exercises (p. 266)

1. $\dfrac{5}{6}$ **2.** $\dfrac{7}{11}$ **3.** $\dfrac{7}{11}$ **4.** $\dfrac{3}{4x^2}$ **5.** $\dfrac{x-3}{x+5}$ **6.** $\dfrac{y}{2(y-6)}$ **7.** $\dfrac{1}{2y}$ **8.** $\dfrac{-1}{2y}$ **9.** $\dfrac{1}{a-b}$

10. $\dfrac{-1}{a+b}$ **11.** $\dfrac{7}{15}$ **12.** $\dfrac{15}{112}$ **13.** $\dfrac{10}{9}$ **14.** 3   **15.** $\dfrac{-35}{6}$ **16.** $\dfrac{a+2}{2(a-7)}$ **17.** $\dfrac{-(y+2)}{2(y-7)}$

18. $\dfrac{x(x-4)}{(x+4)(x+3)}$ **19.** $\dfrac{1}{3}$ **20.** 1   **21.** -1 **22.** 105   **23.** 210   **24.** 180

25. $x^3 y^4$ **26.** $(y+5)(y-5)(y-4)$ **27.** $(x^2+1)(x+1)(x-1)$ **28.** $3(2x+5)(2x-5)(x-1)$

29. $\dfrac{73}{180}$ **30.** $\dfrac{79}{180}$ **31.** $\dfrac{79}{150}$ **32.** $\dfrac{11}{x^2 y}$ **33.** $\dfrac{4a-1}{(2a+1)(2a-1)}$ **34.** $\dfrac{5x+7}{x^2}$

35. $\dfrac{5y-24}{(y+6)(y-8)}$ **36.** $\dfrac{-y+24}{(y+6)(y-8)}$ **37.** $\dfrac{2(x^2+5x-7)}{(x+2)(x-2)(x+5)}$ **38.** $\dfrac{-2x(x-2)}{(2x-1)(x+1)(x-1)}$

39. $\dfrac{-6(5y+3)}{(y+3)^2(y-3)^2}$ **40.** $\dfrac{a+b}{a^2 b}$ **41.** $\dfrac{b-a}{b}$ **42.** $\dfrac{-2x+9}{2x}$ **43.** $\dfrac{(2x+1)(4x-3)}{4x}$

44. $\dfrac{-(a+3)}{2}$ **45.** $\dfrac{a(a+3)}{2}$ **46.** $x=\dfrac{9}{2}$ **47.** $x=-4$ **48.** $x=\dfrac{5}{2}, x=-1$ **49.** $x=-2, x=1$

50. $\dfrac{ay}{a+y+2ay}$

Cumulative Review Exercises (p. 267)

1. 6   **2.** $-24x^4 y^6$ **3.** 48 m^2 **4.** $x=-\dfrac{8}{3}$ **5.** 3 kg   **6.** $\dfrac{y^5}{6x^7}$ **7.** $3x^3-4x^2+7x+5$

8. $2x-16+\dfrac{65}{x+3}$ **9.** $2x(x-5)(x+3)$ **10.** $x=4, x=\dfrac{-3}{2}$

Chapter 7 Test (p. 269)

1. $\dfrac{8}{9}$ **2.** $\dfrac{5b}{7a}$ **3.** $\dfrac{x-4}{x+5}$ **4.** $\dfrac{-(3y-2)}{2y+1}$ **5.** $\dfrac{9}{4x}$ **6.** $\dfrac{4(x+1)}{x+2}$ **7.** $\dfrac{(x-5)(x^2+2x+4)}{(x+2)(x-2)(x+5)}$

8. $\dfrac{2y-3}{y^2(y+4)}$ **9.** $\dfrac{41}{75}$ **10.** $\dfrac{7a+5b}{a^2 b^4}$ **11.** $\dfrac{5x-38}{(x-2)(x-8)}$ **12.** $\dfrac{z+16}{(z-8)(z+2)}$ **13.** (a) 225

(b) $x(x+1)^2(x-1)$ **14.** $\dfrac{2xy}{3}$ **15.** $\dfrac{7x+8}{8x}$ **16.** $y=40$ **17.** $x=7, x=2$ **18.** No solution

19. $x=2$ **20.** $\dfrac{fs_2}{s_2-f}$

Chapter 8

Section 8.1 Exercises (p. 275)

1.   **3.**

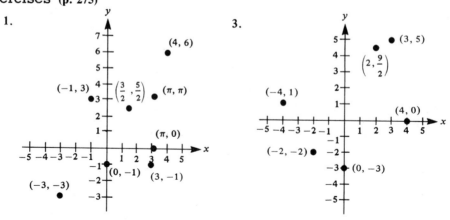

5. $A\,(1,1), B\,(-2,1), C\,(-5,-5), D\,(3,-3), E\,(4,0), F\,(0,5), G\,\left(-\dfrac{9}{2},0\right)$

7. $A\,(2,4), B\,(3,-2), C\,(-3,-2), D\,(-2,4), E\,(-4,2), F\,(-3,0), G\,\left(\dfrac{11}{2},-1\right)$ **9.** 3   **11.** 4

13. 0   **15.** II

17. (a)

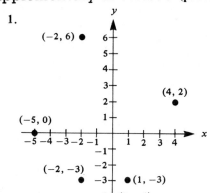

(b) On a line parallel to the x-axis, one unit below it   **19.** IV

Section 8.1 Supplementary Exercises (p. 277)

1.

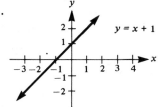

3. $A\ (3, 0), B\ (0, 3), C\ (-2, 0), D\ (0, -1), E\ (1, 2), F\ (-1, 3),$ $G\ (2, -2)$

5. On a line parallel to the y-axis, 7 units to the left of it   **7.** On a line parallel to the x-axis, 8 units below it   **9.** They lie on the x-axis

---

Section 8.2 Exercises (p. 283)

1.

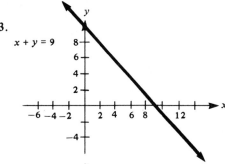

3.

5.

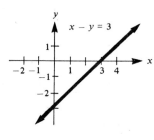

7.

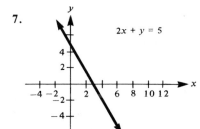

9.

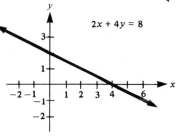

11.

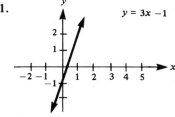

13.

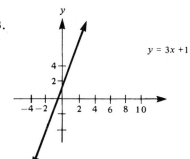

15.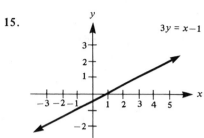

$3y = x - 1$

17.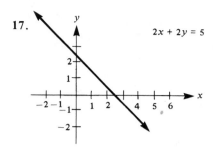

$2x + 2y = 5$

19.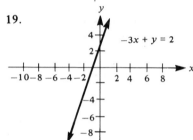

$-3x + y = 2$

21.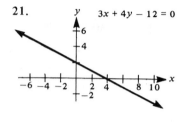

$3x + 4y - 12 = 0$

23.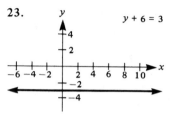

$y + 6 = 3$

25.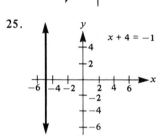

$x + 4 = -1$

27.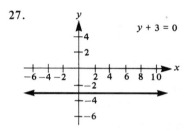

$y + 3 = 0$

29.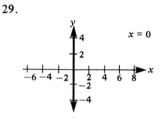

$x = 0$

31. (a), (b), and (c)   33. False   35. False   37. $x = -1$

Section 8.2 Supplementary Exercises (p. 284)

1.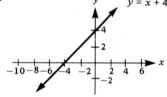

$y = x + 4$

3.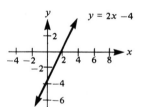

$y = 2x - 4$

5.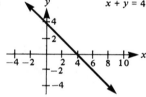

$x + y = 4$

7.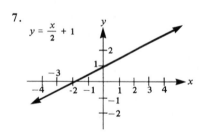

$y = \dfrac{x}{2} + 1$

9.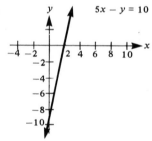

$5x - y = 10$

11.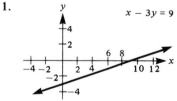

$x - 3y = 9$

13.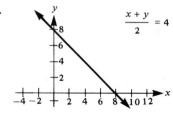

$\dfrac{x + y}{2} = 4$

15.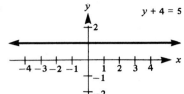

17.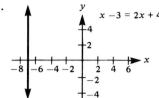

19.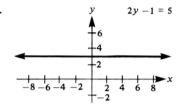

Section 8.3 Exercises (p. 290)

1. x-intercept: $(-2, 0)$
y-intercept: $(0, 4)$
slope: 2

3. x-intercept: $(7, 0)$
y-intercept: $(0, 7)$
slope: -1

5. x-intercept: $(4, 0)$
y-intercept: $(0, 2)$
slope: $-\dfrac{1}{2}$

7. x-intercept: $\left(\dfrac{5}{3}, 0\right)$
y-intercept: $(0, 5)$
slope: -3

9. x-intercept: $(0, 0)$
y-intercept: $(0, 0)$
slope: 2

11. x-intercept: $(0, 0)$
y-intercept: $(0, 0)$
slope: -1

13. x-intercept: $(2, 0)$
y-intercept: $(0, 2)$
slope: -1

15. x-intercept: none
y-intercept: $(0, -3)$
slope: 0

17. x-intercept: $(1, 0)$
y-intercept: $\left(0, -\dfrac{1}{3}\right)$
slope: $\dfrac{1}{3}$

19. x-intercept: $(1, 0)$
y-intercept: $(0, -1)$
slope: 1

21. x-intercept: $(7, 0)$
y-intercept: none
slope: none

23. x-intercept; $\left(\dfrac{-3}{2}, 0\right)$
y-intercept: $(0, -0.3)$
slope: -0.2

25. 2 **27.** 4 **29.** 0 **31.** $\dfrac{-3}{2}$ **33.** undefined   **35.** $\dfrac{1}{2}$ **37.** $\dfrac{20}{9}$ **39.** $\dfrac{-16}{15}$

Section 8.3 Supplementary Exercises (p. 290)

1. x-intercept: $(2, 0)$
y-intercept: $(0, -8)$
slope: 4

3. x-intercept: $(3, 0)$
y-intercept: $(0, -2)$
slope: $\dfrac{2}{3}$

5. x-intercept: $(2, 0)$
y-intercept: none
slope: none

7. $m = \dfrac{4 - 12}{-1 - (-5)} = \dfrac{-8}{4} = -2$ **9.** $m = \dfrac{6 - 6}{-2 - 4} = \dfrac{0}{-6} = 0$

Section 8.4 Exercises (p. 294)

1. $m = 5, b = 3$
$y = mx + b$
$y = 5x + 3$

3. $m = -3, b = -5$
$y = mx + b$
$y = -3x - 5$

5. $m = \dfrac{2}{3}, b = -2$
$y = mx + b$
$y = \dfrac{2}{3}x - 2$

7. $m = 3, b = -6$
$y = mx + b$
$y = 3x - 6$

9. $y - y_1 = m(x - x_1)$
$y - 4 = 3(x - 2)$
$y - 4 = 3x - 6$
$y = 3x - 2$

11. $y - y_1 = m(x - x_1)$
$y - 8 = -4(x + 1)$
$y - 8 = -4x - 4$
$y = -4x + 4$

13. $y - y_1 = m(x - x_1)$
$y + 4 = 0(x - 1)$
$y = -4$

15. slope undefined means a vertical line: $x = -3$

17.
$$y - y_1 = m(x - x_1)$$
$$y + 4 = -3(x + 1)$$
$$y + 4 = -3x - 3$$
$$y = -3x - 7$$

19.
$$y - y_1 = m(x - x_1)$$
$$y + \frac{3}{2} = -\frac{1}{2}(x + 1)$$
$$y + \frac{3}{2} = -\frac{1}{2}x - \frac{1}{2}$$
$$y = -\frac{1}{2}x - 2$$

21. $y = \frac{3}{2}x - 6$

23. $y = 4x + 1$ **25.** $y = 2x - 5$ **27.** $y = \frac{1}{16}x + \frac{1}{2}$ or $16y = x + 8$

29. $15y = 8x + 1$ or $y = \frac{8}{15}x + \frac{1}{15}$ **31.** (a) Solve equation in Step 3 for b.   (b) Substitute $y_1 - mx_1$ for b in equation in Step 2.   (c) Subtract y_1 from both sides of equation in Step 5.   (d) Factor the righthand side of equation in Step 6.   **33.** $y = 6$ **35.** $x = 0$

Section 8.4 Supplementary Exercises (p. 294)

1. $m = 5, b = -4$
$$y = mx + b$$
$$y = 5x - 4$$

3. $m = 0, b = \frac{1}{2}$
$$y = mx + b$$
$$y = \frac{1}{2}$$

5. undefined slope means vertical line: $x = -2$

7. $m = \frac{1}{3}, (x_1, y_1) = (6, -1)$
$$y - y_1 = m(x - x_1)$$
$$y + 1 = \frac{1}{3}(x - 6)$$
$$y + 1 = \frac{1}{3}x - 2$$
$$y = \frac{1}{3}x - 3$$

9. slope $= \dfrac{0 - 2}{-4 - 0} = \dfrac{1}{2}$
$$b = 2$$
$$y = mx + b$$
$$y = \frac{1}{2}x + 2$$

11. $m = \dfrac{3 - 6}{2 + 1} = \dfrac{-3}{3} = -1$
$$y - y_1 = m(x - x_1)$$
$$y - 3 = -1(x - 2)$$
$$y - 3 = -x + 2$$
$$y = -x + 5$$

13. $m = \dfrac{4 - 15}{5 + 6} = \dfrac{-11}{11} = -1$
$$y - y_1 = m(x - x_1)$$
$$y - 4 = -1(x - 5)$$
$$y - 4 = -x + 5$$
$$y = -x + 9$$

Section 8.5 Exercises (p. 299)

1.
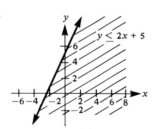
$y \leq 2x + 5$

3.
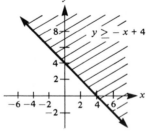
$y \geq -x + 4$

5.
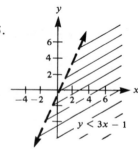
$y < 3x - 1$

7.
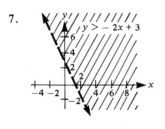
$y > -2x + 3$

9.
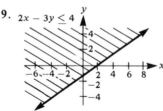
$2x - 3y \leq 4$

11.
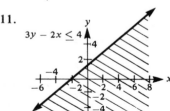
$3y - 2x \leq 4$

13.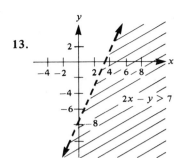

15.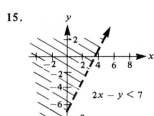

17.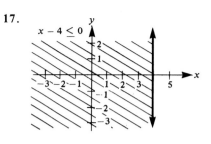

19.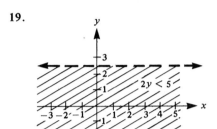

Section 8.5 Supplementary Exercises (p. 299)

1.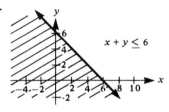

3.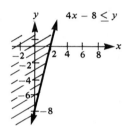

5.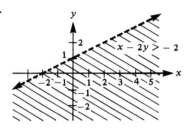

7.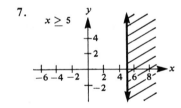

9.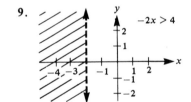

Vocabulary Quiz (p. 300)

1. g   2. d   3. b   4. j   5. c   6. l   7. n   8. h   9. m   10. k   11. a   12. f
13. i   14. e

Chapter 8 Review Exercises (p. 301)

1. Quadrant IV   2. Quadrant III   3. Quadrant I   4. Quadrant II   5. Quadrant IV

6.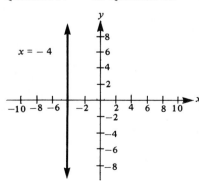

7.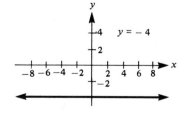

8.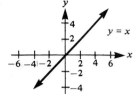

9.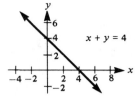

10.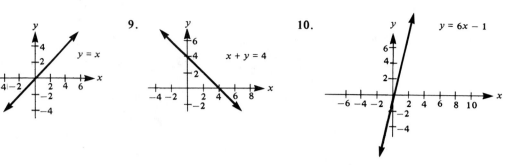

11.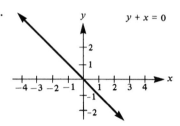

12.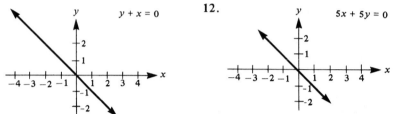

13.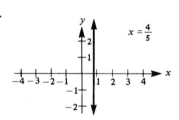

14.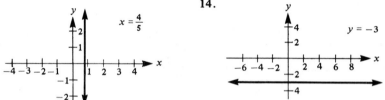

15.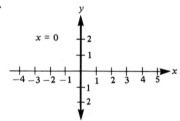

16. 1   **17.** 0   **18.** 0   **19.** undefined   **20.** $\dfrac{-12}{119}$

21. 3   **22.** $-\dfrac{1}{2}$ **23.** -1 **24.** $\dfrac{1}{3}$ **25.** undefined

	26.	**27.**	**28.**	**29.**	**30.**	**31.**	**32.**	**33.**	**34.**	**35.**
x-intercept	$(-1,0)$	none	none	$(5,0)$	$\left(\dfrac{16}{3},0\right)$	$\left(\dfrac{-2}{3},0\right)$	$\left(\dfrac{5}{2},0\right)$	$\left(\dfrac{-2}{3},0\right)$	$\left(\dfrac{-7}{3},0\right)$	$(-1,0)$
y-intercept	$(0,1)$	$(0,4)$	$(0,5)$	none	$\left(0,\dfrac{64}{119}\right)$	$(0,2)$	$\left(0,\dfrac{5}{4}\right)$	$\left(0,\dfrac{-2}{3}\right)$	$\left(0,\dfrac{7}{9}\right)$	none

36. $y = 5x - 3$ **37.** $y = -4x - 2$ **38.** $y = -\dfrac{2}{3}x + 3$ **39.** $y = \dfrac{1}{2}x - \dfrac{3}{4}$ **40.** $y = -3x$

41. $y = 2x + 10$ **42.** $y = -6x - 6$ **43.** $y = -5$ **44.** $y = -\dfrac{1}{4}x + \dfrac{15}{8}$ **45.** $y = -\dfrac{3}{5}x - \dfrac{63}{20}$

46. $y = x + 1$ **47.** $y = 3x - 6$ **48.** $y = \dfrac{1}{4}x + \dfrac{9}{4}$ **49.** $y = \dfrac{5}{3}x - \dfrac{1}{12}$ **50.** $y = 6$

51.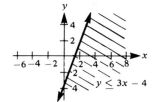

52.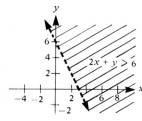

53.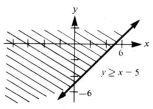

54.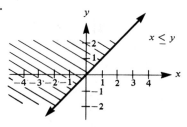

55.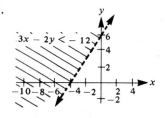

Cumulative Review Exercises (p. 302)

1. $-21x + 8$ **2.** $-y^5 + 6y^4 - 10y^3$ **3.** $t = \dfrac{3xy - 5}{2}$ **4.** 147   **5.** $-24x^9 y^{13}$ **6.** $8x^3 - 27y^3$

7. $3(x - 5y)(x + 5y)$ **8.** $4t(2t - 1)(t + 4)$ **9.** $\dfrac{1}{8}$ **10.** $\dfrac{5x^2 - 16x - 12}{(x - 3)(x - 2)(x - 4)(x + 2)}$

Chapter 8 Test (p. 303)

1. 3   **2.** $A\,(-4, -3), B\,(0, 0), C\,(4, -3)$

3.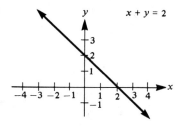

4.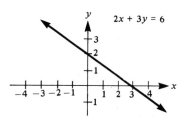

5.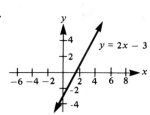

6.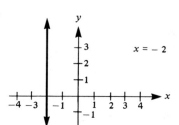

7. x-intercept: $(2, 0)$
y-intercept: $(0, -3)$

8. 7   **9.** -1 **10.** 0   **11.** $y = 3x - 5$ **12.** $y = -2x - 4$ **13.** $y = -x + 1$

14.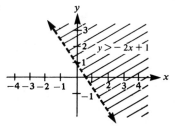

15.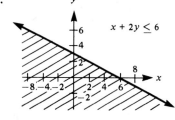

Chapter 9

Section 9.1 Exercises (p. 311)

1.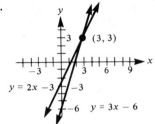

3.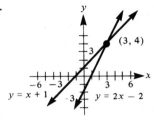

5.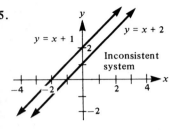

7.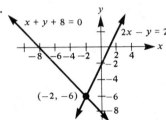

9.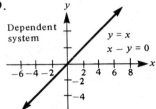

11.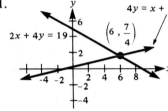

13.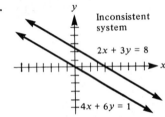

15.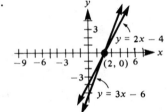

Section 9.1 Supplementary Exercises (p. 311)

1.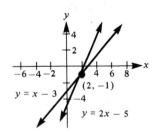

3.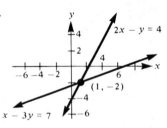

5.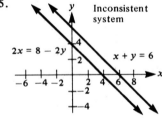

7.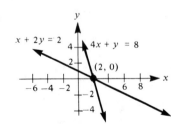

Section 9.2 Exercises (p. 315)

1.
$$x + y = 9$$
$$\underline{x - y = 5}$$
$$2x \quad\;\; = 14$$
$$x = 7$$
$$x + y = 9$$
$$7 + y = 9$$
$$y = 2$$
Solution: $x = 7, y = 2$

3. $2x - 4y = -4$
$\quad x + 2y = 1$ (Multiply by 2) ⌐
$$2x - 4y = -4$$
$$\underline{2x + 4y = 2}$$
$$4x \quad\;\; = -2$$
$$x = -\frac{1}{2}$$

$$2x - 4y = -4$$
$$2\left(-\frac{1}{2}\right) - 4y = -4$$
$$-4y = -3$$
$$y = \frac{3}{4}$$
Solution: $x = -\frac{1}{2}, y = \frac{3}{4}$

5.
$$2x + y = 1$$
$$\underline{-3x - y = 0}$$
$$-x \quad\;\; = 1$$
$$x = -1$$

$$2x + y = 1$$
$$2(-1) + y = 1$$
$$y = 3$$
Solution: $x = -1, y = 3$

7. $2x + 3y = 11$ (Multiply by 10)
$\quad 5x - 10y = 10$ (Multiply by 3)
$$20x + 30y = 110$$
$$\underline{15x - 30y = 30}$$
$$35x \quad\quad\; = 140$$
$$x = 4$$
$$2x + 3y = 11$$
$$2(4) + 3y = 11$$
$$3y = 3$$
$$y = 1$$
Solution: $x = 4, y = 1$

9. $3x + 2y = -2$ (Multiply by -2) ⌐
$\quad 3x + 4y = -3$
$$-6x - 4y = 4$$
$$\underline{3x + 4y = -3}$$
$$-3x \quad\quad = 1$$
$$x = -\frac{1}{3}$$

$$3x + 2y = -2$$
$$3\left(-\frac{1}{3}\right) + 2y = -2$$
$$y = -\frac{1}{2}$$
Solution: $x = -\frac{1}{3}, y = -\frac{1}{2}$

11. $5x + 7y = 12$
$\quad\; 5x + 7y = 12$

Same equation, therefore it is a dependent system.

13. $3x - 2y = 5$ (Multiply by 2) ⌐
$\quad -6x + 4y = -10$
$$6x - 4y = 10$$
$$\underline{-6x + 4y = -10}$$
$$0 = 0$$
Dependent system

15. $0.2x + 0.2y = 1$ (Multiply by 5)⌐
$\quad\quad x - y = 1$
$$x + y = 5$$
$$\underline{x - y = 1}$$
$$2x \quad\;\; = 6$$
$$x = 3$$
$$x - y = 1$$
$$3 - y = 1$$
$$y = 2$$
Solution: $x = 3, y = 2$

17. $3x + 4y = 9$ (Multiply by 2) ⌐
$\quad -6x - 8y = 3$
$$6x + 8y = 18$$
$$\underline{-6x - 8y = 3}$$
$$0 = 21$$
Inconsistent system

19.
$$\begin{array}{rcl} x - y &=& 3 \\ 3x + 3y &=& -3 \end{array}$$ (Multiply by 3)

$$\begin{array}{rcl} 3x - 3y &=& 9 \\ 3x + 3y &=& -3 \\ \hline 6x &=& 6 \\ x &=& 1 \end{array}$$

$$\begin{array}{rcl} x - y &=& 3 \\ 1 - y &=& 3 \\ y &=& -2 \end{array}$$

Solution: $x = 1, y = -2$

21. Dependent   23. $x = -\dfrac{3}{2}, y = -3$ 25. Dependent

27. (b) $x = 4, y = 0$

Section 9.2 Supplementary Exercises (p. 316)

1.
$$\begin{array}{rcl} x - y &=& 6 \\ x + y &=& 10 \\ \hline 2x &=& 16 \\ x &=& 8 \end{array}$$

$$\begin{array}{rcl} x - y &=& 6 \\ 8 - y &=& 6 \\ y &=& 2 \end{array}$$

Solution: $x = 8, y = 2$

3.
$$\begin{array}{rcl} x + y &=& 10 \\ x + y &=& 7 \end{array}$$ (Multiply by -1)

$$\begin{array}{rcl} -x - y &=& -10 \\ x + y &=& 7 \\ \hline 0 &=& -3 \end{array}$$

Inconsistent system

5.
$$\begin{array}{rcl} 2x - y &=& 4 \\ x - y &=& 5 \end{array}$$ (Multiply by -1)

$$\begin{array}{rcl} -2x + y &=& -4 \\ x - y &=& 5 \\ \hline -x &=& 1 \\ x &=& -1 \end{array}$$

$$\begin{array}{rcl} 2x - y &=& 4 \\ 2(-1) - y &=& 4 \\ y &=& -6 \end{array}$$

Solution: $x = -1, y = -6$

7.
$$\begin{array}{rcl} x + 3y &=& 1 \\ 2x - y &=& -5 \end{array}$$ (Multiply by 3)

$$\begin{array}{rcl} x + 3y &=& 1 \\ 6x - 3y &=& -15 \\ \hline 7x &=& -14 \\ x &=& -2 \end{array}$$

$$\begin{array}{rcl} x + 3y &=& 1 \\ -2 + 3y &=& 1 \\ y &=& 1 \end{array}$$

Solution: $x = -2, y = 1$

9.
$$\begin{array}{rcl} \frac{1}{2}x + y &=& 2 \\ x - y &=& 3 \\ \hline \frac{3}{2}x &=& 5 \\ x &=& \frac{10}{3} \end{array}$$

$$\begin{array}{rcl} x - y &=& 3 \\ \frac{10}{3} - y &=& 3 \\ y &=& \frac{1}{3} \end{array}$$

Solution: $x = \dfrac{10}{3}, y = \dfrac{1}{3}$

11.
$$\begin{array}{rcl} -3x + 2y &=& -2 \\ 2x - 3y &=& -2 \end{array}$$ (Multiply by 3)
(Multiply by 2)

$$\begin{array}{rcl} -9x + 6y &=& -6 \\ 4x - 6y &=& -4 \\ \hline -5x &=& -10 \\ x &=& 2 \end{array}$$

$$\begin{array}{rcl} -3x + 2y &=& -2 \\ -3(2) + 2y &=& -2 \\ 2y &=& 4 \\ y &=& 2 \end{array}$$

Solution: $x = 2, y = 2$

13. $0.3x - 0.2y = 0.1$ (Multiply by 5)
    $x + \ \ \ y = 2$

$$1.5x - \ \ \ y = 0.5$$
$$\underline{\ \ x + \ \ \ y = 2\ \ }$$
$$2.5x \ \ \ \ \ \ \ \ = 2.5$$
$$x = 1$$

$$x + y = 2$$
$$1 + y = 2$$
$$y = 1$$

Solution: $x = 1, y = 1$

15. $x = 2y$
   $y = 2x - 3$

$$x - 2y = 0$$
$$2x - \ y = 3 \quad \text{(Multiply by } -2)$$

$$x - 2y = 0$$
$$\underline{-4x + 2y = -6\ }$$
$$-3x \ \ \ \ \ \ = -6$$
$$x = 2$$

$$x = 2y$$
$$2 = 2y$$
$$y = 1$$

Solution: $x = 2, y = 1$

Section 9.3 Exercises (p. 320)

1. $y = 3x - 1$
$2x + 3y = 8$

Substitute $3x - 1$ for y in the second equation:

$$2x + 3(3x - 1) = 8$$
$$2x + 9x - 3 = 8$$
$$11x = 11$$
$$x = 1$$

Now find y:
$$y = 3x - 1$$
$$y = 3(1) - 1$$
$$y = 2$$

Solution: $x = 1, y = 2$

3. $x = -y + 1$
$3x - y = 23$

Substitute $-y + 1$ for x in the second equation:

$$3(-y + 1) - y = 23$$
$$-3y + 3 - y = 23$$
$$-4y = 20$$
$$y = -5$$

Now find y:
$$x = -y + 1$$
$$x = 5 + 1$$
$$x = 6$$

Solution: $x = 6, y = -5$

5. $x = -y + 1$
$x + y = 2$

Substitute $-y + 1$ for x in the second equation:

$$(-y + 1) + y = 2$$
$$1 = 2$$

Inconsistent system

7. $2x - 3y = -13$
$y = 2x + 11$

Substitute $2x + 11$ for y in the first equation:

$$2x - 3(2x + 11) = -13$$
$$2x - 6x - 33 = -13$$
$$-4x = 20$$
$$x = -5$$

Now find y:
$$y = 2x + 11$$
$$y = 2(-5) + 11$$
$$y = 1$$

Solution: $x = -5, y = 1$

9. $x = 3 - 2y$
$3x - 4y = 14$

Substitute $3 - 2y$ for x in the second equation:

$$3(3 - 2y) - 4y = 14$$
$$9 - 6y - 4y = 14$$
$$-10y = 5$$
$$y = -\frac{1}{2}$$

Now find x:
$$x = 3 - 2y$$
$$x = 3 - 2\left(-\frac{1}{2}\right)$$
$$x = 4$$

Solution: $x = 4, y = -\frac{1}{2}$

11. $x = 3 - 2y$
$-5x - 10y = -15$

Substitute $3 - 2y$ for x in the second equation:

$$-5(3 - 2y) - 10y = -15$$
$$-15 + 10y - 10y = -15$$
$$0 = 0$$

Dependent system

13. $y = 3x - 4$
$6x + 5y = 29$

Substitute $3x - 4$ for y
in the second equation:

$6x + 5(3x - 4) = 29$
$6x + 15x - 20 = 29$
$21x = 49$
$x = \dfrac{49}{21} = \dfrac{7}{3}$

Now find y:

$y = 3x - 4$
$y = 3\left(\dfrac{7}{3}\right) - 4$
$y = 3$

Solution: $x = \dfrac{7}{3}, y = 3$

15. $8x + 12y = -20$

$x = \dfrac{(7y - 10)}{4}$

Substitute $\dfrac{(7y - 10)}{4}$ for x in the
first equation:

$8\left[\dfrac{(7y - 10)}{4}\right] + 12y = -20$
$14y - 20 + 12y = -20$
$y = 0$

Now find x:

$4x - 7y = -10$
$4x - 7(0) = -10$
$x = -\dfrac{5}{2}$

Solution: $x = -\dfrac{5}{2}, y = 0$

17. $x = \dfrac{(3y + 12)}{2}$

$5x + 4y = -3$

Substitute $\dfrac{3y + 12}{2}$ for x in the
second equation:

$5\left(\dfrac{3y + 12}{2}\right) + 4y = -3$
$15y + 60 + 8y = -6$
$23y = -66$
$y = \dfrac{-66}{23}$

Now find x:

$x = \dfrac{3y + 12}{2}$

$x = \dfrac{3\left(\dfrac{-66}{23}\right) + 12}{2}$

$x = \dfrac{39}{23}$

Solution: $x = \dfrac{39}{23}, y = \dfrac{-66}{23}$

19. $x = y + 1$
$2x - 3y = -2$

Substitute $y + 1$ for x in the
second equation:

$2(y + 1) - 3y = -2$
$2y + 2 - 3y = -2$
$-y = -4$
$y = 4$

Now find x:

$x = y + 1$
$x = 4 + 1$
$x = 5$

Solution: $x = 5, y = 4$

Section 9.3 Supplementary Exercises (p. 320)

1. $y = 2x - 1$

$x + y = -4$

Substitute $2x - 1$ for y in the second equation:

$x + y = -4$

$x + (2x - 1) = -4$

$3x = -3$

$x = -1$

Now find y:

$y = 2x - 1$

$y = 2(-1) - 1$

$y = -3$

Solution: $x = -1, y = -3$

3. $x = y + 2$

$2y - x = 0$

Substitute $y + 2$ for x in the second equation:

$2y - x = 0$

$2y - (y + 2) = 0$

$y = 2$

Now find x:

$x = y + 2$

$x = 4$

Solution: $x = 4, y = 2$

5. $y = 3x$

$y = \dfrac{x}{4}$

Substitute $3x$ for y in the second equation:

$y = \dfrac{x}{4}$

$3x = \dfrac{x}{4}$

$12x = x$

$11x = 0$

$x = 0$

Now find y:

$y = 3x$

$y = 3(0)$

$y = 0$

Solution: $x = 0, y = 0$

7. $x = y + 6$

$x - 5y = 2$

Substitute $y + 6$ for x in the second equation:

$x - 5y = 2$

$(y + 6) - 5y = 2$

$-4y = -4$

$y = 1$

Now find x:

$x = y + 6$

$x = 1 + 6$

$x = 7$

Solution: $x = 7, y = 1$

9. $x = 2y - 2$

$3x - y = 4$

Substitute $2y - 2$ for x in the second equation:

$3x - y = 4$

$3(2y - 2) - y = 4$

$6y - 6 - y = 4$

$5y = 10$

$y = 2$

Now find x:

$x = 2y - 2$

$x = 2(2) - 2$

$x = 2$

Solution: $x = 2, y = 2$

Section 9.4 Exercises (p. 324)

1. Let x = one number

y = the other number

$x + y = 124$

$\underline{x - y = 24}$

$2x = 148$

$x = 74$

$x + y = 124$

$74 + y = 124$

$y = 50$

The numbers are 74 and 50.

3. Let x = the first number

y = the second number

$2x + 3y = 18$ (Multiply by 2)—

$3x - 2y = 1$ (Multiply by 3)—

$4x + 6y = 36$ ◄——

$\underline{9x - 6y = 3}$ ◄——

$13x = 39$

$x = 3$

$2x + 3y = 18$

$2(3) + 3y = 18$

$y = 4$

The numbers are 3 and 4.

5. Let q = the number of quarters
h = the number of half-dollars

$$
\begin{aligned}
q + h &= 30 \\
.25q + .5h &= 14 \quad \text{(Multiply by } -2)\text{⌐} \\
q + h &= 30 \\
-.5q - h &= -28 \quad \longleftarrow \\
\hline
.5q &= 2 \\
q &= 4 \\
q + h &= 30 \\
h &= 26
\end{aligned}
$$

There are 4 quarters and 26 half-dollars.

7. Let x = cost of 1 ton of material
y = cost of 1 work-hour

$$
\begin{aligned}
3x + 100y &= 2500 \quad \text{(Multiply by } -4)\text{⌐} \\
4x + 150y &= 3500 \quad \text{(Multiply by } 3)\text{⌐} \\
-12x - 400y &= -10000 \quad \longleftarrow \\
12x + 450y &= 10500 \quad \longleftarrow \\
\hline
50y &= 500 \\
y &= 10 \\
3x + 100y &= 2500 \\
3x + 100(10) &= 2500 \\
3x &= 1500 \\
x &= 500
\end{aligned}
$$

One ton of material cost \$500 and 1 work-hour of labor cost \$10.00.

9. Let x = amount invested at 12%
y = amount invested at 9%

$$
\begin{aligned}
x + y &= 6000 \quad \text{(Multiply by } -9)\text{⌐} \\
.12x + .09y &= 648 \quad \text{(Multiply by } 100)\text{⌐} \\
-9x - 9y &= -54000 \quad \longleftarrow \\
12x + 9y &= 64800 \quad \longleftarrow \\
\hline
3x &= 10800 \\
x &= 3600 \\
x + y &= 6000 \\
y &= 2400
\end{aligned}
$$

Hank invests \$3600 at 12% and \$2400 at 9%.

11. Let x = the number of \$5 bills
y = the number of \$10 bills

$$
\begin{aligned}
x + y &= 47 \quad \text{(Multiply by } -5)\text{⌐} \\
5x + 10y &= 335 \\
-5x - 5y &= -235 \quad \longleftarrow \\
5x + 10y &= 335 \\
\hline
5y &= 100 \\
y &= 20 \\
x + y &= 47 \\
x &= 27
\end{aligned}
$$

There are twenty-seven \$5 bills and twenty \$10 bills.

13. Let x = Tony's age
y = Lea's age

$$
\begin{aligned}
x &= 3y + 6 \\
x + y &= 50 \\
3y + 6 + y &= 50 \\
4y &= 44 \\
y &= 11 \\
x &= 3y + 6 \\
x &= 33 + 6 \\
x &= 39
\end{aligned}
$$

Thus, Tony is 39 and Lea is 11.

15. Let x = the number of 20-cent stamps
y = the number of 18-cent stamps

$$
\begin{aligned}
x + y &= 23 \\
.2x + .18y &= 4.56 \quad \text{(Multiply by } -5)\text{⌐} \\
x + y &= 23 \\
-x - .9y &= -22.80 \quad \longleftarrow \\
\hline
.1y &= .2 \\
y &= 2 \\
x + y &= 23 \\
x &= 21
\end{aligned}
$$

Ruby purchased 21 twenty-cent stamps and 2 eighteen-cent stamps.

17. Let x = the time the first car travels
y = the time the second car travels

$$
\begin{aligned}
x - y &= 3 \\
48x &= 64y \\
48(y + 3) &= 64y \\
48y + 144 &= 64y \\
16y &= 144 \\
y &= 9
\end{aligned}
$$

The second car needs to travel for 9 hours.
It will cover $64 \cdot 9 = 576$ miles.

Section 9.4 Supplementary Exercises (p. 325)

1. Let x = the first number
  y = the second number

$$x + y = 4$$
$$\underline{x - y = -2}$$
$$2x \quad\; = 2$$
$$x \;\; = 1$$

$$x + y = 4$$
$$1 + y = 4$$
$$y = 3$$

The numbers are 1 and 3.

3. Let x = the cost of 1 ton of material
  y = the cost of 1 work-hour

$$2x + 100y = 2800 \quad \text{(Multiply by } -3)$$
$$3x + 250y = 5000 \quad \text{(Multiply by 2)}$$
$$-6x - 300y = -8400$$
$$\underline{6x + 500y = 10000}$$
$$200y = 1600$$
$$y = 8$$
$$2x + 100y = 2800$$
$$2x + 800 = 2800$$
$$x = 1000$$

Thus, one ton of material cost $1000 and 1 work-hour of labor cost $8.00.

5. Let x = the amount invested at 12%
  y = the amount invested at 9%

$$x = 2y + 500$$
$$.12x + .09y = 1380$$
$$.12(2y + 500) + .09y = 1380$$
$$.24y + 60 + .09y = 1380$$
$$.33y = 1320$$
$$y = 4000$$

$$x = 2y + 500$$
$$x = 8500$$

Walt invested $8500 at 12% and $4000 at 9%.

7. Let x = the farmer's age
  y = the son's age

$$x = 2y - 4$$
$$x - y = 35$$
$$2y - 4 - y = 35$$
$$y = 39$$
$$x = 2y - 4$$
$$x = 2(39) - 4$$
$$x = 74$$

The farmer is 74; his son is 39.

Vocabulary Quiz (p. 326)

1. b **2.** c **3.** a

Chapter 9 Review Exercises (p. 326)

1.
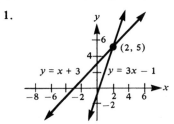

2.
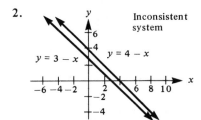

3.
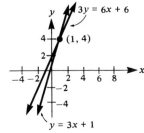

4.
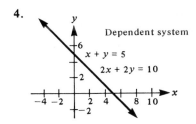

5.
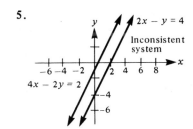

6.
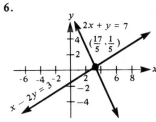

7. $x = 2, y = \dfrac{1}{3}$ **8.** $x = 10, y = 0$ **9.** $x = \dfrac{-5}{3}, y = 5$ **10.** $x = -3, y = 4$ **11.** $x = -1, y = -3$

12. $x = -1, y = 1$ **13.** $x = 18, y = 10$ **14.** $x = 5, y = -2$ **15.** Dependent **16.** Dependent

17. $x = 1, y = 1$ **18.** $x = -4, y = 3$ **19.** $x = 2, y = 1$ **20.** $x = 5, y = -2$ **21.** Inconsistent

22. Dependent   **23.** $x = \dfrac{59}{29}, y = \dfrac{-7}{29}$ **24.** $x = 7, y = -5$ **25.** -3 and 12

26. Sugar is $1 per kilogram and flour is 50¢ per kilogram.   **27.** Bill earns $3.25 per hour and Sandy earns $5.25 per hour.   **28.** $2500 at 12% and $1500 at 10%.   **29.** 4 records and 11 tapes
30. 10 mph and 14 mph

Cumulative Review Exercises (p. 327)

1. $-6a^6 b^5$ **2.** $t = 2$ **3.** $x > 1$ 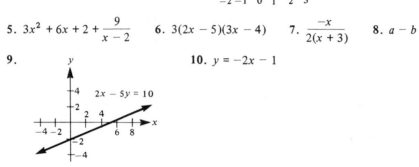   **4.** $216

5. $3x^2 + 6x + 2 + \dfrac{9}{x - 2}$ **6.** $3(2x - 5)(3x - 4)$ **7.** $\dfrac{-x}{2(x + 3)}$ **8.** $a - b$

9.   **10.** $y = -2x - 1$

Chapter 9 Test (p. 328)

1.   **2.** Inconsistent   **3.** $x = -3, y = 5$ **4.** $x = 5, y = 3$

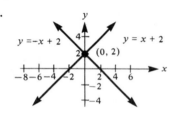

5. Inconsistent   **6.** $x = 2, y = -3$ **7.** $x = 5, y = -3$ **8.** 13.8 and 19.2   **9.** 75¢
10. $15,000 at 10% and $7500 at 8%

Chapter 10

Section 10.1 Exercises (p. 334)

1. 5   **3.** 7 and -7 **5.** -9 **7.** 11   **9.** True   **11.** 3.742   **13.** 3.162   **15.** -9.539
17. 2.646   **19.** ± 2.449 **21.** $2.646 - 3.317 = -0.671$ **23.** 0   **25.** Yes
27. 50.24 square feet

29. (a) $C = 8\pi$ meters $A = 16\pi$ square meters
  (b) $C \approx 25.12$ meters $A \approx 50.24$ square meters
  (c) $C \approx \dfrac{176}{7}$ meters $A \approx \dfrac{352}{7}$ square meters

31. (a) $\pi \approx 3.1415926, 3.14, \dfrac{22}{7} \approx 3.142857$:

  they all differ in the third decimal place

  (b) $\sqrt{10}$ (c) 3.14   (d) 3.14 and $\dfrac{22}{7}$

Section 10.1 Supplementary Exercises (p. 336)

1. 3   **3.** -2 **5.** ±12 **7.** 4.123   **9.** -6.245 **11.** ±9.11 **13.** $3 - 2 = 1$
15. $4.583 - 3.162 = 1.421$ **17.** 3 feet

Section 10.2 Exercises (p. 341)

1. $\sqrt{220} = \sqrt{4} \cdot \sqrt{55} = 2\sqrt{55} \approx 14.832$ **3.** $\sqrt{500} = \sqrt{100} \cdot \sqrt{5} = 10\sqrt{5} \approx 22.36$
5. $\sqrt{450} = \sqrt{225} \cdot \sqrt{2} = 15\sqrt{2} \approx 21.21$ **7.** $\sqrt{1000} = \sqrt{100} \cdot \sqrt{10} = 10\sqrt{10} \approx 31.62$
9. $\sqrt{10,400} = \sqrt{100} \cdot \sqrt{104} = \sqrt{100} \cdot \sqrt{4} \cdot \sqrt{26} = 20\sqrt{26} \approx 101.98$ **11.** $\sqrt{\dfrac{81}{16}} = \dfrac{9}{4}$

13. $\sqrt{0.16} = \sqrt{16} \cdot \sqrt{10^{-2}} = 4 \cdot 10^{-1} = 0.4$ **15.** $\sqrt{\dfrac{-4}{25}}$ is not a real number   **17.** $-\sqrt{\dfrac{16}{9}} = \dfrac{-4}{3}$

19. $6x$ **21.** $4x$ **23.** $x^2 y\sqrt{y}$ **25.** x^{300} **27.** $9xyz^2\sqrt{xz}$ **29.** $\dfrac{9x}{y}\sqrt{3x}$ **31.** $2x^4\sqrt{30}$
33. $\pm xy^2 z^8\sqrt{z}$ **35.** ±1

Section 10.2 Supplementary Exercises (p. 341)

1. $\sqrt{200} = \sqrt{100} \cdot \sqrt{2} = 10\sqrt{2} \approx 14.14$ **3.** $-\sqrt{432} = -\sqrt{144} \cdot \sqrt{3} = -12\sqrt{3} \approx 20.784$
5. $\sqrt{0.04} = 0.2$ **7.** $\sqrt{0.0009} = 0.03$ **9.** $-\sqrt{125} = -\sqrt{25} \cdot \sqrt{5} = -5\sqrt{5} \approx -11.18$ **11.** $4x^3$
13. x^{200} **15.** $5xy^3\sqrt{5xy}$ **17.** $\dfrac{11x}{y}\sqrt{x}$

Section 10.3 Exercises (p. 344)

1. $\sqrt{6} + \sqrt{24} = \sqrt{6} + 2\sqrt{6} = 3\sqrt{6}$ **3.** $\sqrt{18} + \sqrt{16} = 3\sqrt{2} + 4$ **5.** $\sqrt{144} - \sqrt{100} = 12 - 10 = 2$
7. $\sqrt{3} + \sqrt{12} = \sqrt{3} + 2\sqrt{3} = 3\sqrt{3}$ **9.** $\sqrt{8} - \sqrt{18} + \sqrt{2} = 2\sqrt{2} - 3\sqrt{2} + \sqrt{2} = 0$
11. $\sqrt{44} + \sqrt{200} - \sqrt{99} - \sqrt{50} = 2\sqrt{11} + 10\sqrt{2} - 3\sqrt{11} - 5\sqrt{2} = 5\sqrt{2} - \sqrt{11}$
13. $2\sqrt{3} + \sqrt{13}$ cannot be simplified   **15.** $\sqrt{9x} + \sqrt{16x} = 3\sqrt{x} + 4\sqrt{x} = 7\sqrt{x}$
17. $\dfrac{-5 \pm \sqrt{50}}{5} = \dfrac{-5 \pm 5\sqrt{2}}{5} = \dfrac{\cancel{5}(-1 \pm \sqrt{2})}{\cancel{5}} = -1 \pm \sqrt{2}$ **19.** $\dfrac{2 \pm 2\sqrt{2}}{2} = \dfrac{\cancel{2}(1 \pm \sqrt{2})}{\cancel{2}} = 1 \pm \sqrt{2}$
21. $1 + \sqrt{2}$ **23.** $3 - 3\sqrt{3}$ **25.** $\dfrac{-1 \pm 2\sqrt{2}}{2}$ **27.** $-1 \pm \sqrt{15}$ **29.** $\dfrac{-4 \pm 2\sqrt{7}}{7}$

Section 10.3 Supplementary Exercises (p. 344)

1. $\sqrt{8} + \sqrt{18} = 2\sqrt{2} + 3\sqrt{2} = 5\sqrt{2}$ **3.** $\sqrt{32} - \sqrt{8} = 4\sqrt{2} - 2\sqrt{2} = 2\sqrt{2}$
5. $\sqrt{112} + \sqrt{18} - \sqrt{28} + \sqrt{27} = 4\sqrt{7} + 3\sqrt{2} - 2\sqrt{7} + 3\sqrt{3} = 2\sqrt{7} + 3\sqrt{2} + 3\sqrt{3}$
7. $\sqrt{500} + 2\sqrt{48} - \sqrt{300} = 10\sqrt{5} + 8\sqrt{3} - 10\sqrt{3} = 10\sqrt{5} - 2\sqrt{3}$
9. $\dfrac{-6 \pm 12\sqrt{7}}{2} = \dfrac{\cancel{2}(-3 \pm 6\sqrt{7})}{\cancel{2}} = -3 \pm 6\sqrt{7}$ **11.** $\dfrac{-8 \pm 4\sqrt{6}}{4} = \dfrac{\cancel{4}(-2 \pm \sqrt{6})}{\cancel{4}} = -2 \pm \sqrt{6}$
13. $\dfrac{-2 \pm \sqrt{8}}{2} = \dfrac{-2 \pm 2\sqrt{2}}{2} = \dfrac{\cancel{2}(-1 \pm \sqrt{2})}{\cancel{2}} = -1 \pm \sqrt{2}$ **15.** $-7 \pm \sqrt{49} = -7 \pm 7 = 0$ or -14

Section 10.4 Exercises (p. 347)

1. $\dfrac{2}{\sqrt{2}} \cdot \dfrac{\sqrt{2}}{\sqrt{2}} = \dfrac{2\sqrt{2}}{2} = \sqrt{2} \approx 1.414$ **3.** $\dfrac{20}{\sqrt{8}} \cdot \dfrac{\sqrt{8}}{\sqrt{8}} = \dfrac{20\sqrt{8}}{8} = \dfrac{40\sqrt{2}}{8} = 5\sqrt{2} \approx 7.070$

5. $\dfrac{9}{\sqrt{3}} \cdot \dfrac{\sqrt{3}}{\sqrt{3}} = \dfrac{9\sqrt{3}}{3} = 3\sqrt{3} \approx 5.196$ **7.** $\dfrac{2}{\sqrt{3} - \sqrt{2}} \cdot \dfrac{\sqrt{3} + \sqrt{2}}{\sqrt{3} + \sqrt{2}} = \dfrac{2(\sqrt{3} + \sqrt{2})}{1} = 2(\sqrt{3} + \sqrt{2}) \approx 6.292$

9. $\dfrac{12-\sqrt{3}}{\sqrt{3}} \cdot \dfrac{\sqrt{3}}{\sqrt{3}} = \dfrac{12\sqrt{3}-3}{3} = 4\sqrt{3}-1 \approx 5.928$

11. $\dfrac{\sqrt{5}+\sqrt{10}}{\sqrt{20}} \cdot \dfrac{\sqrt{20}}{\sqrt{20}} = \dfrac{10+10\sqrt{2}}{20} = \dfrac{10(1+\sqrt{2})}{20} = \dfrac{1}{2}(1+\sqrt{2}) \approx 1.207$

13. $\dfrac{1}{\sqrt{8}-\sqrt{6}} \cdot \dfrac{\sqrt{8}+\sqrt{6}}{\sqrt{8}+\sqrt{6}} = \dfrac{\sqrt{8}+\sqrt{6}}{2} \approx 2.639$ 15. $\dfrac{\sqrt{12}}{\sqrt{8}+\sqrt{3}} \cdot \dfrac{\sqrt{8}-\sqrt{3}}{\sqrt{8}-\sqrt{3}} = \dfrac{4\sqrt{6}-6}{5} \approx 0.759$

17. $\dfrac{3}{\sqrt{8}-\sqrt{2}} = \dfrac{3}{2\sqrt{2}-\sqrt{2}} = \dfrac{3}{\sqrt{2}} \cdot \dfrac{\sqrt{2}}{\sqrt{2}} = \dfrac{3\sqrt{2}}{2}$

19. $\dfrac{\sqrt{3}}{6+\sqrt{3}} \cdot \dfrac{6-\sqrt{3}}{6-\sqrt{3}} = \dfrac{6\sqrt{3}-3}{33} = \dfrac{3(2\sqrt{3}-1)}{33} = \dfrac{2\sqrt{3}-1}{11}$

Section 10.4 Supplementary Exercises (p. 348)

1. $\dfrac{5}{\sqrt{5}} \cdot \dfrac{\sqrt{5}}{\sqrt{5}} = \dfrac{5\sqrt{5}}{5} = \sqrt{5} \approx 2.236$ 3. $\dfrac{-6}{\sqrt{3}} \cdot \dfrac{\sqrt{3}}{\sqrt{3}} = \dfrac{-6\sqrt{3}}{3} = -2\sqrt{3} \approx 3.464$

5. $\dfrac{15}{\sqrt{3}} \cdot \dfrac{3}{\sqrt{3}} = \dfrac{15\sqrt{3}}{3} = 5\sqrt{3} \approx 8.660$ 7. $\dfrac{-2}{\sqrt{5}-\sqrt{2}} \cdot \dfrac{\sqrt{5}+\sqrt{2}}{\sqrt{5}+\sqrt{2}} = \dfrac{-2(\sqrt{5}+\sqrt{2})}{3} = -2.434$

9. $\dfrac{1}{3-\sqrt{2}} \cdot \dfrac{3+\sqrt{2}}{3+\sqrt{2}} = \dfrac{3+\sqrt{2}}{7} \approx 0.631$

Section 10.5 Exercises (p. 355)

1. $\sqrt{x} = 10$ 3. $\sqrt{3x} = 10$ 5. $\sqrt{6x} = 0.1$ 7. $1-\sqrt{x} = 3$
   $x = 100$ $3x = 100$ $6x = 0.01$ $-\sqrt{x} = 2$
                   $x = \dfrac{100}{3}$ $x = 0.001\overline{6}$ $x = 4$

   Does not check.
   No solution

9. $1-\sqrt{2x} = 3$ 11. $\sqrt{x-7}+2 = 4$ 13. $\sqrt{5x} = \sqrt{3x+1}$
   $-\sqrt{2x} = 2$ $\sqrt{x-7} = 2$ $5x = 3x+1$
   $2x = 4$ $x-7 = 4$ $2x = 1$
   $x = 2$ $x = 11$ $x = \dfrac{1}{2}$

   Does not check.
   No solution

15. $1+\sqrt{2x-3} = 8$ 17. $\sqrt{x} = 7x$ 19. $1+\sqrt{5x-1} = 1+\sqrt{19}$
    $\sqrt{2x-3} = 7$ $x = 49x^2$ $\sqrt{5x-1} = \sqrt{19}$
    $2x-3 = 49$ $49x^2 - x = 0$ $5x-1 = 19$
    $2x = 52$ $x(49x-1) = 0$ $5x = 20$
    $x = 26$ $x = 0$ or $x = \dfrac{1}{49}$ $x = 4$

                              $\downarrow$ $\downarrow$

                             checks   checks

21. (a) $v = \sqrt{\dfrac{5r}{2}}$ 23. (a) 711 ft/sec   (b) $h = \dfrac{v^2}{32}$ (c) 6050 ft

    (b) 5 mph

25. (a) $r = \sqrt{\dfrac{V}{\pi h}}$ (b) approximately 2.45 cm   27. $\sqrt{13}$ 29. $\dfrac{\sqrt{3}}{4}$ 31. 10

33. $\sqrt{108} = 6\sqrt{3} \approx 10.4$ 35. 13   37. 84.85 feet   39. (a) 12 sq in.   (b) base: $4\sqrt{3}$ m, height: $2\sqrt{3}$ m   (c) base: 12 m, height: 2 m

---

Section 10.5 Supplementary Exercises (p. 357)

1. $\sqrt{y} = 9$
   $y = 81$

3. $\sqrt{2y} = 8$
   $2y = 64$
   $y = 32$

5. $\sqrt{\dfrac{y}{5}} = 3$
   $\dfrac{y}{5} = 9$
   $y = 45$

7. $2\sqrt{3x} = 6$
   $\sqrt{3x} = 3$
   $3x = 9$
   $x = 3$

9. $\sqrt{x} - 2 = 3$
   $\sqrt{x} = 5$
   $x = 25$

11. $\sqrt{2x} = \sqrt{x-1}$
    $2x = x - 1$
    $x = -1$
    Does not check.
    No solution

13. $2 + \sqrt{3x+1} = 6$
    $\sqrt{3x+1} = 4$
    $3x + 1 = 16$
    $3x = 15$
    $x = 5$

15. $\sqrt{3x-1} + 4 = 5$
    $\sqrt{3x-1} = 1$
    $3x - 1 = 1$
    $3x = 2$
    $x = \dfrac{2}{3}$

17. $5 - \sqrt{1-x} = 3$
    $-\sqrt{1-x} = -2$
    $1 - x = 4$
    $x = -3$

19. $\sqrt{5x} = \sqrt{2x-3}$
    $5x = 2x - 3$
    $3x = -3$
    $x = -1$
    Does not check.
    No solution

Vocabulary Quiz (p. 358)

1. c   2. g   3. e   4. h   5. j   6. d   7. i   8. f   9. k   10. l   11. a   12. b

---

Chapter 10 Review Exercises (p. 359)

1. True   2. False   3. False   4. True   5. True   6. True   7. 4.123   8. 8.246
9. 14.14   10. 0.968   11. 0.6325   12. ±13.416   13. −34.641   14. 11.944, −5.944

15. 5.071, −9.071   16. y^2 17. $2x\sqrt{x}$ 18. $7xyz$ 19. $\dfrac{7x}{4y}\sqrt{x}$ 20. $\dfrac{x\sqrt{y}}{5y}$ 21. $7\sqrt{5}$

22. $3\sqrt{6}$ 23. $-2\sqrt{7}$ 24. $6\sqrt{x}$ 25. $2\sqrt{2} - \sqrt{3}$ 26. $1 - \sqrt{2}$ 27. $\dfrac{-3 \pm \sqrt{2}}{2}$

28. $-3 \pm \sqrt{5}$ 29. $\dfrac{2 \pm \sqrt{2}}{2}$ 30. $\dfrac{2 \pm \sqrt{2}}{2}$ 31. $\sqrt{5}$ 32. $2\sqrt{5} + 1$ 33. $-2 - 2\sqrt{2}$

34. $\sqrt{2} - 1$ 35. $\dfrac{-\sqrt{6}}{2}$ 36. $x = 25$ 37. $x = 25$ 38. No solution   39. $x = 81$ 40. $x = 8$

41. $x = \dfrac{50}{3}$ 42. $x = 0$ 43. $x = 17$ 44. $x = 8$ 45. $x = \dfrac{15}{2}$ 46. (a) 80 ft/sec   (b) $h = \dfrac{v^2}{64}$

(c) 156.25 feet   47. 59.36 km   48. 9 cm by 12 cm   49. 11.62 inches

Cumulative Review Exercises (p. 360)

1. $x^5 - x^4 - 5x^3 + 6x^2 + 3x$ **2.** 3   **3.** $2x^3 - 3x^2 + 5x + 21$ **4.** $4x(x^2 + 9)$

5. 60 feet by 125 feet   **6.** $t = \dfrac{3}{2}$ **7.** x-intercept: $(-6, 0)$ y-intercept: $(0, 2)$ slope: $\dfrac{1}{3}$

8. 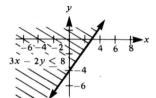   **9.** $x = 5, y = 3$ **10.** Inconsistent system

Chapter 10 Test (p. 361)

1. 7 and -7 **2.** True   **3.** $6\sqrt{5}$ **4.** 0.4   **5.** $3xy\sqrt{2y}$ **6.** $3\sqrt{5}$ **7.** $-\sqrt{6x}$ **8.** $1 + \sqrt{2}$

9. $\dfrac{-1 \pm \sqrt{2}}{2}$ **10.** $\dfrac{8\sqrt{5}}{5}$ **11.** $\sqrt{7} + \sqrt{2}$ **12.** $t = 20$ **13.** $x = 14$ **14.** $x = \pm \sqrt{\dfrac{8J}{7y}}$

15. 12 inches

Chapter 11

Section 11.1 Exercises (p. 366)

1. $x^2 - 5x - 6 = 0$
$(x - 6)(x + 1) = 0$
$x - 6 = 0$ $x + 1 = 0$
$x = 6$ $x = -1$

3. $x^2 + 5x - 6 = 0$
$(x + 6)(x - 1) = 0$
$x + 6 = 0$ $x - 1 = 0$
$x = -6$ $x = 1$

5. $x^2 - 7x + 6 = 0$
$(x - 6)(x - 1) = 0$
$x - 6 = 0$ $x - 1 = 0$
$x = 6$ $x = 1$

7. $x^2 - 6x - 7 = 0$
$(x - 7)(x + 1) = 0$
$x - 7 = 0$ $x + 1 = 0$
$x = 7$ $x = -1$

9. $2x^2 - 5x + 2 = 0$
$(2x - 1)(x - 2) = 0$
$2x - 1 = 0$ $x - 2 = 0$
$x = \dfrac{1}{2}$ $x = 2$

11. $2x^2 - 3x - 2 = 0$
$(2x + 1)(x - 2) = 0$
$2x + 1 = 0$ $x - 2 = 0$
$x = -\dfrac{1}{2}$ $x = 2$

13. $4x^2 - 13x + 3 = 0$
$(4x - 1)(x - 3) = 0$
$4x - 1 = 0$ $x - 3 = 0$
$x = \dfrac{1}{4}$ $x = 3$

15. $6x^2 + 17x - 3 = 0$
$(6x - 1)(x + 3) = 0$
$6x - 1 = 0$ $x + 3 = 0$
$x = \dfrac{1}{6}$ $x = -3$

17. $6x^2 + 19x + 3 = 0$
$(6x + 1)(x + 3) = 0$
$6x + 1 = 0$ $x + 3 = 0$
$x = -\dfrac{1}{6}$ $x = -3$

19. $6x^2 + 7x - 3 = 0$
$(3x - 1)(2x + 3) = 0$
$3x - 1 = 0$ $2x + 3 = 0$
$x = \dfrac{1}{3}$ $x = -\dfrac{3}{2}$

21. $x = -\dfrac{1}{2}, x = \dfrac{2}{3}$ **23.** $x = 1, x = \dfrac{1}{3}$ **25.** $x = 0, x = -\dfrac{7}{2}$

27. $x = \dfrac{9}{2}, x = -\dfrac{9}{2}$ **29.** $x = -\dfrac{3}{2}$ **31.** 5

Section 11.1 Supplementary Exercises (p. 366)

1. $x^2 - 5x - 36 = 0$
$(x - 9)(x + 4) = 0$
$x - 9 = 0 \quad x + 4 = 0$
$x = 9 \quad\quad x = -4$

3. $x^2 - 6x + 8 = 0$
$(x - 4)(x - 2) = 0$
$x - 4 = 0 \quad x - 2 = 0$
$x = 4 \quad\quad x = 2$

5. $2x^2 - 3x - 2 = 0$
$(x - 2)(2x + 1) = 0$
$x - 2 = 0 \quad 2x + 1 = 0$
$x = 2 \quad\quad x = -\dfrac{1}{2}$

7. $5x^2 + 7x + 2 = 0$
$(5x + 2)(x + 1) = 0$
$5x + 2 = 0 \quad x + 1 = 0$
$x = -\dfrac{2}{5} \quad\quad x = -1$

9. $x^2 - 2x = 0$
$x(x - 2) = 0$
$x = 0 \quad x - 2 = 0$
$x = 2$

11. $5x^2 - 25x = 0$
$5x(x - 5) = 0$
$x = 0 \quad x = 5$

13. $x^2 - 9 = 0$
$(x + 3)(x - 3) = 0$
$x = -3 \quad x = 3$

15. $x^2 - x - 6 = 0$
$(x - 3)(x + 2) = 0$
$x - 3 = 0 \quad x + 2 = 0$
$x = 3 \quad\quad x = -2$

Section 11.2 Exercises (p. 369)

1. $x^2 + 2x - 7 = 0$
$x^2 + 2x + [\ ] = 7 + [\ ]$
$x^2 + 2x + 1 = 8$
$(x + 1)^2 = 8$
$x + 1 = \pm\sqrt{8}$
$x = -1 \pm \sqrt{8}$
$x = -1 \pm 2\sqrt{2}$

3. $x^2 - 6x + 2 = 0$
$x^2 - 6x + [\ ] = -2 + [\ ]$
$x^2 - 6x + 9 = 7$
$(x - 3)^2 = 7$
$x - 3 = \pm\sqrt{7}$
$x = 3 \pm \sqrt{7}$

5. $x^2 - 6x + 4 = 0$
$x^2 - 6x + [\ ] = -4 + [\ ]$
$x^2 - 6x + 9 = 5$
$(x - 3)^2 = 5$
$x - 3 = \pm\sqrt{5}$
$x = 3 \pm \sqrt{5}$

7. $x^2 - 6x + 6 = 0$
$x^2 - 6x + [\ ] = -6 + [\ ]$
$x^2 - 6x + 9 = 3$
$(x - 3)^2 = 3$
$x - 3 = \pm\sqrt{3}$
$x = 3 \pm \sqrt{3}$

9. $x^2 - 8x + 1 = 0$
$x^2 - 8x + [\ ] = -1 + [\ ]$
$x^2 - 8x + 16 = 15$
$(x - 4)^2 = 15$
$x - 4 = \pm\sqrt{15}$
$x = 4 \pm \sqrt{15}$

11. $x^2 - 7x + 1 = 0$
$x^2 - 7x + [\ ] = -1 + [\ ]$
$x^2 - 7x + \dfrac{49}{4} = \dfrac{45}{4}$
$\left(x - \dfrac{7}{2}\right)^2 = \dfrac{45}{4}$
$x - \dfrac{7}{2} = \dfrac{\pm 3\sqrt{5}}{2}$
$x = \dfrac{7}{2} \pm \dfrac{3\sqrt{5}}{2}$
$x = \dfrac{7 \pm 3\sqrt{5}}{2}$

13. $x^2 - 7x + 2 = 0$
$x^2 - 7x + [\ ] = -2 + [\ ]$
$x^2 - 7x + \dfrac{49}{4} = \dfrac{41}{4}$
$\left(x - \dfrac{7}{2}\right)^2 = \dfrac{41}{4}$
$x - \dfrac{7}{2} = \pm\dfrac{\sqrt{41}}{2}$
$x = \dfrac{7}{2} \pm \dfrac{\sqrt{41}}{2}$
$x = \dfrac{7 \pm \sqrt{41}}{2}$

15. $2x^2 - 8x - 1 = 0$
$x^2 - 4x - \dfrac{1}{2} = 0$
$x^2 - 4x + [\ ] = \dfrac{1}{2} + [\ ]$
$x^2 - 4x + 4 = \dfrac{9}{2}$
$(x - 2)^2 = \dfrac{9}{2}$
$x - 2 = \pm\dfrac{3\sqrt{2}}{2}$
$x = 2 \pm \dfrac{3\sqrt{2}}{2}$
$x = \dfrac{4 \pm 3\sqrt{2}}{2}$

17. $2x^2 - 6x - 1 = 0$
$x^2 - 3x - \dfrac{1}{2} = 0$
$x^2 - 3x + [\ ] = \dfrac{1}{2} + [\ ]$
$x^2 - 3x + \dfrac{9}{4} = \dfrac{11}{4}$
$\left(x - \dfrac{3}{2}\right)^2 = \dfrac{11}{4}$
$x - \dfrac{3}{2} = \pm\dfrac{\sqrt{11}}{2}$
$x = \dfrac{3}{2} \pm \dfrac{\sqrt{11}}{2}$
$x = \dfrac{3 \pm \sqrt{11}}{2}$

19. $2x^2 - 6x - 8 = 0$

$x^2 - 3x - 4 = 0$

$x^2 - 3x + [\ ] = 4 + [\ ]$

$x^2 - 3x + \dfrac{9}{4} = \dfrac{25}{4}$

$\left(x - \dfrac{3}{2}\right)^2 = \dfrac{25}{4}$

$x - \dfrac{3}{2} = \pm\dfrac{5}{2}$

$x = \dfrac{3}{2} \pm \dfrac{5}{2}$

$x = 4, x = -1$

23. $x = .95 \pm \sqrt{5.0025}$

$x \approx 3.187, x \approx -1.287$

25. $x = -0.63 \pm \sqrt{8.73023}$

$x \approx 2.325, x \approx -3.585$

Section 11.2 Supplementary Exercises (p. 370)

1. $x^2 - 4x + 1 = 0$

$x^2 - 4x + [\ ] = -1 + [\ ]$

$x^2 - 4x + 4 = -1 + 4$

$(x - 2)^2 = 3$

$x - 2 = \pm\sqrt{3}$

$x = 2 \pm \sqrt{3}$

3. $x^2 - 4x + 2 = 0$

$x^2 - 4x + [\ ] = -2 + [\ ]$

$x^2 - 4x + 4 = -2 + 4$

$(x - 2)^2 = 2$

$x - 2 = \pm\sqrt{2}$

$x = 2 \pm \sqrt{2}$

5. $x^2 - 8x + 6 = 0$

$x^2 - 8x + [\ ] = -6 + [\ ]$

$x^2 - 8x + 16 = -6 + 16$

$(x - 4)^2 = 10$

$x - 4 = \pm\sqrt{10}$

$x = 4 \pm \sqrt{10}$

7. $x^2 + 2x - 1 = 0$

$x^2 + 2x + [\ ] = 1 + [\ ]$

$x^2 + 2x + 1 = 1 + 1$

$(x + 1)^2 = 2$

$x + 1 = \pm\sqrt{2}$

$x = -1 \pm \sqrt{2}$

9. $3x^2 - 9x + 6 = 0$

$x^2 - 3x + 2 = 0$

$x^2 - 3x + [\ ] = -2 + [\ ]$

$x^2 - 3x + \dfrac{9}{4} = -2 + \dfrac{9}{4}$

$\left(x - \dfrac{3}{2}\right)^2 = \dfrac{1}{4}$

$x - \dfrac{3}{2} = \pm\dfrac{1}{2}$

$x = \dfrac{3}{2} \pm \dfrac{1}{2}$

$x = 2, x = 1$

11. $\dfrac{x^2}{2} - x - 1 = 0$

$x^2 - 2x - 2 = 0$

$x^2 - 2x + [\ ] = 2 + [\ ]$

$x^2 - 2x + 1 = 2 + 1$

$(x - 1)^2 = 3$

$x - 1 = \pm\sqrt{3}$

$x = 1 \pm \sqrt{3}$

13. $\dfrac{x^2}{4} + 2x - 1 = 0$

$x^2 + 8x - 4 = 0$

$x^2 + 8x + [\ ] = 4 + [\ ]$

$x^2 + 8x + 16 = 4 + 16$

$(x + 4)^2 = 20$

$x + 4 = \pm 2\sqrt{5}$

$x = -4 \pm 2\sqrt{5}$

Section 11.3 Exercises (p. 374)

1. $x^2 + x - 12 = 0$
$a = 1, b = 1, c = -12$

$$x = \frac{-1 \pm \sqrt{1 - 4(1)(-12)}}{2}$$

$$x = \frac{-1 \pm \sqrt{49}}{2}$$

$$x = \frac{-1 + 7}{2} = 3, x = \frac{-1 - 7}{2} = -4$$

3. $x^2 - 7x + 10 = 0$
$a = 1, b = -7, c = 10$

$$x = \frac{7 \pm \sqrt{49 - 4(1)(10)}}{2}$$

$$x = \frac{7 \pm \sqrt{9}}{2}$$

$$x = \frac{7 + 3}{2} = 5, x = \frac{7 - 3}{2} = 2$$

5. $x^2 - 6x + 5 = 0$
$a = 1, b = -6, c = 5$

$$x = \frac{6 \pm \sqrt{36 - 4(1)(5)}}{2}$$

$$x = \frac{6 \pm \sqrt{16}}{2}$$

$$x = \frac{6 + 4}{2} = 5, x = \frac{6 - 4}{2} = 1$$

7. $y^2 - 7y - 12 = 0$
$a = 1, b = -7, c = -12$

$$y = \frac{7 \pm \sqrt{49 - 4(1)(-12)}}{2}$$

$$y = \frac{7 \pm \sqrt{97}}{2}$$

9. $x^2 + 3x - 1 = 0$
$a = 1, b = 3, c = -1$

$$x = \frac{-3 \pm \sqrt{9 - 4(1)(-1)}}{2}$$

$$x = \frac{-3 \pm \sqrt{13}}{2}$$

11. $3y^2 - 5y - 2 = 0$
$a = 3, b = -5, c = -2$

$$y = \frac{5 \pm \sqrt{25 - 4(3)(-2)}}{6}$$

$$y = \frac{5 \pm \sqrt{49}}{6}$$

$$y = \frac{5 + 7}{6} = 2, y = \frac{5 - 7}{6} = -\frac{1}{3}$$

13. $x^2 - 3x - 2 = 0$
$a = 1, b = -3, c = -2$

$$x = \frac{3 \pm \sqrt{9 - 4(1)(-2)}}{2}$$

$$x = \frac{3 \pm \sqrt{17}}{2}$$

15. $x^2 = x$
$x^2 - x = 0$
$a = 1, b = -1, c = 0$

$$x = \frac{1 \pm \sqrt{1 - 4(1)(0)}}{2}$$

$$x = \frac{1 \pm 1}{2}$$

$$x = \frac{1 + 1}{2} = 1, x = \frac{1 - 1}{2} = 0$$

17. $3x^2 - 6x - 6 = 0$
$x^2 - 2x - 2 = 0$
$a = 1, b = -2, c = -2$

$$x = \frac{2 \pm \sqrt{4 - 4(1)(-2)}}{2}$$

$$x = \frac{2 \pm \sqrt{12}}{2}$$

$$x = \frac{2 \pm 2\sqrt{3}}{2}$$

$$x = 1 \pm \sqrt{3}$$

19. $3m^2 - 5m + 1 = 0$
$a = 3, b = -5, c = 1$

$$m = \frac{5 \pm \sqrt{25 - 4(3)(1)}}{6}$$

$$m = \frac{5 \pm \sqrt{13}}{6}$$

21. $x \approx 10.5944, x \approx -0.0944$

23. $x \approx 86{,}899.8, x \approx 0.23014$ **25.** $x \approx 0.80902, x \approx -0.30902$

Section 11.3 Supplementary Exercises (p. 374)

1. $x^2 - 3x + 1 = 0$
$a = 1, b = -3, c = 1$

$$x = \frac{3 \pm \sqrt{9 - 4(1)(1)}}{2}$$

$$x = \frac{3 \pm \sqrt{5}}{2}$$

3. $x^2 + 5x - 2 = 0$
$a = 1, b = 5, c = -2$

$$x = \frac{-5 \pm \sqrt{25 - 4(1)(-2)}}{2}$$

$$x = \frac{-5 \pm \sqrt{33}}{2}$$

5. $2x^2 - 5x + 1 = 0$
$a = 2, b = -5, c = 1$

$$x = \frac{5 \pm \sqrt{25 - 4(2)(1)}}{4}$$

$$x = \frac{5 \pm \sqrt{17}}{4}$$

7. $y^2 + 3y + 7 = 0$
$a = 1, b = 3, c = 7$

$$y = \frac{-3 \pm \sqrt{9 - 4(1)(7)}}{2}$$

$$y = \frac{-3 \pm \sqrt{-19}}{2}$$

No real solutions

9. $-x^2 + 3x + 1 = 0$
$x^2 - 3x - 1 = 0$
$a = 1, b = -3, c = -1$

$$x = \frac{3 \pm \sqrt{9 - 4(1)(-1)}}{2}$$

$$x = \frac{3 \pm \sqrt{13}}{2}$$

11. $\dfrac{x^2}{2} - 2x + 1 = 0$

$x^2 - 4x + 2 = 0$
$a = 1, b = -4, c = 2$

$$x = \frac{4 \pm \sqrt{16 - 4(1)(2)}}{2}$$

$$x = \frac{4 \pm \sqrt{8}}{2}$$

$$x = \frac{4 \pm 2\sqrt{2}}{2}$$

$$x = 2 \pm \sqrt{2}$$

13. $-t^2 + 4t + 2 = 0$
$t^2 - 4t - 2 = 0$
$a = 1, b = -4, c = -2$

$$t = \frac{4 \pm \sqrt{16 - 4(1)(-2)}}{2}$$

$$t = \frac{4 \pm \sqrt{24}}{2}$$

$$t = \frac{4 \pm 2\sqrt{6}}{2}$$

$$t = 2 \pm \sqrt{6}$$

Section 11.4 Exercises (p. 378)

1. (a) upward   (b) $(3, -4)$
(c)

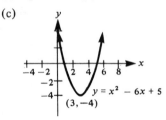
$y = x^2 - 6x + 5$
$(3, -4)$

3. (a) upward   (b) $(0, -1)$
(c)

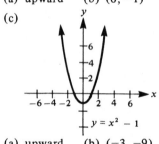
$y = x^2 - 1$

5. (a) upward   (b) $(-3, -4)$
(c)

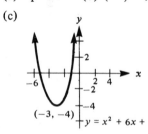
$(-3, -4)$
$y = x^2 + 6x + 5$

7. (a) upward   (b) $(-3, -9)$
(c)

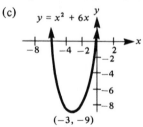
$y = x^2 + 6x$
$(-3, -9)$

9. (a) downward   (b) $(3, 9)$
(c)

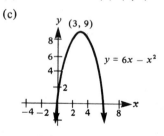
y $(3, 9)$
$y = 6x - x^2$

11. (a) upward   (b) $(0, 0)$
(c)

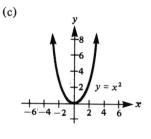
$y = x^2$

13. (a) upward   (b) $\left(\dfrac{1}{4}, \dfrac{-81}{8} \right)$
(c)

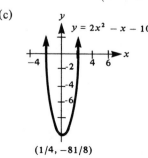
$y = 2x^2 - x - 10$
$(1/4, -81/8)$

15. (a) upward   (b) $(2, -2)$
(c)

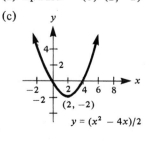
$(2, -2)$
$y = (x^2 - 4x)/2$

19. (a)   (b) 6.6 and 1.4   (c) 6.64575, 1.354249

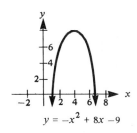

$y = -x^2 + 8x - 9$

Section 11.4 Supplementary Exercises (p. 379)

1. (a) upward   (b) $(3, -16)$ **3.** (a) downward   (b) $\left(\dfrac{-1}{2}, \dfrac{9}{4}\right)$ **5.** (a) upward   (b) $(1, -1)$

(c)

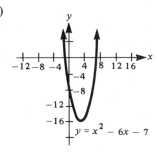

$y = x^2 - 6x - 7$

(c)

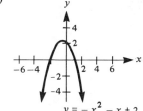

$y = -x^2 - x + 2$

(c)

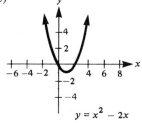

$y = x^2 - 2x$

7. (a) downward   (b) $(1, 1)$ **9.** (a) upward   (b) $\left(\dfrac{3}{4}, \dfrac{-25}{8}\right)$

(c)

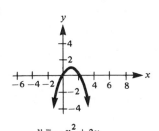

$y = -x^2 + 2x$

(c)

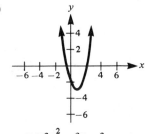

$y = 2x^2 - 3x - 2$

Section 11.5 Exercises (p. 388)

1.

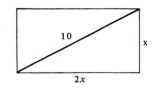

$$x^2 + 4x^2 = 100$$
$$5x^2 = 100$$
$$x^2 - 20 = 0$$
$$x = \pm 2\sqrt{5}$$

reject $-2\sqrt{5}$

Answer:

width is $2\sqrt{5}$ yards ≈ 4.47 yards

length is $4\sqrt{5}$ yards ≈ 8.94 yards

3.

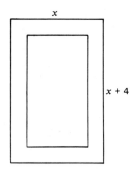

$$(x - 3)(x + 4 - 3) = 45$$
$$x^2 - 2x - 3 = 45$$
$$x^2 - 2x - 48 = 0$$
$$(x - 8)(x + 6) = 0$$
$$x = 8 \qquad x = -6 \text{ (reject)}$$

Thus, the page is 8 inches by 12 inches.

5.

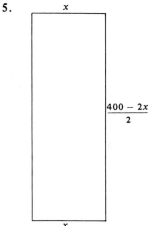

$$x(200 - x) = 7500$$
$$x^2 - 200x + 7500 = 0$$
$$(x - 150)(x - 50) = 0$$
$$x = 150 \quad \text{or} \quad x = 50$$

The dimensions are 50 feet by 150 feet.

7.

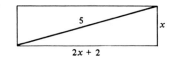

$$x^2 + (2x + 2)^2 = 25$$
$$x^2 + 4x^2 + 8x + 4 = 25$$
$$5x^2 + 8x - 21 = 0$$
$$(5x - 7)(x + 3) = 0$$
$$x = \frac{7}{5} = 1.4 \text{ feet}$$

The dimensions are 1.4 feet by 4.8 feet.

9.

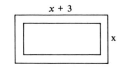

$$(x + 1)(x - 2) = 46.75$$
$$x^2 - x - 48.75 = 0$$
$$(x - 7.5)(x + 6.5) = 0$$
$$x = 7.5$$

(a) 7.5 m by 10.5 m

(b) 8.5 m by 5.5 m

11.

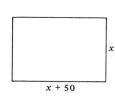

$x(x + 50) = 12600$
$x^2 + 50x - 12600 = 0$
$(x - 90)(x + 140) = 0$
$x - 90 = 0 \qquad x + 140 = 0$
$\qquad x = 90 \qquad\qquad x = -140 \text{ (reject)}$

If the dimensions are x (90 feet) by $x + 50$ (140 feet), then he will need 460 feet of fencing. Thus, he will have 40 feet left over.

13. (a) $n = \dfrac{5000(5001)}{2} = 12{,}502{,}500$

(b) $n = \dfrac{10000(10001)}{2} = 50{,}005{,}000$

15. (a) $D = 0.05(50)^2 + 1.1(50)$
$\qquad = 180$ feet

(b) $\qquad\qquad 100 = 0.05v^2 + 1.1v$
$\qquad 0.05v^2 + 1.1v - 100 = 0$
$\qquad\qquad v^2 + 22v - 2000 = 0$

$$v = \frac{-22 \pm \sqrt{484 + 8000}}{2}$$

$v \approx 35.05 \qquad v \approx -51.1 \text{ (reject)}$
$\qquad 35.05$ mph

17. $d^2 = (4001)^2 - (4000)^2$
$\quad d^2 = 8001$
$\quad d = \sqrt{8001} \approx 89.45$ miles

19. (a)

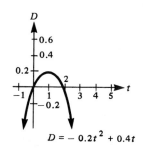

$D = -0.2t^2 + 0.4t$

(b) The first quadrant   (c) $t = 1$

21. (a)

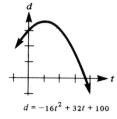

$d = -16t^2 + 32t + 100$

(b) The first quadrant   (c) Approximately 3.7 seconds   (d) $t = 1$

Section 11.5 Supplementary Exercises (p. 391)

1. width: 2 feet, length: 5 feet   **3.** 7 feet   **5.** 150 feet by 200 feet

Section 11.6 Exercises (p. 392)

1. (a) $b^2 - 4ac = 100 - 4(25)$
$\qquad\qquad\qquad = 0$

(b) 1

3. (a) $b^2 - 4ac = 25 - 4(2)(-3)$
$\qquad\qquad\qquad = 25 + 24$
$\qquad\qquad\qquad = 49$

(b) 2

5. (a) $b^2 - 4ac = 1 - 4(2)(4)$ **7.** (a) $b^2 - 4ac = 1 - 4(4)(-2)$ **9.** (a) $b^2 - 4ac = 1 - 4(1)(-2)$
$$= 1 - 32$$ $$= 1 + 32$$ $$= 1 + 8$$
$$= -31$$ $$= 33$$ $$= 9$$

(b) 0   (b) 2   (b) 2

11. False   **13.** True

Section 11.6 Supplementary Exercises (p. 393)

1. (a) $b^2 - 4ac = 25 - 4(1)(7)$ **3.** (a) $b^2 - 4ac = 0 - 4(1)(-9)$
$$= 25 - 28$$ $$= 36$$
$$= -3$$ (b) 2

(b) 0

5. (a) $b^2 - 4ac = 9 - 4(2)(-5)$ **7.** (a) $b^2 - 4ac = 4 - 4(1)(-5)$
$$= 9 + 40$$ $$= 4 + 20$$
$$= 49$$ $$= 24$$

(b) 2   (b) 2

Vocabulary Quiz (p. 393)

1. a   **2.** f   **3.** b   **4.** e   **5.** g   **6.** c   **7.** d   **8.** i   **9.** h

---

Chapter 11 Review Exercises (p. 394)

1. $x = 3, x = 4$ **2.** $x = 9, x = -1$ **3.** $x = 0, x = 8$ **4.** $x = 2$ **5.** $x = \frac{1}{2}, x = -1$

6. $x = \dfrac{-3 \pm \sqrt{13}}{2}$ $(x \approx 0.30278, x \approx -3.30278)$

7. $x = -1 \pm \sqrt{2}$ $(x \approx 0.41421, x \approx -2.41421)$

8. $x = \dfrac{-5 \pm \sqrt{13}}{6}$ $(x \approx -0.23241, x \approx -1.43426)$

9. $x = \pm \dfrac{\sqrt{6}}{6}$ $(x \approx 0.40825, x \approx -0.40825)$

10. $x = \dfrac{-5 \pm 3\sqrt{5}}{10}$ $(x \approx 0.17082, x \approx -1.17082)$ **11.** $x = 9, x = -1$

12.

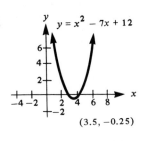

$y = x^2 - 7x + 12$

$(3.5, -0.25)$

13.

14.

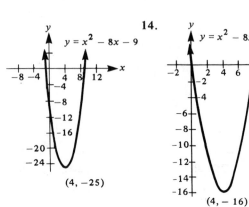

$y = x^2 - 8x - 9$

$(4, -25)$

$y = x^2 - 8x$

$(4, -16)$

15.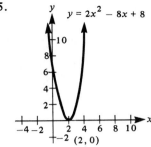
$y = 2x^2 - 8x + 8$
$(2, 0)$

16.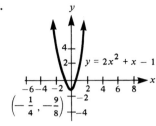
$y = 2x^2 + x - 1$
$\left(-\dfrac{1}{4}, -\dfrac{9}{8}\right)$

17.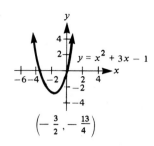
$y = x^2 + 3x - 1$
$\left(-\dfrac{3}{2}, -\dfrac{13}{4}\right)$

18.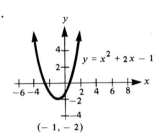
$y = x^2 + 2x - 1$
$(-1, -2)$

19.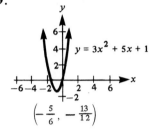
$y = 3x^2 + 5x + 1$
$\left(-\dfrac{5}{6}, -\dfrac{13}{12}\right)$

20.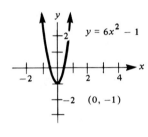
$y = 6x^2 - 1$
$(0, -1)$

21.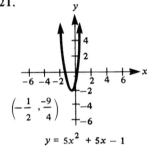
$\left(-\dfrac{1}{2}, \dfrac{-9}{4}\right)$
$y = 5x^2 + 5x - 1$

22.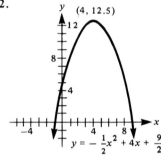
$(4, 12.5)$
$y = -\dfrac{1}{2}x^2 + 4x + \dfrac{9}{2}$

24. (a) $\sqrt{10}$ cm and $3\sqrt{10}$ cm
(b) 15 sq cm

25. $-1 + \sqrt{17}$ inches by $1 + \sqrt{17}$ inches (3.123 inches by 5.123 inches)

Cumulative Review Exercises (p. 394)

1. 100.48 cu in. **2.** $y = 4$ **3.** $\dfrac{3x^2 y z^5}{2}$ **4.** $2(x^3 - 6y^2 + 9xy^2)$ **5.** $\dfrac{5x + 38}{(x + 7)(x - 3)}$

6. $y = -3x + 8$ **7.** $x = -3, y = 4$ **8.** \$2500 at 10% and \$7500 at 12% **9.** $7xy\sqrt{2x}$ **10.** $x = 6$

Chapter 11 Test (p. 396)

1. $x = \dfrac{7}{2}, x = -2$ **2.** $x = 4 \pm \sqrt{13}$ **3.** $y = 2 \pm 2\sqrt{2}$ **4.** $x = \dfrac{3 \pm \sqrt{57}}{4}$

5.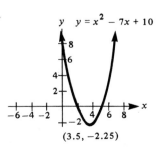
$y = x^2 - 7x + 10$
$(3.5, -2.25)$

6. Downward with vertex
$\left(-\dfrac{1}{2}, \dfrac{49}{4}\right)$

7. (a)

$$s = -16t^2 + 32t + 160$$

(b) The first quadrant

8. 4.3 seconds   **9.** (a) -11 (b) complex   **10.** True

Index